Hydraulic Conductivity Handbook

Hydraulic Conductivity Handbook

Edited by **William Sobol**

Callisto Reference
New York

Published by Callisto Reference,
106 Park Avenue, Suite 200,
New York, NY 10016, USA
www.callistoreference.com

Hydraulic Conductivity Handbook
Edited by William Sobol

International Standard Book Number: 978-1-63239-421-7 (Hardback)

Printed in China.

Contents

Preface

This book consists of a complete analysis of the latest measurement methods for estimating hydraulic conductivity. This has been accomplished by covering several aspects of hydraulic conductivity. For instance, the topics covered within the book include the importance of hydraulic conductivity, hydraulic conductivity and plant systems, conclusions arrived at by field methods, conclusions arrived at by mathematical and laboratory techniques, and hydraulic conductivity and modeling. These topics have been elaborately explained with the help of case studies and facts. Hence, the book contains a mixture of varied and important aspects of this field as noted by many experts from different countries.

After months of intensive research and writing, this book is the end result of all who devoted their time and efforts in the initiation and progress of this book. It will surely be a source of reference in enhancing the required knowledge of the new developments in the area. During the course of developing this book, certain measures such as accuracy, authenticity and research focused analytical studies were given preference in order to produce a comprehensive book in the area of study.

This book would not have been possible without the efforts of the authors and the publisher. I extend my sincere thanks to them. Secondly, I express my gratitude to my family and well-wishers. And most importantly, I thank my students for constantly expressing their willingness and curiosity in enhancing their knowledge in the field, which encourages me to take up further research projects for the advancement of the area.

Editor

Part 1

Hydraulic Conductivity and Its Importance

1

Dynamics of Hydraulic Properties of Puddled Soils

K. B. Singh
Krishi Vigyan Kendra, Moga
Punjab Agricultural University, Ludhiana
India

1. Introduction

Rice-wheat system has emerged as the dominating cropping system in South and South East Asia and Indo-Gangetic plains, because of favourable soil and climatic factors, suitable high yielding varieties, subsidized resource availability in terms of power and fertilizers and remunerative support purchase prices backed by the procurement system. The sustainability of this system in many areas is threatened by the rising water table and salinization, declining water tables and deterioration of both the chemical and the physical fertility of soil. Chemical fertilizers, appropriate tillage operations and addition of organic amendments supplement the low nutrient supplying capacity of the soil and can maintain favourable soil physical conditions. But there is no substitute of water for crop production. Rice being grown under submerged conditions requires high amount of irrigation water. In India, about 50% of the total irrigated area is under rice cultivation and hence, 50% of irrigation water is used for rice crop. On an average, farmers apply 32 irrigations to rice in Indian Punjab and assuming each irrigation to be of 7 cm, a total of 224 cm water is required. In many areas of Punjab and Haryana in India withdrawal of ground water exceeds its recharge which leads to lowering of ground water table. In central Punjab having good quality ground waters, the areas with water table below 10 m depth increased from 3% in 1973 to 76% in 2002 (Hira et al. 2004). Eighty percent area of the Punjab state is facing a rapid decline in water table which is quite alarming to the sustainability of the rice production. Competing demands of water for domestic use, sanitation, industrial and recreational purposes also make it all the more essential to maximize the efficiency of water for crop production. Since, rice is the lowest productive crop per unit of water consumed amongst cereals, therefore, optimum water management and cultural practices need to be followed to ensure minimum losses of water. In India, about 15000 liters of water is used for the production of one kg of rice (Sharma 1992). This low water use efficiency is because of high (55 to 80 percent) water losses through deep percolation (Singh, 1998). Though percolation is gain to water table but high amount of energy is required for lifting the underground water to meet irrigation requirements of rice - wheat system. Singh et al (1990) has shown that irrigation consumes the maximum energy in farm operations for both rice (82%) and wheat (38%). Therefore, water and energy are main elements for the sustainability of the rice-wheat system. Hence, to maintain desired yield levels while conserving the scarce water resources and to make the system energy efficient, it is essential to increase the productivity of rice and wheat per unit of water use. This can be achieved by curtailing the unproductive water losses such as evaporation and deep percolation. Soil evaporation can be manipulated by varying irrigation scheduling in wheat and staggering the date of transplanting of rice to the period of lower climatic evaporative demand (Singh et al, 2001). Most popularly, the deep percolation of water is controlled by manipulating hydraulic properties of soils through puddling.

2. Mechanism of puddling

Puddling (a most common method of land preparation for transplanting of rice seedlings) is associated with the soil disturbance through tillage operations, at or near saturation soil moisture content. The puddling tillage usually comprises one or two ploughing to a depth of 0.15 m and two or more harrowings and a final leveling. In puddling process the soil is submerged under standing water to promote soil chemical reduction, reduce soil mechanical strength and hence reduce the force and energy required for the puddling tillage. Two forces are applied to the soil during the puddling: one is ploughing to loosen and break clods in soils and the other is compaction of the sub-soil due to machinery used for tillage. Puddling leads to break down of soil aggregates into ultimate micro aggregates and individual particles forming a muddy suspension having dispersed fine particles. The degree of dispersion is dependent on the structural stability of the soil. Different research workers have shown that as compared to unpuddled, in puddled soils about 40 % of the aggregates were completely broken down to fractions less than 0.05 mm and all the aggregates were smaller than coarse sand (Naphade and Ghildyal, 1971). Due to differential settling the larger particles of soil in the suspension settle first and the finer ones later. Therefore, the sand fraction settles first from the muddy water and gradually is covered by finer silt and clay, resulting in a thin layer of low permeability due to clogging of macro pores at the surface. The thickness of the layers depends on the original texture. In sandy soils, the clay cover is thin. In fine clay soils there may be no coarse layer. In medium textured mineral soils stratification is well developed, with a fine textured surface layer a few mm thick that overlies 1-2 cm of almost pure sand (Moorman & Van Breemen, 1978).The clogging process is similar to filtration. Three mechanisms may be involved in filtration including surface clogging (large soil particles fail to pass through pore and deposit on the soil surface area), strain filtration (soil particle moves into pore space and is retarded by shear stress among other particles) and physical-chemical filtration (soil particle adsorbed onto the particle surface by surface reaction). If the soil texture is not uniform and contains a certain amount of clay particles, it is easy to develop plough sole by strain and physical-chemical filtrations. A Japanese study showed that the upper 0-15 mm of the puddled layer is composed of fine particles, the middle layer is thin and porous with sandy shingles, and the lowest layer is massive without particle differentiation (Saito and Kawaguchi, 1971). Closely packed parallel particles in puddled soils reduce saturated hydraulic conductivity (Singh et al 2001). The amount of dispersed clay or silt +clay increases with increasing puddling energy applied. Bodman and Rubin (1948) suggested the term puddlability, a measure of susceptibility of soil to puddling, to mean the change in apparent specific volume of soil per unit work extended in causing such change. From a farmer point of view, puddling is mixing soil with water to make it soft for transplanting and impervious to water. The ease and degree of puddling depend on moisture content, soil type, tillage implement, and cultural practices. Maximum puddling occurs at moisture content between field capacity and saturation. High clay content facilitates puddling. Soils with predominantly kaolinitic clay are more difficult to puddle than those with montmorillonitic clay (Sharma and De Datta, 1985). The sodium saturated clays are easier to puddle than calcium saturated clays owing to their dispersed nature in the former type of soil. During the puddling operation the soil layer just below the puddled soil is stamped down by the puddling equipment. This causes crushing and blocking of the pores in the lower unpuddled layer and results in reduction of saturated hydraulic conductivity. A portion of clay fraction from the surface puddled horizon is also deposited as clay-skins along pore surfaces at the top fringe of the compacted unpuddled subsurface layer. These processes reduce macro pore volume in the upper portion of the soil profile while increasing the bulk density in the compacted, anthropogenic subsurface horizon that is alternately termed the plough sole or tillage pan or hardpan. The formation of hardpan in the subsoil below the puddled layer is variable in different soils. It may take 3 to 200 years for a hardpan to form depending on soil type, climate, hydrology and puddling frequency (Moormann and van Breeman, 1978). Subsurface hardpans develop from physical compaction and precipitation of Fe, Mn and Si (Sharma and De Datta, 1985). This hardpan has lower saturated hydraulic conductivity. The non-puddled subsoil beneath this less permeable layer usually has a higher hydraulic conductivity and consequently there can be non-saturated subsoil flow in conjunction with saturated flow in the puddled and compact layers. Therefore a typical soil profile of a puddled rice soil consists of a (i) ponded water layer, (ii) a muddy layer with little resistance to water flow, (iii) a compacted hard layer with large resistance to water flow and (iv) non puddled subsoil with high saturated conductivity having unsaturated flow of water.

3. Effect of puddling on soil hydraulic properties

Puddling results changes in porosity and pore size distribution of the soil. Therefore water retention and movement under saturated and unsaturated conditions are highly affected by puddling.

3.1 Soil water retention

Puddling decreased pores >30 μm (transmission pores) by about 83% and increased pores of 0.6-30 μm (storage pores) and < 0.6 μm (residual pores) by 7% and 52% respectively (Sharma and De Datta, 1985). Eliminating non capillary pores in puddled soils usually lowers water retention for potentials above -0.01 MPa. At lower potentials (-0.01 to -1.5 MPa) water retention in puddled soils was more than non-puddled soils (Gupta *et al*, 1984). Water retention at lower (-0.01 to -1.5 MPa) potentials in puddled soils always exceeds than that of non-puddled soils depending on soil texture and initial aggregation (Taylor 1972 and Yunsheng 1983). The change in water status is defined by the difference between the previous and the current water contents in the paddy system when all the inflows and outflows have been accounted for. It is therefore, the residual amount of water in the paddy system, which can be positive or negative. Mohanty et al. (2003) reported that at harvest, on the average, the puddled soil maintained 25% more water than the unpuddled one. Puddling markedly increases soil water retention in soils, dominated by 2 : 1 swelling clays and the effect was less in soils dominated by kaolinitic clays (Sanchez 1973). Increase in water retention of 0-15 cm soil depth with puddling has also been reported by Sur *et al*

(1981). Puddling increases the water-holding capacity of soils because of compaction, settling, and flocculation of dispersed clay particles (Sharma and De Datta, 1985). The dispersion of soil aggregates during puddling destroys macro pore volume within soil aggregates (Moorman and van Breemen, 1978). Puddling changes the water content-tension relationship with more easy-to-extract water being available in puddled soils. Measurements of water-release characteristics showed that puddling increased storage porosity from 6 to 16 % and residual porosity from 34 to 43 % at the expense of transmission porosity, which declined from 26 to 4 % (Painuli et al 1988).

Porosity contributed to the 5-15 cm layer by pores wider than 30 μm (regardless of shape) was decreased by roto tiller puddling under standing water, and the number of pores per standard area were proportionally more affected (Pagliai et al ., 1987). The distribution of porosity among pores of different shapes and sizes in the 0-5 cm layer of puddled soil showed that elongated pores dominated the porosity, though they were in fact fewer in number. Porosity was less in puddled than in non-puddled soil in all shape categories in 0-20 cm, but below 20 cm there was no statistical or practical difference.

Effect of puddling on soil water retention depends upon climate, soil texture, soil depth, intensity of tillage, use of green and farmyard manures and crop residues incorporated into the soil.

3.1.1 Climate, texture and soil depth

Water retained by different soils at varying suctions in the four agro climatic zones of Punjab where puddling has been continuous from more than 15 years is reported by Singh *et al* (2009). Water retention was higher in zone 1(undulating sub region), followed by zone 2 (piedmont alluvial plains) and zone 3 (central alluvial plains) and least in zone 4 (southwestern alluvial plains) (Figure 1a). These trends are in accordance with amount of organic carbon (OC) and silt + clay contents. More the silt + clay and OC content in the soil, more will be water retention at a given level of suction. The textural class of silty clay loam retained more water at all the suctions which were followed by clay loam, loam and sandy clay loam soils (Figure 1b and 'a' values of soil moisture content (θ) vs. suction (Ψ) relations in Table 1). More water retention in fine textured soils is due to more micro as well as total porosity.

Soil texture	**θ vs. Ψ**		
	a	B	R^2
Sandy clay loam	0.3028	0.0035	0.83
Loam	0.3384	0.0022	0.82
Clay loam	0.3538	0.0020	0.80
Silty clay loam	0.3919	0.0012	0.81
Soil Depth (cm)			
0-7.5	0.3602	0.2132	0.81
7.5-15	0.3415	0.2054	0.80
15-22.5	0.3365	0.1965	0.79
22.5-30	0.3455	0.2051	0.80

Table 1. Parameter's of the equation y = a exp b θ, θ vs. Ψ.

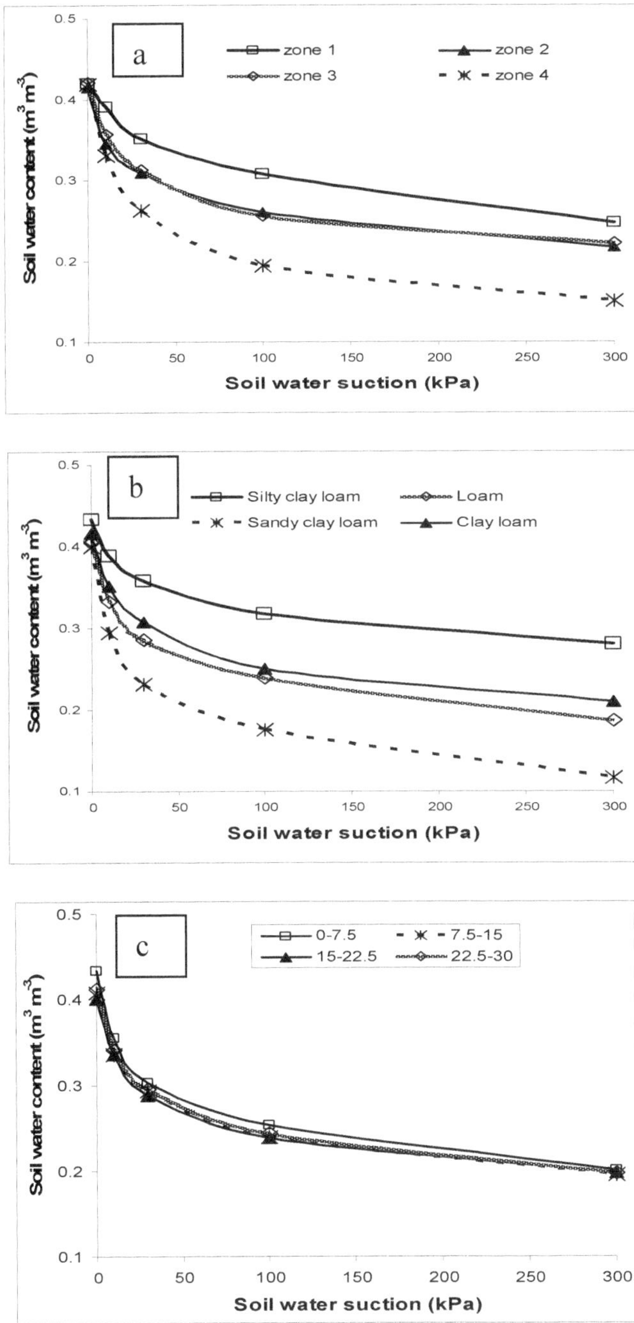

Fig. 1. Soil water retention at different suctions in relation to (a) zone, (b) texture and (c) depth in puddled soils of Punjab.

Water retention in different layers showed that water retention was highest in 0-7.5 cm soil layer and lowest in 15-22.5 cm soil layers at all the suctions applied viz. 1, 3, 10 and 30 k Pa as shown in Figure 1c and 'a' values of (θ) vs. (Ψ) relations Table 1. More volumetric water content in 0-7.5 cm soil layer may be attributed to its higher OC which improves aggregation status of the soil and hence its water retention capacity. Lowest water retention in 15-22.5 cm soil layer was due to its lower porosity (caused by more compaction) and OC contents as compared to other layers.

3.1.2 Intensity of tillage

Yoshida and Adachi (2002) studied the influence of puddling intensity on the water retention characteristics of a clayey paddy soil. Low intensity (one time puddling, P_1), medium intensity (two time puddling, P_2) and high intensity puddling (three time puddling, P_3) were compared. They observed that at the surface layer (0-4 cm), the water content increased over the whole range of suction and the increment of water content at a suction of 100 kPa depended significantly on puddling intensity. Moisture content was reduced linearly with time (Mausavi et al 2009). The line slope for P_0 treatment (no puddling) is more than P_1, P_2 and P_3. This shows that unpuddled soil dried faster than puddled soils. Ten days after puddling, soil moisture was reduced by 11.4, 8.3, 5.2 and 5.1 % respectively in P_0, P_1, P_2 and P_3 treatments. This may be attributed to the greater proportion of pore volume occupied by micro pores in the puddled soils. The effect of puddling intensity on water retention in soil was significant up to a suction of 90 kPa (in laboratory study) and up to 500 kPa in field study. Soil moisture retention in P_1 is much higher than P_2 but P_2 is not very much different from P_3 treatment. Higher number of micro pores in highly puddled soil is an important factor. Puddled soils dry more slowly than unpuddled soils, probably because of higher unsaturated hydraulic conductivity of puddled soils can keep surface soil wet during evaporation by supplying water from lower layers. Also because of increased water retention at a given suction, more energy is required to evaporate the same amount of water from a puddled soil than from an unpuddled soil. Thus a puddled soil may take several weeks or even months to dry and reach workable moisture content (Gupta and Jaggi, 1979).

3.1.3 Organic manures

Singh *et al* (2000) studied the effect of incorporation of green manure (*Sesbania rostrata*) @ 10.4 Mg ha^{-1} and farmyard manure (FYM) @ 6.5 Mg ha^{-1} on dry weight basis before rice transplanting in combination with 100% recommended NPK for 3 years in silty loam soil. They found that in 0-15cm soil layer, water retained at field capacity in green manured plots increased from 28.36% (100% NPK alone treatment) to 30.87%, and in FYM treated plots to 32.33%, when measured after harvest of wheat crop. Tiwari *et al* (2000) reported that incorporation of green manure (*Sesbania rostrata*) before puddling of loam soil for 2 years along with 50% recommended N increased the water holding capacity of the soil from 36.45% (initial) to 37.10%, when measured after harvest of rice crop. Biswas *et al* (1970) in a 10-year study on alluvial sandy loam soil found that, incorporation of green manure (equivalent to 45kg N ha^{-1}) increased water retention from 14.4 % to 18.0% at 33 kPa tension. However, at 1500 kPa tension increase was very small *i. e.* from 3.8% to 4.6%.

3.1.4 Residue incorporation

Pandey *et al* (1985) studied the water retention of the soil at 33 kPa suction in rice residue management experiment. It was 23.2% in control treatment, and increased to 26.3% in rice straw incorporation treatment and up to 26.8% in wheat straw incorporation treatment, when measured at harvest of succeeding wheat crop. Under all treatments, water content decreased abruptly up to 20 kPa suction and beyond this value decrease in water content was gradual. Bhagat and Verma (1991) in a 5-year study on silty clay loam soil found that straw incorporation @ 5Mg ha^{-1} plus farmyard manure treatment retained the highest, whereas in control and straw burnt the lowest water content, at all suction values between 0 and 1500 kPa. At 33 and 1500 kPa, the straw incorporation plus farmyard manure treatment had about 0.055 m^3 m^{-3} and 0.030 m^3 m^{-3} higher water content, respectively, compared to the control and straw burnt. The straw incorporation alone treatment had intermediate water retention compared to the above two sets of treatments, at all the suction values.

3.2 Saturated hydraulic conductivity

In puddled soil some soil physical characteristics which determine water conductivity are the total porosity, the distribution of pore sizes and the pore geometry of the soil. Thus structural destruction due to puddling may not be of much important in reducing water flux through the soil. Saturated hydraulic conductivity (Ks) is a quantitative measure of ability of saturated soil to transmit water under the hydraulic head difference. It defines the linear relationship between flux and hydraulic gradient. So it is the slope of line showing relationship between flux and hydraulic gradient. If the same hydraulic gradient is applied to two different soils, the soil from which the greater quantity of water is collected/discharged, is more conductive means have more flow rate. Mostly sandy soils yield higher flux (i.e. more conductive) than the clayey soils. Saturated hydraulic conductivity is affected by soil and fluid properties. It depends on the soil pore geometry as well as the fluid density. Flux is numerically equal to Ks only when the hydraulic gradient is equal to one. Flux is a dependent variable & hydraulic gradient is the driving force (i.e. independent variable) and Ks is the proportionality constant that defines the relationship between two. Puddling altered the nature of water flow, particularly in silty clay loam (Fig. 2). At lower hydraulic gradients the specific discharge versus hydraulic gradient curve does not remain a straight line, hence Darcy Law does not hold good under puddled conditions. However, as the gradient exceeds 1, the relationship assumes the form of a straight line. It therefore seems that puddling induces non-Darcian flow at lower hydraulic gradients. Evidently in puddled fine-textured soils a critical hydraulic gradient must exist beyond which the water flux is Darcian and before which it remains non-Darcian (Fig. 2). Since puddling reduces the non-capillary pore spaces, a closer packing of soil particles results in higher bulk density. Reduction on the pore volume, in turn, decreases the hydraulic conductivity and free percolation of water. Studies on laterite sandy loam soils indicated that hydraulic conductivity of a field puddle decreased from 0.192 to 0.054 cm h^{-1}, causing a reduction of 72 % from the unpuddled soil.

Climate, soil texture, depth and intensity of puddling, bulk density, type of implement used for puddling, structural regeneration, organic manures applied and crop residue incorporation into the puddled soil etc determine the effect of puddling on soil hydraulic properties.

3.2.1 Climate

Saturated hydraulic conductivity of sandy clay loam soil (Fig 3) generally decreased from agro climatic zone 2 to zone 4 in soil depths because of changes in temperature which resulted into decrease of organic carbon content in the surface soil layers and increase in the

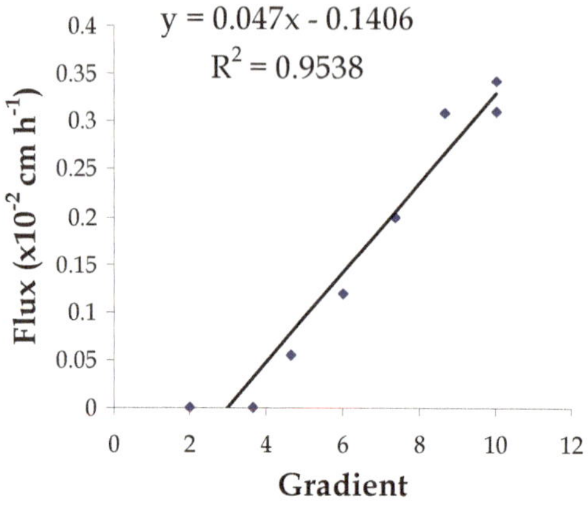

Fig. 2. Relationship between flux and gradient in puddled silty clay loam soil.

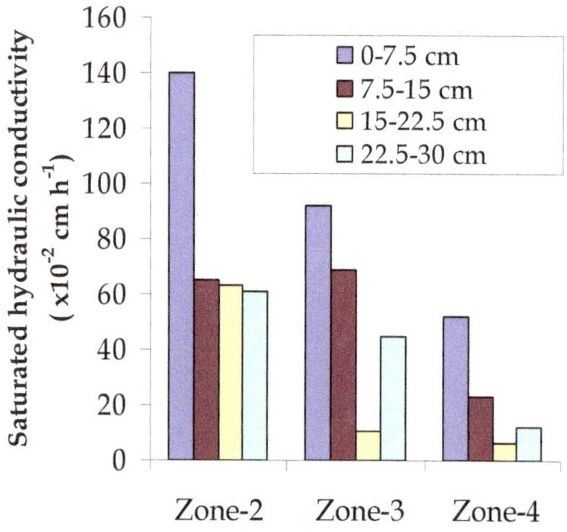

Fig. 3. Saturated hydraulic conductivity of sandy clay loam.

bulk density of subsoil layers (Singh et al, 2009). Hence better puddling for reducing water flux is difficult to achieve in the soil having high organic carbon leading to high structural stability index. Therefore, the high content of organic matter increases resistance of the soil to puddling. Dominance of loam and sandy clay loam textures in agro climatic zone 2 of Punjab compared to other zones resulted in significantly higher Ks (Table 2). Amongst the textural class, saturated hydraulic conductivity was significantly more in sandy clay loam due to more sand content in the soil (66 per cent) than other classes, where it ranged from 31-55 per cent. Saturated hydraulic conductivity was more in the surface soil (due to more organic carbon and less bulk density) and it decreased with depth and was lowest in 15-22.5 cm soil layer, where bulk density and organic carbon were having reverse trend than the surface soil (Singh *et al*, 2009).

Variables		Saturated hydraulic conductivity (mm h^{-1})
Agro climatic Zones	*Zone 1*	0.77[a]
	Zone 2	4.32[b]
	Zone 3	1.77[a]
	Zone 4	1.06[a]
Textures	Silty clay loam	1.05[a]
	Loam	2.06[a]
	Sandy clay loam	4.85[b]
	Clay loam	1.38[a]
Soil depths	0-7.5 cm	4.2[a]
	7.5-15 cm	2.02[b]
	15-22.5 cm	1.20[b]
	22.5-30 cm	1.91[b]
Rows having different letters are significantly different		

Table 2. Averaged effects of agro climatic zones, textures and soil depths on saturated hydraulic conductivity.

3.2.2 Soil texture

The decrease in hydraulic conductivity due to puddling was greater in sandy loam and clay loam soils than in clay soil (Mambani et al., 1989). Thus the benefits of puddling as well as dynamics of pan formation are dependent on soil type. Adachi (1990) reported that with the same intensity of puddling in clayey soils the downward flow rate declined slowly compared to coarse textured soils. In coarse textured soils, flow rate after puddling declined quickly with soil consolidation. In medium textured soils, decrease in flow rate just after puddling was caused by the consolidation in the lower part of the mixed layer and filling of that layer by fine particles. Then from series of field and laboratory studies, Adachi (1992) concluded that depth of least permeable layer and its formation mechanism due to puddling depended upon texture of the soil (Table 3). For fine textured soils, the most important mechanism for decreasing percolation rate was the blocking of layer just below the puddled soil by fine dispersed particles. In medium textured soils the decrease in percolation rate is because of increased bulk density at the lower part of puddled layer. In coarse textured soils the reduction is mainly due to a very clayey layer formed in the top 0-1, 2 cm of puddled layer. A large reduction in Ks of puddled soil was due to the disaggregation induced by puddling. The greater the soil dispersion, the lower is the Ks. High clay content facilitates puddling. Soils with predominantly kaolinitic clay are more difficult to puddle than those with montmorillonitic clay. The sodium saturated clays are easier to puddle (owing to their dispersed nature) than the calcium saturated clays. Saturated hydraulic conductivity was exponentially related with silt content in both silty clay loam($y = 2238e^{-0.1616X}$, $r^2=0.8469$) and sandy clay loam ($y = 355.18\ e^{-0.1858X}$, $r^2 = 0.86$) soils of Punjab (Singh, K.B, Personnel Communication).

3.2.3 Depth and intensity of puddling

The resistance to water flow, as evidenced by specific discharge, was increased in all the soils following puddling. The reduction in specific discharge (cm h^{-1}) after first puddling varied from 6 to 10 times at different hydraulic gradients in sand. In sandy loam it was small, especially at lower gradients. In silty clay loam soil the reduction was marginal. The

Soil layers	Saturated hydraulic conductivity (cm sec-1)		
	Heavy clay	Clay loam	Loamy sand
Initial	$4x10^{-2}$	$4x10^{-3}$	$1x10^{-2}$
Upper part of puddled layer	$2x10^{-6}$	$2x10^{-5}$	$4x10^{-4}$
Lower part of puddled layer	$2x10^{-7}$	$2x10^{-6}$	$3x10^{-3}$
Just below the puddled layer	$8x10^{-8}$	$2x10^{-6}$	$1x10^{-2}$
Lower layer	$4x10^{-6}$	$3x10^{-4}$	$1x10^{-2}$

Table 3. Effect of soil texture on saturated hydraulic conductivity (cm sec-1).

second puddling was less effective than the first in reducing specific discharge except in sandy loam soil where the second puddling might have resulted in more uniform sealing of pores. The results of the third puddling were not any different from the second. The amount of dispersed clay or silt + clay increased with increasing puddling energy applied (Bakti et al 2010). Saturated hydraulic conductivity decreased rapidly when the soil was puddled with only 50-100 Joule energy input. When the soil is puddled with greater energy (100-200 Joule) however, the additional reduction in Ks was small. Sharma and Bhagat (1993) reported that greater the depth of puddling, the smaller is the hydraulic head gradient across the puddled soil layer and hence smaller is the water flux. They further found that puddling was effective in decreasing water permeability of soils, if their sand content was less than 70% and the finer fraction was dominated by clay (13-20%). On texturally different soils about 95% reduction in flux due to per unit increase in puddling depth was explained by the linear function of clay content of the soil. At a given puddling depth, the relative water flux (flux of puddled/non puddled) decreased exponentially with the increase in clay content, the effect was more pronounced at greater puddling depths. They also reported that relatively shallower puddling depth was required in fine textured soils. Effect of intensity of puddling was also found to be significant on hydraulic conductivity. With increased level of puddling there was breakdown of natural aggregates due to which the apparent specific volume of soil decreased which in turn decreased its hydraulic conductivity (Naphade and Ghildyal 1971). Saturated hydraulic conductivity was highest with the moldboard treatment followed by that with mouldboard + puddling and conventional tillage treatments (Akhtar and Qureshi, 1999). Saturated hydraulic conductivity of the native profile, which represented by conventional tillage was 3.0 cm day-1 in the 0-20 cm layer, 1.2 cm day-1 in 20-35 cm layer and 0.7 cm day-1 in 35-75 cm layer whereas the corresponding values of saturated hydraulic conductivity for deep tilled plots were 7.6, 4.1 and 1.4 cm day-1 respectively. Puddling after mouldboard resulted in reduction in saturated hydraulic conductivity to the level of conventional tillage.

Saturated hydraulic conductivity of 0-20 cm layer was 3.0, 1.0 and 0.1 mm h-1 under deep tillage, deep tillage+ puddling and conventional tillage treatments respectively. Singh *et al* (2001) have found that increase in puddling intensity substantially decreased Ks in puddled layer in sandy loam and silty clay loam soils. Generally Ks was more in the surface layers and decreased with soil depth down to 25 cm depth with minimum at 15-20 cm (Table 4).

Texture & Soil depth (cm)	Saturated hydraulic conductivity			$LSD_{(0.05)}$
	No puddling	Puddling 2 times with disc harrow	Puddling 4 times with disc harrow	
Sandy loam				
0-5	7.7(0.6)	1.0(0.58)	0.5(0.12)	2.1
5-10	5.3(1.84)	1.0(0.09)	0.9(0.02)	1.8
10-15	2.5(0.42)	0.8(0.20)	0.7(0.08)	1.0
15-20	2.3(0.89)	0.7(0.18)	0.4(0.17)	1.2
20-25	2.2(0.13)	1.1(0.10)	1.0(0.27)	NS
25-30	1.3(0.54)	1.3(0.54)	1.3(0.54)	NS
Silty clay loam				
0-5	5.2(0.60)	1.1(0.18)	0.8(0.39)	1.4
5-10	2.1(0.72)	1.0(0.05)	0.4(0.04)	0.8
10-15	1.6(0.19)	0.8(0.20)	0.4(0.10)	NS
15-20	0.8(0.43)	0.7(0.07)	0.4(0.20)	NS
20-25	1.0(0.28)	1.0(0.28)	0.4(0.08)	NS
25-30	1.1(0.78)	1.1(0.78)	1.1(0.45)	NS

Figure in parenthesis are standard deviation from mean based on 18 replications.

Table 4. Effect of puddling intensity on laboratory measured saturated hydraulic conductivity (cm day^{-1}).

Saturated hydraulic conductivity was significantly decreased with puddling intensity in all layers upto 20 cm depth in sandy loam and upto 10 cm depth in silty clay loam. The less effect of puddling (upto 10 cm) in decreasing Ks in silty clay loam soil is due to less depth of puddling compared to sandy loam (Table 4). The decrease in Ks with puddling levels was more in surface layers (0-20 in sandy loam and 0-15 in silty clay loam) compared to lower layers. In 0-5 cm soil layer Ks decreased 7.7 times with P_2 (two time puddling) and 15.4 times with P_4 (Four time puddling) in sandy loam and 4.7 times with P_2 and 6.5 times with P_4 in silty clay loam respectively, compared to P_0 (unpuddled control). However differences in Ks between P_2 and P_4 were non significant on both soils, throughout the profile (0-30 cm). The decrease in Ks with puddling levels in due to closely packing of soil particles in parallel orientation because of differential settling, and breakage of soil aggregates and elimination (sealing) of non capillary pores responsible for water transmission through soils (Sharma and De Datta, 1985 and Adachi, 1992).

3.2.4 Bulk density

Since puddling reduces the non-capillary pore spaces, a closer packing of soil particles results in high bulk density. Reduction in the pore volume i.e increase in dry bulk density, in turn, decreases the saturated hydraulic conductivity of water. Studies on laterite sandy loam soils (Naphade and Ghildyal, 1971)indicated that hydraulic conductivity of field puddle decreased from 0.192 to 0.054 cm h-1, causing reduction of 72% from the unpuddled soil. The apparent specific volume (reciprocal of bulk density) decreased from 0.714 to 0.591 cm^3 g^{-1} and hydraulic conductivity from 0.192 to 0.019 cm h^{-1} when the soil was puddled with 0 to 2.700 watt hour energy input. The decrease in the apparent specific volume and saturated hydraulic conductivity with decreased degree of puddling was remarkable

initially. However, the magnitude of decrease in hydraulic conductivity exceeded the apparent specific volume. The apparent specific volume and the hydraulic conductivity of the field puddle were 0.615 cm3 g-1 and 0.054 cm h-1 respectively. The reduction in the apparent specific volume and the hydraulic conductivity of field puddle when compared with the unpuddled soil was 13.9 and 71.9 percent respectively. These values were closely related to those observed for the soil puddle with 0.450 watt hour indicating that puddling the soil to this energy in the laboratory represented field puddle. The results of apparent specific volume and Ks under different degrees of puddling were correlated significantly. The coefficient of correlation (r=0.9292) was significant at 5 percent level and the regression of apparent specific volume (X) on the hydraulic conductivity (Y) was: Y=6.17 X-3.41.
An empirical relationship ($Ks= 0.3266\ bd^{-0.1404}$) was established between bulk density(bd) and hydraulic conductivity (Ks) (Behra et al, 2009). As bulk density increased hydraulic conductivity decreased but at higher bulk density the rate of decrease of hydraulic conductivity was low. The correlation coefficient (R^2 = 0.7821) was found significant at 5 % level of significance. Singh, K.B (Personnel Communication) has observed a linear relationship between bulk density and saturated hydraulic conductivity (Table 5). Increase in bulk density has resulted linear decrease in Ks.

Soil texture	a	b	R^2
Silty clay loam	108.88	60.799	0.904
Clay loam	213.54	112.55	0.893
Loam	521.59	285.87	0.834
Sandy clay loam	554.88	303.81	0.881

Table 5. Saturated hydraulic conductivity ($x10^{-2}$ cm h^{-1}) as function of bulk density (Mg m^{-3}) (Ks = a – b (Bulk density) in texturally different soils.

3.2.5 Type of implement used for puddling

Tyagi et al (1975) found that in loamy sand soil puddling with power tiller having a rotovator resulted in the maximum reduction in saturated hydraulic conductivity when compared with puddling with local plough, the control. The value of hydraulic conductivity was 0.1971×10^{-2} cm h^{-1} in puddling with power tiller compared to 2.871×10^{-2} cm h^{-1} with local plough. When compared with control the reduction in hydraulic conductivity was 76% with power tiller, 74 % with tractor having cage wheel, 34.6% with disc-harrow and only 18% with mould-board plough. Minimum hydraulic conductivity of 0.257 mm hr^{-1} was found in case of two passes of peg type puddler which was significantly lower than that of one pass of peg type puddler (0.315 mm hr^{-1}) but statistically at par with one pass of rotary puddler (0.270 mm hr^{-1}) at 30 days after puddling (Behra et al 2009). At 60 days after puddling, there was no appreciable variation in hydraulic conductivity over that of 30 days after puddling. The hydraulic conductivity depends upon the amount and size of coarse pores (transmission pores) in the soil. Saturated hydraulic conductivity was significantly reduced due to puddling by angular bladed puddler followed by disc harrow, mould board plough and deshi plough and no puddling treatment (Rane and Varade 1972). All the treatments differ significantly from each other with respect to hydraulic conductivity. The percent reduction in hydraulic conductivity over control was 84.87, 74.44, 63.17 and 51.4 percent under angular bladed puddler, disc harrow, mould board plough and deshi plough.

3.2.6 Structural regeneration

Wetting/ drying cycles are known to improve the soil structure. Results clearly showed reduction in the amount of silt+clay (<20μm) and clay (<2 μm) dispersed during wetting/drying cycles in clay and sandy loam soil increased Ks (Bakti et al, 2010). Repeated wetting/drying cycles to air dry water contents increased Ks of the puddled soil significantly in the fine textured soil but had little effect on the sandy soil. However in the fine textured soil the recovery of Ks was not significant. This improvement was partly the consequences of rapid wetting. Rapid wetting causes partial slaking by inducing micro-cracks and these micro cracks has the effect of making the soil easily crumbled. The degree of drying has a strong effect on the structural regeneration shown by the decrease in silt+clay. The hydraulic conductivity increases with time after puddling due to regeneration of soil structure through reflocculation of dispersed clays (Pagliai *et al,* 1990) and greater roots proliferation (Prathapar *et al,* 1989). However, Tyagi *et al* (1975) reported that on loamy sand the hydraulic conductivity was higher at the time of puddling and there was a general decrease in its value after 25 and 50 days of transplanting. This decrease in hydraulic conductivity with time was attributed to the sealing effect caused by the settlement of finer particles in course of time.

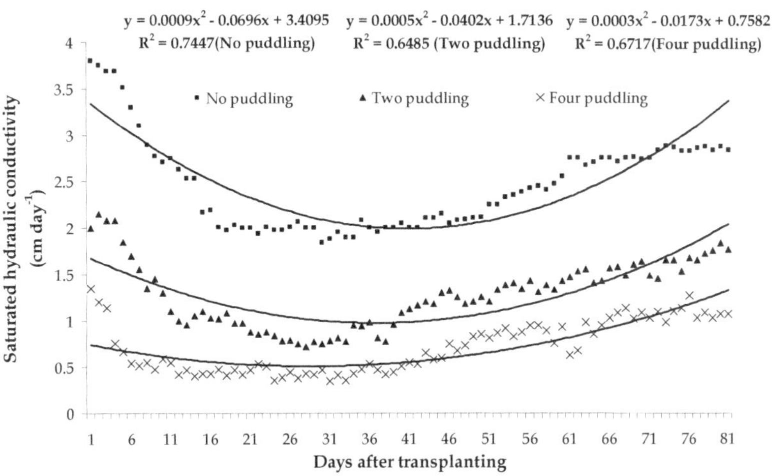

Fig. 4. Effect of puddling on *in situ* saturated hydraulic conductivity of sandy loam.

In field conditions saturated hydraulic conductivity (of the puddled 0-20 cm layers) of sandy loam (Fig 4) at 1 day after transplanting(DAT) was 3.8, 2.1 and 1.3 cm day-1 under unpuddled (P_0), 2 time puddled(P_2) and 4 time puddled(P_4) treatments respectively which decreased to almost half after 15 days of transplanting in all puddling treatments. In silty clay loam (Fig 5) the corresponding values of saturated hydraulic conductivity were 2.9, 1.8 and 1.7 cm day-1 which decreased to 2.0, 1.1 and 0.9 cm day-1 in P_0, P_2 and P_4 treatments after 28 DAT (Singh and Manchanda, 2008). This was because of more time taken for settling of clay particles dispersed in suspension during puddling and then clogging of the pores in the top layers and consolidation of the lower layer in silty clay loam. During puddling of silty clay loam, more clay in suspension takes more time for settling. In the fine textured soil, the soil consolidation development slowly from the lower part of the mixed (puddled) layer to the upper part and consequently, the Ks was slowly reduced. However, within the coarse textured soil, the Ks after puddling was quickly reduced with the settling and the settlement rate was approximately constant. Thus it was considered that the soil settlement phenomenon was sedimentation without soil consolidation (Adachi 1992). In sandy loam (> 50 DAT) and silty clay loam (> 60 DAT), the daily steady water intake rate increased gradually with the increase in root growth in all puddling treatments (Singh and Manchanda, 2008).

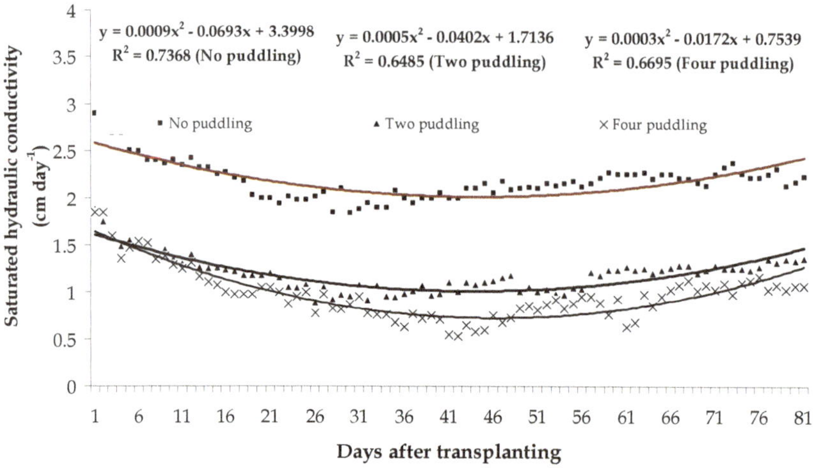

Fig. 5. Effect of puddling on *in situ* saturated hydraulic conductivity of silty clay loam.

3.2.7 Organic manures

Addition of organic amendments (combination of green manures, FYM and residue incorporation) increased in situ saturated hydraulic conductivity (Table 6)of the soil after harvesting of rice (Singh, 2003). Incorporation of rice residues also increased Ks. Incorporation of green manures, farmyard manures and rice straw significantly increased the hydraulic conductivity of puddled soils (Bhagat and Verma, 1991).

Treatments	Ks
Experiment 1	
Control	1.69
Green manure	4.11
FYM applied @ 12 Mg ha^{-1}	4.67
FYM + Green Manure	6.17
Wheat straw incorporated into soil	4.86
Wheat straw + Green manure incorporated	5.50
Experiment 2	
Rice straw removed	2.08
Rice straw burnt	2.17
Rice straw incorporated into soil	2.56

Table 6. Hydraulic conductivity of soil (x 10^{-3} ms^{-1}) as influenced by organic amendments (Experiment 1) and reice residue incorporation (Experiment 2).

3.3 Unsaturated hydraulic conductivity

For the estimation the retention and movement of water under unsaturated conditions, the knowledge of relationship between soil water pressure, water content and hydraulic conductivity is required. Soil moisture characteristics curve obtained through pressure plate apparatus can be fitted to the desired soil water retention model. Once the retention function is estimated, the hydraulic conductivity and moisture content relationship can be evaluated if the saturated hydraulic conductivity is known.

Unsaturated hydraulic conductivity k(θ) as represented by constant 'b' of the equation type k(θ) = a exp bθ, increased with fineness of the textural class (Singh et al, 2009). Similarly, higher unsaturated conductivity was observed in 15-22.5 cm soil layer (having high bulk density) as compared to other layers irrespective of the soil texture (Table 7). Wind (1961) also reported higher unsaturated conductivity in finer textured soils in the dry moisture range due to higher micro porosity.

Soil texture	**K(θ) vs. θ**		
	a (x 10^{-7})	b	R^2
Sandy clay loam	70	39.33	0.96
Loam	0.8	46.70	0.98
Clay loam	0.1	49.00	0.98
Silty clay loam	0.001	56.28	0.99
Soil Depth, cm			
0-7.5	2	42.84	0.99
7.5-15	0.4	48.30	0.98
15-22.5	0.006	52.83	0.98
22.5-30	0.4	47.35	0.98

Table 7. Parameter's of the equation, y = a exp b θ, for K (θ) vs. θ

Wopereis et al (1992) found that unsaturated flow conditions prevail in the non-puddled subsoil. The least permeable layer within the profile, as determined by the greatest gradient in pressure head was found to be at the interface of puddle topsoil (0-15 cm) and non-puddled subsoil. Average thickness of this layer was about 5 cm. Hydraulic conductivity of the least permeable layer was 0.36 mm day^{-1} with 95 % confidence limits at 0.27 and 0.45 mm day^{-1}. Unsaturated downward flow of water prevails below the subsoil layer (15-60 cm) having restriction to water. Unsaturated hydraulic conductivities decreased by 3 to 4 orders of magnitude as the soil matric potential decreased from near zero to -700 cm (Hundal & De Datta 1984). The lowest unsaturated hydraulic conductivity was found in the 45-60 cm soil depth. The differences in conductivities observed in different layers are resulted from differences in the pore size distribution of various layers An equation of the form:

$$K(h)= a(h)^b$$

where 'h' represents the soil matric potential in absolute cm, described the K versus h relation with best fit (Table 8). The power function gave the best fit.

Puddled soils dry more slowly than unpuddled soils probably because the higher unsaturated hydraulic conductivity of puddled soils can keep surface soil wet during evaporation by supplying water from lower layers. Also because of increased water retention at a given suction, more energy is needed to evaporate the same amount of water from puddled than from an unpuddled soil. Thus a puddled soil may take several weeks or months to dry and to reach workable moisture content.

Soil depth (cm)	*a*	*b*	R^2
0-15	6.398 x 10^3	-2.287	0.95
15-30	1.498 x 10^1	-1.360	0.92
30-45	0.742	-1.005	0.95
45-60	0.509	-1.104	0.70
60-75	1.175	-1.037	0.87
75-90	1.624	-1.056	0.87

Table 8. Coefficients a and b in the text equation ($K(h)= a(h)^b$)for different soil depths.

3.4 Preferential flow

If clay soils dry long enough, the soft mud cracks and dries to a stiff paste. When the soil floods again, the cracks do not completely close as a result of partial swelling of clays and may cause preferential flow of water and nutrients. Bypass flow may continue unabated until puddling seals the cracks. Introducing discontinuities in soil cracks by shallow surface tillage (0-5 cm) after the first monsoon rain reduced bypass flow and resulted in 45-60 % water savings, thus increasing the retention of water within the topsoil and within the bunded field (Woperis, 1993).

4. Summary

Puddling increases micro porosity of soil which in turn results increase in water retention at lower potentials than unpuddled soils depending on soil texture, initial aggregation. Water retention in puddled soils is in accordance with amount of organic carbon and silt plus clay content. Silty clay loam soil retained more water than, clay loam, loam and sandy clay loam soils at all suctions. Incorporation of green and farm yard manures, rice and wheat straw also increased water retention. Saturated hydraulic conductivity decreased with decrease in organic carbon and increase in bulk density of soil. The decrease in hydraulic conductivity due to puddling was greater in sandy loam and clay loam soils than in clay soil. Increase in depth and intensity of puddling significantly reduced saturated hydraulic conductivity. Reduction in water flux is directly linked with clay content of soil. Increase in bulk density of soil significantly reduced saturated hydraulic conductivity. However with repeated wetting and drying cycles and addition of organic amendments, the soil structure regenerates which results in increase in saturated hydraulic conductivity. Puddling results in close packing of soil particles which increases soil dry bulk density and unsaturated hydraulic conductivity.

5. Future strategies

Quantitative understanding of the puddled soil system and of the influence that soil management can have on water fluxes and water conservation, would be advanced if the effect of puddling on the hydraulic conductivity of various layers is better defined. Similarly in the simulation of rice production crop-water relations and soil water transmission are crucial components. Many soil-water simulations assume one-dimensional (vertical) water flow through a succession of soil layers, and the hydraulic conductivities of such layers are needed as parameters in the simulation. Soils with 25-50 % clay in the topsoil and a similar or somewhat higher clay percentage in the subsoil produce the highest rice yields (Grant, 1965). Therefore, more understanding on soil hydraulic fluxes is required for the development of mechanistic water flow models to simulate root water uptake and yield response for increasing water use efficiency of puddled rice system.

6. References

Adachi, K. (1990). Effects of rice-soil puddling on water percolation. In: Proceedings of the Transactions of the 14th International Congress of Soil Science, Kyoto, Japan, pp. 146-151.

Adachi, K (1992) Effect of puddling on rice soil physics : Softness of puddled soil and percolation. In: Murty, V.V.N., Koga K. (Eds.). Soil and Water Engineering for Paddy Field Management. *Proceedings of the International Workshop on Soil and Water Engineering for Paddy Field Management.* Asian Institute of Technology, Bangkok, pp. 220-231

Akhtar, M.S. and Qureshi, S. (1999). Soil hydraulic properties and rice root development as influenced by tillage. *Pakistan J. Biological Sci. 2(4):1245-1251*

Bakti, L.A.A., Kirchhof, G. and So, H.B. (2010). Effect of wetting and drying on structural regeneration of puddled soil. Proceedings of 19th World Congress of soil Science, Soil Solutions for changing World, 1-6 Agust 2010, Brisbane, Australia, Published on DVD, pp, 17-20.

Behera, B.K, Varshney, B.P, and Goel , A.K. (2009). Effect of puddling on puddled soil characteristics and performance of self propelled transplanter in rice crop. Agricultural Engineering International: the CIGR Ejournal, Vol. X. Manuscript PM 08 020, September, 2009.

Bhagat, R. M. and Verma, T. S. (1991) Impact of rice straw management on soil physical properties and wheat yield. *Soil Science* 152 (1): 108-14.

Biswas, T. D., Roy, M. R. and Sahu, B. N. (1970) Effect of different sources of organic manures on the physical properties of the soil growing rice. *J Indian Soc Soil Sci* 18 (3): 233-42.

Bodman, G.B and Rubin, J. (1948) Soil Puddling. *Soil Sci. Soc. Amer. Proc. 13: 27-36*

Grant, C. J. (1965). Soil characteristics associated with wet cultivation of rice. In: The mineral nutrition of the rice plant. Proceedings of a symposium at International Rice Research Institute, Johns Hopkins Press, Baltimore, Maryland, pp: 15-28

Gupta, R.K and Jaggi, I.K. (1979). Soil physical conditions and paddy yield as influenced by depth of puddling. *J Agron. Crop Sci. 148:329-336*

Gupta, R. P., Kumar, S. and Singh, T. (1984) Soil management to increase crop production. A Consolidated Report 1967-82. ICAR, New Delhi.

Hira, G.S., Jalota, S.K., Arora, V.K. (2004). Efficient management of water resources for sustainable cropping in Punjab. Technical Bulletin, Department of Soils, Punjab Agricultural University, Ludhiana, India, pp, 20.

Hundal, S.S. and De Datta, S K, (1984) In situ water transmission characteristics of a tropical soil inder rice based cropping systems. *Agricultural Water Management 8: 387-396*

Mambani, B., De Datta, S.K. and Redulla, A.C (1989) Soilphysical behaviour and crop responses to tillage in lowland rice soil of varying clay content. Plant Soil 126(2):227-235

Mohanty, M, Painuli, D.K. and Mandal, K.G. (2003). Effect of puddling intensity on temporal variation in soil physical conditions and yield of rice (*Oryza sati*va L.) in vertisol of central India. *Soil Tillage Res. 76:83-94*

Moormann, F.R and Van Breemen, N (1978) Rice: Soil, water, land. International Rice Research Institute, Los Banos, Phillipines.

Mousavi, S.F. Yousefi-Moghadam, S., Mostafazadeh-Fard, B. Hemmat, A. and Yazdani, M.R. (2009). Effect of puddling intensity on physical properties of a silty clay soil under laboratory and field conditions. *Paddy Water Environ 7: 45-54*

Naphade, J. D. and Ghyldyal, B. D. (1971) Effect of puddling on physical properties of rice soils. *Indian J Agric Sci* 41 : 1065-67.

Pagliai, M., Painuli, D. K. and Woodhead, T. (1990) Soil Management for Sustainable Agriculture in the Tropics. *Trans 14th Intl congr Soil, Kyoto, Japan,* p. 192.

Pagliai, M, Woodhead, T. and Painuli, D.K. (1987). Towards more efficient use of water and energy in rice-soil puddling: can micromorphometric studies help? Trans XIII Congress of I.S.S.S. Hamburg, 5:436-448

Painuli, D.K., Woodhead, T and Pagliai. (1988). Effective use of energy and water in rice soil puddling. *Soil Tillage Res. 12: 149-161.*

Pandey, S.P., Shanker, H. and Sharma, U. K. (1985) Efficiency of some organic and inorganic residues in relation to crop yield and soil characteristics. *J Indian Soc Soil Sci* 33: 179-81.

Prathapar, S. A., Meyer, W. S. and Cook, F. J. (1989). Effect of cultivation on the relationship between root length density and unsaturated hydraulic conductivity in a moderately swelling soil. *Aust J Soil Res* 27 : 645-650.

Rane, D.B, and Varade, S.B. (1972). Hydraulic conductivity as an index for evaluating performance of different puddlers. J Agrci. Engg. 9(1):11-16

Saito, M, and Kawaguchi, K. (1971). Flocculating tendency of paddy soils. IV Soil structure of paddy plow layers. J. *Sci. Soil Manure, Japan. 42:95-96,* Also in *Soil Sci. Plant Nutrition. 18:202(English Abstract).*

Sanchez, P. A. (1973) Puddling tropical rice soils II Effects of water losses. *Soil Sci* 115 : 303-8.

Sharma, B. R. (1992) Water saving techniques for rice production in eastern region. *Indian Farming* 42 : 19-21.

Sharma, P. K. and De Datta, S. K. (1985) Puddling influence on soil, rice development and yield. *Soil Sci Soc Am J* 49 : 1451-57.

Sharma, P.K. and Bhagat, R.M. (1993). Puddling and compaction effect on water permeability of texturally different soils. *J Indian Soc. Soil Sci. 41:1-6*

Singh, G. (2003) Physical and hydraulic properties of soil as influenced by organic amendments in rice-wheat cropping system. M.Sc Thesis, Department of Soils, Punjab Agricultural University, Ludhiana, Punjab, India

Singh, A. K., Amgain, L. P. and Sharma, S. K. (2000) Root characteristics, soil physical properties and yield of rice (*Oryza sativa*) as influenced by integrated nutrient management in rice-wheat (*Triticum aestivum*) system. *Indian J Agron* 45 (2): 217-22.

Singh, K. B. (1998) Quantification of water balance components in rice-wheat system as influenced by soil and water management. Ph D Thesis, Department of Soils, Punjab Agricultural University, Ludhiana, Punjab, India

Singh, K. B., Gajri, P. R. and Arora, V. K. (2001) Modelling the effects of soil and water management practices on the water balance and performance of rice. *Agricultural Water Management* 49:77-95

Singh, K. B. and Manchanda, J. S. (2008) Effect of puddling on water intake rate and root growth of rice and wheat. *Environment & Ecology* 26(3):1046-1050

Singh, K. B., Jalota, S. K. and Sharma, B. D. (2009) Effect of continuous rice-wheat rotation on soil properties from four agro-ecosystems of Indian Punjab. *Communications in Soil Science and Plant Analysis* 40:2945-2958

Singh, S., Singh, M. P. and Bakshi, R. (1990) Unit energy consumption for paddy -wheat rotation. *Energy Conserv Magmt* 30 : 121.

Sur, H. S., Prihar, S. S. and Jalota, S. K. (1981) Effect of rice-wheat and maize-wheat rotations on water transmission and wheat root development in a Sandy loam of the Punjab, India. *Soil Tillage Res* 1:361-71.

Tiwari, V. N., Tiwari, K. N. and Awasthi, P. N. (2000) Role of *Sesbania rostrata* and phosphomicrobe at varying levels of N in sustaining the production and productivity of soil under rice-wheat/chickpea cropping sequence. *J Indian Soc Soil Sci* 48 (2): 257-62.

Taylor, H. M. (1972) Effect of drying on water retention of a puddled soil. *Soil Sci Soc Am Proc* 36 : 972-73.

Tyagi, N. K., Acharya, N. and Mohanty, P. C. (1975) Effect of puddling implements on percolation losses and water use efficiency in rice field. *Indian J Agric Sci* 45 : 132-35.

Wind, G.P. (1961). Capillary rise and some applications of the theory of moisture in unsaturated soil. Institute of Land and Water Management Research Technical Bulletin 22. Wageningen, the Netherlands.

Wopereis, M. C. S., Wosten, J.H.M, Bouman, J., and Woodhead, T. (1992). Hydraulic resistance in peddle rice soils: measurement and effects on water movement. *Soil Tillage Res. 24 (3):199-209*

Wopereis, M.C.S. , Wosten, J.H.M, ten Berge, H.F.M, Woodhead, T. and Agustin, E.M.D.S. (1993). Comparing the performance of soil water balance model using measured and calibrated hydraulic conductivity data; A case study for dryland rice. Soil Sci. 156(3):133-140

Yoshida, S. Adachi, K. (2002). Influence of puddling intensity on the water retention characteristics of clayey paddy soil. 17 th WCSS, 14-21 August, Thailand, Symposium No. 53, Paper No. 235, pp,1-8

Yunsheng, C. (1983) Drainage of paddy soils in Taihu lake region and its effects. Soil Res. Rept. No. 81. Institute of Soil Science, Academia Sinica, Nanjing, China.

2

Role of Hydraulic Conductivity on Surface and Groundwater Interaction in Wetlands

Cevza Melek Kazezyılmaz-Alhan
Istanbul University, Civil Engineering Department
Turkey

1. Introduction

There has been a growing interest in understanding the mechanisms involved in surface and groundwater interactions since these interactions play a crucial role in the behavior of hydrology and contaminant transport in streams, lakes, wetlands, and groundwater (Hakenkamp et al, 1993; Winter, 1995; Packman & Bencala, 2000; Bencala, 2000; Medina et al, 2002). Wetlands are an important part of water resources since they control peak flow of surface runoff and clean polluted water as downstream receiving water bodies and therefore have been recognized as one of the best management practices (Mitsch & Gosselink, 2000; Moore et al., 2002; Mitchell et al., 2002). Wetlands are located in transitional zones between uplands and downstream flooded systems. Surface and groundwater interactions, which occur in these critical zones, result in a change in surface and groundwater depth. Moreover, pollutants in either surface water or groundwater are mixed and the quality of both sources is affected by each other. Therefore, it is important to understand the role of surface and groundwater interactions on wetland sites and incorporate them into the wetland models in order to obtain accurate solutions.

The definition of a wetland is difficult since there is no definite boundary for wetlands over the landscape and wetland characteristics change. Different definitions have resulted from government agencies that take either legal or ecological criteria as a basis for wetlands within their jurisdiction. In Section 404 of the Clean Water Act of Environmental Protection Agency (EPA), wetlands are defined as "areas that are inundated or saturated by surface or groundwater at a frequency and duration sufficient to support a prevalence of vegetation typically adapted for life in saturated soil conditions". From a hydrologic point of view, the change of surface water level or subsurface water table level through time is important. Usually, areas where the depth of standing water is less than 2 m are considered as wetlands. The amount of water present in wetlands is important to support water supply and water quality. It also affects the type of animals and plants living in these areas. Wetlands are classified according to their ecological and hydrological similarities (Mitsch & Gosselink, 2000).

The type of interaction between groundwater and wetlands depends on the geomorphological location of the wetland. Wetlands gain water if they are located on seepage faces where there is an abrupt change in landscape slope (Figure 1A), or if there is a stream near the wetland location (Figure 1B). Water level in wetlands is changed usually by direct precipitation or runoff. Especially in riverine wetlands, water level changes periodically and very often, since its source comes from rivers. Due to this fact, this type of wetland has more complex interactions which affect its hydraulic/hydrologic characteristics. The water and chemical balances determine the principal characteristics and functions of wetlands. Wetlands are very sensitive to changing hydrological conditions. Since interactions between wetland and groundwater affect the water and chemical balances, it is important to include these interactions into the wetland models.

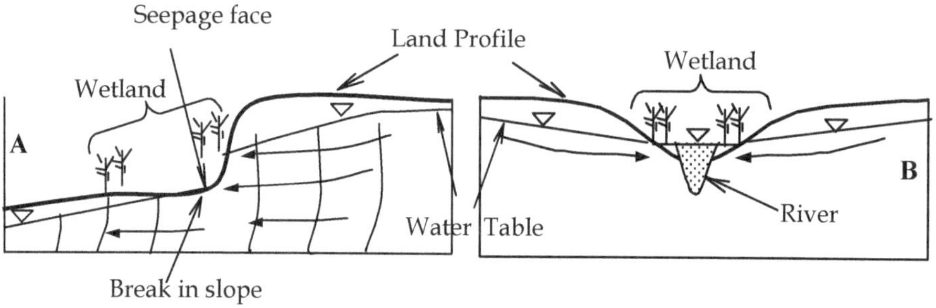

Fig. 1. Wetland and groundwater interactions: (A) Inflow from seepage faces and break in slope of water table. (B) Inflow to streams. (*modified after Winter et al., 1998*)

There have been previous studies reported in the literature that investigate various aspects of surface and groundwater interactions in wetlands. The importance of modeling interactions between groundwater and wetlands and their effect on wetland functions are discussed in detail by Winter et al. (1998), Winter (1999), and Price & Wadington (2000). Experiments are conducted in order to observe the effect of surface and groundwater interactions on wetland hydrology and contaminant transport at different wetland sites (Winter & Rosenberry, 1995; Devito & Hill, 1997; Choi & Harvey, 2000; McHale et al 2004). In addition to these studies, many researchers worked on developing numerical models of wetland hydrology and wetland water quality incorporating surface and groundwater interactions (Restrepo et al., 1998; Krasnostein & Oldham, 2004; Keefe et al., 2004; Crowe et al., 2004; Kazezyılmaz-Alhan et al., 2007).

Examples of recent studies include Harvey et al (2006) who modeled interactions between surface water and groundwater in the wetland area located in central Everglades, Florida, USA in order to quantify recharge and discharge in the basin's vast interior areas. Kazezyılmaz-Alhan & Medina (2008) discussed the effect of surface and groundwater interactions on wetland sites with different characteristics. He et al (2008) developed a coupled finite volume model by using depth averaged two dimensional surface flow and three dimensional subsurface flow for wetlands incorporating surface-subsurface Water‾Resources Investigations, Book 6, Chap A1, U.S. Geological Survey.

McHale, M.R.; Cirmo, C.P.; Mitchell, M.J. & McDonnell, J.J. (2004). Wetland Nitrogen Dynamics in an Adirondack Forested Watershed. Hydrological Processerical model of subsurface vertical flow constructed wetlands called as FITOVERT. Min & Wise (2010) developed a two-dimensional hydrodynamic and solute transport modeling of a large-scaled, subtropical, free water surface constructed wetland in the Everglades of Florida, USA. In this chapter, the role of hydraulic conductivity on surface and groundwater interactions in wetlands is discussed. Both wetland hydrology and wetland water quality are investigated and particularly, the behavior of surface water and groundwater depths and the flux between surface water and groundwater are observed. For this purpose, several models are employed which incorporate surface and groundwater interactions and handle the interactions from different points of view. Among these models are WETland Solute TrANsport Dynamics (WETSAND), Visual MODular Finite-Difference FLOW model (MODFLOW) and EPA Storm Water Management Model (SWMM). WETSAND is a wetland model which has both surface flow and solute transport components, and accounts for upstream contributions from urbanized areas. Visual MODFLOW is a three-dimensional groundwater flow and contaminant transport simulation model. EPA SWMM is a dynamic rainfall-runoff model and calculates surface runoff, channel flow, groundwater flow and depth in aquifer underlying each subcatchment, and water quality. Applications are presented by simulating a conceptual wetland-aquifer system with Visual MODFLOW, the Duke University restored wetland site in the Sandy Creek watershed of Durham, North Carolina in USA with WETSAND and Büyükçekmece wetland site located around Büyükçekmece Lake in Istanbul, Turkey with EPA SWMM.

2. Numerical modelling

This section discusses three numerical models on surface water and groundwater hydrology and contaminant transport. The common point of these models is incorporating surface and groundwater interactions but each model approaches the mechanism and the consequence of these interactions from a different point of view.

2.1 WETland Solute TrANsport Dynamics (WETSAND)

WETland Solute TrANsport Dynamics (WETSAND) is a general comprehensive dynamic wetland model developed by Kazezyılmaz-Alhan et al (2007) which has both water quantity and water quality components, and incorporates the effects of surface and groundwater interactions. While the water quantity component computes water level and velocity distribution as a function of time and space, the water quality component computes Phosphorus and Nitrogen compounds also as a function of time and space. WETSAND also takes into account the effect of flow generated from upstream areas. Figure 2 shows the graphical representation of the conceptual wetland model. During a storm event, overland flow develops on urbanized areas and flows into the wetland area and streams located downstream of the watershed. Overland flow washes off the pollutants which build up on the surface during dry days and these pollutants also reach the wetland site with the overland flow. Besides the overland flow, rainfall and groundwater discharge also contribute to the surface water of the wetland site. Evapotranspiration, infiltration, and groundwater recharge are the water sink terms of the wetland site.

2.1.1 Wetland hydrology

The surface water depth, velocity, and flow through the wetland area are calculated by the diffusion wave equation that applies to the milder slopes (% 0.1 to % 0.01) which is the case in wetlands. The one-dimensional diffusion wave equation is given as follows:

$$\frac{\partial y}{\partial t} + c\frac{\partial y}{\partial x} = K_1 \frac{\partial^2 y}{\partial x^2} + \bar{q} \tag{1a}$$

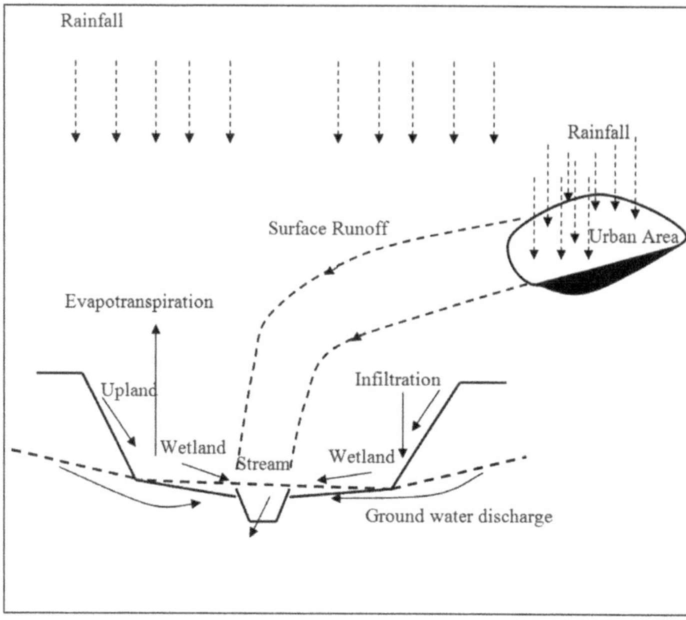

Fig. 2. Schematic of the WETSAND model (Kazezyılmaz-Alhan et al, 2007).

$$\bar{q} = q_r - q_{\inf} - q_{et} + q_{drch} + q_l \qquad c = mV \qquad K_1 = \frac{Vy}{2S_0} \tag{1b}$$

where y is the surface water depth (L), t is time (T), x is the distance (L), c is the wave celerity (L/T), K_1 is the hydraulic diffusivity (L^2/T), $\bar{q}$ is the water source/sink term (L/T), V is the water velocity (L/T), S_0 is the bottom slope (L/L) and m is given according to the flow rate-friction slope relationship (Ponce, 1989). While rainfall (q_r), groundwater discharge (q_{drch}), and lateral inflow (q_l) occupy as water source terms; infiltration (q_{inf}), evapotranspiration (q_{et}), and groundwater recharge (q_{drch}) occupy as water sink terms in the term $\bar{q}$. Infiltration is calculated by the modified version of the Green-Ampt method during unsteady rainfall (Chu, 1978) and evapotranspiration is calculated by the Thornthwaite (1948) method. The groundwater recharge and groundwater discharge terms represent surface and groundwater interaction at the wetland site and are calculated by using the Darcy's Law as follows:

$$q_{drch} = -K_x \frac{\partial H}{\partial x} \begin{cases} < 0 & \text{groundwater recharge} \\ > 0 & \text{groundwater discharge} \end{cases} \tag{2}$$

where H is total head (L), and K_x is the horizontal hydraulic conductivity (L/T). The exchange between surface water and groundwater is calculated in the lateral direction at the banks of the wetland. Overland flow generated over both upland and wetland sites becomes the lateral inflow of the stream. The flow on the wetland site is calculated by the power law for velocity in terms of depth and the friction slope (Kadlec, 1990). This law employs both the effect of a vertical vegetation stem density gradient and a bottom-elevation distribution. The flow rate on a wetland site is given by (Kadlec and Knight, 1996):

$$Q = \begin{cases} K_d \, W y^3 S_0 & \text{dense vegetation} \\ K_s \, W y^3 S_0 & \text{sparse vegetation} \end{cases} \tag{3}$$

where Q is the flow rate in (m^3/day), W is the wetland width (L), and K_d and K_s are the coefficients which reflect the vegetation density with $K_d=1\times10^7$ $m^{-1}day^{-1}$ and $K_s=5\times10^7$ $m^{-1}day^{-1}$. In diffusion wave theory, the term S_0 is replaced by $(S_0 - \partial y / \partial x)$. Therefore, the surface water velocity V on a wetland with a cross-sectional area $A=Wy$ is calculated using both the continuity $Q=VA$ and Equation (3) as follows:

$$V = \begin{cases} K_d y^2 (S_0 - \partial y / \partial x) \\ K_s y^2 (S_0 - \partial y / \partial x) \end{cases} \tag{4}$$

The upper boundary condition of the stream flowing through the wetland site is defined as the upstream surface runoff flowing from urbanized areas and the flow rate in the stream is calculated by using the diffusion wave equation as follows:

$$\frac{\partial Q}{\partial t} + c\frac{\partial Q}{\partial x} = K_1 \frac{\partial^2 Q}{\partial x^2} \qquad c = mV \qquad K_1 = \frac{Q}{2BS_0} \tag{5}$$

where B is the channel width (L) and c and K_1 are the wave celerity (L/T) and the hydraulic diffusivity (L^2/T) in stream, respectively.

2.1.2 Wetland water quality

The water quality component of the WETSAND model calculates the concentration distribution of both total Nitrogen and total Phosphorus through the wetland and along the stream. WETSAND has also the capability to calculate each compound of nitrogen, namely, organic nitrogen, ammonium nitrogen, and nitrate nitrogen, individually. For each constituent, one dimensional advection-dispersion-reaction equation is solved. The equations for nitrogen compounds are coupled through the first order loss rate constants K_{ON} and K_{AN}, which represent ammonification of organic nitrogen into ammonium and nitrification of ammonium into nitrate, respectively. The equations also take into account the vegetation effect of a wetland site represented by plant uptake/release terms as sources and sinks. Finally, the influence of surface and groundwater interactions on contaminant transport is incorporated via the mass flux terms that represent the incoming/outgoing mass due to groundwater recharge/discharge. The surface water velocity in the wetland calculated by the hydrology component of WETSAND is used in the advection term of concentration equations. The concentration formulations of WETSAND are given as follows:
Total Phosphorus (TP):

$$\frac{\partial C_{TP}}{\partial t} = -V\frac{\partial C_{TP}}{\partial x} + \frac{1}{A_x}\frac{\partial}{\partial x}\left(A_x D_x \frac{\partial C_{TP}}{\partial x}\right) + \frac{q_{Lin}}{A_x}\left(C_{TP}^{L} - C_{TP}\right) + \frac{q_{gwd}}{A_x}\left(C_{TP}^{gw} - C_{TP}\right) - K_{TP}C_{TP} \tag{6}$$

Total Nitrogen (TN):

$$\frac{\partial C_{TN}}{\partial t} = -V\frac{\partial C_{TN}}{\partial x} + \frac{1}{A_x}\frac{\partial}{\partial x}\left(A_x D_x \frac{\partial C_{TN}}{\partial x}\right) + \frac{q_{Lin}}{A_x}\left(C_{TN}^{L} - C_{TN}\right) + \frac{q_{gwd}}{A_x}\left(C_{TN}^{gw} - C_{TN}\right) - K_{TN}C_{TN} \tag{7}$$

Organic Nitrogen (ON):

$$\frac{\partial C_{ON}}{\partial t} = -V\frac{\partial C_{ON}}{\partial x} + \frac{1}{A_x}\frac{\partial}{\partial x}\left(A_x D_x \frac{\partial C_{ON}}{\partial x}\right) + \frac{q_{Lin}}{A_x}\left(C_{ON}^{L} - C_{ON}\right) + \frac{q_{gwd}}{A_x}\left(C_{ON}^{gw} - C_{ON}\right) - K_{ON}C_{ON} + J_{RON} \tag{8}$$

Ammonium Nitrogen (AN):

$$\frac{\partial C_{AN}}{\partial t} = -V\frac{\partial C_{AN}}{\partial x} + \frac{1}{A_x}\frac{\partial}{\partial x}\left(A_x D_x \frac{\partial C_{AN}}{\partial x}\right) + \frac{q_{Lin}}{A_x}\left(C_{AN}^{L} - C_{AN}\right) + \frac{q_{gwd}}{A_x}\left(C_{AN}^{gw} - C_{AN}\right) + K_{ON}C_{ON} - K_{AN}C_{AN} - J_{UAN} \tag{9}$$

Nitrate Nitrogen (NN):

$$\frac{\partial C_{NN}}{\partial t} = -V\frac{\partial C_{NN}}{\partial x} + \frac{1}{A_x}\frac{\partial}{\partial x}\left(A_x D_x \frac{\partial C_{NN}}{\partial x}\right) + \frac{q_{Lin}}{A_x}\left(C_{NN}^{L} - C_{NN}\right) + \frac{q_{gwd}}{A_x}\left(C_{NN}^{gw} - C_{NN}\right) + \psi K_{AN}C_{AN} - K_{NN}C_{NN} - J_{UNN} \tag{10}$$

where C is the concentration (M/L^3), C^L is the lateral concentration (M/L^3), C^{gw} is the concentration in groundwater (M/L^3), K is the first order loss rate constant ($1/T$), A_x is the cross-sectional area in x-direction (L^2), D_x is the dispersion coefficient (L^2/T), q_{Lin} is the lateral inflow (L^2/T), q_{gwd} is the groundwater discharge (L^2/T), J_{RON} is the release flux of organic nitrogen from biomass (M/T), J_{UAN} is the uptake flux of ammonium nitrogen absorbed by biomass (M/T), J_{UNN} is the uptake flux of nitrate nitrogen absorbed by biomass (M/T), ψ is the fraction of ammonium that is nitrified, and TP, TN, ON, AN, NN are the subscripts denoting total phosphorus, total nitrogen, organic nitrogen, ammonium nitrogen, and nitrate nitrogen, respectively.

2.2 Visual MODular Finite-Difference FLOW model (MODFLOW)

Visual MODFLOW is a three-dimensional groundwater flow and contaminant transport model that integrates several packages such as MODFLOW-2000, SEAWAT, MODPATH, MT3DMS, MT3D99, RT3D, VMOD 3D-Explorer, WinPEST, Stream Routing, Zone Budget, MGO, SAMG, and PHT3D.

MODFLOW (Modular Three-Dimensional Finite-Difference Ground-Water Flow Model) package solves the three-dimensional ground-water flow equation for a porous medium by using a finite-difference method. MODFLOW is first developed by United States Geological Survey (USGS) (McDonald & Harbaugh, 1988), then continuously improved and enhanced (Harbaugh & McDonald, 1996a; Harbaugh & McDonald, 1996b; Harbaugh et al., 2000; Harbaugh, 2005) and finally integrated into Visual MODFLOW. The three-dimensional movement of groundwater of constant density through porous earth material may be described by the following partial-differential equation (McDonald & Harbaugh, 1988):

$$\frac{\partial}{\partial x}\left\{K_{xx}\frac{\partial h}{\partial x}\right\} + \frac{\partial}{\partial y}\left\{K_{yy}\frac{\partial h}{\partial y}\right\} + \frac{\partial}{\partial z}\left\{K_{zz}\frac{\partial h}{\partial z}\right\} + W = S_s\frac{\partial h}{\partial t} \tag{11}$$

where K_{xx}, K_{yy}, and K_{zz} are the hydraulic conductivities along the x, y, and z coordinate axes, respectively and are assumed to be parallel to the major axes of hydraulic conductivity (L/T), h is the potentiometric head (L), W is a volumetric flux per unit volume and represents sources and/or sinks of water ($1/T$), S_s is the specific storage of the porous material ($1/L$), and t is time (T). MODFLOW takes into account the surface and groundwater interactions in wetlands through the RIVER (RIV) boundary condition via a seepage layer separating the surface water body from the groundwater system as shown in Figure 3. River boundary condition simulates the influence of a surface water body such as rivers, streams, lakes, and wetlands on the groundwater flow. The term, which represents the seepage to or from the surface, is added to the groundwater flow equation in this boundary condition. The flow between the surface water and the groundwater system is given by the following equation:

$$Q_{riv} = \frac{KLW}{M(H_{riv} - h)} \tag{12}$$

where Q_{riv} is the flow between the surface water and the aquifer, taken as positive if it is directed into the aquifer, $Hriv$ is the head in the surface water, L and W are the X-Y dimensions of the River boundary grid cells, M is the thickness of the bed of the surface water body, K is the vertical hydraulic conductivity of the bed material of the surface water body, and h is the groundwater head in the cell underlying the River boundary. The term C_{riv}=KLW/M may be defined as the hydraulic conductance of the surface water-aquifer interconnection which represents the resistance to flow between the surface water body and the groundwater caused by the seepage layer.

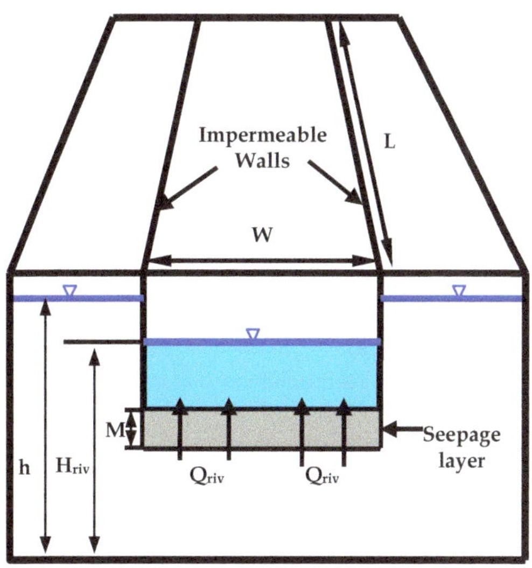

Fig. 3. Schematic of River boundary in MODFLOW (*modified after Visual MODFLOW, 2009*)

MT3DMS (Modular 3-Dimensional Transport Model, Multi-Species) package solves the three-dimensional contaminant transport in groundwater. MT3D is first developed by Zheng (1990) at S. S. Papadopulos & Associates, Inc.; subsequently documented for the Robert S. Kerr Environmental Research Laboratory of the U.S. EPA, then continuously expanded and finally integrated into Visual MODFLOW as a package. MT3DMS employs three different numerical solution techniques, which are the standard finite-difference method, the particle-tracking-based Eulerian-Lagrangian methods, and the higher-order finite-volume TVD method. It has the capability of simulating advection, dispersion/diffusion, and chemical reactions of contaminants in groundwater flow systems under general hydrogeologic conditions. MT3DMS solves the following partial differential equation which describes the fate and transport of contaminants of species k in 3-D:

$$\frac{\partial(\theta C_k)}{\partial t} = \frac{\partial}{\partial x_i}\left(\theta D_{ij}\frac{\partial C_k}{\partial x_j}\right) - \frac{\partial}{\partial x_i}\left(\theta v_i C^k\right) + q_s C_k^s + \sum R_n \tag{13}$$

where θ is the porosity of the subsurface medium (dimensionless), C_k is the dissolved concentration of species k (ML^{-3}), t is time (T), $x_{i,j}$ is the distance along the respective Cartesian coordinate axis (L), D_{ij} is the hydrodynamic dispersion coefficient tensor (L^2T^{-1}), v_i is the seepage or linear pore water velocity (LT^{-1}), q_s is the volumetric flow rate per unit volume of aquifer representing fluid sources (positive) and sinks (negative) (T^{-1}), C_k^s is the concentration of the source or sink flux for species k (ML^{-3}), and $\sum R_n$ is the chemical reaction term ($ML^{-3}T^{-1}$). The transport equation is related to the flow equation through the Darcy's Law:

$$v_i = \frac{q_i}{\theta} = -\frac{K_i}{\theta}\frac{\partial h}{\partial x_i} \tag{14}$$

where K_i is the principal component of the hydraulic conductivity tensor (LT^{-1}) and h is the hydraulic head (L). The hydraulic head is obtained from the solution of the three-dimensional groundwater flow equation (Eqn. 11), which is solved by MODFLOW package.

2.3 Environmental Protection Agency Storm Water Management Model (EPA SWMM)

Environmental Protection Agency Storm Water Management Model (EPA SWMM) is a dynamic rainfall-runoff simulation model of a watershed for a single storm event or for continuous simulation of multiple storms. EPA SWMM also models groundwater flow within the aquifer underlying each subcatchment of the watershed and the interflow between groundwater and the drainage system. The model is extensively used to plan, analyze, and control storm water runoff; to design drainage system components; and to evaluate watershed management of both urban and non-urban areas (Huber and Dickinson, 1988; Rossman, 2010). With the analyses of EPA SWMM, the quantity and quality of surface runoff on each subcatchment; the flow rate, depth, and concentration in each conduit; and groundwater flow and groundwater elevation in each aquifer are obtained. Among the EPA SWMM inputs are precipitation data, subcatchment delineation, pipe system characteristics, and aquifer and soil properties. Change of flow rate (hydrograph), change of groundwater depth, and change of concentration (pollutograph) through time and total simulation summaries are obtained at the end of the analysis.

In EPA SWMM, while precipitation and flow from upstream subcatchments are considered as inflow, infiltration and evaporation are considered as outflow in surface runoff calculation. Flow rate in each conduit is calculated by using the continuity and momentum equations for flood routing. The most general form of flood routing equations is the dynamic wave equations or also known as St. Venant equations which describe unsteady and gradually varied flow. By neglecting the inertial terms in the momentum equation, diffusion wave equations are obtained and by neglecting both inertial and pressure terms, kinematic wave equations are obtained. One can select anyone of these equations as the flood routing option in EPA SWMM according to the characteristics of the modeled watershed. The dynamic wave equations for flow routing in conduits are given as follows (Eagleson, 1970):

$$\frac{\partial Q}{\partial x}+\frac{\partial A}{\partial t}=0 \tag{15}$$

$$\frac{1}{A}\frac{\partial Q}{\partial t}+\frac{1}{A}\frac{\partial Q}{\partial x}\left(\frac{Q^2}{A}\right)+g\frac{\partial y}{\partial x}-g\left(S_0-S_f\right)=0 \tag{16}$$

where Q is flow rate (L^3/T), A is cross-sectional area (L^2), y is water depth (L), S_f is friction slope (L/L), S_0 is bed slope (L/L), g is gravitational acceleration (L/T^2), t is time (T), and x is distance (L). The kinematic wave equation from dynamic wave equations follows (Lighthill and Whitham, 1955):

$$\left.\begin{array}{l}\dfrac{\partial A}{\partial t}+\dfrac{\partial Q}{\partial x}=0\\ Q=\alpha A^m\end{array}\right\}\Rightarrow\frac{\partial A}{\partial t}+\alpha\frac{\partial\left(A^m\right)}{\partial x}=0 \tag{17}$$

where a and m are given according to the flow rate-friction slope relationship. The diffusion wave equation from dynamic wave equations follows (Ponce, 1989):

$$\left.\begin{array}{l}\dfrac{\partial A}{\partial t}+\dfrac{\partial Q}{\partial x}=0\\ S_f=S_0-\dfrac{\partial y}{\partial x}\end{array}\right\}\Rightarrow\frac{\partial Q}{\partial t}+c\frac{\partial Q}{\partial x}=K\frac{\partial^2 Q}{\partial x^2}\qquad c=\frac{\partial Q}{\partial A}\qquad K=\frac{Q}{2BS_0} \tag{18}$$

where c is the diffusion wave celerity (L/T), K is the hydraulic diffusivity (L^2/T), and B is the width (L). EPA SWMM has three options for infiltration calculation which are the Green-Ampt Method, the Integrated Horton Method and the SCS Curve Number Method. The equations for each method are given as follows:
Green-Ampt Method (Huber and Dickinson, 1988):

$$\begin{array}{l}\text{for } \underline{F<F_s}: f=i\\ \quad\text{if } i>K_s \Rightarrow F_s=\dfrac{S_u M}{i/K_s-1}\\ \quad\text{if } i<K_s \Rightarrow F_s \text{ is not calculated.}\end{array} \tag{19}$$

$$\text{for } \underline{F\geq F_s}: f=f_p \text{ and } f_p=K_s\left(1+\frac{S_u M}{F}\right) \tag{20}$$

where F is the cumulative infiltration (L), F_s is the cumulative infiltration of saturated soil (L), i is the rainfall intensity (L/T), K_s is the hydraulic conductivity for saturated soil (L/T), S_u

is the suction head (L), M is the initial moisture deficit (L/L), f is the infiltration rate (L/T), and f_p is the infiltration capacity (L/T).
Integrated Horton Method (Huber and Dickinson, 1988):

$$f_p = f_\infty t + (f_0 - f_\infty) e^{-\alpha t} \qquad f(t) = \min[f_p(t), i(t)] \qquad F(t) = \int_0^t f(\tau) d\tau \tag{21}$$

where f_∞ is minimum infiltration capacity (L/T), f_0 is infiltration capacity for dry soil (L/T), and a is a constant ($1/T$).
SCS Curve Number Method (Ponce and Hawkins, 1996):

$$\frac{F}{S} = \frac{Q}{P - I_a} \tag{22}$$

$$P = Q + I_a + F \tag{23}$$

where F is actual retention (L), S is potential retention (L), Q is actual runoff (L), P is potential runoff (L), and I_a is initial abstraction (L).
The rate of groundwater flow as shown in Figure 4 is calculated as a function of groundwater and surface water levels with the following general equation (Rossman, 2010):

$$Q_{gw} = A_1 (H_{gw} - H^*)^{B_1} - A_2 (H_{sw} - H^*)^{B_2} + A_3 (H_{gw} H_{sw}) \tag{24}$$

where Q_{gw} is the groundwater flow rate per unit area (L^3T^{-1}/L^2), H_{gw} is the height of saturated zone above bottom of aquifer (L), H_{sw} is the height of surface water at receiving node above aquifer bottom (L), H^* is the threshold groundwater height (L), A_1 is the groundwater flow coefficient, B_1 is the groundwater flow exponent, A_2 is the surface water flow coefficient, B_2 is the surface water flow exponent, and A_3 is the surface and groundwater interaction coefficient. If groundwater flow rate per unit area is calculated by using the Darcy's Law, Equation (24) becomes:

$$Q_{gw} = k \frac{(H_{gw} - H_{sw})}{L_a} \tag{25}$$

where $A_1 = A_2 = k/L_a$, k is the hydraulic conductivity (L/T) and L_a is the length of the aquifer, $B_1 = B_2 = 1$, and $H^* = A_3 = 0$. Dupuit-Forcheimer leakage equation is used in groundwater flow calculation, in order to take into account surface and groundwater interactions in watershed modeling:

$$q_{dupuit} = \frac{k}{2L_a}(h_1^2 - h_2^2) \tag{26}$$

where h_1 is the elevation of the highest point of the water table (L), h_2 is the elevation of the water surface in the channel (L) and q_{dupuit} is the flow rate per unit length (L^2/T). If we substitute for $h_1 = 2H_{gw} - h_2$ by assuming that H_{gw} is an average value over the entire horizontal extent of the saturated zone of the aquifer and therefore $H_{gw} = (h_1 + h_2)/2$; $H_{sw} = h_2$; and $Q_{gw} = q_{dupuit}/B$, B being the aquifer thickness (L) in Dupuit-Forcheimer equation, Equation (26) becomes as follows:

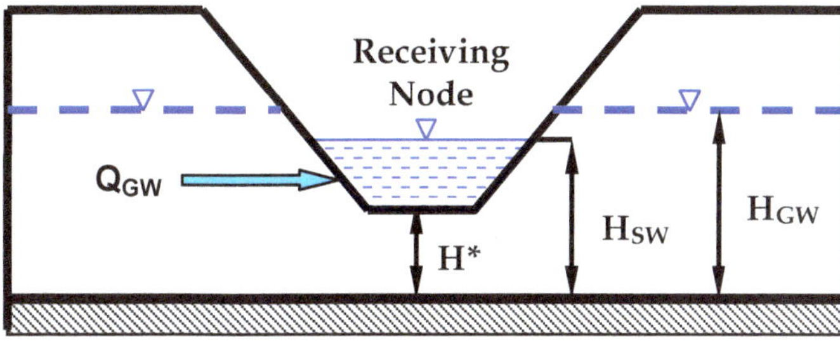

Fig. 4. Schematic of groundwater flow in EPA SWMM (*modified after Rossman, 2010*)

$$Q_{gw} = \frac{2k}{BL_a}(H_{gw}^2 - H_{gw}H_{sw}) \tag{27}$$

When Equation (27) is compared with Equation (24), we see that $A_1 = 2k/BL_a$, B_1=2, A_2=B_2=H^*=0 and A_3= $-2k/BL_a$.

3. Applications

Applications of the models discussed in the previous section are presented in this section. Each model is used to simulate a different case study and shows different aspects of surface and groundwater interactions and the impact of hydraulic conductivity for different scenarios. For comparison purposes, the same set of lateral and vertical hydraulic conductivity values are used in each application. The simulations are conducted under four different combinations of the conductivity values: (A) K_x=K_z=0.01 m/hr, (B) K_x=0.01 m/hr and K_z=0.001 m/hr, (C) K_x=0.1 m/hr and K_z=0.01 m/hr, and (D) K_x=K_z=0.001 m/hr.

3.1 Case study using WETSAND

An application of WETSAND model is presented for Duke University restored wetland site located in North Carolina, USA. The model is simulated to show the importance of surface and groundwater interactions on surface water and nitrogen concentration in wetland and the role of lateral and vertical hydraulic conductivity on surface and groundwater interactions.
The study site is located in the Sandy Creek watershed, in the southern section of Durham County in North Carolina, United States with an area of 554.41 ha (1,370 acres). Storm water runoff generated over part of the Duke University campus and part of the City of Durham flows into the wetland area; its peak flow decreases and its water quality improves after reaching the wetland site. The stream restoration project within the wetland area is completed by closing part of the original streambed of Sandy Creek and opening a new streambed with more meanders. Over 579 m (1900 ft) of stream restoration aims enhancing water flow over the floodplain and removal of nutrients and sediments. Figure 5A shows the position of the wetland site, the boundary of Duke University campus and the tributaries of the Sandy Creek within the Duke University campus area. Figure 5B shows the topography of the restored wetland site and restored part of the Sandy Creek, a total of 20 groundwater sampling well locations and the flooded area behind the earthen dam. The earthen dam was completed also as part of the wetland restoration project which allows for altering the water level in the stream and wetlands.

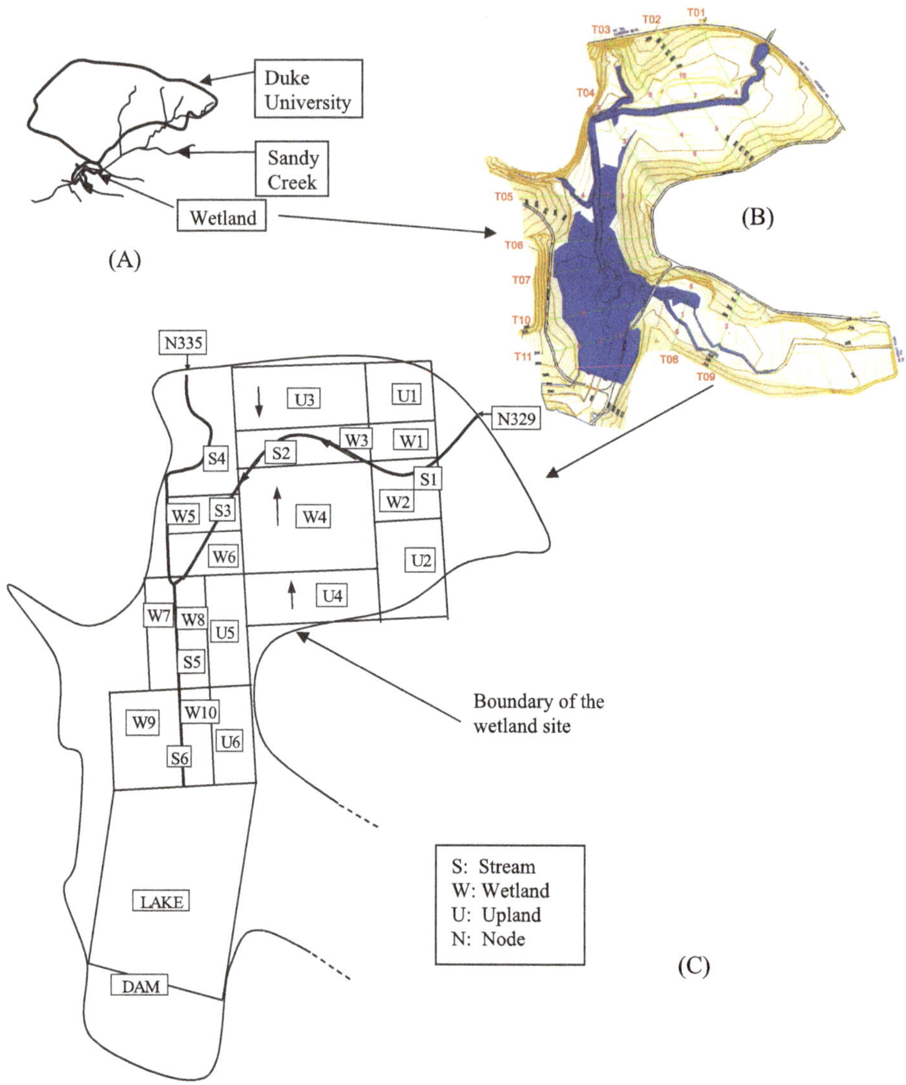

Fig. 5. (A) Boundary of Duke University and wetland site. (B) Restored wetland and stream site at Sandy Creek in Duke Forest (Duke Wetland Center). Contour lines (shown with yellow color) are shown at 30 cm. The stream and lake restoration areas are shown with blue color in the map. Numbers along green lines (T) indicate water well locations. (C) Discretization of the Duke University wetland site (Kazezyılmaz-Alhan et al, 2007). Maps shown are not to scale.

3.1.1 Results and discussion

The WETSAND model has been applied to the Duke University restored wetland site to investigate the role of lateral and vertical hydraulic conductivity on surface and groundwater interactions in terms of wetland hydrology and wetland water quality. For this purpose, first, the study site is discretized into six upland (U) and ten wetland (W) sections and six stream (S) segments (Figure 5C). Nodes N329 and N335 are the receiving nodes of upstream surface runoff. The simulations are conducted by using the rainfall data collected at the nearby Duke Forest Site and groundwater level data recorded by Duke University Wetland Center investigators during year 2002. The average monthly temperature for Durham, NC is obtained from the NOAA National Climatic Data Center (NCDC). Change of surface water depth and concentration of total nitrogen through time are obtained for the case with interaction effect included and with no interaction effect included on each wetland section. Note that, WETSAND provides us with the opportunity to compare interaction and no interaction cases. Here, the results obtained on wetland section four (W4) is presented where we observe groundwater recharge for the major part of the simulation and groundwater discharge towards the end of the simulation.

Figure 6 shows the change of surface water depth on wetland section W4 for different lateral and vertical hydraulic conductivity values with surface and groundwater interaction and with no surface and groundwater interaction. First, we observe that low vertical hydraulic conductivity value (K_z=0.001 m/hr) in Figures 6B and 6D results in higher water depths on the surface as compared to high vertical conductivity value (K_z=0.01 m/hr) in Figures 6A and 6C. Then, when Figures 6B and 6D are compared, we observe that the difference between the surface water depth with interaction effect included and with no interaction effect included is higher in Figure 6B where lateral hydraulic conductivity is K_x = 0.01 m/hr than in Figure 6D where lateral hydraulic conductivity is K_x = 0.001 m/hr. Thus, we conclude that as lateral hydraulic conductivity increases, the effect of surface and groundwater interaction also increases. Moreover, the surface water depth is lower with the surface and groundwater interaction effect for the most part of the simulation as mostly groundwater recharge is observed throughout the simulation on wetland section W4 (Figure 7). When Figures 6A and 6B are compared, eventhough the lateral hydraulic conductivity values are the same, we observe higher difference between the surface water depth with interaction effect and with no interaction effect in Figure 6B. We link this result to different vertical conductivity values: Since the vertical hydraulic conductivity in Figure 6A is higher than the one in Figure 6B, most of the surface water infiltrates into ground in Figure 6A and therefore for both with interaction and no interaction, surface water depth takes the value of about zero. Thus, the vertical hydraulic conductivity plays an indirect role on surface and groundwater interactions especially for the parts where groundwater recharge is dominant. Finally, when Figures 6A and 6C are compared, in the last portion of Figure 6C, we observe a relatively large difference between the surface water depths with interaction and no interaction and the surface water is higher for the case with interaction this time. The reason is that the groundwater discharge comes to the stage in the last part of the simulation (Figure 7) and high lateral hydraulic conductivity (K_x = 0.1 m/hr) in Figure 6C results in an increase in difference of surface water for interaction and no interaction.

Figure 8 shows the change of total Nitrogen concentration on wetland section W4 again for the same set of lateral and vertical hydraulic conductivity values with surface and groundwater interaction and with no surface and groundwater interaction. When Figure 8B and 8D are compared, we observe that high lateral hydraulic conductivity value in Figure 8B (K_x = 0.01 m/hr) results in higher difference between the concentration with interaction and with no interaction. Further, we observe that the concentration for the case with interaction is in general lower than the one with no interaction except for the last part where groundwater discharges in this portion. In the last part, the concentration value reaches about 1.5 mg/l in all figures (Figures 8A-8D), because the total nitrogen concentration in groundwater is defined as 1.5 mg/l. In other words, for the time periods where there is a groundwater discharge, the concentration on the wetland takes the value of groundwater concentration if the surface and groundwater interaction is incorporated into the simulation. For the case with no interaction, the concentration at wetland site reaches a value of only about 0.5 mg/l. Thus, we conclude that it is extremely important to incorporate surface and groundwater interactions into the simulation models as neglecting this physical situation may cause a huge difference in the analysis for certain cases. Moreover, the role of lateral hydraulic conductivity on surface and groundwater interaction is important also in terms of concentration and as lateral hydraulic conductivity increases, the effect of surface and groundwater interaction on concentration also increases. When Figures 8A and 8B are compared, we observe that high vertical hydraulic conductivity (K_z = 0.01 m/hr) results in less oscillation in concentration values and concentration in general reaches a steady state

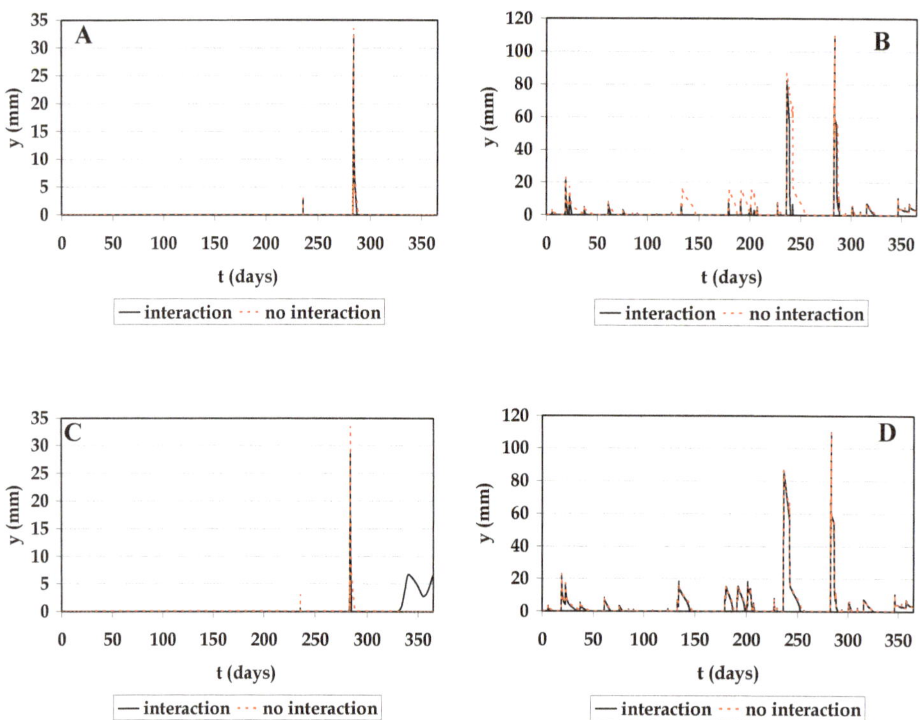

Fig. 6. Comparison of surface water depth on wetland section W4 with surface and groundwater interaction and with no surface and groundwater interaction (A) K_x=K_z=0.01 m/hr, (B) K_x=0.01 m/hr and K_z=0.001 m/hr, (C) K_x=0.1 m/hr and K_z=0.01 m/hr, (D) K_x=K_z=0.001 m/hr.

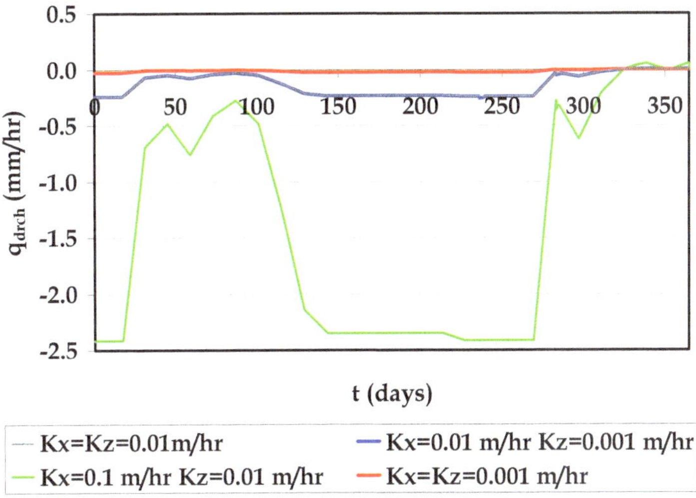

Fig. 7. Water flux between surface and ground on wetland section W4.

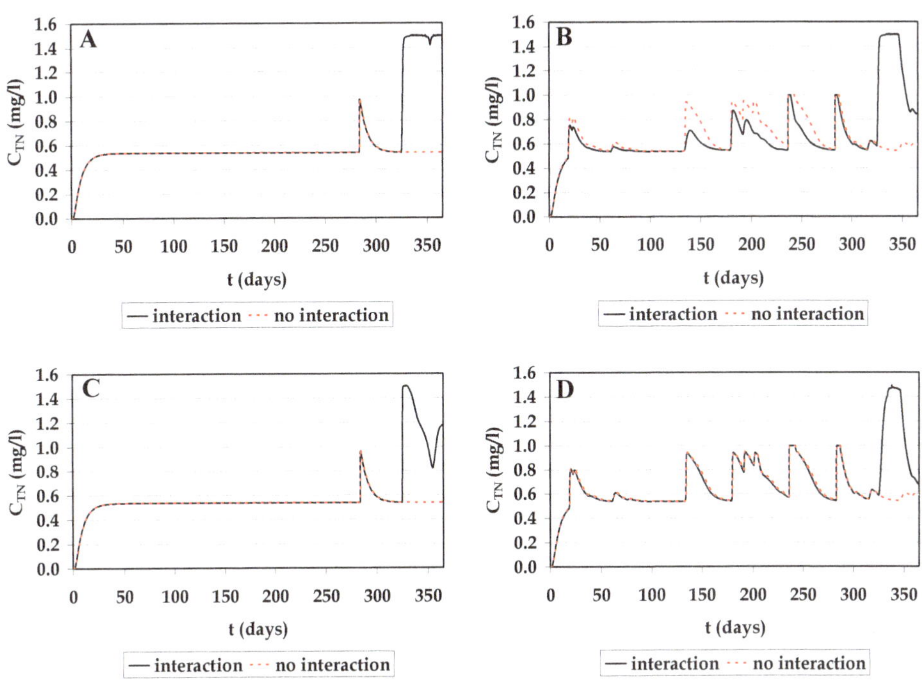

Fig. 8. Comparison of total nitrogen concentration in surface water on wetland section W4 with surface and groundwater interaction and with no surface and groundwater interaction (A) K_x=K_z=0.01 m/hr, (B) K_x=0.01 m/hr and K_z=0.001 m/hr, (C) K_x=0.1 m/hr and K_z=0.01 m/hr, (D) K_x=K_z=0.001 m/hr.

value of about 0.5 mg/l for both with interaction and no interaction. Thus, the vertical hydraulic conductivity plays an indirect role on surface and groundwater interactions especially for the parts where groundwater recharge is dominant. As the vertical hydraulic conductivity increases, the effect of surface and groundwater interaction on concentration decreases.

3.2 Case study using MODFLOW

An application of Visual MODFLOW is presented by using a wetland-aquifer conceptual model. The model is simulated to show different fluxes between wetland surface water and groundwater for different hydraulic conductivities and their effects on concentration distribution in groundwater.

3.2.1 Conceptual model

A model domain of 2000 m × 2000 m with 20 × 20 cells is selected. The aquifer has one layer with a thickness of 10 m; the porosity and the specific storage are selected as θ=0.5 and S_s=10^{-5} m^{-1}, respectively. A wetland site of 300 m × 1300 m is defined within the model domain. The surface water depth at the wetland site is selected as 1m and 10 mg/l of contaminant is assigned to the wetland site for a duration of 100,000 sec (Figure 9).

3.2.2 Results and discussion

The influence of hydraulic conductivity on surface and groundwater interactions in groundwater flow and groundwater contaminant transport modeling is simulated with different conductivity values at a wetland site. Figure 10 shows the water flux between surface water and groundwater along the wetland. Positive values stand for groundwater recharge and negative values stand for groundwater discharge. As it can be seen from the figure, as the lateral hydraulic conductivity increases, the flow between surface water and groundwater also increases in both directions. On the other hand, we don't observe a significant difference between the cases where K_x=K_z=0.01 m/hr and K_x=0.01 m/hr, K_z=0.001 m/hr. Thus, vertical hydraulic conductivity does not play a significant role on surface and groundwater interactions when Visual MODFLOW results are considered.

Figure 11 shows the concentration distribution in groundwater due to the contaminant defined in surface water at the wetland site for the four set of lateral and vertical hydraulic conductivities. As it can be seen from the figure, for the cases where vertical hydraulic conductivity is K_z=0.01 m/hr (Figure 11A) and K_z=0.001 m/hr (Figure 11B), the concentration distribution in groundwater differs slightly and reaches the value of 1×10^{-5} mg/l towards the mid-portion of the wetland site. When Figures 11A and 11C are compared, where the lateral hydraulic conductivity is K_x=0.01 m/hr and K_x=0.1 m/hr, respectively, we see a significant difference in concentration distributions. For the case of K_x=0.1 m/hr, the concentration reaches a value of 5×10^{-4} mg/l. Thus, when the lateral hydraulic conductivity increases, the mass flux and therefore the concentration in groundwater also increases. When Figure 11D is considered, we observe a small fraction of pollutant passed to the groundwater.

We observe that the concentration distribution is observed only in the upper portion of the wetland site where groundwater recharges (see Figure 10). On the other hand, we don't see any contaminant in groundwater in the lower portion of the wetland site where groundwater discharges. Since Visual MODFLOW simulates contaminant transport in groundwater, the results show only pollutants in groundwater due to the mass flux from surface water to groundwater. If contaminant transport in surface water could be simulated, we would expect to see a decrease in surface water concentration in the lower portion of the wetland due to the mass flux of pure water from ground to surface.

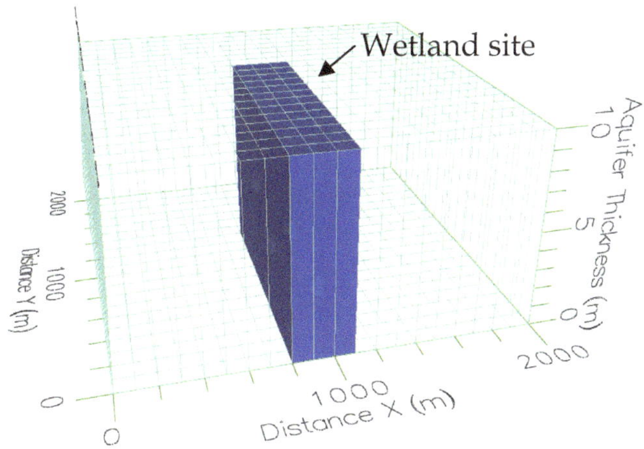

Fig. 9. Schematic of the conceptual wetland-aquifer system.

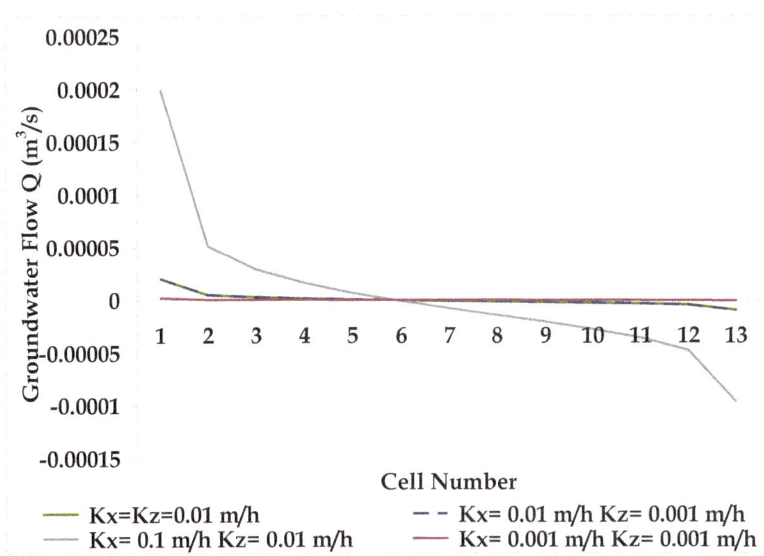

Fig. 10. Water flux between surface and ground on wetland-aquifer system.

3.3 Case study using EPA SWMM

An application of EPA SWMM is presented for Büyükçekmece wetland site located in Istanbul, Turkey. The model is simulated to show the influence of surface and groundwater interactions on groundwater depth and flow and surface runoff concentration and the role of lateral and vertical hydraulic conductivities on surface and groundwater interactions. Büyükçekmece wetland site is located at downstream point of Büyükçekmece Lake in Büyükçekmece Watershed, Istanbul and is one of the most important wetlands of Turkey. Büyükçekmece Watershed has a drainage area of 622 km^2 and supplies a major part of Istanbul's drinking water. A lagoon connects Büyükçekmece Lake with the Marmara Sea. In order to protect the environmental habitat of the lake, a dam was constructed at lake-lagoon-sea interface. About 18 streams gather flow generated over the catchment which are connected to 3 rivers and the rivers flow through the wetland site and reaches the lake. There exist three types of aquifers under Büyükçekmece Watershed: local spaced and cracked Kırklareli limestone, local cracked metamorphic units classified as Istıranca group, and local granular aquifer specified as Pınarhisar formation (Birpınar et al, 2006). The boundaries of Büyükçekmece Watershed and Büyükçekmece wetland site are shown in Figure 12.

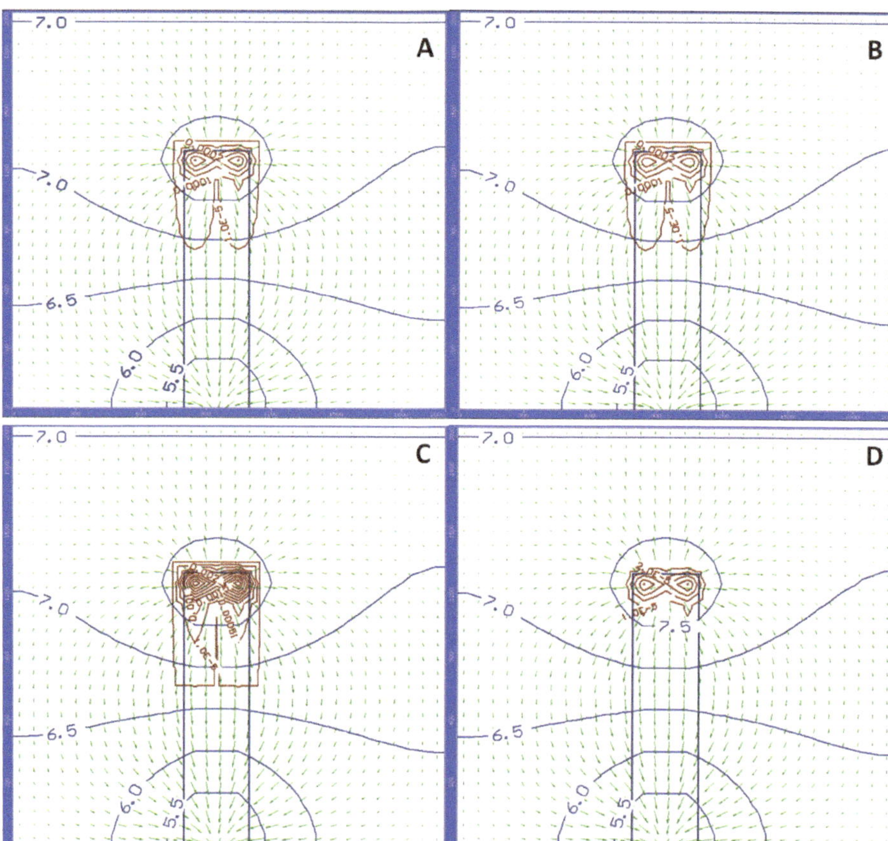

Fig. 11. Concentration distribution (shown with red color in mg/l) in groundwater (A) K_x=K_z=0.01 m/hr, (B) K_x=0.01 m/hr and K_z=0.001 m/hr, (C) K_x=0.1 m/hr and K_z=0.01 m/hr, (D) K_x=K_z=0.001 m/hr.

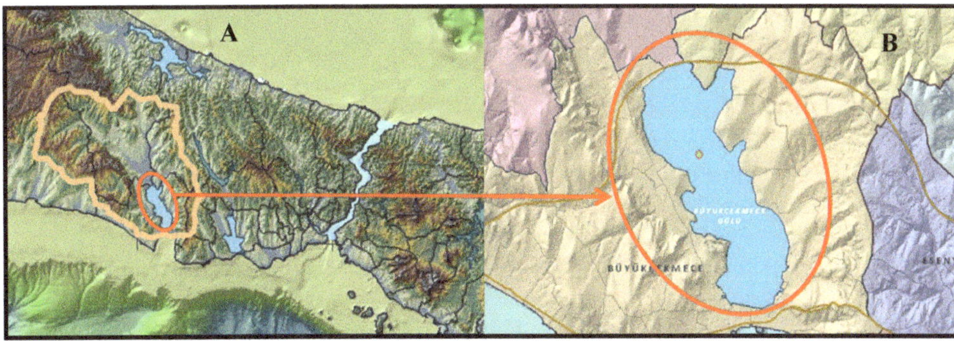

Fig. 12. Büyükçekmece watershed (A) and Büyükçekmece wetland site around Büyükçekmece Lake (B) (Google Earth).

3.3.1 Results and discussion

EPA SWMM has been applied to Büyükçekmece wetland site in order to observe surface and groundwater interactions in groundwater. For this purpose, first a hydrological model for the site is developed by means of discretizing the region into subcatchments and describing the channels and junctions. Slope, area, and width are input data for each subcatchment; length, cross sectional area, and roughness are input data for each conduit; invert elevation, maximum depth, and inflow are input data for each junction. A total number of 167 subcatchments, 118 conduits, and 157 junctions are defined for the site. In addition, a rain gauge is defined in the model in order to introduce the rainfall data in terms of intensity, volume, or cumulative precipitation.

For groundwater flow and groundwater depth simulation, an aquifer is defined under each subcatchment by porosity, wilting point, field capacity, hydraulic conductivity, conductivity slope, and tension slope. In addition, infiltration is calculated with the Green-Ampt option of EPA SWMM and suction head, hydraulic conductivity, and initial soil moisture deficit are defined as Green-Ampt parameters for each subcatchment. The groundwater flow is simulated according to both Dupuit-Forcheimer leakage equation (Eqn. 27) and Darcy's Law (Eqn. 25) in order to see the difference between the cases with surface and groundwater interaction effect included and not included. Figure 13 shows the hydrological model of Büyükçekmece wetland site. Here, the blue lines represent the subcatchment borders and the red lines represent the conduits, namely open channels.

Figure 14 shows the change of groundwater depth through time on wetland section 44s for the cases with interaction effect and with no interaction effect by using the four set of lateral and vertical hydraulic conductivities. As it can be seen from this figure, groundwater depth is affected from both lateral and vertical hydraulic conductivities. For the case with no interaction, the maximum steady state with a value of 3.98 m has been reached with $K_x = K_z = 0.01$ m/hr and the minimum steady state with a value of 3.91 m has been reached with $K_x= 0.01$ m/hr, $K_z = 0.001$ m/hr. The aquifer type with the highest conductivity values, i.e., $K_x= 0.1$ m/hr and $K_z = 0.01$ m/hr reaches the steady state earliest, whereas the aquifer type with the lowest conductivity values, i.e., $K_x = K_z = 0.001$ m/hr reaches the steady state latest. We observe a clear difference between the cases with interaction effect included and with no interaction effect included for each combination of lateral and vertical hydraulic conductivity value. This difference is greater in the rising part with the lowest conductivity values K_x=K_z = 0.001 m/hr and in the descending part with the highest conductivity values K_x= 0.1 m/hr and K_z = 0.01 m/hr. Thus, we conclude that both lateral and vertical hydraulic conductivity have a significant impact on surface and groundwater interactions.

Figure 15A shows the groundwater flow through time for the cases with interaction effect and with no interaction effect for the four set of lateral and vertical hydraulic conductivities. As it can be seen from this figure, we observe the largest difference between interaction effect included and not included with the highest conductivity values K_x= 0.1 m/hr and K_z = 0.01 m/hr. Although it is minor, there is a difference for other combinations of conductivity values, too. Thus, the impact of both lateral and vertical hydraulic conductivity is significant also on groundwater flow. Figure 15B shows the change of surface runoff through time over the study subcatchment (44s), under which the study aquifer lies. As it can be seen from this figure, surface runoff is also affected due to the exchange of water between surface and subsurface.

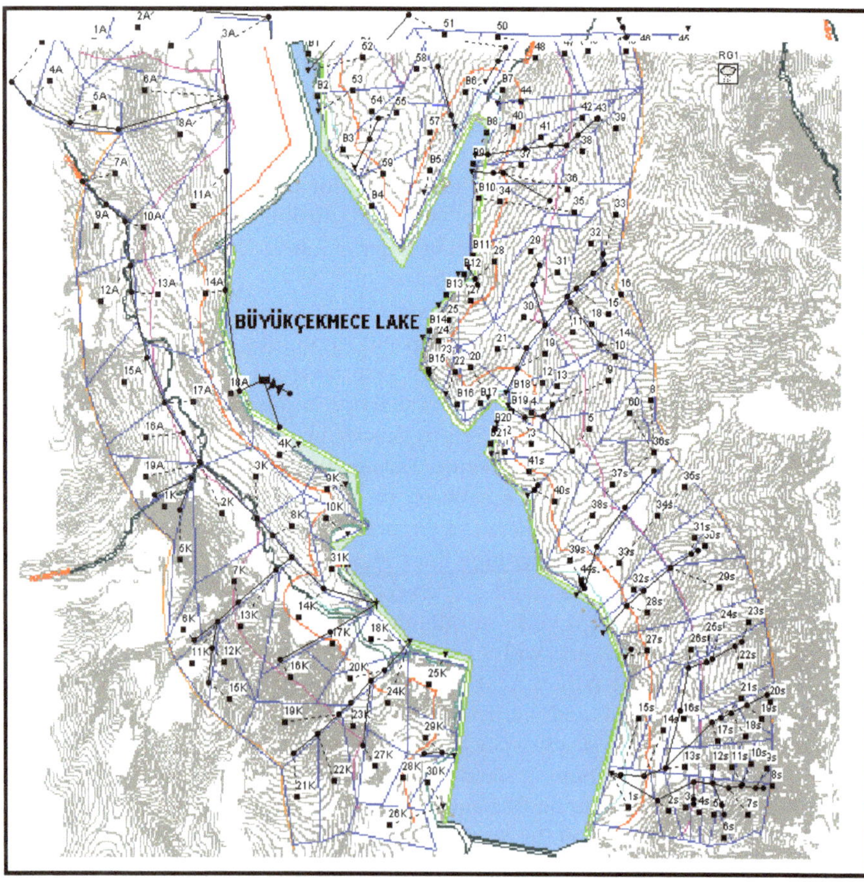

Fig. 13. Hydrological Model of Büyükçekmece wetland site developed by EPA SWMM. Subcatchment borders are shown with blue and conduits are shown with red (not to scale).

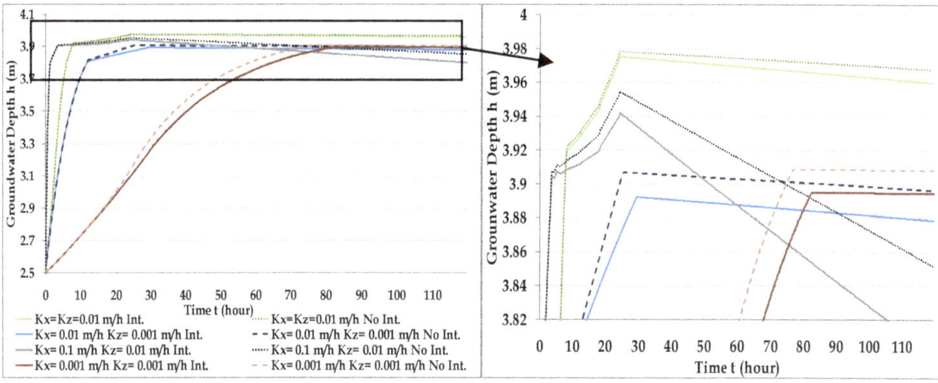

Fig. 14. Comparison of groundwater depth on wetland section 44s with surface and groundwater interaction and with no surface and groundwater interaction.

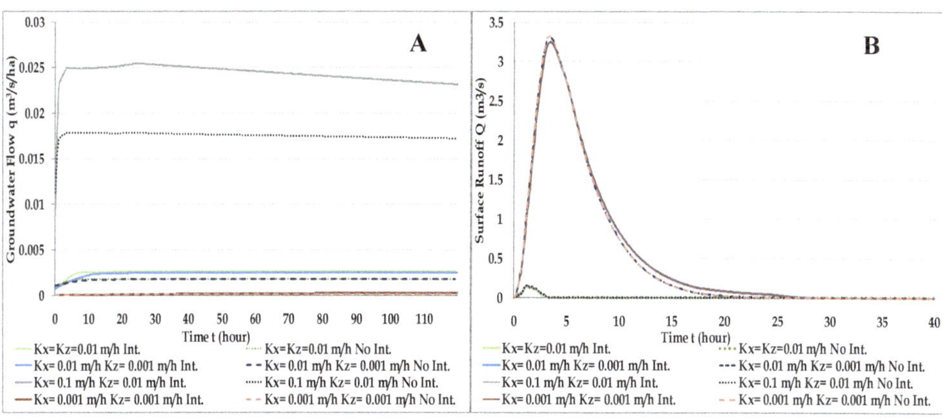

Fig. 15. Comparison of (A) groundwater flow and (B) surface runoff on wetland section 44s with surface and groundwater interaction and with no surface and groundwater interaction.

Figure 16 shows the change of concentration of total nitrogen through time at the outlet of the wetland site which builds up on the catchment during the dry days and is washed off by the surface runoff during a rainfall event. As it can be seen from this figure, the arrival time of the peak concentration to the outlet changes according to different lateral and vertical hydraulic conductivity values. Moreover, we observe different concentration curves for the cases with interaction effect and with no interaction effect. Thus, we conclude that hydraulic conductivity affects both the concentration curve and the arrival time of the peak concentration of the pollutants in the surface runoff significantly.

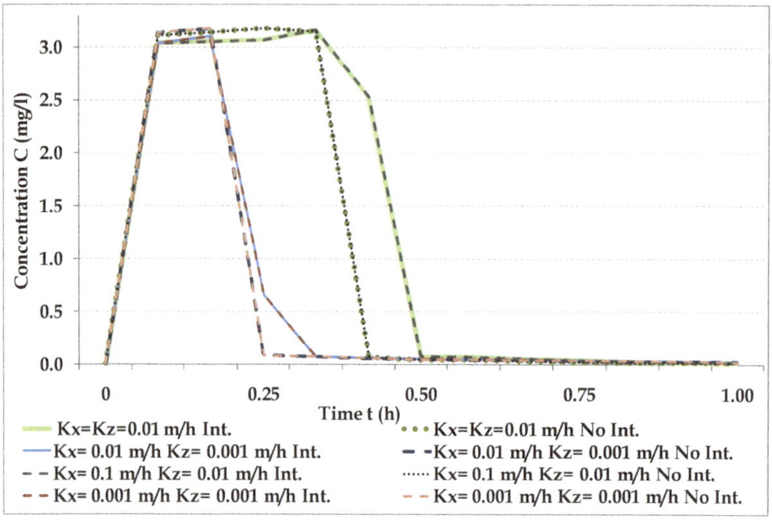

Fig. 16. Comparison of total nitrogen concentration in surface runoff on wetland site with surface and groundwater interaction and with no surface and groundwater interaction.

4. Conclusion

Surface and groundwater interactions play a crucial role in the behavior of hydrology and contaminant transport in streams, lakes, and wetlands. Therefore, it is important to take these interactions into account when modeling water resources. In this chapter, wetland hydrology, wetland water quality, and surface and groundwater interactions are presented. Then, several models, which incorporate surface and groundwater interactions in their hydrological and contaminant transport simulations, are illustrated. These models are WETSAND, Visual MODFLOW and EPA SWMM. Particularly, the role of hydraulic conductivity on surface and groundwater interactions in wetlands is investigated in detail by using these models.

An example study is given for each model. Each model presents the influence of surface and groundwater interactions from a different point of view: WETSAND shows the effect of the interactions on surface water depth and surface water contaminant at a wetland site; Visual MODFLOW shows the effect of the interactions on groundwater flow and groundwater contaminant; EPA SWMM shows the effect of interactions on surface runoff, groundwater depth and flow, and wash off of the contaminant on land surface. Thus, each model has a different feature and therefore is used in presenting different aspects and characteristics of surface and groundwater interactions. Simulations are conducted for a conceptual wetland-aquifer system with Visual MODFLOW, the Duke University restored wetland site in the Sandy Creek watershed of Durham, North Carolina in USA with WETSAND and Büyükçekmece wetland site located around Büyükçekmece Lake in Istanbul, Turkey with EPA SWMM.

The results clearly show that both lateral and vertical hydraulic conductivity influence surface and groundwater interactions. The effects of surface and groundwater interactions play a significant role on wetland dynamics and therefore should be taken into account when modeling wetland hydrology and wetland solute transport.

5. Acknowledgment

The author would like to express her gratitude to General Directorate of State Hydraulics Works (DSI), Turkish State Meteorological Service (DMI), Istanbul Metropolitan Municipality (IBB) and Istanbul Municipality Waterworks (ISKI) for providing rainfall data, site map, soil formation, and valuable discussions on the Büyükçekmece wetland site. Rainfall data for Duke University restored wetland site were obtained from a project supported by the Office of Science (BER), U.S. Department of Energy, Grant No. DE-FG02-95ER62083. The author wishes to acknowledge also the Facilities Management Department of Duke University for partial funding and valuable discussions on stormwater and wetland hydrologic and water quality modeling needs on Duke University Campus, and the Duke University Wetland Center for providing water level data.

6. References

Bencala, K.E. (2000). Hyporheic Zone Hydrological Processes. *Hydrological Processes*, Vol.14, No.15, pp . 2797-2798.

Birpınar M.E. ; Özkılıç N. ; Aktürk M.A. ; Mumcuoğlu H. ; Pirim S. ; Kurtuluş S. ; Aykırı S. ; Yaman M. ; Vardar A. ; Kuzlu A. ; Çakmak, B. ; Akdağ, B. ; Cihan, F. ; Selvi, G.M. ; Şanlımeşhur, İ. ; Sadıkel, İ. ; Özdoğan, J. ; Karabulut, A. ; Akkaş, S. Ö. ; Bukni, R.; Tezcan, Ş.; Erdoğan, T. & Karaaslan, Y. (2006). *Environmental Report of Istanbul for year 2005,* Available from http://www.cedgm.gov.tr.

Choi, J. & Harvey, J.W. (2000). Quantifying Time-Varying Ground-Water Discharge and Recharge in Wetlands of the Northern Florida Everglades. *Wetlands*, Vol.20, No.3, pp. 500-511.

Chu, S.T. (1978). Infiltration During an Unsteady Rain. *Water Resources Research*, Vol.14, No.3, pp. 461-466.

Crowe, A.S.; Shikaze, S.G. & Ptacek, C.J. (2004). Numerical Modelling of Groundwater Flow and Contaminant Transport to Point Pelee Marsh, Ontario, Canada. *Hydrological Processes*, Vol.18, No.2, pp. 293-314.

Devito, K.J. & Hill, A.R. (1997). Sulphate Dynamics in Relation to Groundwater-Surface Water Interactions in Headwater Wetlands of the Southern Canadian Shield. *Hydrological Processes*, Vol.11, No.5, pp. 485-500.

Eagleson, P.S. (1970). *Dynamic Hydrology*, McGraw-Hill, Inc.

EarthInfo Inc. (2004). USGS Daily Values East. CO: EarthInfo Inc.

Giraldi, D.; Vitturi, M. de M. & Iannelli, R. (2010). FITOVERT: A Dynamic Numerical Model of Surbsurface Vertical Flow Constructed Wetlands. *Environmental Modelling & Software*, Vol.25, pp. 633-640.

Hakenkamp, C.C.; Valett, H.M. & Boulton, A.J. (1993). Perspectives on the Hyporheic Zone: Integrating Hydrology and Biology. Concluding Remarks. *Journal of the North American Benthological Society*, Vol.12, No.1, pp. 94-99.

Harbaugh A.W. & McDonald M.G. (1996a). User's Documentation for MODFLOW-96, an Update to the U.S. Geological Survey Modular Finite-Difference Ground-Water Flow model. *Open-File Report 96-485*, USGS.

Harbaugh A.W. & McDonald M.G. (1996b). Programmer's Documentation for MODFLOW-96 an Update to the U.S. Geological Survey Modular Finite-Difference Ground-Water Flow Model", *Open-File Report 96-486, USGS.*

Harbaugh A.W.; Banta E.R.; Hill M.C. & McDonald M.G. (2000). MODFLOW-2000, the US Geological Survey Modular Ground-Water Model -- User Guide to Modularization Concepts and the Ground-Water Flow Process", *Open-File Report 00-92*, USGS.

Harbaugh, A.W. (2005). MODFLOW-2005, the U.S. Geological Survey Modular Ground-Water Model -- the Ground-Water Flow Process: U.S. Geological Survey Techniques and Methods. 6-A16, variously p.

Harvey, J.W.; Newlin, J.T. & Krupa, S.L. (2006). Modeling Decadal Timescale Interactions Between Surface Water and Ground Water in the Central Everglades, Florida, USA. *Journal of Hydrology*, Vol.320, pp. 400-420.

He, Z.; Wu, W. & Wang, S.S.Y. (2010). Coupled Finite-Volume Model for 2D Surface and 3D Subsurface Flows. *Journal of Hydrologic Engineering, ASCE*, Vol.13, No.9, pp. 835-845.

Herron, N. & Croke, B. (2009). Including the Influence of Groundwater Exchanges in a Lumped Rainfall-Runoff Model. *Mathematics and Computers in Simulation*, Vol. 79, pp. 2689-2700.

Huber, W.C. & Dickinson, R.E. (1988). *Storm Water Management Model, Version 4, User's Manual*, Athens, GA.: Environmental Research Laboratory, Office of Research and Development, U.S. Environmental Protection Agency (EPA).

Kadlec, R.H. (1990). Overland Flow in Wetlands: Vegetation Resistance. *Journal of Hydraulic Engineering, ASCE*, Vol.116, No.5, pp. 691-706.

Kadlec, R.H. & Knight, R.L. (1996). *Treatment Wetlands*, CRC Press, Inc.

Kazezyılmaz-Alhan, C. M. & Medina, M. A. Jr. (2008). The Effect of Surface/Ground Water Interactions on Wetland Sites with Different Characteristics. *Desalination*, Vol.226, pp. 298–305.

Kazezyılmaz-Alhan, C. M., Medina, M. A. Jr. & Richardson, C. (2007). A Wetland Hydrology and Water Quality Model Incorporating Surface Water/Groundwater Interactions. *Water Resources Research*, Vol.43, No.4, W04434, pp. 1-16.

Keefe, S.H.; Barver, L.B.; Runkel, R.L.; Ryan, J.N.; McKnight D.M. & Wass, R.D. (2004). Conservative and Reactive Solute Transport in Constructed Wetlands. *Water Resources Research*, Vol.40, No.1, W01201.

Krasnostein, A.L. & Oldham, C.E. (2004). Predicting Wetland Water Storage. *Water Resources Research*, Vol.40, No.10, W10203.

Lighthill, M.J. & Whitham, G.B. (1955). On Kinematic Waves. I. Flood Movement in Long Rivers. *Proceedings, Royal Society of London, London, England, Series A*, Vol.229, No.1178, pp. 281-316.

McDonald, M.G. & Harbaugh, A.W. (1988). A Modular Three-Dimensional Finite Difference Groundwater Flow Model. Techniques of Water-Resources Investigations, Book 6, Chap A1, *U.S. Geological Survey.*

McHale, M.R.; Cirmo, C.P.; Mitchell, M.J. & McDonnell, J.J. (2004). Wetland Nitrogen Dynamics in an Adirondack Forested Watershed. *Hydrological Processes*, Vol.18, No.10, pp. 1853-1870.

Medina, M.A.; Doneker, R.L.; Grosso, N.; Johns, D.M.; Lung, W.; Mohsen, M.F.N.; Packman, A.I. & Roberts, P.J. (2002). Chapter 1: Surface Water-Ground Water Interactions and Modeling Applications, In: *Environmental Modeling and Management: Theory, Practice and Future Directions,* Chien, C.C.; Medina, M.A. Jr.; Pinder, G.F; Reible, D.R.; Sleep, B.E. & Zheng, C., (Eds.), Published by Today Media, Inc. for the DuPont Company, Wilmington, pp. 1-62.

Min, J.-H. & Wise, W.R. (2010). Depth-Averaged, Spatially Distributed Flow Dynamic and Solute Transport Modeling of a Large-Scaled, Subtropical Constructed Wetland. *Hydrological Processes,* Vol.24, pp. 2724-2737.

Mitchell G.F.; Hunt, C.L. & Su, Y.M. (2002). Mitigating Highway Runoff Constituents via a Wetland. *Soil Mechanics 2002 Transportation Research Record,* Vol.1808, pp. 127-133.

Mitsch, W.J. & Gosselink, J.G. (2000). *Wetlands, Third Edition,* John Wiley, New York.

Moore, M.T.; Schulz, R.; Cooper, C.M. & Rodgers, J.H. (2002). Mitigation of Chlorpyrifos Runoff Using Constructed Wetlands. *Chemosphere,* Vol.46, No.6, pp. 827-835.

Packman, A.I. & Bencala, K.E. (2000). Modeling Surface-Subsurface Hydrologic Interactions, In: *Streams and Ground Waters,* Jones, J.B. & Mulholland, P.J. (Eds.), Academic Press, San Diego, CA, pp. 45-80.

Ponce, V.M. (1989). *Engineering Hydrology: Principles and Practices.* Prentice Hall, Inc, Englewood Cliffs, New Jersey 07632.

Ponce, V.M. & Hawkins, R.H. (1996). Runoff Curve Number: Has It Reached Maturity? *Journal of Hydrologic Engineering-ASCE,*Vol 1, No.1, pp:11-19.

Price, J.S. & Wadington, J.M. (2000). Advances in Canadian Wetland Hydrology and Biochemistry. *Hydrological Processes,* Vol.14, No.9, pp. 1579-1589.

Restrepo, J.I.; Montoya, A.M. & Obeysekera, J. (1998). A Wetland Simulation Module for the MODFLOW Ground Water Model. *Ground Water,* Vol.36, No.5, pp. 764-770.

Rossman, L.A. (2010). *Storm Water Management Model, User's Manual, Version 5.* Water Supply and Water Resources Division National Risk Management Research Laboratory, Cincinnati, OH, U.S. Environmental Protection Agency, EPA/600/R-05/040.

Schlumberger Water Services. (2009). *Visual MODFLOW 2009.1 User's Manual.* For Professional Applications in Three-Dimensional Groundwater Flow and Contaminant Transport Modeling.

Thornthwaite, C.W. (1948). An Approach Toward a Rational Classification of Climate. *Am. Geogr. Rev.,* Vol.38, pp. 55-94.

Winter, T.C. (1995). Recent Advances in Understanding the Interaction of Groundwater and Surface water. *Reviews of Geophysics,* Supplement, pp. 985-994.

Winter, T.C. & Rosenberry, D.O. (1995). The Interaction of Ground Water with Prairie Pathole Wetlands in the Cottonwood Lake Area, East-Central North Dakota, 1979-1990. *Wetlands,* Vol.15, No.3, pp. 193-211.

Winter, T.C.; Harvey, J.W.; Franke, O.L. & Alley, W.M. (1998). Ground Water and Surface Water A Single Resource. *U.S. Geological Survey Circular 1139.*

Winter, T.C. (1999). Relation of Streams, Lakes, and Wetlands to Groundwater Flow Systems. *Hydrogeology Journal*, Vol. 7, pp. 28-45.

Zheng, C. (1990). MT3D, A Modular Three-Dimensional Transport Model for Simulation of Advection, Dispersion and Chemical Reactions of Contaminants in Groundwater Systems. *Report to the U.S. Environmental Protection Agency*, Robert S. Kerr Environmental Research Laboratory, Ada, OK.

3

Variation in Hydraulic Conductivity by the Mobility of Heavy Metals in a Compacted Residual Soil

Rejane Nascentes[1], Izabcl Christina Duarte de Azevedo[2] and Ernani Lopes Possato[3]
[1]*Universidade Federal de Viçosa, Campus de Rio Paranaíba*
[2]*Universidade Federal de Viçosa, Campus Universitário*
[3]*Universidade Federal de Lavras , Campus Universitário*
Brazil

1. Introduction

There is a growing consciousness worldwide that progress must be linked to environmental preservation. However, in order to preserve the environment it is necessary to know it and only with an understanding of the mechanisms that regulate the integration of man with nature is it possible to use the environment resources without degrading or destroying it.

Heavy metals are important environmental contaminants that are toxic above a given concentration. Causes of soil contamination by metals include domestic and industrial solid waste disposal, atmospheric deposition of vehicular and industrial emissions, agricultural use of fertilizers, soil additives and pesticides and disposal of crop wastes (Alloway, 1995). Underground waters may be contaminated when metals levels exceed the maximum soil retention capacity.

Population growth and the consequent increase in waste generation has led to an increased demand for technologies that decrease the environmental impact of these wastes, especially with regard to barrier systems used to minimize the infiltration of waste leachates and contain migration of contaminants through soils and underground water in areas of waste disposal.

Tropical soils are common in Brazil. Compacted soils of this type is either used alone or associated with geomembranes have been used as liners in industrial and urban solid waste disposal areas. However a great deal of uncertainty exists with regard to use of these soils since few studies have been undertaken to evaluate their applicability. More studies on the interactions that occur between tropical soils and contaminant solutions are therefore necessary because these interactions may modify properties, such as hydraulic conductivity, which are important in controlling contaminant transport through soils.

The only legal requirements for liners are hydraulic conductivity limits. However, sensibility to contaminants of the soil used for impermeabilization may affect its structure and modify the liner layer by increasing its hydraulic conductivity, thus favoring contamination. Microscopic mechanisms responsible for thecha nges are of physico-chemical and/or purely chemical nature. Redistribution of the pore space driven by dehydration of clay fraction, and rearrangement of clay particles (flocculation, peptization, andmicro-migration) togetherwith chemical reactions between contaminants and claymineral, such as dissolution of the solids are believed to be the most important causes of the permeability evolution (Kaczmarek, et al., 1997). Laboratory testing of soils used in liner layers should therefore be of long enough duration to allow for long term interactions to occur between soil and leachate.

Some of the most important parameters used to express solute mobility trough soil are diffusion, hydrodynamic dispersion coefficients and the retardation factor. Given the scarcity of information on these parameters in Brazilian tropical soils, especially in subsurface horizons and under compacted soil conditions, the main objectives of this study were to evaluate mobility of six heavy metals (Mn2+, Zn2+, Cd2+, Cu2+, Pb2+ and Cr3+) in residual compacted soil and to study the variation in soil hydraulic conductivity during percolation of a multispecies metal contaminant solution through the soil.

2. Background

Underground water deposits are generally more protected from pollution sources than surface water since in the former the overlying soil layer acts as a chemical and physical filter. The ease with which a contaminant reaches the underground water will depend on whether the aquifer is freatic or confined, on the aeration zone depth, the aeration zone and aquifer permeability, the level of organic matter present in the soil and on the types of oxides and minerals clay existing in the soil. Deeper aeration zones permit a longer filtration time and also increase the exposure time of contaminants to oxidizing and adsorbing agents present in this layer. Soil with a higher organic fraction has a higher capacity to adsorb heavy metals.

A contaminant may go through a series of chemical, biochemical and photochemical reactions and physical interactions with soil constituents before reaching underground water. These reactions may neutralize, modify or retard the polluting effect.

The main functions of liner systems are to minimize infiltration of percolates and contain migration of contaminants to the soil and underground water. To properly design liner systems, not only must contaminant flux be determined but the different physicochemical mechanisms that influence transport in contamination evolution must also be known. Although legal requirements for liner materials only establish limits for hydraulic conductivity (maximum value typically limited to 10-6 or 10-7 cm/s), at least four mechanisms control contaminant transport across impermeable layers: advection, diffusion, dispersion and sorption. For practical and economic reasons natural soils alone or combined with geomembranes are being used in these barrier systems in domestic and industrial waste disposal areas.

Soils in tropical climates have greatly different mineralogical compositions than those of temperate regions since they develop under hot humid conditions and minerals in more advanced stages of weathering are generally present. Kaolinite is the most abundant silicate mineral, goethite and hematite are the most abundant iron oxides and gibsite is almost the only aluminum oxide present in tropical soils.

Mineralogical composition has a profound affect on metals retention in soils. In most tropical soils the adsorption of metals is quite intense due to the oxide composition, since iron and aluminum oxides retain heavy metals with high energy in both specific and non-specific interactions.

Two important classes of tropical soils are the lateritic and saprolitic soils. Lateritic soils are often denominated as mature residual soils or red tropical clays while saprolitic soils are called young residual soils or soils from rock alteration.

The clay soil fraction of lateritic soils is essentially composed of clay minerals of the kaolinite group and of hydrated iron and/or aluminum hydroxides and oxides. A characteristic of lateritic soil structure is that macropores are formed in the empty spaces between soil aggregates and micropores are formed within the aggregates during the laterization process. Consequently, lateritic soils in their natural state have low density, high permeability and high porosity that are associated with the low support capacity of these soils. However, when compacted, these soils exhibit a high support capacity.

The variation in permeability of the impermeabilization layers is as important an aspect of soil and underground water contamination as are the soil components, their physicochemical properties and soil retention mechanisms. Contact between the contaminant solution and soil that can cause spatial redistribution because of clay particle rearrangement (flocculation, dispersion, peptization and micro-migration) together with chemical reactions between contaminants and clay minerals, such as solids dissolution and precipitation, are the most important causes of variations in permeability.

The initial soil structure varies with compaction humidity, energy and degree of compaction. According to Boscov (1997), significant variations in permeability may occur within a relatively small range of compaction humidity and density because of the formation of different structural arrangements.

In order to better understand the interactions that occur between tropical soils and contaminant solutions and how these interactions can alter soil properties it is necessary to perform laboratory tests for sufficient time so that long term interactions between soil and the percolating solution may occur.

3. Heavy metals

The meaning of the term heavy metals is controversial and a variety of definitions based on different criteria can be found in the literature. In the density based definition, heavy metals are high density (≥ 6.0 g/cm3) chemical elements and their ions belonging to the transition and non-transition groups of the periodic table (Matos et al., 1999). According to Guilherme et al. (2005), the term trace element has been preferred over heavy metal in several recent publications since the latter has never been formally defined by an official organization of chemistry professsionals.

Some heavy metals, such as Co, Cu, Fe, Mn, Mo, Ni and Zn, are essential human, animal and plant elements. Other elements, such as Cd, Hg and Pb, have no known biological function (Srivastava & Gupta, 1996). Both essential and non-essential metals can cause metabolic problems in living beings if absorbed above a certain amount (McBride, 1994). These elements´ capacities to accumulate in living tissue and to concentrate along the food chain increases the chance of their causing disturbances in ecosystems that may occur even after the release of the metals is stopped. (Tavares & Carvalho, 1992).

Studies on the behavior of heavy metals in soils have concluded that soil retention of these elements depends on the nature of the solid phase and the proportions of its constituents, the properties of the liquid phase and the metal species present in the soil solution (Sposito, 1984; Yuan & Lavkulich, 1997; Naidu et al., 1998).

The concentration of heavy metals in the soil solution results from equilibrium between precipitation, dissolution, complexation and adsorption reactions and is affected by various factors, such as soil type, climate, vegetative cover, and chemical form of the elements (Cooker & Matthews, 1983). However, given the equilibrium changes and chemicals forms of metals in wastes and soil and the possibility of exceeding the soil pollutant retention capacity, the metals may be leached, especially under acid conditions, and may thus reach the underground water. The specific surface area, texture, apparent density, temperature, pH, redox potential, cation exchange capacity (CTC), organic matter content, amount and type of clay minerals and metals and ion competition are among the soil properties that affect the metals reactions and their mobility in soil (Matos, 1995).
The existence of competition for adsorption sites between ions has been recognized by many researchers (Matos, 1995), and it has been observed that the rate of adsorption of any ionic species decreases with the increase in number of competing species. Factors such as solution pH, concentration and the nature of competing species affect the competitive adsorption.
Soils generally have a large variety of adsorption sites with different bonding properties and contain abundant aqueous ionic and non-ionic complexes capable of participating in adsorption processes and possibly in metals precipitation processes.
The soil CTC is one of the most important indicators of heavy metals retention capacity in the soil solid phase. Soils with higher CTC values generally have greater metals adsorption capacity than those with lower CTC values (Lake, 1987).

4. Mass transport mechanisms in porous media

The accumulation of contaminants in soil is a consequence of soil-solute physicochemical interactions arising from transport through the soil. Physical, chemical and biochemical mechanisms can govern solute transport in hydrogeological environments. The most important mechanisms in contaminant transport across saturated clay soil layers are the physical mechanisms of advection and diffusion. In the case of transport in aquifers, advection and dispersion are usually the most important mechanisms.
The transfer of the solute from the interstitial fluid to the solid soil particles is as important as the physical mechanisms. The transfer processes depend on the chemical composition of the solute, its reactivity and organic and inorganic content as well as on soil constituents and characteristics and system pH. These processes may include ion exchange sorption reactions (adsorption and desorption), precipitation and complexation. Biodegradation and radioactive decay are other mechanisms that may be involved.
According to Rowe et al. (1995), the most important factor in soil-contaminant interaction processes and substance transport through soil is probably the diffuse double layer expansion-contraction phenomenon.

5. Diffuse double layer

The thickness of the diffuse double layer (Equation 1) depends on charge density, surface electric potential, electrolyte concentration, valence, pH, dielectric constant of the medium and temperature. Changes in any of these variables may cause alterations in system behavior since forces of repulsion and attraction depend on interaction between adjacent double layers. In general, an increase in thickness of the diffuse double layer corresponds to a lower tendency of the particles in suspension to flocculate. That is, the thicker the double layer, the thinner and more torturous will be the path of percolating solutions in the soil and consequently, the lower the hydraulic conductivity will be.

$$\frac{1}{K} = \left(\frac{\varepsilon_0 D\, k\, T}{2 n_0 e^2 v^2} \right)^{\frac{1}{2}} \quad (1)$$

where 1/K is the electric double layer thickness; (= 8,8542 x 10-12$C^2J^{-1}m^{-1}$; D is the dielectric constant; k is the Boltzman constant; T is the temperature; n0 is the ionic concentration; v is the cationic valence and e is the electron charge.

For a constant void volume, contraction or flocculation of the diffuse double layer causes an increase in voids between soil particles, and increases the hydraulic conductivity and accelerates the advance of percolating fluid. If, on the contrary, an ion exchange reaction that favors expansion of the diffuse double layer takes place, the hydraulic conductivity will decrease and the percolating fluid will be slowed. At a constant void volume, substitution of monovalent cations by bivalent or trivalent cations on the clay particle surface and an increase in temperature are among the factors that cause diffuse double layer contraction and can lead to dramatic increases in hydraulic conductivity (Rowe et al, 1995).

Hydraulic conductivity is inversely proportional to resistance that the medium offers to fluid flow caused by a hydraulic gradient and depends on the characteristics of both the fluid and the porous medium. According to Lambe (1969), the factors with greatest effect on soil permeability are its composition, void index, structure, degree of saturation as well as fluid characteristics, including chemical composition, since the compounds present in the fluid interact with the minerals that form the soil (Mesri & Olson, 1971; Folkes, 1982).

6. Case study - city of Viscount of Rio Branco - Minas Gerais / Brazil

The material used in this study was collected from the B horizon of a red-yellow latosol classified according to Unified Soil Classification (USC) as high plasticity silt, located on line A (MH/CH), and according to the Highway Review Board (HRB) system as A-7 soil with group index 12 (Nascentes, 2005). The geotechnical soil characterization is presented in Tables 1 and 2.

Granulometry				Atterberg Limits		
Clay (%)	Silt (%)	Sand (%)	Rock (%)	LL (%)	LP (%)	PI (%)
42	10	47.1	0.9	52	30	22

LL- liquid limit; LP- limit of plasticity; PI- Plasticity index.

Table 1. Granulometry and Atterberg limits for the soil used in this study.

γ_S (kN.m^{-3})	Activity	$\gamma_{dmax}{}^1$ (kN.m^{-3})	$w_{optimum}{}^1$ (%)	$\gamma_{dmax}{}^2$ (kN.m^{-3})	$w_{optimum}{}^2$ (%)
27	0.52	16.45	22.3	15.82	24.1

γ_s - specific weight of solids; γ_{dmax} and $w_{optimum}$ – dry soil specific weight and optimum soil humidity: [1] at Normal Proctor energy; [2] at an energy of 233 kJ/m^3.

Table 2. Physical indexes of the soil used in this study.

X-ray analyses were performed on three sample types: (i) natural clay randomly placed in powdered form on a glass well slide, (ii) an oriented sample, prepared with natural clay spread as a paste to orient minerals and clay and (iii) an oriented sample, after treatment to remove ferric oxides in order to better identify clay silicates and aluminum oxides possibly present in the soil sample. Analyses were performed in the Mineralogy Laboratory of the Soils Department of the Federal University of Viçosa using a Rigaku D-Max X-ray difractometer. A cobalt tube and curved graphite monochromator was used to obtain Co-Kα radiation. The difractometer was operated at 40 kV and 30 mA.

The presence of various peaks that permit identification of kaolinite, goethite and small quantities of hematite can be observed in the difractogram of the random natural clay fraction (Figure 1a). Given its yellow color, it was assumed that only a very small amount of hematite was present in this soil since the presence of hematite, even in small quantities imparts a reddish color to the soil (Fontes & Carvalho Jr., 2005), which was not the case for the sample used.

The attempt to orient the clay minerals in the presence of ferric oxides was not successful, as can be seen in Figure 1b. Face to face alignment of kaolinite minerals was not possible due mainly to the presence of goethite, resulting in a large series of peaks characteristic of kaolinite. The difractogram of the clay fraction of the soil after removing ferric oxides is presented in Figure 1c. Only two large kaolinite peaks (1st and 2nd order) were observed proving that when the kaolinite is perfectly aligned only peaks characteristic of this mineral's base atomic planes will appear.

Through combined evaluation of the difractograms the soil's clay fraction composition was defined as kaolinite and goethite with very little hematite.

The Fe content was determined using the dithionite extraction method (Coffin, 1963) to quantify iron oxides. Iron oxides corresponded to 13.3% of the clay fraction and were assigned to goethite. It is important to determine the presence of iron oxides since they have the capacity to retain heavy metals with high energy even when present in small quantities. Important chemical and physicochemical characteristics of the soil are presented in Table 3.

Ca^{2+}	Mg^{2+}	K^+	Al^{3+}	$H^+ + Al^{3+}$	CTC_{ef}	CTC_{pot}	pH	MO
			cmol$_c$	kg^{-1}				dag kg^{-1}
1.23	0.11	0.026	0.0	0.7	1.37	2.07	6.01	0.0

CTC_{ef}- Effective cation exchange capacity at natural soil pH; CTC_{pot}- Cation exchange capacity at pH 7

Table 3. Results of chemical and physicochemical soil analyses.

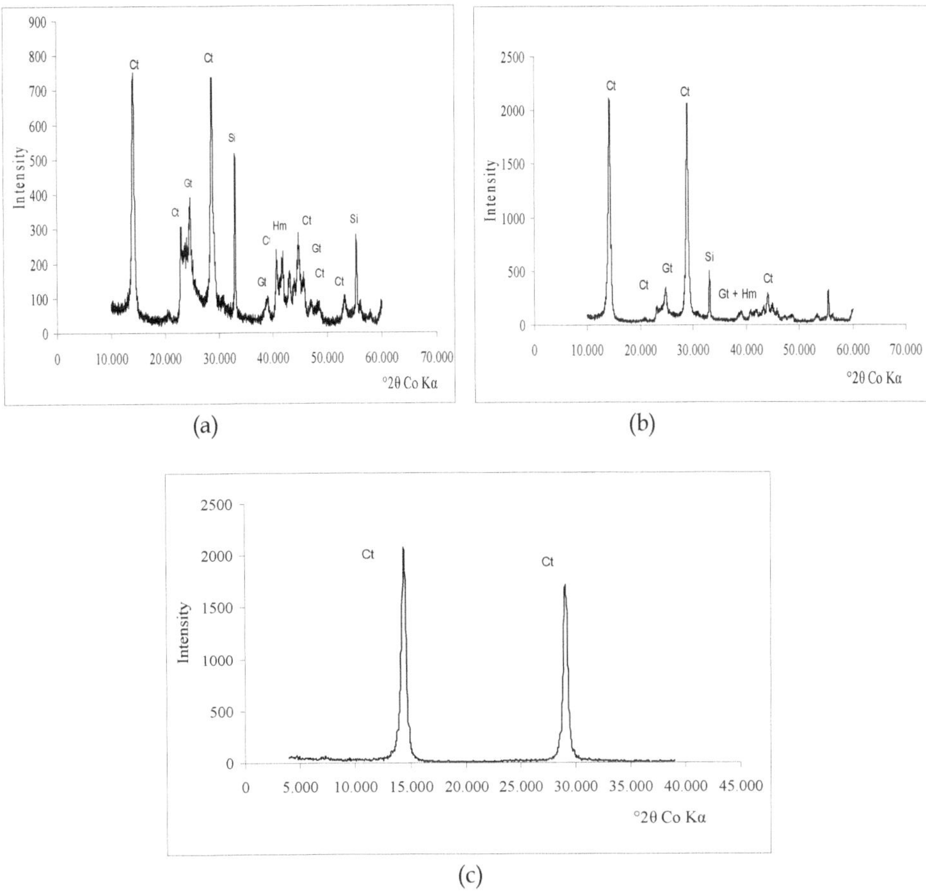

Fig. 1. Difractogram of the clay fraction. (a) Randomly oriented natural clay in a well slide; (b) Oriented natural clay (c) Oriented clay after iron removal.

6.1 Contaminant solution

A contaminant solution was prepared by mixing nitrates of manganese, zinc, cadmium, copper, lead and chrome since these are the metals most commonly found in urban landfill leachates. The pH and concentrations of the heavy metals in the leachate (Table 4) are within the range cited in the literature for Brazilian landfill leachates (Oliveira & Jucá, 1999; Barbosa & Otero, 1999).

Parameter	pH	Cr^{3+} (mgL^{-1})	Cd^{2+} (mgL^{-1})	Pb^{2+} (mgL^{-1})	Cu^{2+} (mgL^{-1})	Mn^{2+} (mgL^{-1})	Zn^{2+} (mgL^{-1})
Value	5.2	0.7	1.6	1.6	5.0	36.0	62.0

Source: Azevedo et al. (2006)

Table 4. Contaminant solution.

6.2 Column tests

Column tests were performed on eight soil samples adjusted to 22.5% humidity to evaluate hydraulic conductivity of compacted soil leached by the contaminant solution and to determine the transport parameters of the metals studied.

All samples were compacted to a specific weight of approximately 15.63 kN/m^3 (95% of the maximum normal Proctor compaction degree). Compaction energy was such that the samples were compacted until reaching 10 cm height and 5 cm diameter (233 kJ/m^3). Compaction data for each sample are presented in Table 5.

Soil columns were saturated with distilled water until a constant flow was obtained before percolating the contaminant solution. A flexible walled permeameter that resembled a triaxial chamber (Azevedo et al., 2003), constructed in the Geotechnical Laboratory of the Civil Engineering Department of the Federal University of Viçosa was used for the tests. Tests were conducted in a temperature controlled room (17 to 20 °C).

	CP 01	CP 02	CP 03	CP 04	CP 05	CP 06	CP 07	CP 08
GC (%)	94.9	94.8	95.1	94.5	94.5	94.9	95.1	94.8
Gradient	13.4	13.4	7.3	7.3	13.4	13.4	7.3	7.3
Void index	0.729	0.731	0.726	0.737	0.737	0.729	0.726	0.731
Porosity	0.422	0.422	0.421	0.424	0.424	0.422	0.421	0.422
Void vol. (mL)	81.4	81.4	80.7	82.0	81.7	81.5	81.0	81.3
Degree of saturation (%)	83.2	83.0	83.6	82.4	82.4	83.3	83.6	83.0
Δh (%)	-1.6	-1.6	-1.6	-1.6	-1.6	-1.6	-1.6	-1.6

Table 5. Compaction assays.

6.2.1 Analyses

Chemical, physical and micromorphological analyses of the samples were performed at the end of the column tests to evaluate soil retention of the heavy metals and variation in hydraulic conductivity caused by the percolating contaminant solution. The following determinations were made: hydraulic conductivity when percolating distilled water; hydraulic conductivity when percolating contaminant solution; leachate cation (Na^+, Ca^{2+}, Mg^{2+}) concentrations; effluent pH; effluent electric conductivity; metals retardation factors of (R_d); metals hydrodynamic dispersion coefficients (D_h) of; sequential extraction; dispersed clay; scanning microphotography.

6.2.2 Chemical analyses

Samples from columns CP01 (control), CP03, CP05 and CP07 were sliced into five 2 cm thick layers, placed in plastic bags to avoid water loss and kept in a humidity chamber for up to 24h for chemical analyses. Sequential extraction was used to determine speciation of each heavy metal in the compacted samples (article in press).

At the end of the column tests three samples of distilled water were collected from each permeameter used to apply the confining pressure and analyzed for the six heavy metals (Mn, Zn, Cd, Cu, Pb and Cr). No differences were detected among the samples indicating that no contaminant migration occurred into the water chamber.

6.2.3 Physical analyses

Percentage of dispersed clay was determined in the upper half of samples CP02, CP04, CP06 (control) and CP08. Slow and fast mechanical dispersion were employed to examine the effect of type of dispersion on the results. The methodology used was described by Ruiz (2005).

6.2.4 Micromorphological analyses

The lower half of samples CP02, CP04, CP06 and CP08 were used for micromorphological analyses. Samples were oven dried at 60°C, for 48h and then impregnated with a mixture of acrylic resin (60%) and styrene (40%) plus 5 mL of catalyst for each 1000 mL of mixture. The samples were left to soak in a well vented environment for 20 days to permit penetration in all the sample pores. Thin slides were prepared for optical and scanning electron microscope observations. The microphotographs were taken in the Microscopy Laboratory of the Geology Department of the Federal University of Ouro Preto. Some microphotographs were also taken of carefully withdrawn 1 cm^3 subsamples from the compacted samples. These subsamples were oven dried at 50°C and fixed on aluminum stubs using double faced adhesive tape, covered with about 20 nm of a gold-palladium mix using a model FDU 010 (Balzers, Inc, USA) metalizer and observed under a VP 1430 scanning electron microscope (LEO Electron Microscopy, Oberkochen, Germany) operated at a voltage of 15 kV. This procedure was performed at the Center for Microscopy and Microanalysis of the Federal University of Viçosa. The images obtained were used to observe changes in soil microstructure.

6.3 Results

6.3.1 Column tests

6.3.1.1 Distilled water percolation

Soil hydraulic conductivity curves (corrected to 20°C) for distilled water percolation are presented in Figure 2. Concentrations of Na^+, Ca^{2+} and Mg^{2+} in the column leachates are presented in Figures 3a, 3b and 3c. Electrical conductivity and pH values measured in the column leachates are presented in Figures 4 and 5, respectively. T is a dimensionless parameter corresponding to the ratio of effluent volume to column void volume and is equivalent to the pore volume.

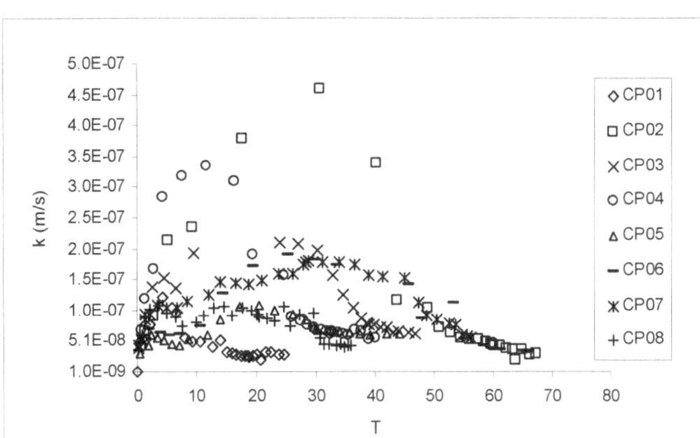

Fig. 2. Hydraulic conductivity in samples percolated with distilled water.

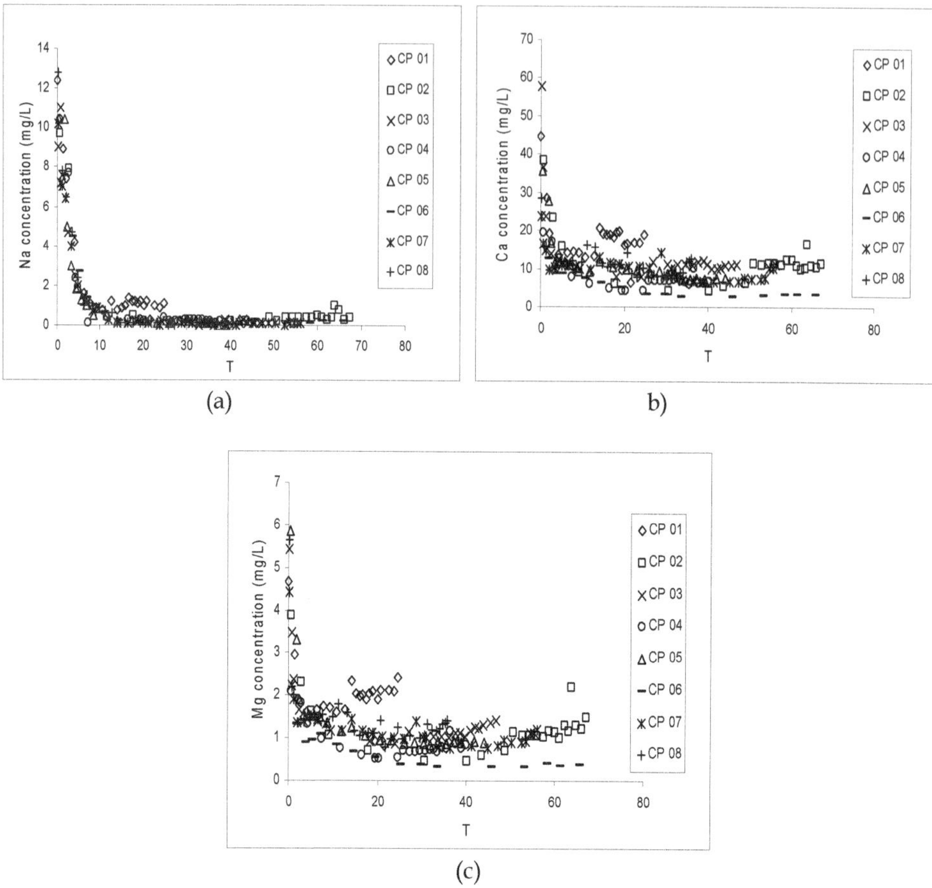

Fig. 3. Cations in soil column leachates percolated with distilled water (a) sodium, (b) calcium, (c) magnesium.

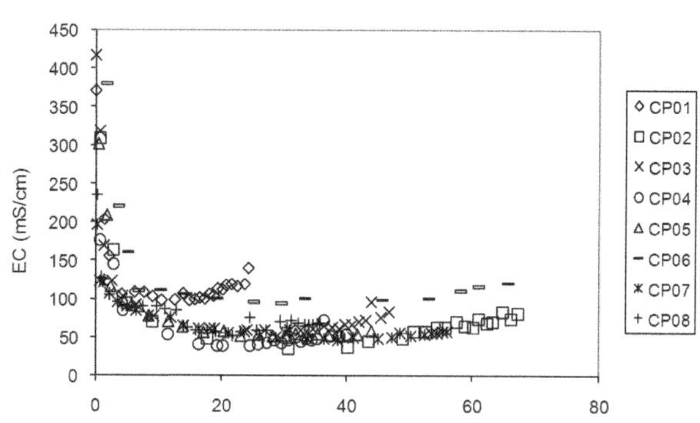

Fig. 4. Electrical conductivity (EC) in soil columns percolated with distilled water.

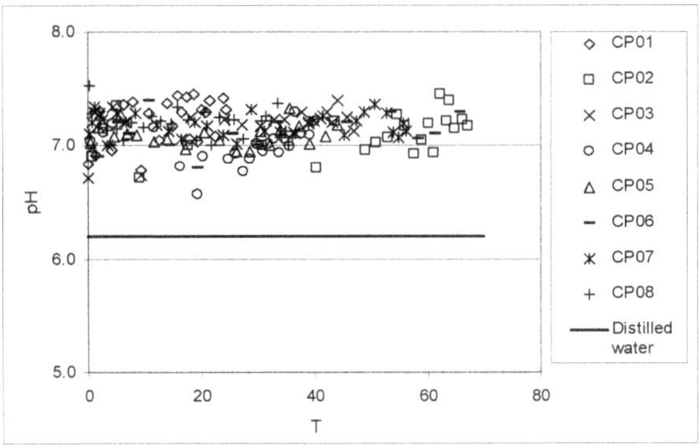

Fig. 5. Leachate pH of soil column percolated with distilled water.

The significant variation (up to one order of magnitude) in hydraulic conductivity with the number of pore volumes percolated (T) is shown in Figure 2. The increase in hydraulic conductivity at the beginning of the test may have been caused by the gradual increase in degree of sample saturation due to expulsion of air from the voids.

Distilled water percolated through the soil led to a decrease in ionic concentration (Na^+, Ca^{2+}, Mg^{2+} etc.) of the soil solution (Figures 3 (a), (b), (c)) as a result of expansion of the diffuse double layer. This expansion may have contributed to colloidal dispersion including that of material that acted as cementing agent of primary particles. The dispersion of cementing agents led to partial disaggregation of the soil structure, causing the greatest effect on soil macroporosity. Given the high correlation between soil hydraulic conductivity and macroporosity, it is clear that a decrease in macroporosity would lead to a proportional reduction in hydraulic conductivity.

Values of electrical conductivity were higher at the beginning of the column test (Figure 4) when greater leaching of basic cations occurred.

Leachate pH values varied somewhat during soil saturation with distilled water. All sample leachate pH values were higher than the influent water (pH=6.2) due to leaching of the bases Na^+, Mg^{2+}, Ca^{2+} adsorbed on the soil exchange complex. Their leaching to the aqueous solution resulted in an increase in hydroxide concentration (OH^-), and consequently in leachate pH.

6.3.1.2 Percolation of contaminant solution

Soil hydraulic conductivity curves for all samples percolated with the contaminant solution are presented in Figure 6. Control samples CP01 and CP06 were percolated with distilled water and served as references. Concentration curves for the leached cations versus pore volumes are presented in Figures 7a, 7b and 7c. Graphs of effluent electrical conductivity and pH versus pore volumes are presented in Figures 8 and 9, respectively.

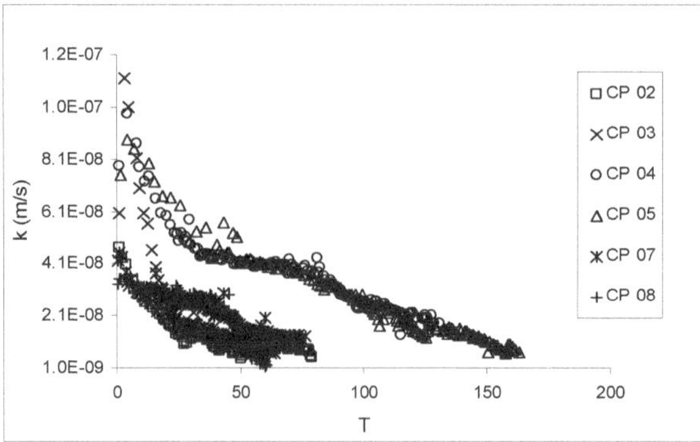

Fig. 6. Hydraulic conductivity in soil samples percolated with contaminant solution.

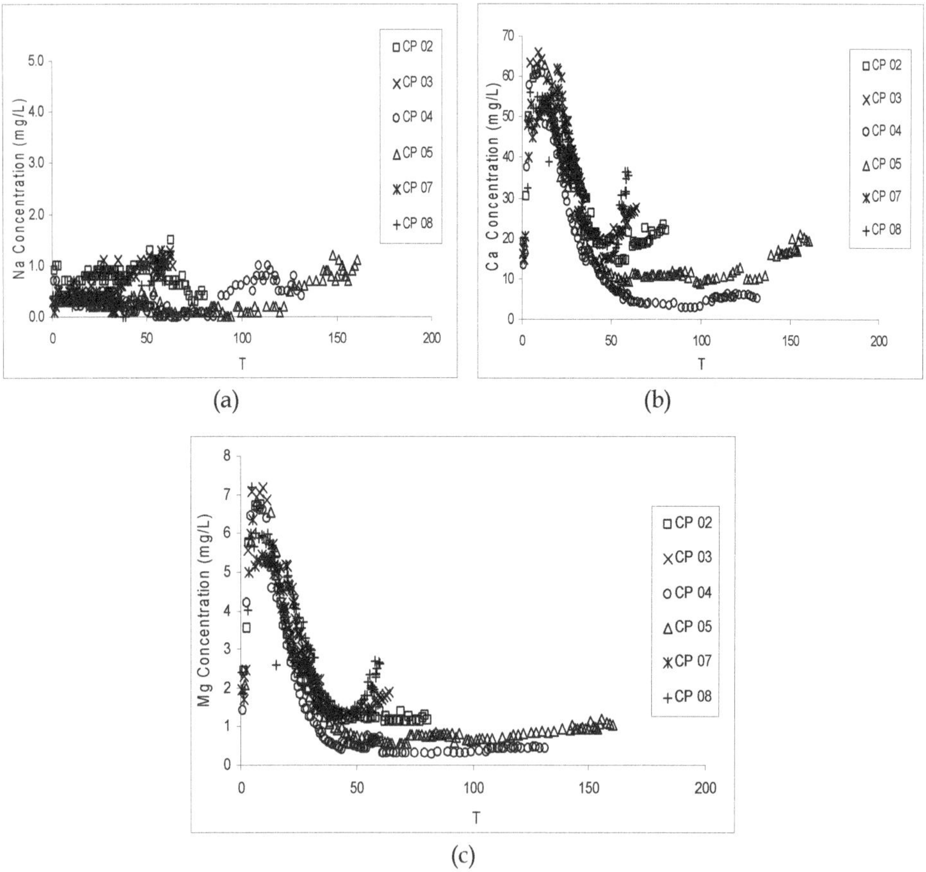

Fig. 7. Cation concentrations concentration in leachates from soil columns percolated with contaminant solution: a) sodium, b) calcium e c) magnesium.

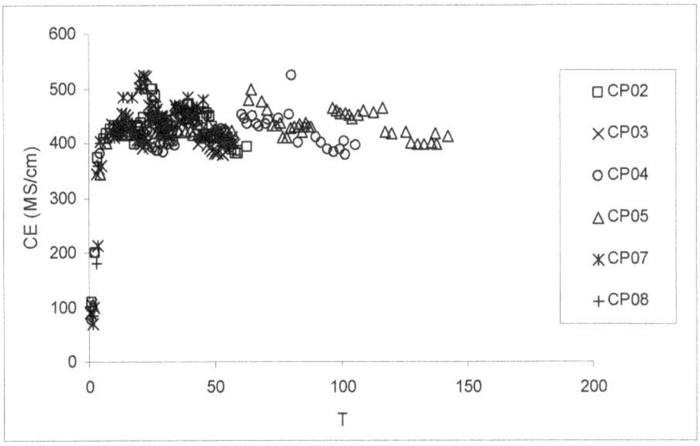

Fig. 8. Electrical conductivity in leachate from soils columns percolated with contaminant solution.

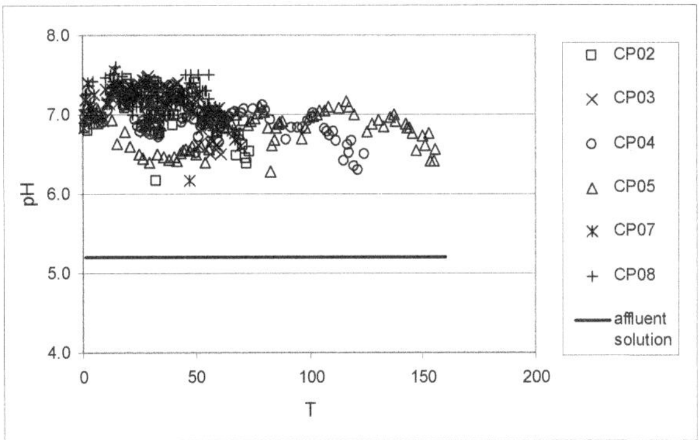

Fig. 9. pH of leachate from soil columns percolated with contaminant solution.

A significant decrease in hydraulic conductivity with percolation of contaminant solution was seen for all samples (Figure 6). The decrease was not homogeneous among samples due to the different structures formed after the initial percolation with distilled water (Figure 2). The large difference in water volume percolated directly influenced the hydraulic conductivity behavior when percolating the contaminant solution. In samples CP04 and CP05, the small difference observed between number of pore volumes and amount of leached cations when percolating water led to a similar hydraulic conductivity behavior when the contaminant solution was percolated through these columns.

The decrease in hydraulic conductivity in all samples was caused by obstruction of soil pores by heavy metals precipitation. According to Alloway (1995), the solubility of Cr^{3+} decreases at pH values greater than 4, with complete precipitation of the metal at values above 5.5. Evidence of metals precipitation from the contaminant solution are indicated by the pH versus T curves in Figure 9. Curves obtained for samples CP04 and CP05 are shown in Figure 10 because of the large amount of data overlay in Figure 9. Hydraulic conductivity *versus* T curves presented the same trend in variation as the pH *versus* T curves in all samples tested.

Column CP04 presented lower hydraulic conductivity than CP05 for number of pore volumes between 13 and 50 (Figure 6). CP04 leachate pH was greater than that of CP05 over this pore volume interval, indicating greater precipitation in CP04. From that point on the pH values approached the hydraulic conductivity values up to T=104. At that point CP05 leachate pH increased compared to that of CP04 and consequently the hydraulic conductivity of CP05 decreased more than that of CP04. This occurred in all samples.
Column CP08 presented the highest leachate pH value (Figure 11) and lowest final hydraulic conductivity (Figure 6) of all samples. Higher pH values were observed at the beginning of the test (maximum of 7.6) for T values up to about 17, decreasing afterwards to T=38. From that point on, the pH increased indicating an increase in metals precipitation and consequent decrease in hydraulic conductivity.
More colloidal dispersion also occurred in the samples after percolation with contaminant solution than after percolation with distilled water. Microphotographs of sample CP06 (control) and CP02 are presented in Figures 12 and 13. It can be seen that the soil mass was more uniform in sample CP02 than in sample CP06, indicating more clay dispersion in the former. It is therefore possible to conclude that clay dispersion occurred during contaminant solution percolation and the dispersion contributed to the decrease in hydraulic conductivity.

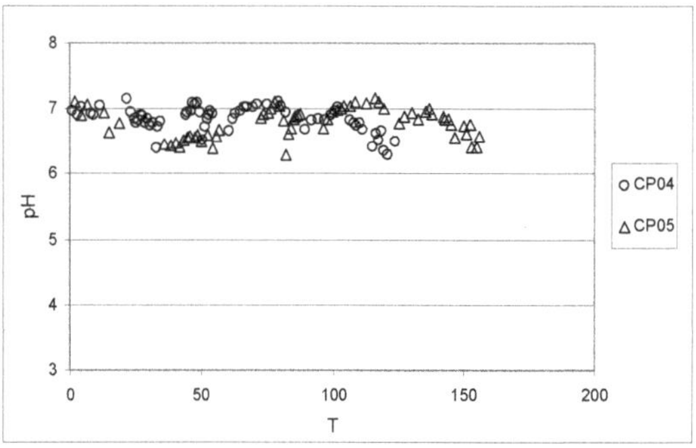

Fig. 10. Comparison of pH values of leachate collected from soil columns CP04 and CP05 percolated with contaminant solution.

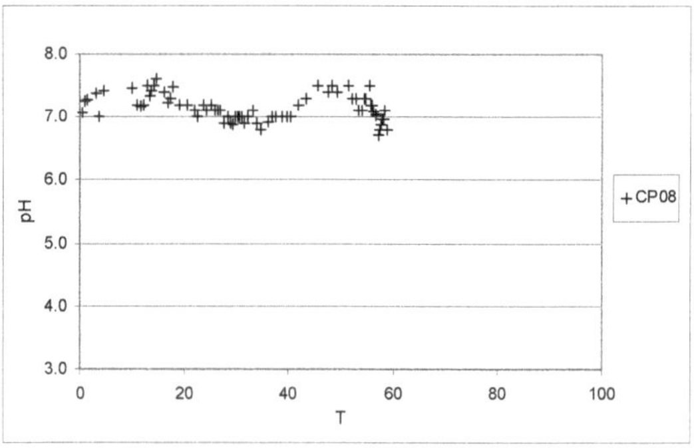

Fig. 11. pH values of leachate collected from soil column CP08 percolated with contaminant solution.

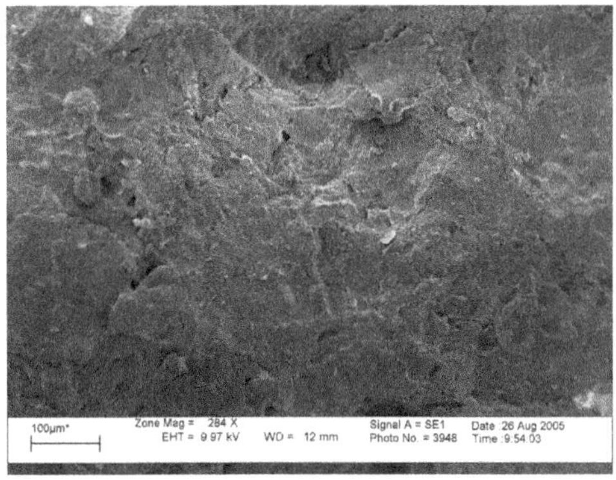

Fig. 12. Microphotograph of sample CP02.

Fig. 13. Microphotograph of sample CP06 (control).

Alterations in of all sample leachate pH values were attributed to an initial washing of the bases Na^+, Mg^{2+}, Ca^{2+} (distilled water percolation increased the concentration of OH^- in the leachate because the bases are accompanied by hydroxyl) present in the soil solution exchange complex (Figures 7a, 7b and 7c).

In aqueous solution the bases promoted an increase in hydroxide concentration (OH^-), and thus an increase in solution pH. The fact that the leachate pH value remained above that of the influent solution may be due to the fact that ion exchange of metals by the bases adsorbed in the soil solid phase continued over the course of the test. It is believed that when sample saturation with all the metals is reached the leachate pH will tend to equal the influent pH, as was observed by Azevedo et al. (2005).

Cation leaching occurred due to substitution by heavy metals added to the soil. The initially high exchange of Ca^{2+} and Mg^{2+} decreased as the exchange sites occupied by these cations were occupied and their soil concentration decreased. Na^+ was leached in small concentrations during almost the entire test. The concentrations of Ca^{2+}, Mg^{2+} and Na^+ in the leachate of some samples tended to increase at the end of the test, indicating greater leaching of these cations caused by adsorption of metals with greater affinity for the soil matrix.

Electrical conductivity values presented fluctuations. The final portion of the curve in Figure 8 presented a decline that indicates that a greater adsorption and/or precipitation of metals occurred.

6.4 Determination of transport parameters

Only the metals Mn (in all samples) and Zn (only in CP04) reached a steady state concentration (Figures 16 and 17c) in the column tests, that is they saturated the soil adsorption sites. Therefore it was only possible to determine the retardation factor (R_d) and hydrodynamic dispersion coefficient (D_h) using the cumulative mass method (Shackelford, 1995) for these two these metals. The curves of cumulative mass ration (CMR) *versus* theoretical and experimental T values are presented in Figures 14 and 15.

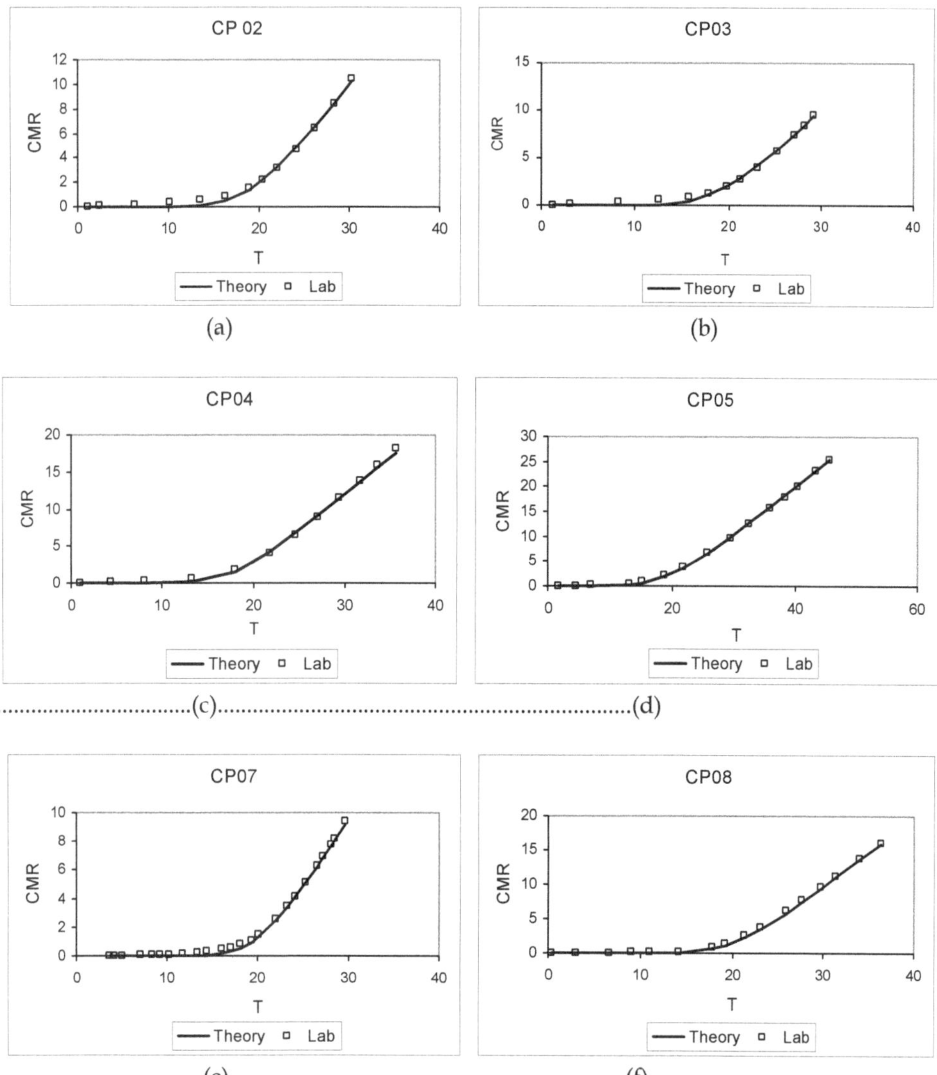

Fig. 14. Manganese cumultative mass ratio curves. CP02, (b) CP03, (c) CP04, (d) CP05, (e) CP07 e (f) CP08.

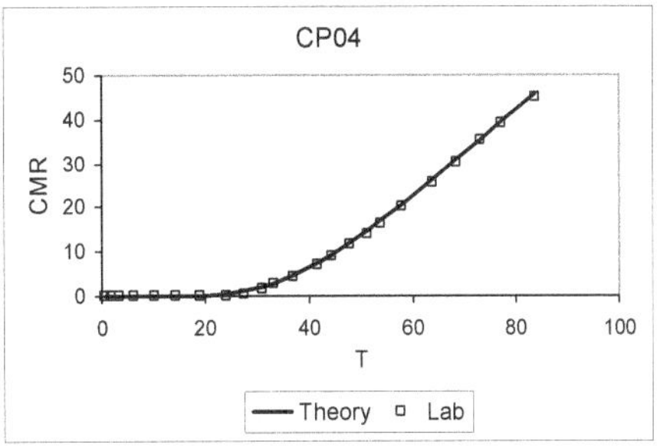

Fig. 15. Zinc cumulative mass ratio curve (CP04).

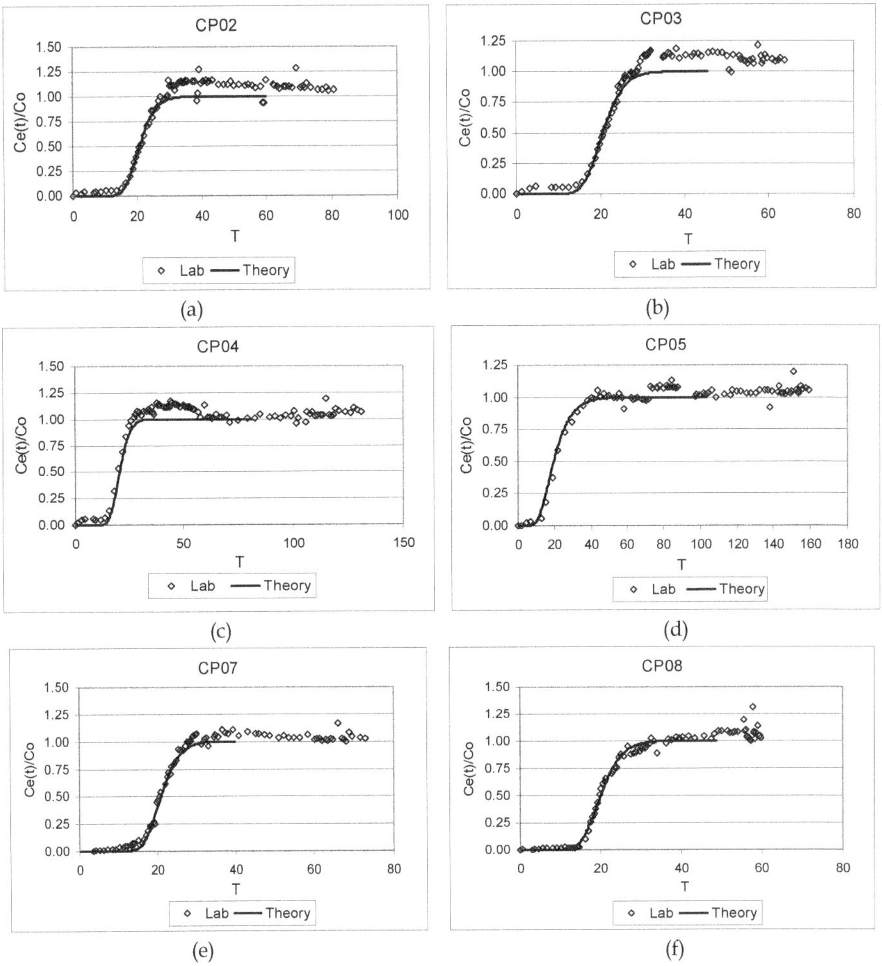

Fig. 16. Manganese elution curves from soil columns: (a) CP02, (b) CP03, (c) CP04, (d) CP05, (e) CP07, (f) CP08.

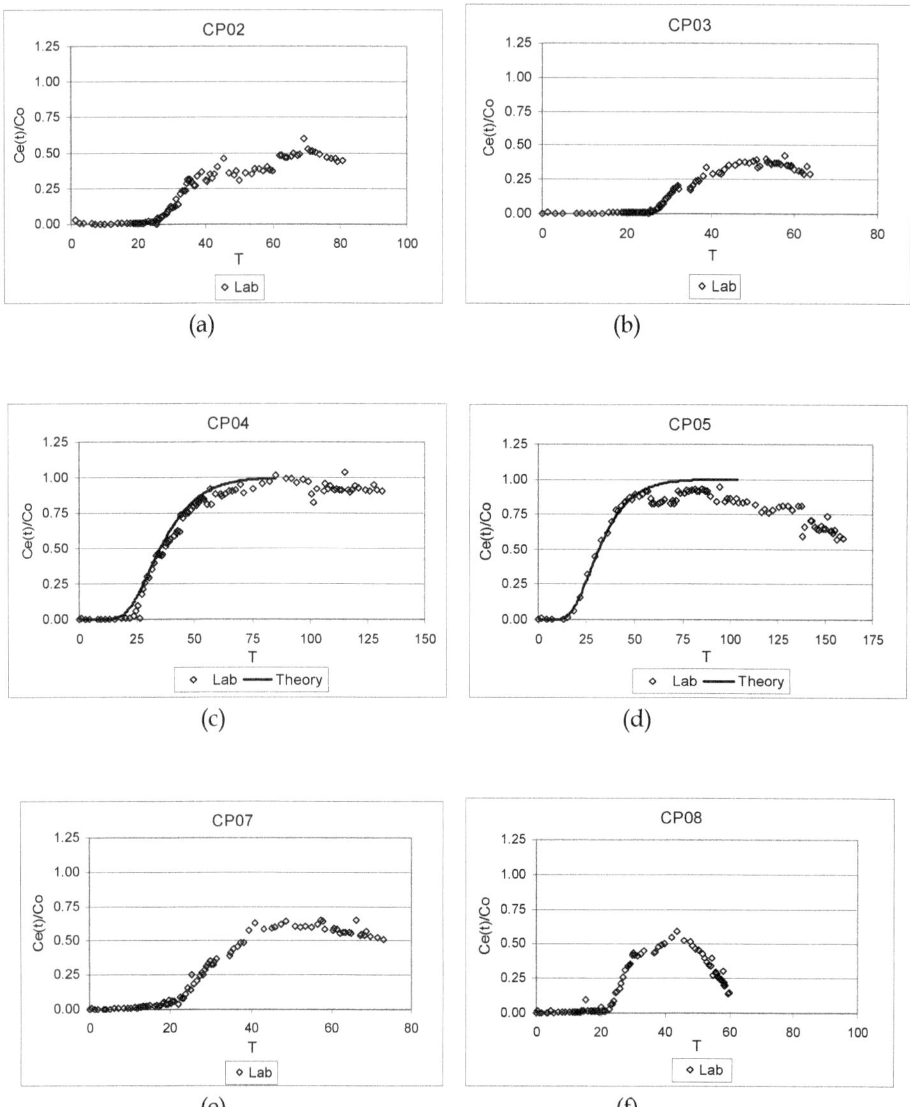

Fig. 17. Zinc elution curves from soil columns: (a) CP02, (b) CP03, (c) CP04, (d) CP05, (e) CP07, (f) CP08.

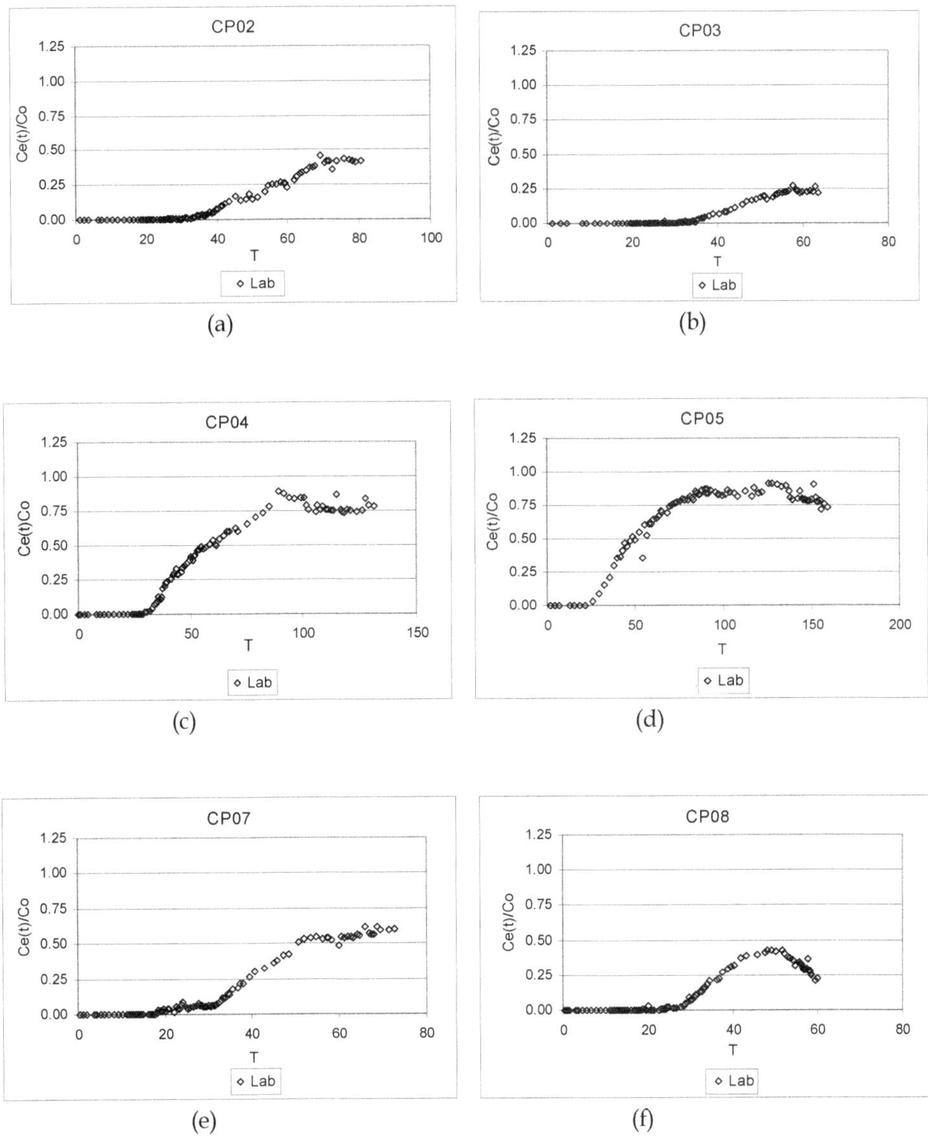

Fig. 18. Cadmium elution curves from soil columns: (a) CP02, (b) CP03, (c) CP04, (d) CP05, (e) CP07, (f) CP08.

Mn, Zn and Cd elution curves are presented in Figures 16 to 18. The desorption of Mn (Figure 16) and the increase in adsorption of Zn (Figures 17a, b, d, e, f) and Cd (Figures 18c, d, f) at the end of the column tests were observed in some samples.

The metals Cu, Pb and Cr remained totally retained in the soil, since these metals have low mobility and high affinity for iron oxides present in the soil. These elements can also form precipitates depending on their concentrations as well as on soil and solution pH.

No large data dispersion was observed in the CMR *versus* T curves (Figures 14 and 15) and the experimental curves adjusted well to the theoretical curves. The transport parameters R_d and D_h are presented in Table 6.

The relative Mn concentration (C/Co) in the elution curve (Figure 16) was greater than unity, evidence of desorption. For Zn (except CP04) and Cd, the C/Co ratio never reached unity (and decreased in some cases), due to the decrease in hydraulic conductivity, principally in CP08. It is thus possible to conclude that Zn and Cd replaced the desorbed Mn on the soil adsorption sites.

A Mn retardation factor of 18.8 was reported by Azevedo et al. (2005), close to the average value presented in Table 6. Given the greater hydraulic conductivity values (10^{-5} to 10^{-6} cm s^{-1}) in the work by Azevedo et al. (2005) as compared to those in the present study (10^{-6} to 10^{-7} cm s^{-1}), the mobility of Mn was found to be practically independent of soil hydraulic conductivity. The average D_h value equal to 4.3E-04 cm^2 min^{-1} (Table 6) was a little lower than the 8.64E-03 cm^2 min^{-1} presented in Azevedo et al. (2006), probably because the lower average percolation velocity used in that study caused less Mn dispersion.

Metal	Sample	R_d	D_h (cm^2 min^{-1})
	CP02	20	4.6E-04
	CP03	20	3.9E-04
	CP04	18	4.4E-04
Manganese	CP05	20	8.0E-04
	CP07	20.5	2.0E-04
	CP08	20.4	2.7E-04
	Average	19.8	4.3E-04
Zinc	CP04	38	1.4E-03

Table 6. R_d and D_h obtained by the cumulative mass method.

6.5 Soil physical analyses

Results of dispersed clay in samples withdrawn from the soil columns were quite similar for both fast and slow dispersal methods. Average results are indicated in Table 7. The values of dispersed clay indicated little dispersion of fine material in the soil samples analyzed. Greater values were expected since the hydraulic conductivity results indicated that clay dispersion occurred when distilled water was percolated through the columns. However the samples were stored for two months and 23 days (justified by the time necessary to decide on the appropriate sample analyses) and it is believed that they suffered a tixotropic effect. Compacted clays may exhibit a considerable tixotropic effect (increase in resistance and rigidity with time) leading to a natural tendency to flocculate during storage. According to Boscov (1997), the structure of adsorbed water may change in stored compacted samples and can be detected by measurements that show the decrease in neutral pressure with time after compaction.

Sample	CP02	CP04	CP06	CP08
Dispersed clay (kg/kg)	0.002	0.002	0.005	0.003
FI*	0.99	0.99	0.988	0.99

*Floculation index

Table 7. Dispersed clay in samples withdrawn from the soil columns.

6.6 Micromorphological analyses

Mosaics were prepared from 50% of the photos of slides taken in the optical microscope to better visualize pores. The photos taken with the scanning electron microscope (SEM) permitted visualization of a greater quantity of sample fissures in the control (CP06) than in the columns percolated with contaminant solution (CP02, CP04 and CP08). Mosaics of samples CP02, CP04 and CP06 are presented in Figure 19. No mosaic was made of CP08 due to insufficient sample quantity for evaluation.

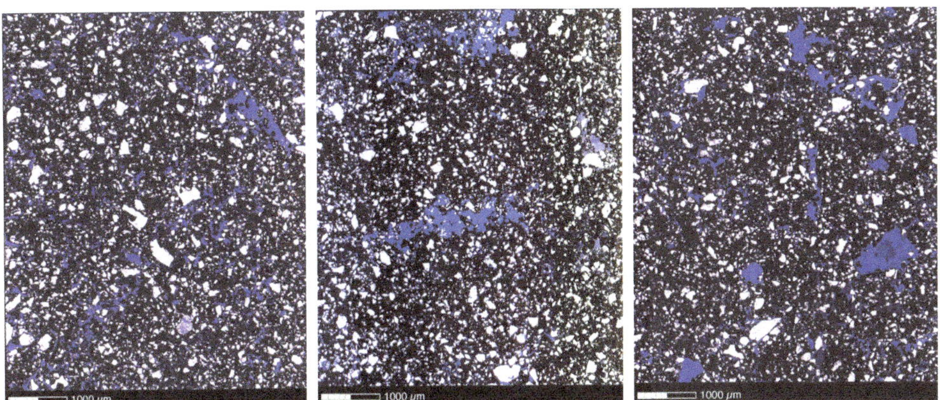

(a) (b) (c)

Fig. 19. Optical microscope mosaic of samples withdrawn from soil column: (a) CP02; (b) CP04; (c) CP06 (control)

A greater quantity of dark stained macropores can be observed in control column CP06 that had the greatest hydraulic conductivity. Sample CP04 presented larger pores than sample CP02, consistent with the larger final hydraulic conductivity of sample CP04, which could be attributed to less plugging of macropores.

The presence of fissures in photos of samples CP06 (control) and CP08 obtained by SEM are indicated by arrows in Figures 20 and 21. More fissures are present in CP06 (Figure 20) than in the other samples. The fissures in the other samples may possibly have been blocked by precipitates formed during the percolation of the contaminant solution in the soil columns and also by particle washout during the test (Figure 21).

Fig. 20. Various fissures present - CP06

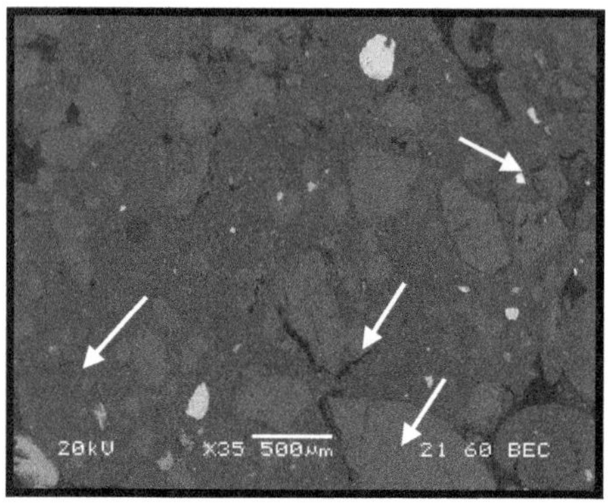

Fig. 21. Few fissures present - CP08.

7. Conclusions

Of the various conclusions to be drawn, it should be emphasized that initial percolation of distilled water to saturate the soil without counter pressure influenced the column test results since the soil structure was altered, especially when a large pore volume was percolated.

Hydraulic conductivity values decreased significantly in all soil columns, although the difference differed among the samples.

The difference in number of pore volumes percolated (T) and the amount of cations leached during percolation of distilled water in the soil columns directly affected hydraulic conductivity when the contaminant solution was percolated possibly because of an alteration in soil structure caused by initial percolation with distilled water.

Evidence of pore obstruction caused by heavy metal precipitation was observed, explaining in part the decrease in hydraulic conductivity. The decrease was also partially attributed to dispersion of colloidal material

Greater leaching of Ca^{2+}, Mg^{2+} and Na^{+} was observed at the end of the soil column tests resulting in greater adsorption of some metals.

At the end of the column test, Mn was desorbed and a proportionally greater amount of Zn (except for CP04) and Cd were adsorbed due to decrease in hydraulic conductivity (especially in CP08), suggesting that Zn and Cd dislocated Mn from the adsorption sites. Dispersed clay measurements indicated high flocculation indexes but the results may have been influenced by a trixotropic effect occurring in the stored samples.

Comparison of Mn transport parameters determined in the present study with those obtained in a previous study showed that Mn mobility was practically independent of soil hydraulic conductivity when all other factors were held constant. Hydraulic conductivity affected metal mobility of the other five metals studied.

8. Acknowledgments

The authors would like to acknowledge CNPq and CAPES for financial support and the Civil Engineering, Agricultural Engineering, and Soil Departments of the Federal University of Viçosa, MG, Brazil, for carrying out tests.

9. References

Alloway, B.J. (1995). Heavy metals in soils. 2nd ed. John Wiley & Sons, New York, p. 3-10.

Azevedo, I.C.D.A.D., Nascentes, R., Azevedo, R.F., Matos, A.T., Guimarães, L.M. (2003). Hydrodynamic dispersion coefficient and retardation factor for heavy metals in residual compacted soils. (In Portuguese). Solos e Rochas - Revista Brasileira de Geotecnia. São Paulo: v. 26:3, p229-249.

Azevedo, I.C.D.; Nascentes, C.R.; Matos, A.T.; Azevedo, R.F. (2005). Determination of heavy metal transport parameters in residual compacted soil. (In Portuguese). Revista Brasileira de Engenharia Agrícola e Ambiental, v.9, n.4, p.623-630. Campina Grande, PB.

Azevedo, I.C.D.A.D., Nascentes, C. R., Matos, A.T., Azevedo, R.F. (2006). Determination of transport parameters for heavy metal in residual compacted soil using two methodologies. *Canadian Journal of Civil Engineering*. 33(7), 912-917.

Barbosa, R.M.; Otero, O.M.F. (1999). Characterization of a pollution plume caused by urban waste disposal. (In Portuguese). Geochimica Brasiliensis, Rio de Janeiro, v.13, n.1, p. 51-65.

Boscov, M.E.G. (1997). Contribution of design of systems of hazardous waste containment using lateritic soils. PhD Dissertation (In Portuguese). Universidade de São Paulo, Escola Politécnica. São Paulo, 259p.

Coffin, D.E. (1963). A method for determination of free iron oxides in soils and clays. Can. Journal of Soil Science 43, p.9-17.

Cooker, E.G., Matthews, P.J. (1983). Metals in sewage sludge and their potential effects in agriculture. Water Sci. Technol., v.15, p.209-225.

Folkes, D.J. (1982). Fifth Canadian Geotechnical Colloquium: Control of contaminant migration by the use of liners. Canadian Geotechnical Journal. V.19. p. 320-344.

Fontes, M.P.F; Carvalho Jr., I. (2005). Color attributes and mineralogical characteristics, evaluated by radiometry, of highly weathered tropical soils. Soil Science Society of America Journal, 69: 1162-1172.

Guilherme, L.R.G, Marques, J.J., Pierangeli, M.A.P, Zuliani, D.Q., Campos, M.L., Marchi, G. (2005). Trace elements in soils and aquatic systems. (In Portuguese). Topics in Soil Science. Soc. Bras. de Ciência do Solo. v. 5. p 345-390.

Kaczmarek, M., Hueckel, T., Chawla, V., Imperiali, P. (1997). Transport Through a Clay Barrier with the Contaminant Concentration Dependent Permeability. Transport in Porous Media 29: 159-178.

Lake, D.J. Sludge disposal to land. (1987). Heavy metals in wastewater and sludge tretment process. Florida. CRC. V.2: treatment and disposal. p.92-124.

Lambe, T.W. (1979). Soil Mechanics. John Wiley & Sons Inc. New York. 553 p.

McBride, M.B. (1994). Environmental chemistry of soils. New York: Oxford University Press. 406 p.

Matos A.T. (1995). Retardation factors and dispersion-diffusion coefficients for zinc, cadmium, copper and lead in soils from the county of Viçosa, Minas Gerais. Doctoral Thesis in Soils and Plant Nutrition. (In Portuguese). Universidade Federal de Viçosa, Viçosa, MG, 183 p.

Matos, A.T.; Costa, L.M.; Fontes, M.P.F.; Martinez, M.A. (1999). Retardation factors and the dispersion-diffusion coefficients of Zn, Cd, Cu and Pb in soils from Viçosa - MG, Brazil, Transaction of the ASAE, American Society of Agricultural Engineers, V.42 (4), p. 903-910.

Mesri, G. & Olson, R.E. 1971. Mechanisms controlling the permeability of clays. Clay and Clay Minerals. V. 19. p. 151-158.

Naidu, R., Sumner, M.E., Harter, R.D. 1998. Sorption of heavy metals in strongly weathered soils: an overview. Environ. Geochem. Health. v.20. n.1. p.5-9.

Oliveira, F.J.S.& Jucá, J.F.T. (1999). Study on contamination of subsoil of the landfill region in Muribeca, Pernambuco. (In Portuguese). In: Brazilian Environmental Geotechnology Congress, REGEO IV, 1999, Salvador. Proceedings. Salvador: REGEO, p. 455-460.

Rowe, R.K., Quigley, R.M. & Booker, J.R. (1995). Clayey Barrier Systems for Waste Disposal Facilities, E&FN Spon, London. 390 p.

Ruiz, H.A. (2005). Incremento da exatidão da análise granulométrica do solo por meio da coleta da suspensão (silte+argila). Rev. Bras. Ciência do Solo. Viçosa, MG. 29(2). p. 297-300.

Shackelford, C.D. (1995). Cumulative Mass Approach for Column Testing. Journal of Geotechnical Engineering, American Society for Civil Engineers, 121: 696-703.

Sposito, G. 1984. The surface chemistry of soils. New York: Oxford University Press. 234 p.

Srivastava, P.C., Gupta, U.C. (1996). Trace elements in crop production. Lebanon: Science Publishers. 356p.

Tavares, T.M., Carvalho, F.M. (1992). Evaluation of human exposure to heavy metals in the environment: examples from the Recôncavo Baiano. (In Portuguese). Química Nova, v.15, n.2, p.147-153.

Yuan, G., Lavkulich, L.M. (1997). Sorption behavior of copper, zinc and cadmium in response to simulated changes in soil properties. Comm. Soil Sci. Plant Anal., v.28. n.6-8. p. 571-587.

4

Evaluation of Cover Systems for the Remediation of Mineral Wastes

Francis D. Udoh
Department of Chemical & Petroleum Engineering
University of Uyo, Uyo - Akwa Ibom State
Nigeria

1. Introduction

Lean ore and waste rock stockpiles, unless controlled, may pose significant environmental problems. Precipitation which enters a mining stockpile is a potential source of surface and groundwater contamination. Minerals present in the stockpile will dissolve in the presence of oxygen and water. Precipitation which percolates through the rock subsequently transports the dissolved minerals from the stockpiles downstream. The degree of transport of the dissolved minerals is dependent upon the chemistry of the component released, the chemistry of the transporting solution and the solids and biota which come in contact with the flow.

Often drainage from mineral wastes can be reduced by proper sitting or diversion of surface and groundwater. Further reduction can only be achieved by minimizing the rate of water infiltration into the waste itself. Infiltration reduction is generally the first step in stemming the water quality problem associate with stockpile drainage. One method of minimizing infiltration into mineral stockpiles is to cover the pile with a low permeability material and route the water off the pile before it becomes contaminated. The purpose of this study was to use the EPA HELP (Hydrologic Evaluation of Landfill Performance) model to simulate field conditions in order to identify the capping options that could be used to stem infiltration into mineral waste stockpiles.

2. Materials and methods

Numerous materials were sourced and screened for use as potential stockpile capping systems (Eger et. al, 1990). Laboratory tests were performed on the selected materials (which included, among others, glacial till, glacial till plus bentonite, fine tailings plus bentonite, paint rock and silty clay). Each material was subjected to a variety of tests using ASTM standards (Table 1). Material property criteria were proposed by the Minnesota Pollution Control Agency (MPCA) and are listed in Table 2. Additionally, each material was selected based on availability, cost, workability, expected hydraulic conductivity, and any potential environmental problems which could result from the use of that material. A summary of the physical properties of the materials for the stockpile capping study is shown in Table 3. Based on the final evaluation of laboratory data, cost and other potential environmental problems, glacial till, glacial till mixed with 5 per cent bentonite, and a 20 mil PVC membrane were chosen for field evaluation. Fine tailings were rejected since the hydraulic conductivity was greater than 2 x 10^{-6} cm/sec., and they might contain asbestiform fibers. Paint rock, although having a suitable hydraulic conductivity, produces "red" water (suspended iron oxides), and was eliminated due to its prohibitively high cost. While the glacial till had an acceptable hydraulic conductivity, the till contained large boulders which would not be suitable for a barrier layer. Therefore, the glacial till was screened through a Read Screen-All to produce a more uniform sized material.

Test or Classification	Procedure	
Description of Soils	ASTM*	D-2488
Classification of Soils	ASTM	D-2487
Water Content Determination	ASTM	D-2216
Specific Gravity Determination	ASTM	D-854
Particle Size Analysis	ASTM	D-421, D-422
Including Sieve and Hydrometer	ASTM	D-1140, and D-4217
Modified Proctor Moisture - Density Relationship	ASTM	D-1557
Permeability Testing in Conjunction with the Falling Head Procedure	ASTM	D-1557
Atterberg Limits	ASTM	D-4318

*ASTM – American Society for Testing and Materials

Table 1. Material Testing Procedures.

Component of Cap	Specifications
Cover	1. Minimum thickness - 12 inches. 2. Must be capable of sustaining vegetation.
Barrier	1. Soil barriers must be at least 12 inches thick. 2. Each layer must be placed in 6-inch lifts and compacted at or above optimum moisture content to achieve greater than 90% Standard Proctor Density.
	3. Barrier material should not contain more than 1% by weight coarse sand and gravel.
	4. At least 3% dry mass bentonite must be used in bentonite-soil barriers.
	5. The hydraulic conductivity of the barrier must be less than or equal to 2 x 10^{-6} cm/sec.
Buffer	Buffers serve to protect the barrier from tears, cracks, punctures and other deteriorations. The buffer can not contain any coarse fragment greater than 6 inches. 12-inch thickness was chosen as a suitable buffer.

Table 2. Material Specifications for Stockpile Capping Program.

In order to study the effectiveness of the selected capping systems to stem infiltration, the HELP (Hydrologic Evaluation of Landfill Performance) model was used to simulate field conditions. To accomplish this objective, a three layer capping system consisting of a cover, a barrier and a buffer was required (Table 2). Laboratory results in Table 3 along with synthetic materials such as polyvinyl chloride (PVC) were used to simulate the field

Soil Type	Permeability* (cm/sec)	Natural Wc	Atterberg Limits (LL-PL)	Gradation				Modified Proctor		
				%Gravel	%Sand	%Silt	%Clay	Dry Density (pcf)	Moisture Content(%)	Specific Gravity
Glacial Till	1.69×10^{-6} 1.55×10^{-6}	2.6	NP	15.5	63.4	17.2	3.9	134.4	8.0	2.74
Glacial Till + 3% Bentonite #	5.60×10^{-8} 1.96×10^{-7}	N/A	N/A	N/A	N/A	N/A	134.8	8.0	N/A	+ 3%
Glacial Till + 5% Bentonite #	6.0×10^{-9} 4.12×10^{-9}		N/A	N/A	N/A	N/A	N/A	135.0	8.0	N/A
Glacial Till +7% Bentonite #	3.17×10^{-9} 4.73×10^{-9}	N/A	N/A	N/A	N/A	N/A	134.6	8.0	N/A	
Paint Rock	6.67×10^{-7} 6.00×10^{-7}	35.1	NP	24.4	41.1	25.8	8.7	123.6	23.0	3.93
Fine Tailings	4.50×10^{-5} 3.66×10^{-5}	7.7	NP	0.1	43.9	50.7	5.3	116.7	13.4	2.98
Fine Tailings**	4.4×10^{-6} 6.6×10^{-6}									
Fine Tailings +1/2% Bentonite #	1.47×10^{-5} 2.02×10^{-5}		N/A	N/A	N/A	N/A	N/A	120.8	12.2	N/A
Fine Tailings +1% Bentonite #	2.39×10^{-6} 1.61×10^{-6}		N/A	N/A	N/A	N/A	N/A	122.2	12.3	N/A
Aurora Sity clay	20×10^{-9} 2.0×10^{-9}	31.4	43.5-16.2	0.0	17.8	40.2	42.0	114.0	17.0	2.67

*Tests performed at approximately 95% of the maximum Modified Proctor dry density (ASTM - 1557).
**Test result is from previous STS testing.
#Wyo-Ben Bentonite used
NP - Non-plastic
N/A - Not appropriate

Table 3. Summary of Physical Properties of Materials for the Stockpile Capping Study.

conditions for the various stockpile capping options. Simulations were also performed for a control (untreated) stockpile, standard reclamation (two feet of cover), and the MPCA hydraulic conductivity barrier requirement of 2 x 10^{-6} cm/sec. The HELP model is a sophisticated water balanced model that can model multilayered capping systems. The HELP model uses climatologic, soil and design data to produce daily estimates of water movement across, into, through and out of mineral stockpiles considered in this study. The climatologic data, which included daily precipitation and mean monthly temperatures in ^{0}F were from Babbitt, Minnesota (Udoh, 1993). The solar radiation data in langleys, were the monthly averages from Winton, Minnesota (Eger et. al, 1990). Other climatologic data such as leaf area indices, evaporative zone depth, and winter cover factors were selected from the HELP model built-in default data files. Leaf area index (which is the area of leaves per unit area of ground) affects the total evaporation from the stockpile capping systems. Maximum leaf area index ranges from about 1.5 for grass up to about 5 for a plant like soybeans. The maximum leaf area index used in the simulations ranged from 1 to 1.5. Typical default values for evaporative zone depth (which is related to root depth) range from 4 inches for bare ground to 18 inches for excellent grass. Fair grass, which is the general cover class found at most landfills (Eger et. al, 1990), has an evaporative zone depth of 10 inches, the default value used in the HELP model simulation. The soil data used in the simulation also came from the built-in default data files for soil texture classes 3, 6 and 20 for the top, drainage and barrier layers respectively. However, the hydraulic conductivities for each soil class were estimated to reflect the hydraulic conductivities required for typical mineral capping projects. The hydraulic conductivity of the buffer layer was computed as 70% of that of the top layer since the layer was assumed to be partially compacted.

The HELP program models a number of hydrologic processes by performing daily, sequential analysis using a quasi-two-dimensional, deterministic approach. The model computes surface runoff using the Soil Conservation Service (SCS) runoff curve number method. The equation developed relates daily runoff, Q, to daily precipitation, P and a watershed retention parameter, S, thus:

$$Qp = \frac{(P-0.2S)^2}{(P+0.8S)} \tag{1}$$

Where

Q, P, and S are in inches

Infiltration, I, is computed in the HELP model as:

$$I = P - Q - SE \tag{2}$$

Where:

- I = infiltration
- P = daily precipitation
- Q = daily runoff
- SE = surface evaporation

Potential evapotranspiration, E_0 is computed as:

$$E_o = \frac{1.28AH}{A+0.68} \tag{3}$$

Where:
E_o = potential evapotranspiration
A = slope of saturation vapour pressure curve
H = net solar radiation in langleys
Percolation, Qp, modeled as Darcian flow, is computed as:

$$Qp = Kp\frac{TH + T_c}{T_c} \quad (4)$$

Where:
Qp = the rate of percolation through the barrier soil layer
Kp = the saturated hydraulic conductivity of the barrier soil layer
TH = the total head in the profile above the barrier soil layer
T_c = the thickness of the barrier soil layer
The lateral drainage rate, Q_D, based on a linearization of the steady-state Boussinesq equation is computed as:

$$Q_D = \frac{2K_D yho}{L^2} \quad (5)$$

Where:
- Q_D = lateral drainage rate
- K_D = hydraulic conductivity for lateral flow
- y = average thickness of flow
- ho = elevation of water surface
- L = maximum length to drain

With a correction factor, the lateral drainage equation rewritten as:

$$Q_D = \frac{2K_D y(0.510 + 0.00205\alpha L)[Y(y/l)^{0.16} + \alpha L]}{L^2} \quad (6)$$

Where: α = fractional slope at surface of cover
The surface vegetation was assumed to be fair grass. A default runoff curve number of 69.76 was determined by the HELP model based on surface vegetation and the minimum infiltration rate of the top soil, and this number was used in the simulations. Additionally, the total area of cover was considered to be 40,000 square feet with a drainage distance of 200 feet. The input parameter values along with the results obtained are summarized in Table 4.

3. Results and discussion

The HELP model is a water balanced model that models multilayered capping systems. Simulations are most accurate when actual field and laboratory data are available for many of the input parameters that are needed to simulate field conditions. However, since data were not available for all of the parameters, various estimates were made for some of the input parameters to the HELP model. By using the materials presented in Table 3 along with polyvinyl chloride (PVC) liners, simulations were carried out (using different scenarios) to evaluate the effectiveness of various capping design systems in minimizing water

Case	Cover		Barrier		Buffer		Drainage			Results (in.)			
	Thickness (in.)	Kh (cm/sec.)	Thickness (in.)	Kh (cm/sec.)	Thickness (in.)	Kh (cm/sec.)	Thickness (in.)	TSlope (z)	Kh (cm/sec.)	ET	Surface Runoff	Lateral Drainage	Infiltration
PCA Reference Case													
# 1	12	7.1×10^{-4}	12	2×10^{-6}	6	5×10^{-4}	0	NAP	NAP	1.2	19.1	7.0	0
# 2	12	4.7×10^{-3}	12	2×10^{-6}	6	3.3×10^{-3}	0	NAP	NAP	1.2	17.0	9.3	0
# 3	12	1×10^{-2}	12	2×10^{-6}	6	$.7\times10^{-2}$	0	NAP	NAP	1.2	16.9	9.5	0
No Reclamation													
# 1	0	NAP	0	NAP	6	7.1×10^{-4}	0	NAP	NAP	0.1	16.6	10.7	0
# 2	0	NAP	0	NAP	6	4.7×10^{-3}	0	NAP	NAP	0	13.3	14.1	0
# 3	0	NAP	0	NAP	6	1×10^{-2}	0	NAP	NAP	0	12.7	14.7	0
Standard Reclamation													
# 1	24	7.1×10^{-4}	0	NAP	6	5×10^{-4}	0	NAP	NAP	0.1	18.7	8.6	0
# 2	24	4.7×10^{-3}	0	NAP	6	3.2×10^{-3}	0	NAP	NAP	0	14.1	13.3	0
# 3	23	1×10^{-2}	0	NAP	6	$.7\times10^{-2}$	0	NAP	NAP	0	13.2	14.3	0
PCA Reference Case with Drainage Layer At 3% Slope													
# 1	12	7.1×10^{-4}	12	2×10^{-6}	6	5×10^{-4}	12	3	1.0	0.1	18.8	2.8	5.9
# 2	12	4.7×10^{-3}	12	2×10^{-6}	6	3.3×10^{-3}	12	3	1.0	0	14.3	3.3	10.0
# 3	12	1×10^{-2}	12	2×10^{-6}	6	$.7\times10^{-2}$	12	3	1.0	0	13.2	3.3	10.9
At 5% Slope													
# 1	12	7.1×10^{-4}	12	2×10^{-6}	6	5×10^{-4}	12	5	1.0	0.1	18.7	2.5	6.1
# 2	12	4.7×10^{-3}	12	2×10^{-6}	6	3.3×10^{-3}	12	5	1.0	0	14.2	2.8	10.4
# 3	12	1×10^{-2}	12	2×10^{-6}	6	$.7\times10^{-2}$	12	5	1.0	0	13.2	2.9	11.4
PCA Reference Case – Barrier With Liner													
# 1	12	7.1×10^{-4}	12	2×10^{-6}	6	5×10^{-4}	0	NAP	NAP	4.5	22.3	0	0
# 2	12	4.7×10^{-3}	12	2×10^{-6}	6	3.3×10^{-3}	0	NAP	NAP	4.3	22.4	0	0
# 3	12	1×10^{-2}	12	2×10^{-6}	6	$.7\times10^{-2}$	0	NAP	NAP	4.3	22.5	0	0
Lower Permeability Barrier with Liner													
# 1	12	7.1×10^{-4}	12	1×10^{-7}	6	5×10^{-4}	0	NAp	NAp	4.5	22.3	0	0
# 2	12	4.7×10^{-3}	12	1×10^{-7}	6	3.3×10^{-3}	0	NAp	NAp	4.4	22.4	0	0
# 3	12	1×10^{-2}	12	1×10^{-7}	6	$.7\times10^{-2}$	0	NAp	NAp	4.4	22.4	0	0
PVC Liners													
Case # 1*	12	3×10^{-3}	18	2.9×10^{-4}	NAp	NAp	6	3	5.8×10^{-3}	0.3	15.4	11.0	0.7
# 2*	12	3×10^{-3}	18	2.9×10^{-4}	Nap	Nap	6	3	5.8×10^{-3}	0.8	17.0	4.0	5.3
Higher Permeability Barrier with Drainage Layer (Base case)													
# 1	12	7.1×10^{-4}	12	1×10^{-5}	6	5×10^{-4}	0	NAp	NAp	0.5	18.7	8.1	0
# 2	12	4.7×10^{-3}	12	1×10^{-5}	6	3.3×10^{-2}	0	NAp	NAp	05	14.6	12.4	0
# 3	12	1×10^{-2}	12	1×10^{-5}	6	$.7\times10^{-2}$	0	NAp	NAp	0.4	13.9	13.1	0
Higher Permeability Barrier Drainage Layer at 3% Slope													
# 1	12	7.1×10^{-4}	12	1×10^{-5}	6	7.1×10^{-4}	12	3	1.0	0.1	18.7	6.4	2.2
# 2	12	4.7×10^{-3}	12	1×10^{-5}	6	4.7×10^{-3}	12	3	1.0	0	14.2	9.2	4.1
# 3	12	1×10^{-2}	12	1×10^{-5}	6	1×10^{-2}	12	3	1.0	0	13.2	9.6	4.7
At 5% Slope													
# 1	12	7.1×10^{-4}	12	1×10^{-5}	6	7.1×10^{-4}	12	5	1.0	0.1	18.7	5.7	2.9
# 2	12	4.7×10^{-3}	12	1×10^{-5}	6	4.7×10^{-3}	12	5	1.0	0	14.2	8.1	5.2
# 3	12	1×10^{-2}	12	1×10^{-5}	6	1×10^{-2}	12	5	1.0	0	13.2	8.5	5.9

Table 4. Summary of Preliminary Results with HELP Model.

infiltration into mineral stockpiles. Simulations were run for the MPCA barrier requirement of $2x10^{-6}$ cm/sec, 20 mil PVC liner, a control (untreated) stockpile, standard reclamation (2 feet of cover), MPCA reference case with drainage layer at 3% and 5% slope, and lower permeability with liner, etc. All the materials (Table 3) except fine tailings alone and mixed with 1/2% bentonite had permeability which were equal to or less than $2x10^{-6}$ cm/sec, which was the maximum value established by Minnesota Pollution Control Agency.

The first scenario involved the Minnesota Pollution Control Agency (MPCA) case which was a barrier with hydraulic conductivity of 2 x 10^{-6} cm/sec. Results from model simulations indicated an average infiltration of 8.6 inches with a surface runoff of 1.2 inches and no lateral drainage. With a 3% drainage slope, the MPCA case registered an average infiltration of 3.1 inches with lateral drainage of 8.4 inches. With a 5% drainage slope, the MPCA case recorded an average infiltration of 2.7 inches with drainage of 9.3 inches. With the PVC liner, the MPCA case had neither infiltration nor drainage but the surface runoff was 4.4 inches.

With no reclamation, the average infiltration from model simulations was 13.7 inches with neither surface runoff nor lateral drainage. The standard reclamation, which required a cover thickness of at least 24 inches, had an average infiltration of 12.1 inches with neither surface runoff nor lateral drainage. The lower permeability barrier with liner with hydraulic conductivity of 1 x 10^{-7} cm/sec had a surface runoff of 4.4 inches with neither infiltration nor lateral drainage. The results obtained from the HELP model are presented in Table 4, and the simulated annual infiltration into stockpiles using the various capping options is graphically depicted in Figure 1. Thus far, synthetic liners appear to be the perfect cover systems, since if intact, they would not transmit any water. Regrettably, a leak-proof liner does not really exist. In general, the thicker the liner system and the better the installation, the smaller the leakage. For the synthetic liner barrier system used in this study, the effective hydraulic conductivity is a function of the leakage factor, f. A leakage factor, f is directly proportional to the area of opening and inversely proportional to the area of the liner system. Typical values for liners range from 0.01 for a 20 mil PVC liner poorly installed to 0.00001 for an 80 mil HDPE with a perfect installation (Eger et. al, 1990). The results from the HELP model simulation imply that the 20 mil PVC liner system has a leakage factor of about 0.001, which is within the expected range. When the flow from a stockpile has been reduced, more efficient use can be made of additional passive treatment systems such as alkaline and wetland treatment. Thus, uncontaminated surface and barrier flow from a stockpile capping system could be collected and used to augment flow downstream of additional passive treatment systems.

From the foregoing results, the three variables that have the greatest effect on the amount of water that infiltrate the cap are the hydraulic conductivity of the barrier layer, the hydraulic conductivity of the cover, and the type and rooting depth (evaporative zone depth) of the vegetation. From the results of the HELP model simulation, none of the barriers reduced flow to a level consistent with a barrier with hydraulic conductivity of 1 x 10^{-7} cm/sec. The United States Environmental Protection Agency's guidelines for capping landfills require a barrier layer with an effective hydraulic conductivity of 1 x 10^{-7} cm/sec. This value is also required for new landfills by present MPCA solid waste rules. Simulations conducted with the HELP model showed that when the hydraulic conductivity of the barrier was reduced from $2x10^{-6}$ cm/sec to 1 x 10^{-7} cm/sec, infiltration decreased by over 90 per cent. Therefore, to minimize the volume of contaminated flow in any stockpile capping system, the hydraulic conductivity of the barrier should be less than or equal to 1 x 10^{-7} cm/sec.

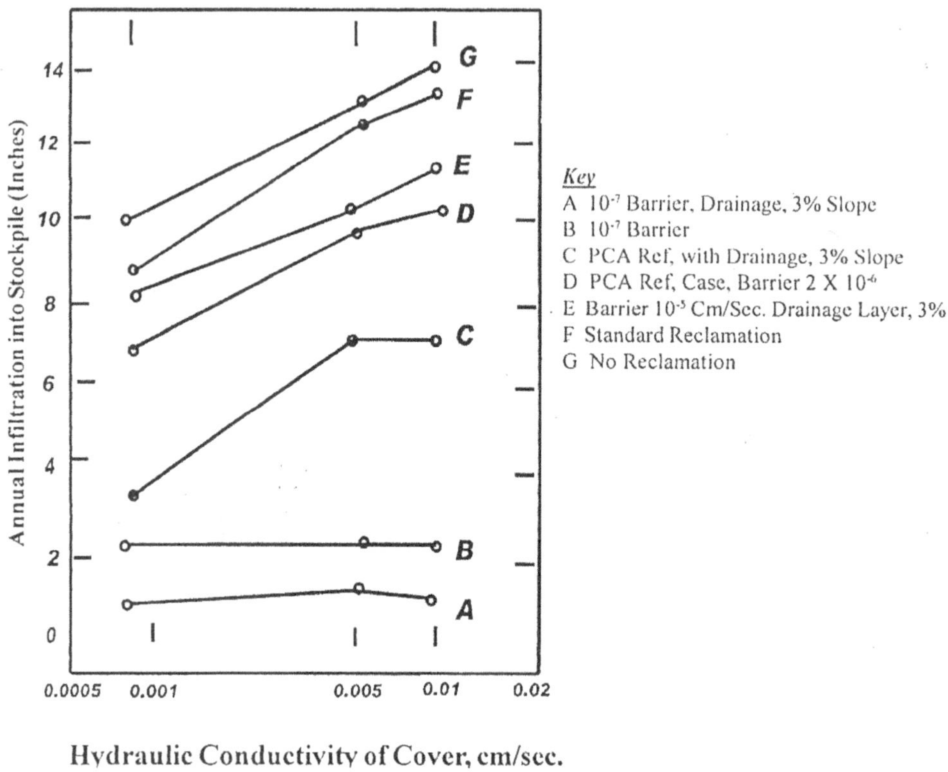

Fig. 1. Simulated annual infiltration into stockpiles using various capping options.

4. Conclusion and recommendations

One key component in mitigating the water quality problem associated with stockpile drainage is the reduction of the amount of water which infiltrates into the stockpiles. While a reduction in infiltration will not change the drainage water quality, the overall mass of contaminants released per year will be reduced as the drainage flow is reduced. Based on the results from the HELP model simulations, the three variables that have the greatest effect on the amount of water that will infiltrate the cap are the hydraulic conductivity of the barrier layer, the hydraulic conductivity of the cover, and the type and rooting depth (evaporative zone depth) of the vegetation. The results from the model simulations showed that, when the hydraulic conductivity of the barrier was reduced from 2×10^{-6} cm/sec to 1×10^{-7} cm/sec, the infiltration reduced by over 90%. Therefore, to minimize bottom flow, the hydraulic conductivity of the barrier layer should be less than or equal to 1×10^{-7} cm/sec.

As earlier alluded to, simulations are most accurate when actual field and laboratory data are available for the many input parameters needed to run the HELP model. Unfortunately, data was not available for all the parameters and various estimates had to be made. Additional field and laboratory data are needed to better determine and model the effectiveness of the various capping alternatives to stem infiltration. Generally, infiltration parameters are often established based on samples which are not representative of field profiles. In other words, laboratory test samples are homogeneous, and thus lack the variability that is associated with similar samples in the field (Udoh, 2008). Since field permeability tests are more likely to yield accurate estimates of hydraulic conductivity than laboratory test, they are recommended as part of either the final design process or construction verification.

Based on the results obtained, a cap design consisting of a three-layer soil barrier is recommended for final capping of any mineral stockpile capping project. Therefore, the selection of materials for the capping of any mineral stockpile and/or waste disposal site should be based on optimizing those properties that have the greatest influence on the long-term performance of the material.

5. References

Barr Engineering Company (1986). Feasibility Assessment of Mitigation Measures for Gabbro and Waste Rock Stockpiles: Dunka Pit Area. Prepared for Erie Mining Company, Hoyt Lakes, Minnesota.

Eger, P.; Antonson, D. and Udoh, F. (1990). *Stockpile Capping Report*. Minnesota Department of Natural Resources, Division of Minerals, St. Paul, Minnesota.

Hauser, V. L. and Jones, O. R. (1991). Runoff Curve Number for the Southern High Plains. *Transactions of the American Society of Agricultural Engineers, ASAE Vol. 34(1): 142-148.*

Ogunlela, A. O. (2001). Predicting Effects of Land Use Changes on Runoff Using the Curve Number Method. *Nigerian Journal of Tropical Engineering. Vol. 2 (1):25-32.*

Ogunlela, A. O. and Kasali, M. Y. (2002). Evaluation of Four Methods of Storm Hydrograph Development for an Ungauged Watershed. *Nigerian Journal of Technological Development Vol. 2 (1): 15-24.*

Ritter, J. B. and Gardner, T. W. (1991). Runoff Curve Numbers for Reclaimed Surface Mines in Pennsylvania. *Journal of Irrigation and Drainage Engineering. Vol. 117(5):656-666.*

Schroeder, P. R. et al (1984). The Hydrologic Evaluation of Landfill Performance (HELP) Model. Volume 1. User's Guide for Version I. EPA/530-SW-84-009, U.S. Environmental Protection Agency, Office of Solid Waste and Emergency Response, Washington, D. C.

Schroeder, P. R. and Peyton, R. L. (1987). Verification of the Hydrologic Evaluation of Landfill Performance (HELP) Model Using Field Data. U.S. Department of Commerce, National Technical Information Service, Springfield, Virginia.

Soil Conservation Service (1972). *National Engineering Handbook* Section 4: Hydrology. U.S. Printing Office, Washington, D. C.

Udoh, F. D. (1993). Minimization of Infiltration into Mining Stockpiles Using low Permeability covers. Doctoral Dissertation. University of Wisconsin - Madison.

Udoh, F. D. (2008). An Integrated Program for Infiltration Control of Gabbro and Waste Rock Stockpiles at the Dunka Mine Site. Journal of Ecology, Environment and Conservation, Vol. 14, No. 4, p. 589-594.

Part 2

Hydraulic Conductivity and Plant Systems

5

Plant Hydraulic Conductivity: The Aquaporins Contribution

María del Carmen Martínez-Ballesta, María del Carmen Rodríguez-Hernández, Carlos Alcaraz-López, César Mota-Cadenas, Beatriz Muries and Micaela Carvajal

Dpto. Nutrición Vegetal. Centro de Edafología y Biología Aplicada del Segura (CEBAS-CSIC)
Campus de Espinardo, Murcia
Spain

1. Introduction

In the soil-plant-atmosphere continuum the major resistance to water flux is the leaf-atmosphere interface which determine the rate of transpiration for a specific evaporative demand. In this scenario, the hydraulic resistance of the different plant tissues is minor; however, the hydraulic conductivity of the whole plant is subjected to a tight physiological regulation in which the aquaporins role may result fundamental. The expression of a large number of aquaporins occurs predominantly in roots and different experimental procedures have demonstrated that aquaporins activity is linked to the hydraulics of some species during abiotic stress. However, the plants roots hydraulic properties also depend on the morphology and anatomy of roots and the length of the absorbing region in addition to the influence of aquaporins. These features change during the plant development and in response to environmental stimuli by altering the hydraulic conductivity of the root. To fully understand root system hydraulics and the contribution of native aquaporins, comprehensive studies at different scales are required. In this chapter the definitions used to describe the plant hydraulic resistances are mentioned and the influence of the root anatomy and morphology on hydraulic conductivity is reviewed. Also, the variations in the hydraulic resistances under different abiotic stresses and distinct environmental conditions have been explored. Finally, the different properties and characteristics among various measuring methods are reviewed.

2. Hydraulic resistances: the physiological significance

The water pass from soil though plant to atmosphere has been classically described as a system of hydraulic resistors arranged in series (van den Honert, 1948). However, plants can vary this resistance (and conductance) of the pathways to maintain the water balance of the shoot (Steudle, 2000). When water exists in the vapour phase, the greatest resistance is the stomatal aperture. However, in the liquid phase, the root system constitutes a highly significant and important resistance to overall flow of water in the plant (Steudle, 2000).

A common error in plant water relations is the interchangeable use of hydraulic conductance and conductivity although both parameters are related. Hydraulic conductance is a measure of the ability of an entity to conduct water, independent of the specific entity's dimensions, whereas hydraulic conductivity is a property of an entity with specified dimensions, usually surface area. Conductance or conductivity may be normalized to various dimensions of the particular organ or system relevant to the flow-path, thus conductance of the root system may be normalized to root length or root weight (Tyree et al., 2002). When water transport across a surface is considered (a cell or a root) the concept is expressed by surface area to give what is also termed hydraulic conductivity (*Lp*, m s^{-1} MPa^{-1}). Conductivity of a stem segment or petiole can be normalized to xylem cross sectional area of a stem segment. Leaf hydraulic conductivity is generally measured as the flux (mmol s^{-1}) divided by the gradient in water potential, $\Delta\Psi$ (MPa), and leaf surface area to give units of mmol m^{-2} s^{-1} MPa^{-1} (Sack & Holbrook, 2006).

Root hydraulic conductivity (Lp_r), is one of the major parameters reflecting root water uptake ability. It has a close correlation with plant water relations under both normal and stressed conditions. The root ability to respond rapidly to fluctuating conditions suggests that Lp_r may participate in plant adaptation to diverse environments (Steudle, 2000). The study of root water uptake has been made progress recently from the anatomical structure of the root to molecular level, i.e., aquaporins activity (Steudle, 2000; 2001). Aquaporins are transmembrane proteins that belong to the Mayor Intrinsic Proteins (MIP), a large family of water channel proteins located in plasma and intracellular membranes and are the main determinants of water flow across plant cells and tissues. Aquaporins can be divided into different subfamilies depending on the sequence homology and subcellular localization, the plasma membrane proteins (PIP, with two phylogenic subgroups PIP1 and PIP2), the tonoplast intrinsic proteins (TIP) the nodulin- 26-like intrinsic membrane proteins (NIPs) and the small basic intrinsic proteins (SIPs) (Javot et al., 2003; Maurel et al., 2008).

The dynamic changes in Lp_r in response to chemical or environmental stimuli may result from modifications of aquaporin abundance or activity (Carvajal et al., 1996; Tournaire-Roux et al., 2003; Boursiac et al., 2005). However, due to the high plasticity of plant root systems both in architecture and metabolism (Liang et al., 1997; Joslin et al., 2000; Linkohr et al., 2002; López-Bucio et al., 2003), and the different properties among various measuring methods and experimental conditions, the root hydraulic conductivity could be highly variable even for the same plant.

Also, leaves contribute to a substantial part of the hydraulic resistance in whole plants (Sack et al., 2003; Sack & Holbrook 2006). Leaf hydraulic conductance may also be linked to transpiration efficiency through regulation of water transport by aquaporins and effects on mesophyll cell water status (Zwieniecki et al., 2007). In fact, ABA (abcisic acid) controls aquaporin PIP levels in the leaf (Morillon & Chrispeels, 2001; Aroca, 2006; Lian et al., 2006; Parent et al., 2009), thereby contributing to the leaf hydraulic conductivity (Morillon & Chrispeels 2001). However, this is not always the case and it has been reported that an antisense inhibition of PIP1 and PIP2 expression did not affect the leaf hydraulic conductance in *Arabidopsis* (Martre et al., 2002).

3. The root anatomy influence on the hydraulic conductivity

In the radial pathway, the water absorbed by the roots has to pass through living tissue, through the walls of the root before reaching the xylem vessels. In the axial plane, the flow of water occurs along the xylem vessels and tracheids. The relationship between radial and axial resistances determines the resistance of the whole root and distribution of water uptake (Zwieniecki et al., 2003). The composite transport model of water proposed in the roots (Steudle & Frensch, 1996; Steudle & Peterson, 1998) accounts for variable contributions of transmembrane (where aquaporins may exert a control) and apoplastic (independent of aquaporin activity) pathways to the overall root water uptake, depending on the nature and the intensity of the driving force. The model explains why hydrostatic gradients may result in higher root hydraulic conductivity (Lp_r) than for osmotic gradients (Steudle, 2000). However, higher root Lp_r for hydrostatic than for osmotic gradients is not always observed (Bramley et al., 2007b). For example, Lpc of epidermal and cortical cells was much greater than Lpr in *Hordeum distichon* and *Phaseolus coccineus* roots, indicating that water flow mainly via the cell-to-cell pathway (Steudle & Brinckmann, 1989). By contrast, analogous measurements on maize (*Zea mays*) roots revealed a predominantly apoplastic flow (Steudle et al., 1987). Comparing the measured values of Lp_r and Lp_c for each cortical cell layer indicated that radial water flow through wheat (*Triticum aestivum*) roots occurs by a similar contribution of the parallel pathways, but radial water flow in the roots of narrow-leafed lupin (*Lupinus angustifolius*) and yellow lupin (*Lupinu luteus*) appears to be predominantly apoplastic (Bramley, 2006).

In addition, the dynamics of root permeability to water has been also associated with the anatomical and morphological features (Kramer, 1983; Moreshet & Huck, 1991). In cereal roots, a maximum of water absorption in the region of less than 100 mm from the root apex has been observed (Sanderson, 1983) since the water flow resistance is higher in both the radial and the axial pathway in the root zones where there is a developed xylem (Steudle, 2001). Also, the Lp_r of wheat roots decreased with the distance from the root tip, indicating that water absorption occurs preferentially in the apical region (Bramley, 2006).

Rieger and Litvin (1999) found that the root diameter was negatively correlated with Lp_r in five species and that drought stimulated the suberisation and other anatomical changes that reduced the Lp_r. Thus, the apoplastic pathway can be inhibited by the presence of Casparian bands, which are deposits of suberin or lignin in the cell wall (Steudle, 2000). Casparian bands occur in radial and transverse walls of the endodermis and exodermis (Steudle & Peterson, 1998). Hydraulic conductivity uses to decline with root age which is likely due to suberization and loss of the cortex reducing surface area available for water uptake (Wells & Eissenstat, 2002). Frequently, suberized layers may assist in reducing water loss to soil during water deficits. Huang and Eissenstat (2000) determined that structural differences in the radial pathway were the main factors that determined the Lp_r in the roots of citrus rootstocks. In maize plants the development of an exodermis in the roots reduced the radial hydraulic conductivity (Hose et al, 2000; Zimmermann et al, 2000). By contrast, Steudle et al. (1993) demonstrated that the endodermis of young maize roots did not affect the Lp_r. In a similar way, Barrowclough et al. (2000) found that in the roots of onion plants (*Allium cepa*), the highest values of radial hydraulic conductivity were correlated with the presence of exodermis. Thus, the anatomical changes are slow and depend on the plant growth and the genotype, and can act as a survival strategy to reduce long-term Lp_r when environmental changes are slow.

In addition, depending on the length of the root species the absorption region can change (Kramer, 1983). For example, the wheat root length is two to ten times higher than the lupino plants (Gallardo et al., 1996). However, despite these differences in root length, the roots of eudicotyledon species tend to have a higher specific rate of water uptake than

cereals. Moreover, these higher rates of water absorption appear to be due to greater hydraulic conductivity (Bramley, 2006; Gallardo et al., 1996) as occurred in wheat roots where aquaporin activity increased causing an overshoot in Lp_r (Bramley et al., 2010).
Although it was believed that the relation between root water uptake and Lp_r was due only to differences in axial and radial anatomy (Hamza & Aylmore, 1992a; Gallardo et al, 1996), the discovering of the aquaporins supposed a tight regulation mechanism of water flux. Since a significant proportion of radial flow of water occurs from cell to cell through the cell membrane, Lp_r can be controlled by the activity of aquaporins. Thus, measurements of the radial hydraulic conductivity after removal of tissue layers and the application of mercury have shown variable activity of aquaporins in different regions of *Agave deserti*, where aquaporins were to be active in regions associated with living cells with high metabolic activity (Martre et al., 2001; North et al., 2004). Also, in *Arabidopsis thaliana* roots the relative contribution of the apoplastic pathway increased when aquaporin activity was inhibited by mercury treatment which was reflected in L_0 (Martinez-Ballesta et al., 2003).
Finally, the absorption of water for several or all of the individual roots can contribute to Lp_r of the entire root system (Bramley, 2006). There is also evidence that individual roots are capable of varying its hydraulic conductivity. In several experiments Vysotskaya et al. (2004a, 2004b) removed four of the seminal roots of durum wheat (*Triticum durum*) and an increased Lp_r of the remaining roots was observed maintaining the shoot water supply.

4. Hydraulic conductivity and environmental stress

It is known that roots offer the greatest resistance to water flow and that the hydraulic conductivity of the root (Lp_r), may be affected by diverse forms of abiotic stress. Although the hydraulic conductivity of the tissues could be regulated by changes in the level of specific aquaporins, regulation could also occur by changing the activity of these proteins (Carvajal et al., 2000; Zimmermann et al., 2000). Thus, the ability to increase or decrease the water permeability of a cell seems to justify the enormous effort in expressing large amounts of these proteins (Schäffner, 1998). However, water uptake by roots is a variable process that depends on the structure and anatomy of roots which, in turn, is affected by environmental factors such as drought, temperature and heavy metals (Azaizeh et al., 1992; North and Nobel, 1996; Peyrano et al., 1997; Schreiber et al., 1999).

4.1 Water stress

Some stresses, such as drought, could be perceived by the roots and transduced to the aerial part as a hydraulic signal reducing cell turgor in the leaves (Christmann et al., 2007). This change increases the leaf ABA levels and subsequently induces the stomatal closure. The Lp_r drop due to the water deficit could amplify this root-shoot signal, decreasing finally the plant transpiration rate. It has been found a correlation between Lp_r and the transpiration rate for eucalyptus (Franks et al., 2007) or grapevine (Vanderleur et al., 2004). However, under specific physiological context, stomatal regulation and Lp_r are uncoupled and more research about the root-shoot hydraulic signalling is necessary. Thus, it has been reported than under low evaporative demand the stomatal conductance was not affected by the application of aquaporin inhibitors on roots. Consequently, transpiration was steady and the water potential gradient between the root medium and the xylem at the leaf base was increased and counteracted the Lp_r reduction. However, this chemical manipulation of root hydraulic conductivity caused simultaneous effects on leaf growth rate and on cell turgor in the growing zone suggesting that turgor and growth are coupled (Ehlert et al., 2009). By contrast, under higher evaporative demand, which induced a dramatic decrease in leaf water potential, Lp_r was reduced to values similar to those observed in maize in field conditions (Tardieu & Simonneau, 1998). Also, previous studies of aspen (*Populus species*) (Wan & Zwiazek, 1999) and pepper (*Capsicum annuum*) (Martinez-Ballesta et al., 2003a) also reported that, on a slightly longer term exposure of the roots to $HgCl_2$ induced a significant decrease in stomatal conductance.

The effects of drought on the root hydraulic conductivity depend on the stress level (Siemens & Zwiazeck, 2004) and plant genotype (Saliendra & Meinzer, 1992). If water uptake becomes limiting, the up-regulation of aquaporins expression could enhance cellular water permeability, increasing root hydraulic conductivity, relieves osmotic pumps, and supports the survival during dry periods (Siefritz et al., 2002).

Gene expression studies in various plant species have shown variable responses of aquaporin isoforms to water stress, with both up- and down-regulation of genes evident (Alexandersson et al., 2005; Jang et al., 2004; Sarda et al., 1999; Suga et al., 2002; Yamada et al., 1997). The down-regulation of PIP gene expression and Lp_r by drought stress may result in reduced membrane water permeability, and may promote cellular water conservation during periods of dehydration stress (Jang et al., 2004). In desert plants, the closure of water channels during drought would help prevent root water loss to a soil that generally has a lower water potential than does the plant (North et al., 2004). In leaves, roots, and twigs of olive (*Olea europaea*), OePIP1;1, OePIP2;1, and OeTIP1;1 were significantly reduced at 3 and 4 weeks after water was withheld (Secchi et al., 2007). Overexpression of AtPIP1b in transgenic tobacco (*Nicotiana tabacum*) caused plants wilting faster when water was withheld (Aharon et al., 2003). In contrast, Siefritz et al. (2002) observed reduced resistance to water stress in antisense tobacco plants with reduced expression of NtAQP1, the homologous aquaporin. Recently, Sade et al. (2009) showed that the tonoplast aquaporin SlTIP2;2, is a key to isohydric to anisohydric behaviour conversion, increasing transpiration under normal growth conditions and limiting the reduction in transpiration under drought and salt stresses. This characteristic attributable to overexpression of the TIP isoform SlTIP2;2 do not appear to exist in many PIP-overexpressing plants.

Effects of drought on root hydraulic conductivity will then have different consequences on whole hydraulic resistance and on leaf water potential depending on species. Isohydric cultivars are those that keep their leaf water potential above a certain threshold regardless of soil water availability or atmospheric water demand. The finding of no variation in transcript level of most important root PIP aquaporins and suberisation implies a lower hydraulic conductance in water deficit conditions. This supports the hypothesis of tight control on stomatal regulation that is typical of isohydric cultivars, which aims to avoid excessively negative xylematic water potential and, therefore, cavitation (Schultz, 2003*a*; Soar et al., 2006; Vandeleur et al., 2009). Anisohydric cultivars are those in which leaf water potential drops with decreasing soil water availability or increasing atmospheric water demand.

4.2 Salinity

It has been reported that salinity affect negatively to the hydraulic conductivity (Munns & Passioura, 1984; Joly, 1989). Although the reductions in root hydraulic conductivity or hydraulic conductance of salinised plants have been suggested as being due to the hyperosmotic stress and ionic imbalance caused by the high apoplastic concentrations of

Na^+ and Cl^- (Evlagon et al., 1990), it has been suggested that these reductions could be due to changes either in the aquaporins functionality or in the amount of this protein in the plasma membrane (Carvajal et al., 2000).
In any case, the L_0 results for the plant roots cannot be always explained in terms of aquaporins abundance in the plasma membrane, as shown in several reports (López-Pérez et al., 2007; Muries et al., 2011). Thus, in root cells of *Brassica oleracea* plants grown with NaCl, apparent disagreement between L_0 values and PIP protein abundance has been observed (Muries et al., 2011). In these plants the most-important modification in the anatomy of the root was phi thickening, which increased in salinity-stressed plants and could be a physical barrier to apoplastic water transport (López-Pérez et al., 2007). The down-regulation of L_0 under saline conditions and the increased protein amount observed could be interpreted as a mechanism to restore and compensate water uptake by roots. Other explanations for the disagreement between L_0 values and PIP protein abundance under stressing conditions could be differences in the contribution of PIP isoforms to the L_0 values or different PIP localisation along the root axis (Benabdellah et al., 2009) or among cellular membranes (Boursiac et al., 2005; Zelazny et al., 2007). Furthermore, a reduction of the phosphorylation state of PIP proteins could cause the observed reduction in L_0 (Wilder et al., 2008) and this and other post-translational modifications as gating control mechanism may be considered.

4.3 Anoxia

Soil compaction or flooding which restrict oxygen diffusion in the soil, result in root anoxia which, in turn, down-regulates Lp_r in certain plant species. Thus, Zhang & Tyerman (1991) using the cell pressure probe showed a 10-fold decrease in the hydraulic conductivity of root cortical cells of wheat under anoxia conditions. These changes in the root hydraulics largely reflected the variations in the transport properties of root cell membranes. Thus, anoxia may reduce the rate of active pumping of nutrients without affecting the passive permeability of roots. Because of the reduction of root hydraulic conductivity, anaerobic conditions should have great consequences for the supply of the shoot with water and, hence, for the whole plant water status. Aquaporins that are highly expressed in roots and facilitate water transport across membranes tended to be down regulated after a few hours of hypoxia (Bramley et al., 2007b). Also, the closure of aquaporins in membranes decreased the hydraulic conductivity and hence increased the half-time of the rate of water exchange across the cell (Bramley et al., 2010).
In *Arabidopsis* plants, hydraulic conductivity inhibition under anoxia or O_2 deprivation by the gating of aquaporins was related to cytosolic acidosis (Tournaire-Roux et al., 2003). Thus, the closure of the plant plasma membrane aquaporin was triggered by the protonation of a conserved hystidine residue under anoxia conditions (Tournaire-Roux et al., 2003). Similarly, in spinach an acidification of the cytosol due to anoxia, would cause a protonation of His 193 in loop D of SoPIP2;1 thereby closing the channel (Törnroth-Horsefield et al., 2006). Also, it has been characterized two protein kinases phosphorylating Ser 115 and Ser 274 in SoPIP2;1 which optima pH reflects the normal cytosolic pH (Sjövall-Larsen et al., 2006). Thus, inactivation of these kinases due to an acidification of the cytosol would lead to a dephosphorylation of Ser115 and Ser274 of SoPIP2;1 and represent an alternative mechanism for aquaporin closing (Törnroth-Horsefield et al., 2006).
These changes in cytosolic pH and H_2O_2 have recently emerged as cellular signals triggered by various external stimuli and mediating pronounced and rapid changes in Lp_r (Aroca et al., 2005; Lee et al., 2004a; Tournaire-Roux et al., 2003).

4.4 Low temperatures

Also, the root system can respond very quickly to changes produced by low temperatures through the variations in its root hydraulic conductivity (Fennell & Markhart, 1998). Chilling induced water stress in plants and it was initiated by the decreased of 17-23% in the root hydraulic conductance followed by a large decrease in leaf water and turgor potential (Aroca et al., 2001). Thus, the water deficit is caused by a reduction in the root water uptake greater than the leaf transpiration rate during chilling (Aroca et al., 2001). The effect of chilling on the root hydraulic conductivity can be attributed to changes in abundance and/or activity of aquaporins (Aroca et al., 2004; Cochard et al., 2007). Thus, in tulip Azad et al. (2004) identified the temperature as an environmental stimulus that induced phosphorylation or dephosphorilation of aquaporins accompanied by changes in the cells water permeability. Reversible phosphorylation is considered as a potent mechanism for plant aquaporin regulation, during development and in the response of plants to environmental stimuli (Luu & Maurel, 2005).

Aroca et al. (2001) reported that chilling-tolerant maize genotype showed an acclimation of Lp_r and root hydraulic conductance was recovered in chilling-tolerant but not in chilling-sensitive varieties. Lee et al. (2004b) showed that in cucumber (*Cucumis sativus*), a cold sensitive species, a brief exposure to low temperature reduces root pressure, hydraulic conductivity, and active nutrient transport. These authors also postulated that changes in the activity of aquaporins underlie the changes in hydraulic conductivity (Aroca et al., 2005; Lee et al., 2004a). Thus, it was concluded that water permeability of cucumber root cell membranes was related to changes in the activity (open/closed state) of aquaporins that were effectively at low temperature (Lee et al., 2005).

In addition to the aquaporins, increased water viscosity accounted for part of the decrease in the L_0 early during chilling (Matzner & Comstock, 2001). Thus, it has been suggested that it is the result of low-temperature-induced alteration of membrane properties that lowers the hydraulic conductance of the symplastic component of radial root water flux (Sanders & Markhart, 2001).

5. Effect of light intensity on hydraulic conductivity

The plant water status is constantly changed by diurnal variations of light intensity as the stoma opening to fix CO_2 is occurring. Therefore, as plants lose substantial amounts of water using the same pathway, they have to develop strategies to optimize the use of water efficiently in response to changes in the light regime, such as variations in hydraulic conductivities of the root (Lp_r) and hydraulic conductance of the leaf (K_{leaf}) (Postaire et al., 2010).

It is well-described that plant leaves respond to light in a short time scale by adjusting leaf hydraulic efficiency. There is also a general consensus that up- or down-regulation of water channels in the plasma membrane of leaf cells plays a central role in the underlying mechanisms. In many plant species it has been reported that the K_{leaf}, can be increased several folds by high irradiance (Nardini et al., 2005, 2010; Lo Gullo et al., 2005; Sack & Holbrook, 2006; Cochard et al., 2007) and can depend on both light duration and intensity (Sellin et al., 2008). Other experiments showed no effect of light conditions on K_{leaf} as it occurs in laurel in laboratory experiments (*Laurus nobilis*) (Cochard *et al.*, 2004) or on K_{lam} (leaf lamina hydraulic conductance) of trembling aspen trees (Voicu et al., 2009). Despite of it, there is current agreement that aquaporin activation and/or expression plays a role in the underlying mechanisms as the increase in K_{leaf} (Nardini et al., 2005; Voicu et al., 2008). In addition, other studies suggest that this light-induced enhancement involves expression or activation of plasma membrane aquaporins in mesophyll or bundle sheath cells (Tyree et al., 2005; Cochard et al.; 2007, Voicu et al., 2008). This idea is supported by the results obtained by Cochard et al. (2007) who found a very good kinetic correlation between the increase in K_{leaf} and the increase in two walnut aquaporin (*Jr*PIP2,1 and *Jr*PIP2,2) expression during a transition from dark to high light. In the same way, pressure probe measurements revealed that the effect of light on leaf water transport was mediated in part through changes in cell hydraulic conductivity (Lp_c) in midrib parenchyma cells of maize leaves (Kim & Steudle, 2007) where an increasing light intensity increased both Lp_c and aquaporin activity. However higher light intensities (800 and 1800 µmol m^{-2} s^{-1}) dramatically decrease Lp_c probably due to an oxidative gating of aquaporins by ROS (Kim & Steudle, 2009). There should be an optimal light intensity to maximize water flow across leaf cells, but enhanced water flow could be inhibited at a certain light intensity.

Although recent findings showed an inhibition of aquaporin-mediated water transport in tobacco and bur oak leaves exposed to high irradiance (Lee et al., 2009; Voicu et al., 2009), these papers rather indicate that regulation of the aquaporin-mediated water transport processes is more complicated and can not always be explained merely by changes in the transcript level. On the other hand, it has also been shown that exposure of *Arabidopsis* plants to darkness increased the hydraulic conductivity of excised rosettes (K_{ros}) by up to 90% and enhanced the transcript abundance of several PIP genes, including AtPIP1;2 which represent a key component of whole-plant hydraulics (Postaire et al., 2010).

The impact of high irradiance on stem (K_{stem}) and leaf lamina (K_{lam}) hydraulic conductance has also been demonstrated with an increase in K_{stem} (field-grown laurel plants - Nardini et al, 2010; silver birch - Sellin et al., 2010) and in K_{lam} (Voicu et al., 2008) whereas some data suggest that petiole hydraulic conductance (K_{pet}) was unchanged upon illumination (Voicu et al., 2008). The quality of light was also found to have an effect in K_{lam} with a higher increase ranked in descending order as follows, white, blue and green, red and amber light, after a 30-min exposure to high irradiance (Voicu et al., 2008) but not in K_{pet}. Neither of these studies demonstrated a direct involvement of the aquaporins on hydraulic conductance changes.

6. Hydraulic conductivity and biological rhythm

The plant water status is not only challenged by light intensity or darkness but also by diurnal variations (biological rhythm). Since the transport of water and certain other small solutes is facilitated by the function of aquaporins, whose expression and functionality follows the changing demands of the plant physiology during the day or night, it is not surprising that root hydraulic conductivity which is indicative of plant water uptake may also be regulated in a day/night-dependent manner and modified by aquaporin activity (Siefritz et al. 2002). In classic experiments, it was observed that the root hydraulic conductivity declined towards the end of the light period and rose again at the end of the dark period (Parsons & Kramer, 1974). In addition, a diurnal variation of Lp_r was shown in young roots of *Phaseolus coccineus* (Peters & Steudle, 1999) and in excised roots of the legume *Lotus japonicus* grown in aeroponic (mist of nutrient reservoir around the plant roots) or in sand culture (Henzler et al., 1999) where Lp_r was found to vary over a 5-fold range during a day/night cycle, with a maximum around noon. This was correlated to the expression of a putative PIP1 aquaporin (Henzler et al., 1999) probably due to the conductivity of membranes of endodermal and stellar cells rather than first four cell layers of the cortex where there was no evidence of any diurnal fluctuation.

Diurnal changes in K_{leaf} have been reported in numerous species, but, in most cases, K_{leaf} was increased during the day, concomitantly to a higher transpiration demand (Nardini et al., 2005; Tyree et al., 2005; Sack & Holbrook, 2006; Cochard et al., 2007). A midday depression of K_{leaf} has been reported in the tropical tree species *Simarouba glauca* (Brodribb & Holbrook, 2004), but in this case, it was due to a vulnerability of the vascular system to cavitation rather than aquaporin regulation. Contrary as it occurs in roots, a higher expression of most of the *Zm*PIP genes during the first hours of the light period than at the end of the day or at night (Hachez et al., 2008) was correlated with changes in the membrane water permeability measured using a cell pressure probe in maize leaves (Heinen et al., 2009).

7. Effect of plant nutrition on hydraulic conductivity

An excess or absence of the main elements in plant nutrition can cause disorders in some parameters of water relations such as hydraulic conductance of roots and the activity of aquaporins at the cellular level (Clarkson et al., 2000).

Several works revealed that both N- and P- deficient conditions decreased the Lp_r (Carvajal et al., 1996; Shaw et al., 2002; Shangguan et al., 2005; Fan et al., 2007) and the Lp_c (Radin & Matthews, 1989) in many plant species. This suggests that the lowered root Lp_r of N-, or P-deficient plants may be due to the decreaseds water channel activity or abundance on the plasma membrane (Carvajal et al. 1996, 1998; Clarkson et al., 2000; Shangguan et al., 2005). Such a decrease has also been observed in SO_4^{2-}-deprived barley (*Hordeum vulgare*) roots, where Lp_r decreased to 20% of controls over a 4-d period (Karmoker et al. 1991) On the other hand, Mg^{2+} and K^+ starvation produced a positive effect on L_0 (Cabañero & Carvajal, 2007) and Lp_r (Benlloch-González et al., 2010) respectively. Nevertheless, available data regarding the effect of K^+ deprivation on aquaporin activity are sparse and contradictory. Prolonged deprivation is reported not to lead to any increase in the activity of mercury-sensitive aquaporins in plant roots of several plant species (Maathuis et al., 2003; Cabañero & Carvajal, 2007; Benlloch-Gonzalez, 2009), even though a greater PIP and MIP aquaporin activity has been observed in the early stages of deprivation (Maathuis et al., 2003). This would suggest that transcriptional regulation of aquaporins by low external K^+ at early stages could provide a potential means of preventing osmotic stress during long-term K^+ deprivation.

Concerning nutrient supply or excess, Adler et al. (1996) were among the first to suggest that lower Lp_r under NH^{4+} supply was due to an effect on aquaporin activity. In addition, the excess of nutrients such as of K^+ and Ca^+ produced a toxic effect on L_0 in agreement with aquaporin functionality in pepper plants (*Capsicum annuum* L.) (Cabañero & Carvajal, 2007) while NO_3^- induction of root Lp_r in maize was not correlated with aquaporin expression (Gorska et al., 2008).

8. Different methods for root hydraulic conductivity measurements

Root resistance is an important parameter in determining plant water relations and influencing whole plant responses to multitude of environmental changes and stress situations, as it accounts for a significant fraction of the entire hydraulic resistance in most plants. Methods for determining the hydraulic conductivity of the entire root system relate the ratio of xylem sap flow, or change in flow, to the difference in water potential, hydrostatic pressure or osmotic potential gradient across the root system, or change of it. These methods include transpirational water flow, osmotically induced flow and hydrostatic pressure-induced flow through the root xylem. Hydrostatic pressure may be applied either to the soil or root medium to induce root exudation (Martínez-Ballesta et al., 2003) or the root xylem through the cut stem surface following excision of the shoot, to induce reverse flow through the roots to the surrounding medium (Frensch & Steudle, 1989; Zhu & Steudle, 1991; Garthwaite et al., 2006; Knipfer et al., 2007).

8.1 Evaporative water flow method

Determination of the root hydraulic conductivity by means of the transpirational water flow method involves the ratio of transpiration to the water potential difference induced across the xylem (root surface to xylem) of the root system. By this method, the measurements are carried out under undisturbed conditions, since the use of transpiration require that the hydraulic pathway is followed by transpiration (Tsuda & Tyree, 2000) without imposed gradients. This method is very practical in the field conditions but its accuracy is limited by the relatively low precision by which the water potential and transpiration can be measured in the field, particularly with large plants. However, under controlled environment conditions with adequate evaporative demand, steady-state transpiration and differences in the osmotic pressure may be readily attained, preventing changes in tissue water content.

8.2 Hydrostatic pressure-induced root exudation method

Measurements of root hydraulic conductivity by pressurising roots are one of the methods most frequently used under laboratory conditions. The entire root system of a detached pant is sealed in a pressure vessel with the cut stem surface exposed to ambient pressure through a seal in the top. Hydrostatic pressure is applied to the root system inducing nutrient solution to flow through the root to the unpressurised cut stem surface (Martinez-Ballesta et al., 2003). The method imposes a unnatural gradient in water potential, and could lead to irreversible changes in the soil-root interface, so that root conductance determined by this method may not accurately reflect the true value under natural conditions, particularly with root systems grown in soil. However, this method, in plants grown in hydroponic solution, the root exudation is more readily attained and more linearly related to applied pressure.

8.3 Natural exudation method

Collecting root exudates under natural root pressure for measuring hydraulic conductance is also widely used (López-Perez et al., 2007). However the flow rate detected by this method hardly represents the natural status of transpiring plants (Emery and Salon, 2002). Using the hydrostatic pressure chamber to force xylem sap out from decapitated plants, it is difficult to know exactly how high the applied pressure should be, because different values of over pressure can result in different xylem water fluxes (Else et al., 1995). In any case, the flow rate is influenced by the inherent hydraulic conductivity of root systems, measured as a conductance. Therefore different values will be obtained, those using the pressurising chamber will be higher as a consequence of pressurizing the roots. In this case, water movement will occur through the apoplast to a greater extent than when the measurements are obtained by natural exudation. Therefore, the resulting root hydraulic conductance will be higher.

8.4 Root pressure probe method

Root pressure probe (RPP) is one of the most reliable techniques able to measure hydraulic conductivity of plant roots. RPP have been used to measure root pressure and water and solute flows (Steudle, 1993). Other important issues is the ability for separating the axial hydraulic resistance of xylem vessels from that related to flow across the root cylinder and to measure the radial hydraulic resistance of individual root zone (Lee et al., 2004b). In this method, the excised segment of the root or whole root system (excised close to its base) from plants is fixed to pressure probe for continuously recording of the root pressures with the aid of a pressure transducer. Water flow across the root could be induced either by changing the hydrostatic pressure in the probe by moving a metal rod with the aid of a micrometer screw or by exchanging the root medium by a medium containing a test solute of known osmotic pressure (Frensch and Steudle, 1989; Lee et al., 2004b). Transient responses in root pressure allow Lp_r to be calculated from rate constant or half-times of pressure relaxations (Steudle et al., 1987).

Root pressure probe has been used to work out the water and solute permeability of roots. As for some species, the results indicate a considerable cell-to-cell component, whereas in others, the apoplast seemed to be preferred (López-Perez et al., 2007). However, the relative contribution of pathways also depended on the nature of the forces applied. In osmotic experiments, the cell-to-cell path was preferred, whereas in hydrostatic experiments the flow was predominantly in the apoplast. The results obtained with the pressure probe measurements indicated the consistence of the technique since the extended osmometer model in which the osmotic barrier in the root is looked at a composite membrane system.

However, in spite of all these methodologies, the mechanism of water ascent has been the subject of much controversy during years. The development of thermocouple psychrometers and of the pressure chamber technique permitted indirect estimates of the xylem pressure on a large number of species.

A major difficulty with the use of psychormeters approach is the extreme sensitivity of the measurement to temperature fluctuations. For example, a change in temperature of 0.01°C corresponds to a change in water potential of about 0.1 MPa. Thus, psychrometers must be operated under constant temperature conditions. For this reason, the method is used primarily in laboratory settings. Because of its feasibility and its simplicity, the pressure chamber technique is widely used by plant physiologists, but also by farmers to measure plant water stress and schedule irrigation. For many species, hydraulic conductance, as determined with these techniques, typically ranges between −1 and −2MPa. Also, direct measurements of hydraulic conductance have been attempted by the pressure probe. In this case, the pressures that were recorded with this technique were much less negative (in the range of 0 to −0.5MPa) than the values produced by the pressure chamber, although new experiments have recently been conducted with the pressure probe (Wei et al., 1999) and were found to agree with the pressure chamber.

Thus by the pressure probe the hydrostatic pressure of individual cells may be measured directly. However, the primary limitation of this method is that some cells are too small to measure. Furthermore, some cells tend to leak after being stabbed with the capillary, and others plug up the tip of the capillary, thereby preventing valid measurements. However, technical problems with cavitation limit the measurement of negative pressures by this technique.

9. Conclusion

Root hydraulic conductivity is one of the main parameters that reflect the capacity of the root to uptake water. It confers to the root the ability to respond rapidly to fluctuating conditions suggesting that this parameter may be involved in the plant adaptation to diverse environments. After the aquaporins discovery the dynamic changes in the hydraulic conductivity were attributable to the modifications of the abundance or activity of these water channels. However, root plasticity and its ability to adapt the water uptake to the variable environment is also the consequence of root architecture and metabolism. Thus, the anatomical and morphological features of the roots, such as the diameter or length, the cell layer and its degree of suberisation and the radial and axial water transport pathway have a great influence on the hydraulic conductivity. Thus, the heterogeneity of aquaporins and of root hydraulic properties feed each other and play critical roles in the integrated root functions.

Several abiotic stresses such as drought, salinity, soil compaction or flooding and low temperatures as well as the light intensity, diurnal variations and the nutritional status affect the hydraulic conductivity of the tissues changing their resistance to water flow and where the role of aquaporins may be essential. The combination of aquaporin genetics with integrated plant physiology will provide critical insights into the hyadraulic conductance architecture in response to these stresses.

Regarding hydraulic conductance methodologies the Scholander chamber is the best option for field measurements, however, the validity of the pressure chamber technique has been seriously challenged and new experimental evidences are needed to rehabilitate the technique.

10. References

Adler, P. R.; Wilcox, G. E. & Markhart, A. H. (1996). Ammonium decreases muskmelon root system hydraulic conductivity. *Journal of Plant Nutrition*, Vol. 19, No. 1-2, pp. 1395-1403

Aharon, R.; Shahak, Y.; Wininger, S.; Bendov, R.; Kapulnik, Y. & Galili, G. (2003). Overexpression of a plasma membrane aquaporin in transgenic tobacco improves plant vigor under favorable growth conditions but not under drought or salt stress. *Plant Cell*, Vol.15, pp. 439-447

Alexandersson, E.; Fraysse, L.; Sjovall-Larsen, S., Gustavsson, S., Fellert, M., Karlsson, M., Johanson, U. & Kjellbom, P. (2005). Whole gene family expression and drought stress regulation of aquaporins. *Plant Molecular Biology*, Vol.59, pp. 469-484

Aroca, R. (2006). Exogenous catalase and ascorbate modify the effects of abscisic acid (ABA) on root hydraulic properties in *Phaseolus vulgaris* L. plants. *Journal of Plant Growth Regulation*, Vol.25, pp. 10-17

Aroca, R.; Amodeo, G.; Fernández-Illescas, S.; Herman, E. M.; Chaumont, F. & Chrispeels M. J. (2005). The Role of Aquaporins and Membrane Damage in Chilling and Hydrogen Peroxide Induced Changes in the Hydraulic Conductance of Maize Roots. *Plant Physiology*, Vol.137, pp. 341-353

Aroca, R.; Tognoni, F.; Irigoyen, J.J.; Sánchez-Diaz, M. & Pardossi, A. (2001). Difference in root low temperature response of two maize genotypes differing in chilling sensitivity. *Plant Physiology and Biochemistry*, Vol.39, pp. 1067-1075

Azad, A. K.; Sawa, Y.; Ishikawa, T. & Shibata, H. (2004). Phosphorylation of Plasma Membrane Aquaporin Regulates Temperature-Dependent Opening of Tulip Petals. *Plant and Cell Physiology*, Vol.45, No.5, pp. 608-617

Azaizeh, H.; Gunse, B. & Steudle, E. (1992). Effects of NaCl and $CaCl_2$ on water transport across root cells of maize (*Zea mays* L.) seedlings. *Plant Physiology*, Vol.99, pp. 886-894

Barrowclough, D. E.; Peterson, C. A. & Steudle, E. (2000). Radial hydraulic conductivity along developing onion roots. *Journal of Experimental Botany*, Vol.51, pp. 547–557

Benabdellah, K.; Ruiz-Lozano J. M. & Aroca, R. (2009). Hydrogen peroxide effects on root hydraulic properties and plasma membrane aquaporin regulation in *Phaseolus vulgaris*. *Plant Molecular Biology*, Vol.70, pp. 647-661

Benlloch-González, M. (2009). Efecto del ayuno en K^+ sobre el transporte de agua en girasol. Mecanismos implicados. PhD thesis, Department of Agronomy of Cordoba University (Spain)

Benlloch-González, M.; Fournier, J. M. & Benlloch, M. (2010). K^+ deprivation induces xylem water and K^+ transport in unflower: evidence for a co-ordinated control. *Journal of Experimental Botany*, Vol. 61, No. 1, pp. 157-164

Boursiac, Y.; Chen, S. ; Luu, D. T. ; Sorieul, M. ; van den Dries, N. & Maurel, C. (2005). Early effects of salinity on water transport in Arabidopsis roots-molecular and cellular features of aquaporin expression. *Plant Physiology*, Vol.139, pp. 790-805

Bramley, H. (2006). Water flow in the roots of three crop species: The influence of root structure, aquaporin activity and waterlogging, *PhD thesis*, The University of Western Australia, Perth

Bramley, H.; Turner, D. W.; Tyerman, S. D. & Turner, N. C. (2007b). Water flow in the roots of crop species: the influence of root structure, aquaporin activity, and waterlogging. *Advances in Agronomy*, Vol.96, pp.133–196

Bramley, H.; Turner, N. C.; Turner, D. W. & Tyerman, S. D. (2010). The contrasting influence of short-term hypoxia on the hydraulic properties of cells and roots of wheat and lupin. *Functional Plant Biology*, Vol.37, No.3, 183–193

Brodribb, T. J. & Holbrook, N. M. (2004). Diurnal depression of leaf hydraulic conductance in a tropical tree species, *Plant, Cell and Environment*, Vol. 27, No. 7, pp. 820-827

Cabañero, F. J. & Carvajal, M. (2007). Different cation stresses affect specifically osmotic root hydraulic conductance, involving aquaporins, ATPase and xylem loading of ions in Capsicum annuum, L. plants. *Journal of Plant Physiology*, Vol. 164, No. 10, pp. 1300-1310

Carvajal, M.; Cerdá, A. & Martínez, V. (2000). Does calcium ameliorate the negative effect of NaCl on melon root water transport by regulating water channel activity?. *New Phytologist*, Vol.145, pp. 439–447

Carvajal, M.; Cooke, D. T. & Clarkson, D. T. (1996). Responses of wheat plants to nutrient deprivation may involve the regulation of water-channel function. *Planta*, Vol. 199, No. 3, pp. 372-81

Carvajal, M.; Cooke, D. T. & Clarkson, D. T. (1998). The lipid bilayer and aquaporins; parallel pathways for water movement into plant cells. *Plant Growth Regulation*, Vol. 25, pp. 89-95

Christmann, A.; Weiler, E. W., Steudle, E. & Grill, E. (2007). A hydraulic signal in root-to-shoot signalling of water shortage. *The Plant Journal*, Vol.52, pp. 167–174

Clarkson, D. T.; Carvajal, M.; Henzler, T.; Waterhouse, R. N.; Smyth, A. J. & Cooke D. T. (2000). Root hydraulic conductance: diurnal aquaporin expression and the effects of nutrient stress. *Journal of Experimental Botany*, Vol. 51, No. 342, pp. 61-70

Cochard, H.; Nardini, A. & Coll, L. (2004). Hydraulic architecture of leaf blades: where is the main resistance?. *Plant Cell and Environment*, Vol. 27, pp. 1257-1267.

Cochard, H.; Venisse, J. S.; Barigah, T. S.; Brunel, N.; Herbette, S.; Guilliot, A.; Tyree, M. T. & Sakr, S. (2007). Putative role of aquaporins in variable hydraulic conductance of leaves in response to light. *Plant Physiology*, Vol. 143, pp. 122-133

Ehlert, C. ; Maurel, C. ; Tardieu, F. & Simonneau, T. (2009). Aquaporin mediated reduction in maize root hydraulic conductivity impacts cell turgor and leaf elongation even without changing transpiration. *Plant Physiology*, Vol.150, pp. 1093–1104

Else, M. A.; Hall, K. C.; Arnold G. M.; Davies, W. J. & Jackson M. B. (1995).Export of abscisic acid, 1-aminocyclopropane-1-carboxylic acid, phosphate, and nitrate from roots to shoots of flooded tomato plants. *Plant Physiology*, Vol.107, pp. 377-384

Emery, R. J. N. & Salon, C. (2002). Water entry into detached root systems saturates with increasing externally applied pressure; a result inconsistent with models of simple passive diffusion. *Physiologia Plantarum*, Vol.115, pp. 406-416

Evlagon, D.; Ravina, Y. & Neumann, P.M. (1990). Interactive effects of salinity and calcium on hydraulic conductivity, osmotic adjusment and growth in primary roots of maize seedlings. *Israel Journal Bot*any, Vol. 39, pp. 239–247

Fan, M.; Bai, R.; Zhao, X. & Zhang, J. (2007). Aerenchyma formed under phosphorus deficiency contributes to the reduced root hydraulic conductivity in maize roots. *Journal of Integrative Plant Biology*, Vol. 49, No. 5, pp. 598-604

Fennell, A. & Markhart, A.H. (1998). Rapid acclimation of root hydraulic conductivity to low temperature. *Journal of Experimental Botany*, Vol.49, No.322, pp. 879-884

Franks, P. J.; Drake, P. L. & Froend, R. H. (2007). Anisohydric but isohydrodynamic: seasonally constant plant water potential gradient explained by a stomatal control mechanism incorporating plant hydraulic conductance. *Plant Cell and Environment*, Vol. 30, pp. 19–30

Frensch, J. & Steudle, E. (1989). Axial and radial hydraulic resistance to roots of maize (Zea mays L.). *Plant Physiology*, Vol.91, pp. 719-726

Gallardo, M.; Eastham, J.; Gregory, P. J. & Turner, N. C. (1996). A comparison of plant hydraulic conductances in wheat and lupins. *Journal of Experimental Botany*, Vol.47, pp. 233–239

Garthwaite, A. J.; Steudle, E. & Colmer, T. D. (2006).Water uptake by roots of Hordeum marinum: formation of a barrier to radial O2 loss does not affect root hydraulic conductivity. *Journal of Experimental Botany*, Vol.57, pp. 655-664

Gorska, A.; Zwieniecka, A.; Holbrook, N. M. & Zwieniecki, M.A. (2008). Nitrate induction of root hydraulic conductivity in maize is not correlated with aquaporin expression. *Planta*, Vol. 228, No. 6, pp. 989-998

Hachez, C.; Heinen, R.; Draye, X. & Chaumont, F. (2008). The expression pattern of plasma membrane aquaporins in maize leaf highlights their role in hydraulic regulation. *Plant Molecular Biology*, Vol. 68, No.4-5 , pp. 337-353

Hamza, M. A. & Aylmore, L. A. G. (1992a). Soil solute concentration and water uptake by single lupin and radish plant roots. I. Water extraction and solute accumulation. *Plant and Soil*, Vol.145, pp. 187–196

Heinen, R. B.; Ye, Q. & Chaumont, F. (2009). Role of aquaporins in leaf physiology. *Journal of Experimental Botany*, Vol.60, No.11, pp. 2971-2985

Henzler, T.; Waterhouse, R.N.; Smyth, A.J.; Carvajal, M.; Cooke, D. T.; Schaffner, A. R.; Steudle, E. & Clarkson, D. T. (1999). Diurnal variations in hydraulic conductivity and root pressure can be correlated with the expression of putative aquaporins in the roots of *Lotus japonicus*. *Planta*, Vol. 210, pp. 50-60

Hose, E.; Steudle, E. & Hartung, W. (2000). Abscisic acid and hydraulic conductivity of maize roots: A study using cell- and root-pressure probes. *Planta*, Vol.211, pp. 874–882

Huang, B. R. & Eissenstat D. M. (2000). Linking root hydraulic conductivity to anatomy in citrus root stocks that vary in specific root length. *Journal of the American Society of Horticultural Science*, Vol.125, pp. 260-264

Jang, J.Y.; Kim, D. G.; Kim, Y. O.; Kim, J. S. & Kang, H. S. (2004). An expression analysis of a gene family encoding plasma membrane aquaporins in response to abiotic stresses in *Arabidopsis thaliana*. *Plant Molecular Biology*, Vol.54, pp. 713-725

Javot, H.; Lauvergeat, V.; Santoni, V.; Martin-Laurent, F.; Güclü, J.; Vinh, J.; Heyes, J.; Franck, K.; Schäffner, A. R.; Bouchez, D. & Maurel, C. (2003). Role of a single aquaporin isoform in root water uptake. *The Plant Cell*, Vol.15, pp. 509–522

Joly, R. J. (1989). Effects of sodium chloride on the hydraulic conductivity of soybean root systems. *Plant Physiology*, Vol. 91, pp. 1262-1265

Joslin, J. D.; Wolfe, M. H. & Hanson, P. J. (2000). Effects of altered water regimes on forest root systems. *New Phytologist*, Vol.147, pp. 117-129

Karmoker, J. L.; Clarkson, D. T.; Saker, L. R.; Rooney, J. M. & Purves, J. V. (1991). Sulphate deprivation depresses transport of nitrogen to the xylem and hydraulic conductivity of barley roots. *Planta*, Vol. 185, No. 2, pp. 269-278

Kim, Y. X. & Steudle, E. (2007). Light and turgor affect the water permeability (aquaporins) of parenchyma cells in the midrib of *Zea mays*. *Journal of Experimental Botany*, Vol. 58, No. 15-16 , pp. 4119-4129

Kim, Y. X. & Steudle, E. (2009). Gating of aquaporins by light and reactive oxygen species in leaf parenchyma cells of the midrib of *Zea mays*, *Journal of Experimental Botany*, Vol. 60, No. 2, pp. 547-556

Knipfer, T.; Das, D. & Steudle, E. (2007). During measurements of root hydraulics with pressure probes, the contribution of unstirred layers is minimized in the pressure relaxation mode: comparison with pressure clamp and high-pressure flowmeter. *Plant Cell and Environment*, Vol.30, pp. 845–860

Kramer, P. J. (1983). Water Relations of Plants, *Academic Press*, New York

Lee, S. H.; Chung, G. C. & Steudle, E. (2005). Low temperature and mechanical stresses differently gate aquaporins of root cortical cells of chilling-sensitive cucumber and -resistant figleaf gourd. *Plant, Cell and Environment*, Vol.28, No. 9, pp. 1191-1202

Lee, S. H.; Singh, A . P.; Chung, G. C.; Ahn, S. J.; Noh, E. K. & Steudle, E. (2004b). Exposure of roots of cucumber (*Cucummis sativus*) to low temperature severely reduces root pressure, hydraulic conductivity and active transport of nutrients. *Physiologia Plantarum*, Vol.120, pp. 413–420

Lee, S. H.; Singh, A. P. & Chung, G. C. (2004a). Rapid accumulation of hydrogen peroxide in cucumber roots due to exposure to low temperature appears to mediate decreases in water transport. *Journal of Experimental Botany*, Vol.55, pp. 1733–1741

Lee, S.; Chung, G. C. & Zwiazek J. J. (2009). Effects of irradiance on cell water relations in leaf bundle sheath cells of wild-type and transgenic tobacco (Nicotiana tabacum) plants overexpressing aquaporins. *Plant Science*, Vol. 176, No. 2, pp. 248-255

Liang, B. M.; Sharp, R. E. & Baskin, T. I. (1997). Regulation of growth anisotropy in well-watered and water-stressed maize roots (I. Spatial distribution of longitudinal, radial, and tangential expansion rates). *Plant Physiology*, Vol.115, pp. 101-111

Linkohr, B. I.; Williamson, L. C.; Fitter, A. H. & Leyser, O. (2002). Nitrate and phosphate availability and distribution have different effects on root system architecture of Arabidopsis. *The Plant Journal*, Vol.29, pp. 751-760

Lo Gullo, M. A.; Nardini, A.; Trifilo, P. & Salleo, S. (2005). Diurnal and seasonal variations in leaf hydraulic conductance in Evergreen and deciduous trees. *Tree Physiology*, Vol. 25, No. 4, pp. 505-512

López-Bucio, J.; Cruz-Ramírez, A. & Herrera-Estrella, L. (2003). The role of nutrient availability in regulating root architecture. *Current Opinion in Plant Biology*, Vol.6, pp. 280-287

López-Pérez, L.; Fernández-García, N.; Olmos E. & Carvajal, M. (2007). The Phi thickening in roots of broccoli plants: An acclimation mechanism to salinity?. *International Journal of Plant Sciences*, Vol.168, pp. 1141-1149

Luu, D. T. & Maurel, C. (2005). Aquaporins in a challenging environment: molecular gears for adjusting plant water status. *Plant Cell and Environment*, Vol.28, pp. 85-96

Maathuis, F. J. M.; Filatov, V.; Herzyk, P.; Krijger, G. C.; Axelsen, K. B.; Chen, S.; Green, B. J.; Li, Y.; Madagan, K. L.; Sánchez-Fernández, R.; Forde, B. G.; Palmgren, M. G.; Rea, P. A.; Williams, L. E.; Sanders, D. & Amtmann, A. (2003). Transcriptome analysis of root transporters reveals participation of multiple gene families in the response to cation stress. *The Plant Journal*, Vol. 35, No. 6, pp. 675-92

Martínez-Ballesta, M.C.; Aparicio, F.; Pallas, V.; Martínez, V. & Carvajal, M. (2003a). Influence of saline stress on root hydraulic conductance and PIP expression in *Arabidopsis*. *Journal of Plant Physiology*, Vol.160, pp. 689-697

Martre, P.; Morillon, R.; Barrieu, F.; North, G. B.; Nobel, P. S. & Chrispeels, M. J. (2002). Plasma membrane aquaporins play a significant role during recovery from water deficit. *Plant Physiology*, Vol.130, pp. 2101-2110

Martre, P.; North, G. B. & Nobel, P. S. (2001). Hydraulic conductance and mercury-sensitive water transport in roots of *Opuntia acanthocarpa* in relation to soil drying and rewetting. *Plant Physiology*, Vol.126, pp. 352-362

Matzner, S. & Comstock, J. (2001). The temperature dependence of shoot hydraulic resistance: implications for stomatal behaviour and hydraulic limitation. *Plant Cell and Environment*, Vol.24, pp. 1299-1307

Maurel, C.; Verdoucq, L.; Luu, D. T. & Santoni, V. (2008). Plant Aquaporins: Membrane channels with multiple integrated functions. *Annual Review of Plant Biology*, Vol.59, pp. 595-624

Moreshet, S. & Huck, M. G. (1991). Dynamices of water permeability, In: *Plant Roots: The Hidden Half*, Y. Waisel; A. Eshel, & U. Kafkafi, (Ed.), pp. 605-626, Marcel Dekker Inc., New York.

Morillon, R. & Chrispeels, M. J. (2001). The role of ABA and the transpiration stream in the regulation of the osmotic water permeability of leaf cells. *Proceeding of the National Academy of Sciences*, Vol.98, No24, pp. 14138-14143

Munns, R. & Passioura, J. B. (1984). Hydraulic resistances of plants III. Effects of NaCl in barley and lupin. *Australian Journal of Plant Physiology*, Vol.11, pp. 351-359

Muries, B.; Faize, M.; Carvajal, M. & Martinez-Ballesta, M. C. (2011). Identification and differential induction of the expression of aquaporins by salinity in broccoli plants. Molecular Biosystems, Vol.7, No.4, pp. 1322-1335

Nardini, A.; Salleo, S. & Andri, S. (2005). Circadian regulation of leaf hydraulic conductance in sunflower (Helianthus annuus L. cv. Margot). *Plant, Cell and Environment*, Vol.28, pp. 750-759

Nardini, N.; Grego, F.; Trifilò, P.; Salleo, S. (2010). Changes of xylem sap ionic content and stem hydraulics in response to irradiance in *Laurus nobilis*. *Tree Physiology*, Vol. 30, No.5, pp. 628-635.

North, G. B.; Martre, P. & Nobel, P. S. (2004). Aquaporins account for variations in hydraulic conductance for metabolically active root regions of *Agave deserti* in wet, dry, and rewetted soil. *Plant Cell and Environment*, Vol.27, pp. 219–228

North, G.B. & Nobel, P.S. (1996). Radial hydraulic conductivity of individual root tissues of Opuntia-indica (L.) Miller as soil moisture varies. *Annals of Botany*, Vol.77, pp. 133-142

Parent, B. ; Hachez, C. : Redondo, E. ; Simonneau, T. ; Chaumont, F. & Tardieu, F. (2009). Drought and abscisic acid effects on aquaporin content translate into changes in hydraulic conductivity and leaf growth rate: A trans-scale approach. *Plant Physiology*, Vol.149, pp. 2000–2012

Parsons, L. R. & Kramer, P. J. (1974). Diurnal cycling in the root resistance to water movement. *Physiologia Plantarum*, Vol. 30, No. 1, pp. 19-23

Peters, K. C. & Steudle, E. (1999) Effect of temperature and xylem suction on the synchronisation of circadian oscillation in root pressure, hydraulic conductivity and xylem loading of bean roots (Phaseolus coccineus L.). In: *Structure and function of root*, F. Baluska, M. Ciamporova,, O. Gasparikova, P. W. Barlow, (Eds), Kluwer Academic Publishers, London, in press.

Peyrano, G.; Taleisnik, E.; Quiroga, M.; de Forchetti, S. M. & Tigier, H. (1997). Salinity effects on hydraulic conductance, lignin content and peroxidase activity in tomato roots. *Plant Physiology* and *Biochemistry*, Vol.35, pp. 387-393

Postaire, O.; Tournaire-Roux, C.; Grondin, A.; Boursiac, Y.; Morillon, R.; Schäffner, A. R. & Maurel, C. (2010). A PIP1 aquaporin contributes to hydrostatic pressure-induced water transport in both the root and rosette of Arabidopsis. *Plant Physiology*, Vol. 152, pp. 1418-1430

Radin, J. W. & Matthews, M. A. (1989). Water transport properties of cortical cells in roots of nitrogen-and phosphorus-deficient cotton seedlings. *Plant Physiology*, Vol. 89, No. 1, pp. 264-268

Rieger, M. & Litvin, P. (1999). Root system hydraulic conductivity in species with contrasting root anatomy. *Journal of Experimental Botany*, Vol.50, pp. 201–209

Sack, L. & Holbrook, N. M. (2006). Leaf hydraulics. *Annual Review of Plant Physioliogy and Plant Molecular Biology*, Vol.57, pp. 361–381

Sack, L.; Cowan, P. D.; Jaikumar, N. & Holbrook, N.M. (2003). The "hydrology" of leaves: co-ordination of structure and function in temperate woody species. *Plant Cell and Environment, Vol.*26, pp. 1343–1356

Sade, N.; Vinocur, B. J.; Diber, A.; Shatil, A.; Ronen, G. & Nissan, H. (2009). Improving plant stress tolerance and yield production: is the tonoplast aquaporin SlTIP2;2 a key to isohydric to anisohydric conversion?. *New Phytology*, Vol.181, pp. 651-61

Saliendra, N. Z. & Meinzer, F.C. (1992). Genotypic, developmental and drought-induced differences in root hydraulic conductance of contrasting sugarcane cultivars. *Journal of Experimental Botany*, Vol.43, pp. 1209-1217

Sanders, P. L. & Markhart, A. H. (2001). Root system functions during chilling temperatures: injury and acclimation, In: *Crop responses and adaptations to temperature stress, S. Basra,* (Ed.), pp. 77-108, Food Products Press, New York, London, Oxford

Sanderson, J. (1983). Water uptake by different regions of the barley root. Pathways of radial flow in relation to development of the endodermis. *Journal of Experimental Botany*, Vol.34, pp. 240–253

Sarda, X.; Tousch, D.; Ferrare, K.; Cellier, F.; Alcon, C.; Dupuis, J. M.; Casse, F. & Lamaze, T. (1999). Characterization of closely related d-TIP genes encoding aquaporins which are differentially expressed in sunflower roots upon water deprivation through exposure to air. *Plant Molecular Biology*, Vol.40, pp. 179-191

Schäffner, A. R. (1998). Water channel function, structure, and expression: are there more surprises to surface in water relations?. *Planta*, Vol.204, pp. 131-139

Schreiber, L.; Hartmann, K.; Skrabs, M. & Zeier, J. (1999). Apoplastic barriers in roots: chemical composition of endodermal and hypodermal cell walls. *Journal of Experimental Botany*, Vol.50, pp. 1267-1280

Schultz, H. R. (2003*a*). Differences in hydraulic architecture account for near isohydric and anisohydric behaviours of two field-grown *Vitis vinifera* L. cultivars during drought. *Plant Cell and Environment*, Vol.26, pp. 1393-1405

Secchi, F.; Lovisolo, C.; Uehlein, N.; Kaldenhoff, R. & Schubert, A. (2007). Isolation and functional characterization of three aquaporins from olive (Olea europaea L.). *Planta*, Vol.225, pp. 381-392

Sellin, A.; Õunapuu, E. & Karusion, A. (2010) Experimental evidence supporting the concept of light-mediated modulation of stem hydraulic conductance. *Tree Physiology*, Vol. 30, pp. 1528-1535

Sellin, A.; Õunapuu, E. & Kupper P. (2008). Effects of light intensity and duration on leaf hydraulic conductance and distribution of resistance in shoots of silver birch (Betula pendula). *Physiologia Plantarum*, Vol. 134, pp. 412-420

Shangguan, Z.; Lei, T.; Shao, M. & Xue, Q. (2005). Effects of phosphorus nutrient on the hydraulic conductivity of sorghum (*Sorghum vulgare* Pers.) seedling roots under water deficiency. *Journal of Integrative Plant Biology*, Vol. 47, No. 4, pp. 421-427

Shaw, B.; Thomas, T. H. & Cooke, D. T. (2002). Responses of sugar beet (Beta vulgaris L.) to drought and nutrient deficiency stress. *Plant Growth Regulation*, Vol. 37, No. 1, pp. 77-83.

Siefritz, F.; Tyree, M. T.; Lovisolo, C.; Schubert, A. & Kaldenhoff, R. (2002). PIP1 plasma membrane aquaporins in tobacco: from cellular effects to functions in plants. *Plant Cell*, Vol. 14, pp. 869-876

Siemens, J. A. & Zwiazek, J. J. (2004). Changes in water flow properties of solutions culture-grown trembling aspen (*Populus tremuloides*) seedlings under different intensities of water-deficit stress. *Physiologia Plantarum*, Vol.121, pp. 44–49

Sjövall-Larsen, S.; Alexandersson, E.; Johansson, I.; Karlsson, M.; Johanson, U. & Kjellbom, P. (2006). Purification and characterization of two protein kinases acting on the aquaporin SoPIP2;1. *Biochimica et Biophysica Acta*, Vol.1758, pp. 1157-1164

Soar, C. J.; Speirs, J.; Maffei, S. M.; Penrose, A. B.; McCarthy, M. G. & Loveys, B. R. (2006). Grape vine varieties Shiraz and Grenache differ in their stomatal response to VPD: apparent links with ABA physiology and gene expression in leaf tissue. *Australian Journal of Grape and Wine Research*, Vol.12, pp. 2-12

Steudle, E. & Brinckmann, E. (1989). The osmometer model of the root: Water and solute relations of roots of Phaseolus coccineus. *Botanica Acta*, Vol.102, pp. 85–95

Steudle, E. & Frensch, J. (1996). Water transport in plants: role of the apoplast. *Plant and Soil*, Vol.187, pp. 67-79

Steudle, E. & Peterson, A. (1998). How does water get through roots?. *Journal of Experimental Botany*, Vol.49, pp. 775 -788

Steudle, E. (1993). Pressure probe techniques: Basic principles and application to studies of water and solute relations at the cell, tissue and organ level, In: *Water Deficits. Plant Responses from Cell to Community* J. A. C. Smith & H. Griffiths, (Ed.), pp. 5-36, Bios Science Publs, Oxford

Steudle, E. (2000). Water uptake by roots: Effects of water deficit. *Journal of Experimental Botany*, Vol.51, pp.1531-1542

Steudle, E. (2001). The cohesion-tension mechanism and the acquisition of water by plant roots. *Annual Review Plant Physiology and Plant Molecular Biology*, Vol.52, pp. 847-863

Steudle, E.; Oren, R. & Schulze, E. D. (1987). Water transport in maize roots. Measurement of hydraulic conductivity, solute permeability and of reflection coefficients of excised roots using the root pressure probe. *Plant Physiology*, Vol.84, pp 1220-1232

Suga, S., Komatsu, S. & Maeshima, M. (2002). Aquaporin isoforms responsive to salt and water stresses and phytohormones in radish seedlings. *Plant and Cell Physiology*, Vol.43, pp. 1229-1237

Tardieu, F. & Simonneau, T. (1998). Variability among species of stomatal control under fluctuating soil water status and evaporative demand: modelling isohydric and anisohydric behaviours. *Journal of Experimental Botany*, Vol.49, pp. 419-432

Törnroth-Horsefield, S., Wang, Y., Hedfalk, K., Johanson, U., Karlsson, M., Tajkhorshid, E., Neutze, R. & Kjellbom, P. (2006). Structural mechanism of plant aquaporin gating. *Nature*, Vol.439, No. 7077, pp. 688-694

Tournaire-Roux, C. ; Sutka, M. ; Javot, H. ; Gout, E. ; Gerbeau, P. ; Luu, D. T.; Bligny, R. & Maurel, C. (2003). Cytosolic pH regulates root water transport during anoxic stress through gating of aquaporins. *Nature*, Vol.425, pp. 393-397

Tsuda, M. & Tyree, M. T. (2000). Plant hydraulic conductance measured by the high pressure flow meter in crops. *Journal of Experimental Botany*, Vol.51, No.345, pp. 823-828

Tyree, M. T.; Nardini, A.; Salleo, S.; Sack, L. & El Omari, B. (2005). The dependence of leaf hydraulic conductance on irradiance during HPFM measurements: any role for stomatal response? *Journal of Experimental Botany*, Vol. 56, No. 412, pp. 737-744

Tyree, M.T., Zimmermann, M.H. (2002). *Xylem structure and the ascent of sap.* New York: Springer Verlag.

Van den Honert T. H. (1948). Water transport in plants as a catenary process. Discussions of the Faraday Society, Vol.3, pp. 146-153

Vandeleur, R. K.; Mayo, G.; Shelden, M. C.; Gilliham, M.; Kaiser, B. N. & Tyerman, S.D. (2009). The role of plasma membrane intrinsic protein aquaporins in water transport through roots: diurnal and drought stress responses reveal different strategies between isohydric and anisohydric cultivars of grapevine. *Plant Physiology*, Vol.149, pp. 445-460

Vandeleur, R. K.; Niemietz, C.; Tilbrook, J. & Tyerman, S. D. (2004). Roles of aquaporins in root responses to irrigation. *Plant and soil*, Vol.274, pp. 141-161

Voicu, M. C. & Zwiaze, J. J. (2009). Inhibitor studies of leaf lamina hydraulic conductance in trembling aspen (Populus tremuloides Michx.) leaves. *Tree Physiology*, Vol. 30, No. 2, pp. 193-204

Voicu, M. C.; Zwiazek, J. J. & Tyree, M. T. (2008). Light response of hydraulic conductance in bur oak (Quercus macrocarpa) leaves. *Tree Physiology*, Vol. 28, No. 7, pp. 1007-1015

Vysotskaya, L. B.; Arkhipova, T. N.; Timergalina, L. N.; Dedov, A. V.; Veselov, S. Y. & Kudoyarova, G. R. (2004a). Effect of partial root excision on transpiration, root hydraulic conductance and leaf growth in wheat seedlings. *Plant Physiology Biochemistry*, Vol.42, pp. 251-255

Vysotskaya, L. B.; Kudoyarova, G. R. & Jones, H. G. (2004b). Unusual stomatal behaviour on partial root excision in wheat seedlings. *Plant Cell and Environment*, Vol.27, pp. 69-77

Wan, X. & Zwiazek, J. J. (1999), Mercuric chloride effects on root water transport in aspen seedlings. *Plant Physiology*, Vol.121, pp. 939-946

Wei, C.; Tyree, M.T. & Steudle, E. (1999). Direct measurement of xylem pressure in leaves of intact maize plants. A test of the Cohesion-Tension theory taking hydraulic architecture into consideration. *Plant Physiology*, Vol.121, pp. 1191-1205

Wells, C.E. & Eissenstat, D. M. (2002). Defining the fine root: marked differences in survivorship among apple roots of different diameter. *Ecology*, Vol.82, pp. 882 - 892

Wilder, V.V.; Miecielica, U.; Degand, H.; Derua, R.; Waelkens E. & Chaumont, F. (2008). Maize Plasma Membrane Aquaporins Belonging to the PIP1 and PIP2 Subgroups are in vivo Phosphorylated. *Plant and Cell Physiology*, Vol.49, No. 9, pp. 1364-1377

Yamada, S.; Komori, T.; Myers, P. N.; Kuwata, S.; Kubo, T. & Imaseki, H. (1997). Expression of plasma membrane water channel genes under water stress in Nicotiana excelsior. *Plant and Cell Physiology*, Vol.38, pp. 1226-1231

Zelazny, E.; Borst, J. W.; Muylaert, M.; Batoko, H.; Hemminga M. A. & Chaumont, F. (2007). FRET imaging in living maize cells reveals that plasma membrane aquaporins interact to regulate their subcellular localization. *Proceedings of the National Academy of Sciences of the United States of America*, Vol.104, pp. 12359-12364

Zhang, W. H. & Tyerman, S. D. (1991). Effect of low O_2 concentration and azide on hydraulic conductivity and osmotic volume of the cortical cells of wheat roots. *Australian Journal of Plant Physiology*, Vol.18, pp. 603-613

Zhu, G. L. & Steudle, E. (1991). Water transport across maize roots. Simultaneous measurement of flows at the cell and root level by double pressure probe technique. *Plant Physiology*, Vol95, pp 305-315

Zimmermann, H. M.; Hartmann, K.; Schreiber, L. & Steudle, E. (2000). Chemical composition of apoplastic transport barriers in relation to radial hydraulic conductivity of corn roots (*Zea mays* L.). *Planta*, Vol.210, pp. 302-311

Zwieniecki, M. A..; Thompson, M. V. & Holbrook, N. M. (2003). Understanding the hydraulics of porous pipes: Tradeoffs between water uptake and root length utilization. *Journal of Plant Growth Regulation*, Vol.21, pp. 315-323

Zwieniecki, M. A.; Brodribb, T. J. & Holbrook, N. M. (2007). Hydraulic design of leaves: insights from rehydration kinetics. *Plant Cell and Environment, Vol.* 30, pp. 910-921

6

Plant and Soil as Hydraulic Systems

Mirela Tulik and Katarzyna Marciszewska
Warsaw University of Life Science – SGGW (WULS-SGGW)
Poland

1. Introduction

Soil is usually defined as a natural body consisting of mineral constituent layers that are different in structure and variable in thicknesses. It is composed of particles of broken rock that have been affected by chemical and environmental processes as weathering and erosion. Soil particles are packed loosely, forming a soil structure such containing pore spaces. These pores are filled with soil solution (liquid) and air (gas). This chapter shows that soil and plant are similar hydraulically. The pathway for water moving is from soil through plant to the atmosphere and can be described with the Soil-Plant-Atmosphere Continuum (SPAC) model. Hydraulic conductivity is the property of both soils and higher plants and the resulting analogue model for water transport is supported by the similar structure and the same source for water movement. This continuum hypothesis characterizes the state of water in different components of the SPAC as expressions of the energy level or water potential of each. Also the review of methodology of hydraulic conductivity measurement has been presented in detail.

2. Major physical properties of water

The surface of the Earth is covered in ca. 70% by water - a chemical substance including one atom of oxygen and two of hydrogen connected by covalent bonds (Fig. 1).

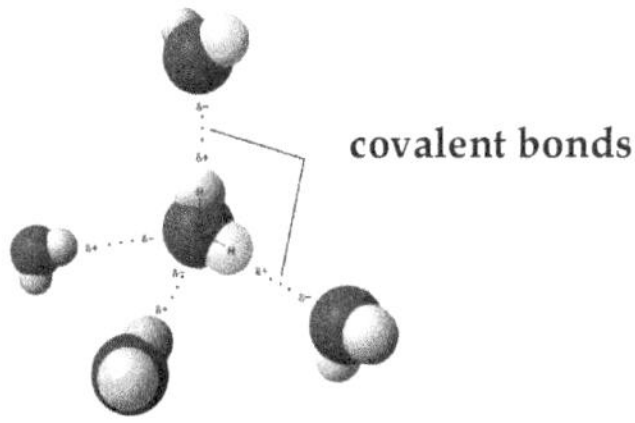

Fig. 1. Model of dipole water structure (http://ffden-2.phys.uaf.edu/212_fall2009.web/Yngve_margaretJ/ice/pg1.htm, modified).

Because water is a polar molecule it has a high surface tension and capillary forces. The capillarity basically makes water move up in narrow tubes against the force of gravity (what occurs in soil and all vascular plants in mechanism of water transport). Polar structure of water causes also the cohesion and the adhesion of water molecules, which contributes to the capillary water threads not to collapse.

The water circulates above and below the surface of the Earth and can change states among liquid, vapour, and ice at various places in its cycle (Fig. 2). Although the balance of water on Earth remains constant over time, individual water molecules can come and go, in and out of the atmosphere. During the physical processes as evaporation, condensation, precipitation, infiltration, runoff, and subsurface flow, water moves from one reservoir to another.

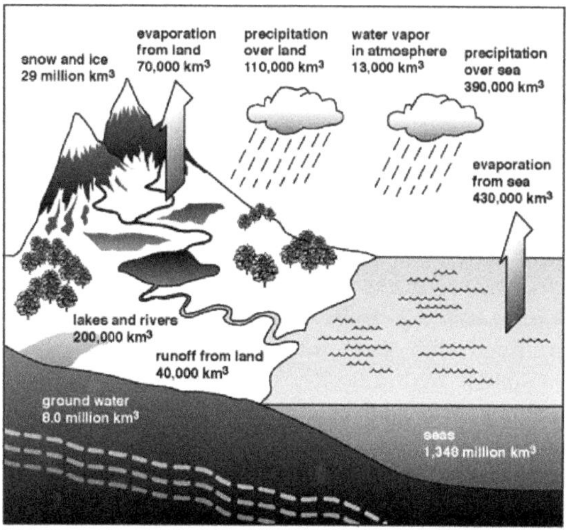

Fig. 2. Water cycle (Clarke, 1991, modified).

According to vertical distribution of water content two main zones in soil profile are distinguished: vadose and saturated (Fig. 3).

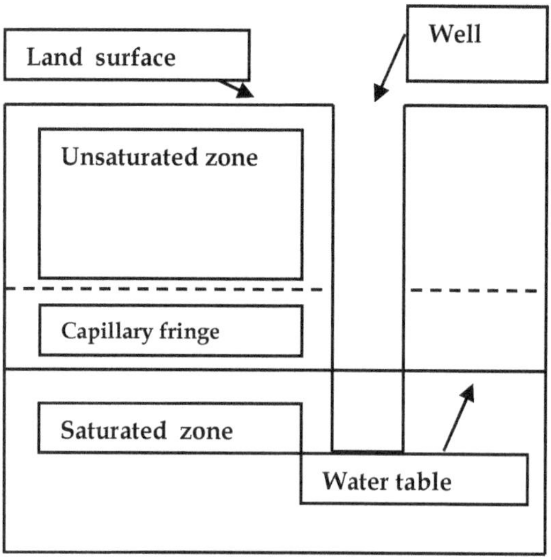

Fig. 3. Vertical distribution of water content in soil profile.

The vadose zone called also unsaturated, is located between the land of Earth surface and zone of saturation and extends from the top of the ground surface to the water table. In this zone water is under less than atmospheric pressure, which is a result of the process of adhesion and capillarity. Water moves predominantly in a vertical direction (Heath, 1983). Within the zone of vadose soil pore spaces usually contain air or other gases. The area below the water table where all open spaces are filled with water under pressure that is equal to or greater than that of the atmosphere makes up the zone of saturation.

3. Hydraulic conductivity of soils and plants

Hydraulic conductivity, which is a property of soils, vascular plants, or rocks describes the ability of water to move through pore spaces or fractures using hydraulic gradient. Saturated hydraulic conductivity expresses water movement *via* saturated media. Symbolically hydraulic conductivity is written as K while saturated conductivity as Ksat. The hydraulic conductivity of soil depends on the soil grain size, the soil matrix structure, the type of soil fluid and is defined by Darcy's law that could be written as follows:

$$U = -K/\eta \times dP/dl \quad (1)$$

where U is the velocity of the soil fluid *via* a geometric cross-sectional area within the soil, K is a hydraulic conductivity, η is a coefficient of the viscosity of water, dP/dl is the pressure gradient (Neuman, 1977). On the basis of the above mentioned equation, the hydraulic conductivity is defined as a ratio of soil fluid velocity (U) to the applied hydraulic gradient (dP/dl), because η is a constant.
Soil and vascular plants are similar hydraulically; the same physical laws might be applied to describe soil and plants hydraulic conductivity (Sperry et al., 2003). Both in the soil and the vascular plants structure the pores filled with water occur and although the pores in plants are highly organized in comparison to soil there is a close analogy between the soil and the vascular plants hydraulics. Additionally soil water potential is the driving force behind water movement. The main advantage of the "potential" concept is that it provides a unified measure by which the water state can be evaluated at any time and everywhere within the soil-plant-atmosphere continuum (Hillel, 1980).
For the theoretical calculation of the volume of water flow in the plant conducting elements the law of Hagen-Poiseuille, a special case of Darcy's law, describing the laminar flow through long cylindrical pipe is applied:

$$dV/dt = -K \times dP/dl = -\pi r^4 x dP/8\eta x dl \text{ beaucause } K = \pi r^4/8\eta \quad (2)$$

where K is hydraulic conductivity, η is the viscosity of water, r the radius of the capillary and - dP/dl is the pressure gradient along the capillary (Tyree et al., 1994; Tyree & Zimmermann, 2002). K can also be considered as the coefficient of the Hagen-Poiseuille law. It is important to note that flow rate dV/dt, is proportional to the fourth power of the capillary diameter. Thus a slight increase in vessel or tracheid diameter causes a considerable increase in conductivity. In a transverse section of the stem, branch or root many capillaries of different diameters d_i are present in parallel thus the aforementioned formula is written as follows:

$$dv/dt = - dp/dl \times K_i = - dp/dl \times \Sigma\pi_r{}^4/8\eta \quad (3)$$

The pressure gradient is the driving force for water flow *via* tracheary elements and is caused by transpiration. Both in the case of soils and plants hydraulics the volume flow of water occurs by reason of decreasing pressure gradient.

4. Methods of hydraulic conductivity determination

To determine the saturated hydraulic conductivity of water in soil both empirical and experimental (field and laboratory) methods could be applied (Jennsen, 1990) (Fig. 4). The empirical approach correlates the hydraulic conductivity with soil properties as: pore and grain size, their distribution and soil texture. The methodology of measurement for laboratory and field experiments is based on Darcy's law and has been described by Klute and Dirkesen (1986) as well as Amoozegar and Warrick (1986).

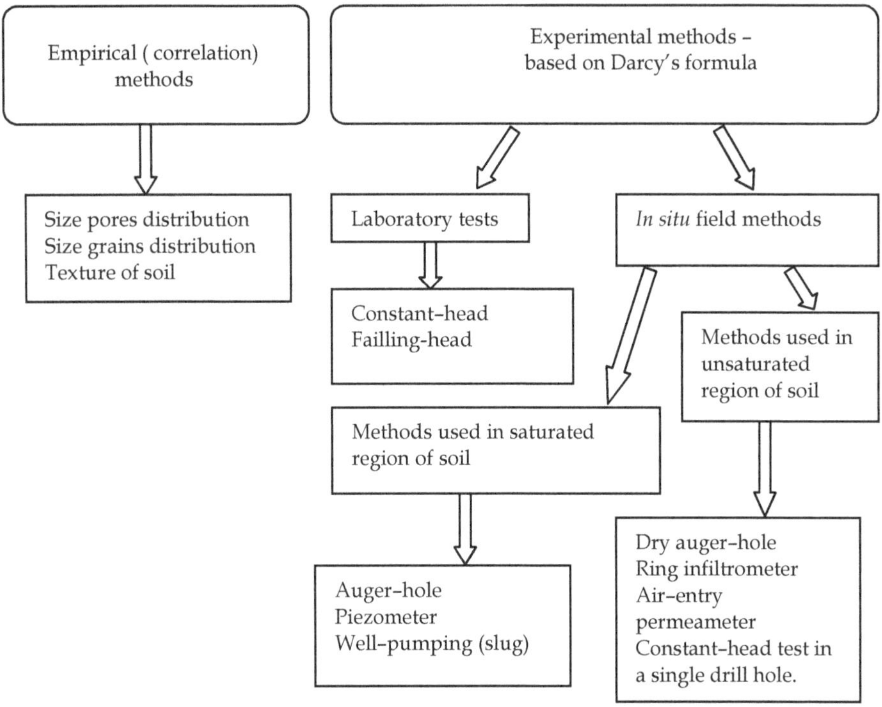

Fig. 4. Scheme of methods for measurement of hydraulic conductivity.

The selection of methods could depend on the objectives to be achieved (Tab. 1).

The laboratory methods are used to delimit the vertical and horizontal hydraulic conductivity in small soil samples collected in accordance with core drilling programs. The results of these methods are considered as point representation of the soil features because of the small sizes of the soil samples. If the structure of soil samples is not disturbed (naturally) the measurement of hydraulic conductivity represents the *in situ* saturated hydraulic conductivity at the particular sampling point.

In the case of the field methods, the evaluation of the hydraulic conductivity is based on a large region of soil, therefore the results of these tests should present the effects of both vertical and horizontal directions and the main value of K. It is important if soil is highly stratified and the values of K measured by means of field methods could reflect the most permeable and dominate layer in soil profile.

4.1 Empirical methods

Empirical approach contains the Shepherd formula (1989) that correlates grain size and hydraulic conductivity. The formula expresses the approximate hydraulic conductivity from grain size analyses:

$$K = a(D_{10})^b \tag{4}$$

where a and b are empirically derived terms based on the soil type, and D_{10} is the diameter of the 10 percentile grain size of the material.

Another specialized empirical estimation of hydraulic conductivity is the pedotransfer function method (PTF). It is described primarily in the soil science literature, but has been increasingly applied in hydrogeology. There are many different PTF methods, however, they all attempt to evaluate soil properties, such as hydraulic conductivity, several given measured soil properties, such as soil particle size, and bulk density.

The hydraulic conductivity is affected not only by grain size but also by the viscosity and quality of the water, the shape of the soil particles, density of the soil, cementation of the soil and the degree of soil saturation. All these factors strongly influence hydraulic conductivity and relationship between these factors and hydraulic conductivity can be expressed by following formula based on Darcy law:

$$K = 2gD^2e^3/vC_s1 + e \tag{5}$$

where:

K - hydraulic conductivity,

g - the acceleration due to gravity,

v - the kinematic viscosity of water,

Cs - particle shape factor,

D - the weighted or characteristic particle diameter,

e - void ratio.

The characteristic particle diameter D is calculated from a grain size distribution analysis using the following equation:

$$D = \Sigma M_i/\Sigma[M_i/D_i] \tag{6}$$

where:

M_i - the mass retained between two adjacent sieves,

D_i - the mean diameter of two adjacent sieves.

Gülser and Candemir (2008) using pedotransfer method to determine the saturated hydraulic conductivity on the base of the soil physical properties concluded that direct effect of some physical properties on K in soils could be expressed in following order: permanent wilting point > bulk density > clay > silt > field capacity. The hydraulic conductivity generally decreases according to soil textural class (Fig. 5) and it may be described as follows: sandy soil > loamy soil > clay soil. If sand and silt contents in soil texture increase the soil bulk density increases generally (Hillel, 1982) while total porosity decreases and ratio of macro porosity in total porosity increases.

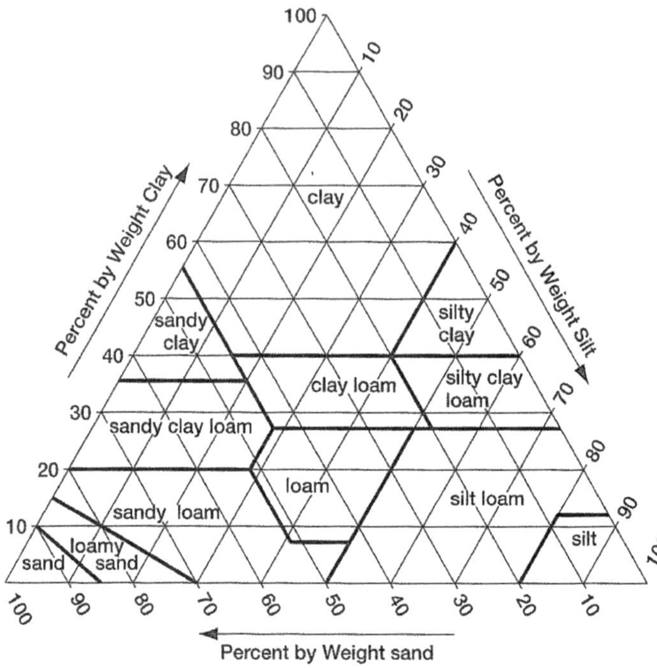

Fig. 5. Diagram of soil textural classes. (http://keys.lucidcentral.org/keys/sweetpotato/key/sweetpotato%20diagnotes/media/html/TheCrop/CropManagement/SoilFertilityManagement/Soil%20structure.htm)

Generally, large particles (sand and stones) pack loosely with large spaces between. Very fine particles (clay) pack very densely with little space. A well structured soil has small particles clumped together in aggregates, so that there are both small spaces (between the particles within the aggregates) and larger spaces (between the aggregates).

Larger spaces allow water to infiltrate easily, and to drain freely enabling air to re-enter after wetting. Smaller spaces are primarily important to hold water, and to ensure contact between soil particles and soil water so that nutrients can dissolve and become available to plant roots.

4.2 Experimental methods

4.2.1 Laboratory tests

While estimating of the hydraulic conductivity of soil by means of laboratory methods, different instruments are used. They contain permeameters, pressure chambers and consolidometers. The samples of soil are placed in a small cylindrical receptacle representing one-dimensional soil configuration which, the circulating liquid is forced to flow through. On the basis of the flow pattern *via* the soil samples some kinds of the laboratory methods for measuring of K are distinguished. The first one is constant-head test with a steady-state flow regimen and the second with unsteady flow regimen called falling-head test. The constant-head method is mainly used on granular soil with an estimated K above 1.0x 10^2 m/yr, while the falling-head method is used on soil samples with K below 1.0x 10^2 m/yr (Freeze & Cherry, 1979). The important considerations concerning the estimation of value of the hydraulic conductivity by means of laboratory methods are the procedure of the soil collection and preparation of the test specimen and circulating liquid. The collection of the soil samples should be performed so as to avoid changes in matrix structure of the soil. It is possible to apply walled tube sampling methods. In this technique undisturbed soil sample is received by pressing a thin walled metal tube into the soil, removing the metal tube filled with soil and then sealing its ends to avoid physical disturbance in the structure of the soil matrix.

4.2.1.1 Constant-head method

The procedure of constant-head method allows water to move through the soil under a steady state head condition while the quantity (volume) of water flowing through the soil specimen is measured over a period of time (Fig. 6). By knowing the quantity Q of water measured, length of specimen L, cross-sectional area of the specimen A, time required for the quantity of water t, and head h, the hydraulic conductivity can be calculated thus:

$$Q = Avt \tag{7}$$

where v is the flow velocity. Using Darcy's Law:

$$v = Ki \tag{8}$$

and expressing the hydraulic gradient i as:

$$i = h/L \tag{9}$$

where h is the difference of hydraulic head over distance L:

$$Q = AKht/L \tag{10}$$

solving for K gives:

$$K = QL/Ath \tag{11}$$

4.2.1.2 Failling-head method

The basis of this test is very similar to the foregoing constant-head method, but it is used for both fine-grained and coarse-grained soils. The soil sample is first saturated under a specific head condition. The water is then allowed to flow through the soil, a constant pressure head not to be maintained. To determine the hydraulic conductivity by use of the failing-head method, a cylindrical soil sample of cross-sectional area A, and length L is placed between two high conductivity plates. The soil sample column is connected to a standpipe of cross-sectional area a, in which the percolating fluid is introduced into the system. Therefore, by measuring the changes in head in the standpipe from h_1 to h_2 during the time t, the hydraulic conductivity can be expressed as follows:

$$K = (aL/At) \ln (h_1/h_2) \tag{12}$$

A common problem of the foregoing, two laboratory methods using permeameter is related to the degree of saturation achieved within the samples of soil during the test. Air bubbles are usually trapped within the pore space, and although they could disappear slowly by

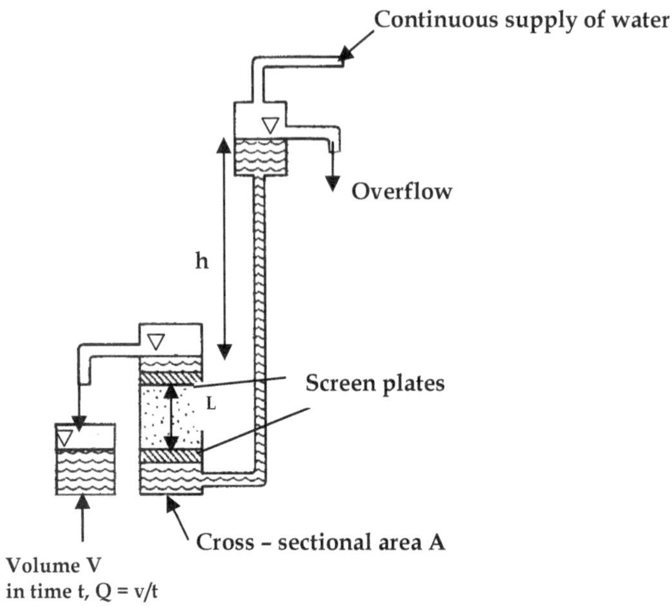

Fig. 6. Constant head permeameter (Beers, 1983, modified).

dissolving into the deaerated water, their presence in the system may influence the results of measurement. For more accurate results of K measurement in soil samples in which the air bubbles could appear and be critical, the conductivity test with back pressure is recommended.

4.2.2 Field methods

The field methods contain several tests for *in situ* determination of hydraulic conductivity that could be divided into two groups: (1) those which refer to sites near or below a shallow water table and (2) those that are applicable to sites above a deep water table or in absence of water table. Similarly to the laboratory procedures, in these groups, to determine the hydraulic conductivity (K), the Darcy's formula is used after the measuring the gradient of hydraulic head at the site and the resulting soil water flux.

4.2.2.1 Field methods used in saturated regions of the soil

These methods will be applied to determine the hydraulic conductivity for saturated zones of soil within a groundwater formation under unconfined and confined conditions. This method contains (1) the auger-hole and piezometer methods, which are used in unconfined shallow water table conditions and (2) well pumping tests, which are used for determination of aquifer properties in confined and unconfined groundwater systems.

4.2.2.1.1 Auger-hole method

If the water table is shallow, the auger-hole method could be used for evaluating the hydraulic conductivity below the water table. This method concerns four steps as follows:

a. drilling of the augerholes

b. removal of water from the augerhole
c. measurement of the rate of rise
d. computation of the value of K from the measurement data.

This method is fast, simple and usually used to design drainage systems in waterlogged land and in canal seepage studies.

4.2.2.1.2 Piezometer method

The piezometer method is designed for applications in layered soil aquifers and for determining either horizontal or vertical components of the saturated hydraulic conductivity. In this method piezometer tube or long pipe are used to penetrate the unconfined system.

4.2.2.1.3 Well-pumping (slug) method

A slug test is a particular type of aquifer test where water is quickly added or removed from a groundwater well, and the change in hydraulic head is monitored over time, to determine the near-well aquifer characteristics. It is a method used by hydrogeologists and civil engineers to determine the transmissivity/hydraulic conductivity.

4.2.2.2 Field methods used in the unsaturated region of the soil

The measurement of the value of hydraulic conductivity of unsaturated soils located above the water table (or if there is no water table) by *in situ* methods is more difficult in comparison to estimation of K for saturated soils because primarily unsaturated soil must be first saturated to perform the measurements. Therefore, the results of these *in situ* measurements are commonly called the field-saturated hydraulic conductivity. The available methods for measuring field saturated K comprise: (1) the shallow-well pump-in or dry auger-hole, (2) the double-tube, (3) the ring infiltrometer, (4) the air-entry permeameter, (5) the constant-head test in a single drill hole.

The shallow-well pump-in, is otherwise known as dry auger-hole method or well permeameter method. It is used to measure the rate of water flow from a cased or uncased auger-hole when a constant height of water is maintained in hole. To maintain a water level with a large water tank providing the water supply a float valve is usually used. The hydraulic conductivity values are calculated by use the steady state outflow rate and a shape factor determined from nomographs or equation. The position of water table or impermeable layer below the bottom of the well must be known. This procedure is easy but is limited by the time requirements needed to reach steady state and to replicate measurements.

Double tube method uses two concentric tubes that are placed in the soil to a given depth. The water flow is manipulated to move from the inner to outer tube at a high and changing rate of hydraulic head. Hydraulic conductivity values are then determined from tables and graphs and express the combination of vertical and horizontal hydraulic conductivities.

Ring infiltrometer method is similar to cylinder permeametr procedure. A large hole is prepared to appropriate depth. A metal sleeve is installed in the center of the hole. The same water level is maintained outside and inside the sleeve. The saturated hydraulic conductivity is taken to be the rate of infiltration when soil suction at the bottom of the ring equals zero (saturated conditions). This method allows measuring the vertical conductivity of layered soils.

Air-entry permeameter method. In this method a small covered cylinder is driven into ground. To this cylinder water is applied until all air is driven out. At the top of cylinder a

Type of methods	Name of method	Application of method	Description of method	Literature
LABORATORY	**A.** Constant-head conductivity with permeameter cylinder	For disturbed soil samples of cohesionless coarse-grained soils with $K>1.0x\ 10^2 m/yr$	It is used to estimate the original, in undisturbed state conductivity in a horizontal direction	Klute & Drikson, 1986
	B. Falling-head conductivity with with permeameter cylinder	For disturbed samples of cohesionless coarse-grained soils with $K<1.0x\ 10^2 m/yr$	As in **A** method	Klute & Drikson, 1986
	C. Conductivity test with sampling tubes	For unchanged cohesionless soil samples that can not be removed from the tube without disturbance	In depends on the estimated conductivity of samples, it is carried out under constant or falling head flow regimen, the conductivity is evaluated in accordance with direction in which samples has been collected (mainly in vertical direction)	Reddi, 2003
	D. Conductivity test with pressure chamber	For cohesive fine-grained soil, with changed, unchanged structure or in compacted state in a fully saturated condition	It is used only for originally fully saturated soil samples under falling-head flow conditions	Reddi, 2003
	E. Conductivity test with back pressure	As in D case, but soil samples are not fully saturated	The back pressure is used to reduce the size of the gas bubbles in the soil sample pores resulting in increase of the degree of water saturated, it is conductive under constant-head flow regimen	Reddi, 2003
	F. Conductivity test with consolidometer	As in D case	As an alternative method to method D	O'Kelly, 2009
FIELD	**A.** Auger-hole	For saturated soil samples located near the ground surface in the presence of a shallow water table	Used for saturated soil samples, gives average value of horizontal conductivity for dominate layers in soil profile	Amoozegar & Warrick, 1986
	B. Piezometer	As above	It is used for determination the hydraulic conductivity in both directions and it allows to measure K in each layer of stratified soils. The piezometer tube or pipe are installed into an aguer - hole with a cavity at the bottom. Water is transferred from the tube and the rate of its rise in tube is measured.	Amoozegar and Warrick, 1986

FIELD	**C**. Single–well (slug) in moderately permeable formations under unconfined conditions	Saturated soil samples of moderate K	Pomp out test, K in horizontal direction is measured. It comprises the removing of a slug of water from a well and then measuring the recovery of the water in the well.	Bouwer & Rice, 1976
	D. Single–well (slug) in moderately permeable formations under confined conditions	As above but in confined conditions, entirely open to the well screen or open borehole	As above, for saturated zone of the soil under confined condition	Kessler & Oosterbaan, 1974
	E. Constant–head conductivity test by the well permeameter method, referred to shallow-well	For measurement field–saturated hydraulic conductivity of soil samples in unsaturated zone near the ground surface	Pump in test, consisting of measuring the rate at which water flows out of an uncased well into the soil under constant–head flow conditions, could be also used for saturated zones	Amoozegar and Warrick, 1986
	F. Pump in or dry–auger–hole	For soil types ranging among sand, silt and clay mixture with K larger than $1{,}0x10^0$ m/yr and relatively clean sand or sandy gravel with $K<1{,}0x10^4$ m/yr	Pump out test, K is measurement in horizontal direction for uniform soil, it is used to determine the conductivity of the most permeable layer of the soil profile	Engqvist et al., 1978
	G. Double–tube	To measure field-saturated K of soil samples in the unsaturated zone near the ground surface	Two concentric cylinders are installed in auger–hole, water is pumped into these cylinders and K is estimated by measuring the flow in the cylinders, K represents conductivity in both directions	Amoozegar & Warrick , 1986
	H. Infiltrometer	As above for soil samples with ranging between $1{,}0x10^{-3}$ m/yr and $1{,}0x10^3$ m/yr	For measurement K in the vertical direction near the ground surface. It consists of pending water within a cylindrical ring placed over the soil surface and measuring the volumetric rate of water needed to maintain a constant head	Amoozegar & Warrick, 1986
	I. Constant–head conductivity test in single drill hole	To measure field-saturated K of soil samples at any depth within unsaturated zone, Soil or rock materials within K ranging between $1{,}0x\ 10^0$ and $1{,}0x\ 10^4$ m/yr	Pump in test consisting of injecting water into an isolated interval of a drill hole in soil or rock under constant–head flow conditions	Amoozegar & Warrick, 1986

Table 1. The review of standard experimental methods applied in measurements of soil saturated hydraulic conductivity (K).

large constant head is kept in a reservoir until saturation is reached at the bottom of the cylinder. The water level is then allowed to fall and the conductivity is calculated by the use of falling head equations. Due to air-entry permeameter method the hydraulic conductivity in vertical direction is obtained.

5. Wood elements as structural basis for hydraulic conductivity of higher plants

Vascular plants also known as tracheophytes or higher plants are those that have developed specialized conductive (vascular) tissues which circulate resources through the plant's body. In these plants, the xylem tissue is a space where long-distance transport of water takes place. The xylem tissue consists primarily of dead, lignified cells named tracheary elements. Either of two types of water conductive cells, tracheids and vessel elements, are found in xylem of vascular plants. Tracheids are found in all vascular plants and are the chief water-conducting elements in most of living gymnosperms and seedless vascular plants (Bailey & Tupper, 1918; Gifford & Foster, 1989), whereas vessel elements are unique to angiosperms and are the chief water-conducting elements for these plants. Both kinds of cells are elongated, die at maturity, but their lignified cell walls remain as the conduits through which water is carried in the xylem. Tracheids are closed at both ends but have pits where the cell wall is modified into a thin membrane, through which water flows from tracheid to tracheid. Vessel elements are stacked one on top of another in long columns, called vessels. In contrast to the tracheids the final walls of the single vessel element are perforated (composed plate) or, completely resolved (simple plate). Water flows almost unimpeded from cell to cell along these columns through perforations in the cell walls.

The size of tracheids is limited as they comprise a single cell. By the end of the Devonian, tracheid diameter had already increased to its maximum of ca. 80 µm (Niklas, 1985). Greater tracheid diameter would be advantageous only if accompanied by increased conduit length. Actually the wider tracheids are longer - up to 10 mm (Schweingruber, 1990). However, further increase in length and diameter of tracheid may be impossible beacause of limits to the maximum cell volume. Vessels, consisting of a number of cells overcame this limit and allowed larger conducts to form, reaching diameters of up to 500 µm, and lengths of up to 10 m (Zimmerman, 1983).

Important feature of the xylem structure is its connectivity i.e. the interconnected conduits form a network (Cruciat et al., 2002; Tyree & Zimmerman, 2002). The spatial arrangement of conduits was investigated by Burgraff (1972), Zimmerman (1971) and more recently by other authors (Steppe et al., 2004; Kittin et al., 2004) giving the support to define vascular system as a network integrating all main parts of the plant's body, i.e. roots, branches and leaves. Any root in the system is more or less directly connected with any branch and not with a single one. Moreover, the xylem network is redundant in two meanings: at a given level of the stem several xylem element are present in parallel and they develop lateral contacts with other tracks of vessels or tracheids.

Scholander et al. (1957) proposed and experimentally tested the hypothesis that the water-conducting xylem in the stem is essentially a flooded, continuous, micropore system, scattered with elongate macrocavities (vessels). The stem may accordingly be compared to a pipe filled with a sinter of fine sand, throughout which large longitudinal cavities are dispersed. If water fills such a system and flows through it, the cavities will offer the paths of least resistance, and through them most of the water will flow. If the hydrostatic pressure is below atmospheric and outside air enters a cavity, this will press the water out of the cavity, but no farther, as the air-water menisci will hang up in the micropores of the cavity walls.

The above way of describing the stem xylem emphasises that there is a great degree of structural similarities between the stem xylem of plants and the soil pore system. Tracheary elements have dimensions of capillaries and thus the same mechanism as for soil hydraulic conductivity might have been applied. Water moves spontaneously through tracheary elements only from places of higher water potential (ψ) to places of lower water potential, i.e. along a decreasing ψ gradient (Fig. 7).

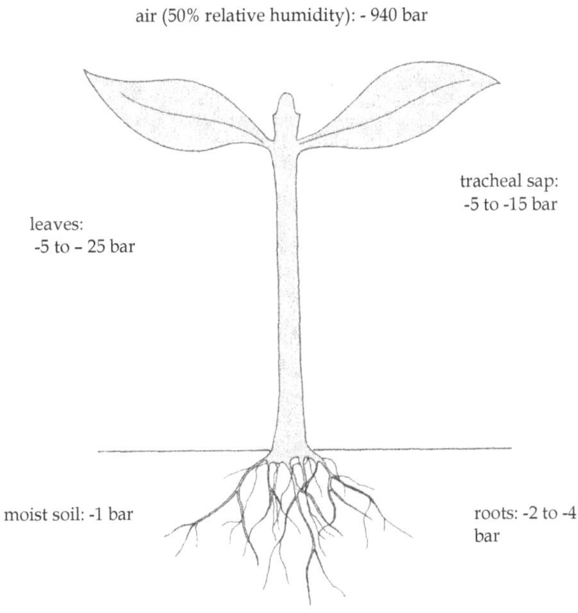

Fig. 7. The scheme of the pathway of water in the soil-plant-atmosphere continuum with the representative values for the water potential ψ (Mohr & Schopfer, 1995, modified).

6. Mechanism of water transport in vascular plants

The water transport in xylem of vascular plants is explained by the cohesion - tension theory formulated by Dixon and Joly (1894) as well as Askenasy (1895). The attractive force between water molecules is one of the principal factors responsible for the occurrence of surface tension in liquid water allowing plant to draw water from the root through the xylem to the leaf. Two phenomena cause xylem sap to flow: transpirational pull and root pressure.

Water is constantly lost by transpiration in the leaf. This creates tension (negative pressure) in the mesophyll cells. Because of this tension water is being pulled up from the roots into leaves, helped by cohesion (the pull between individual water molecules, due to hydrogen bonds) and adhesion (the stickiness between water molecules and the hydrophilic components of cell walls of plants). This mechanism of water flow works because of water potential (waters flows from high to low potential) and rules of diffusion. Transpirational pull requires that conduits transporting the water are small in diameter, otherwise cavitation would break the water column. As water evaporates from leaves, more is drawn up through plant to replace it. When the water pressure within the xylem reaches extreme levels due to low water input from roots, then the gases come out solution and form a bubble - an embolism forms, which will spread to adjacent cells, unless bordered pits are present.

Water potential of the root cells is more negative than that of the soil, usually due to high concentrations of solute, water can move by osmosis into the root from the soil. This causes a positive pressure that forces sap up the xylem towards the leaves. In some circumstances the sap will be forced from the leave through a hydathode in a phenomenon known as guttation. Root pressure is highest in the morning, before stomata open allowing transpiration to begin.

Comparing plant to a hydraulic system evokes the search for basic elements of such a system i.e. a driving force, pipes, reservoirs and regulating systems. In case of plants the driving force is most of the time, the transpiration, which pulls water from the soil to the leaves and creates and maintains a variable gradient of water potential throughout the plant. Pipes in the hydraulic systems correspond to very complex network of fine capillaries (vessels and tracheids) forming plant conducting system.

7. Environmental factors affecting hydraulic conductivity of SPAC

The root hairs play role in water uptake from the soil into plants: having close contact with soil solution they prevent the formation of air-filled cavities between root and soil particles during the process of water uptake. Such spaces could be a barrier zone to the transfer of water from soil to the roots. (Fig. 8.).

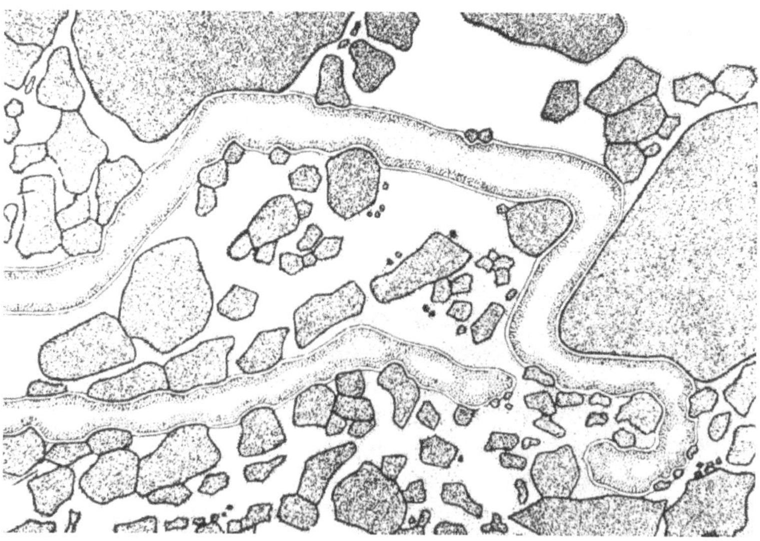

Fig. 8. Longitudinal section of soil profile with root hairs that increase the root surface and have extensive contact with the soil solution (Stocker, 1952 modified).

As mentioned above the continuous movement of water molecule from the soil *via* plant into atmosphere is driven by the differences in water potential between the perirhizal soil and the atmosphere and is maintained by solar energy. Resulting continuity of water columns from soil pores throughout the plant to leaf cells, linked to evaporative flux, is known as the soil-plant-atmosphere continuum (SPAC). Maintenance of this "hydraulic rope" is needed to ensure a continuous water supply to leaves.

The traditional view of plant hydraulics considered stomatal conductance and root conductivity as the main controlling factors of water flow in plants (Jones, 1983). This view is now expanding to include dynamic responses of xylem flow resistance to environment. Xylem conducivity is determined by the structure and size of the vessels (Schultz & Matthews, 1993; Thyree & Ewers, 1991) and by their efficiency, which may be affected by presence of embolism (Tyree & Sperry, 1989). One of the consequence of the cohesion-tension theory of water ascent in plants is the state of tension in the xylem sap and the occurrence of cavitation, which is the abrupt change from liquid water under tension to water vapour. As water is withdrawn from the cavitated conduit, vapour expands filling the entire lumen. Then air diffuses in causing the pressure rising to atmospheric. The conduit becomes embolized i.e. air-blocked. The same occurs in soil, where the larger pores in soil became filled in air, leaving only the smaller pores to hold and transmit water under the drought conditions.

Several reports have shown that water stress induces embolism and loss of function of the vessels (Sperry &Tyree, 1990; Hargrave et al., 1994) contributing to the reduction of water flow across the shoot (Schultz & Matthews, 1988). The resistance to cavitation and embolism is thus an important parameter determining the drought resistance of a plant and its hydraulic conductivity. The relation between the tension of the sap in the xylem and the corresponding degree of embolism is called a vulnerability curve. These curves are measures of the plants drought resistance (Cruziat et al., 2002; Sperry et al., 2003) and to much extend correspond to the unsaturated conductivity of soil while drying and wetting cycles (Sperry et al., 2002).

A negative effect of water stress on vessels size, hypothesized by Zimmermann and Milburn (1982) and implied in the observation that in periods of drought, wood xylem rings are narrower was directly evidenced in experiments with grapevine plants subjected to water stress of different intensity (Lovisolo & Schubert, 1998). It is also suggested that in large - vessel species reduction of vessel size may be an adaptation to a persistent situation of moderate waters stress, while embolism may be induced by a short or more severe water stress. Drought stress is frequently mentioned as an environmental factor implicated in the induction of trees decline recently observed in case of several species in Europe (Lygis et al., 2005; Kowalski & Łukomska, 2005) and North America (Ward et al., 2007; Bricker & Stutz, 2004).

Recently the relevance of xylem network structure for plant hydraulics (Loepfe at all, 2007) has been introduced into the theoretical discussion emphasizing that maximum hydraulic conducivity and vulnerability depend on multiple factors, including the connectivity of the network. The aforementioned authors have stated that connectivity increases both maximum hydraulic conducivity and vulnerability to drought-induced embolism and is therefore an element to be taken into account in any discussion on the efficiency vs. safety trade-off in the xylem. Our own preliminary data on connectivity in *Fraxinus excelsior* L. xylem in relation to the decline process were shown at the 55th Congress of Polish Botanical Society in 2010 and further investigations are currently carried out on that issue.

8. Conclusions

Soil and xylem are similar hydraulically and can be described as systems consisting of the same basic elements: a driving force, pipes, reservoirs, regulating systems. For plants the driving force is mainly the transpiration pulling water from the soil to the leaves and creating and maintaining a variable gradient of water potential throughout the plant. Pipes in plants correspond to a complex and highly organized network of very fine capillaries (vessels and tracheids), which form the xylem conducting system.

The water transport models for soils have been much mechanically based and complete than the corresponding description of plant hydraulics. The physical nature of flow throughout the soil makes it to some extent more amenable to quantitative treatment. An unsaturated conductivity curve for soil corresponds to the vulnerability curve for xylem and the underlying physical basis is the same. Thus any transport model that treats unsaturated soil conducivity provides an opportunity for the SPAC model to incorporate more mechanistic and predictive treatment of plant hydraulics and a better understanding of how the SPAC model is influenced by repeated droughts.

9. Acknowledgments

The publication was, in part, supported by Grant, No N N309 077438 from The Ministry of Science and Higher Education.

10. References

Amoozegar, F. & Warrick, A.W. (1986). Hydraulic conductivity of saturated soils: field methods. In: *Methods of Soil Analysis*, A. Klute, (Ed.), pp. 735-770, ASA and SSSA, Madison, WI, USA

Bailey, I.W. & Tupper, W.W. (1918). Size variation in tracheary cells. I. A comparison between the secondary xylems of vascular cryptogams, gymnnosperms and angiosperms. *Proc. Am. Art Sci. Soc.* Vol. 54, pp. 149-204, ISSN

Beers, W.F.J. (1983). *The auger hole method. A field measurement of the hydraulic conductivity of soil below the water table*. ILRI, ISBN 90 702060 81 6, Wageningen, The Netherlands

Bricker, J.S. & Stutz, J.C. (2004). Phytoplasmas associated with ash decline. *J. Arbor.*, Vol. 30, pp. 193-199, ISSN 0278-5226

Bouwer, H. & Rice, R.C. (1976). A slug test for determining hydraulic conductivity of unconfined aquifers with completely or partially penetrating wells. *Water Resources Res.*, Vol. 13, No. 3, pp. 423-428, ISSN 0043-1397

Burggraaf, P.D. (1972). Some observations on the course of the vessels in the wood of Fraxinus excelsior L. *Acta Bot. Neerl.*, Vol. 21, pp. 32–47, ISNN 0044-5983

Clarke, R. (1991). *Water: The International Crisis.*, Earthscan Publications Limited, ISBN 1-85383-105-0, London

Cruiziat, P.; Cochard, H. & Améglio T. (2002). Hydraulic architecture of trees: main concepts and results, *Ann. For. Sci.*, Vol. 59, pp. 723–725, ISSN 1286-4560

Engqvist, P.; Olsson, T. & Svensson, T. (1978). Pumping and recovery tests in wells sunk in till. Nordic Hydrological Conference, Papers of workshop, pp. 42-51, Haansari Cultural Centre, Finland, 31 July-3 August, 1978

Freeze, A.R. & Cherry, J.A. (1979). *Groundwater. Englewood Cliffs, Prentice-Hal,* ISBN 0133653129, London

Gifford, E. M. & Foster, A. S. (1989). *Morphology and evolution of vascular plants.* W.H. Freeman and Co, ISBN 9780716719465 , New York

Gülser, C. & Candemir, F. (2008). Prediction of saturated hydraulic conductivity using moisture constansts and soil physical properties. *www.balwois.com/balwois/administration/full_paper/ffp-902.pdf*

Hargrave, K.R.; Kolb K.J.; Ewers, F.W. & Davis, S.D. (1994). Conduit diameter and drought-induced embolism in *Salvia mellifera* Greene (Labiateae). *New Phytologist,* Vol. 126, pp. 695-705, ISSN 1469-8137

Hillel, D. (1980). *Fundamentals of soil physics.* Academic Press, ISBN 13: 978-0-12-348525-0, New York, NY

Hillel, D. (1982). Introduction to soil physics. Academic Press, ISBN ISBN: 0123485207, San Diego, California

Jenssen, P.D. (1990). Methods for measuring the saturated hydraulic conductivity of tills. *Nordic Hydrology,* Vol. 21, pp. 95-106, ISSN 0029-1277

Jones, H.G. (1992). *Plants and microclimate.* Cambridge University Press, ISBN 0 521 41502 0

Kitin P.B.; Fujii, T.; Abe H. & Funada, R. (2004). Anatomy of the vessel network within and between tree rings *of Fraxinus lanuginosa* (Oleaceae), *Am. J. Bot.,* Vol. 91, pp. 779-788, ISSN 0002-9122

Kessler, J. & Oosterban, R.J. (1974). Determining hydraulic conductivity of soils. In: Drainage principles and application, IRLI Publ., Vol. 16, No. 3, pp. 255-295, Wageningen, Netherlands

Klute, A., & Dirksen, C. (1986) Hydraulic conductivity and diffusivity: laboratory methods. In: *Methods of soil analysis,* A. Klute, (Ed.), pp. 687 - 724, ASA and SSSA, Madison, WI, USA

Kowalski, T. & Łukomska A. (2005). Badania nad zamieraniem jesionu (*Fraxinus excelsior* L.) w drzewostanach Nadleśnictwa Włoszczowa. *Acta Agrob.* Vol. 58, pp. 429-440, ISSN 0065-0951

Loepfe, L.; Martinez-Vilalta, J.; Piñol, J. & Mencuccini, M. (2007). The relevance of xylem network structure for plant hydraulic efficiency and safety. *Journal of Theoretical Biology,* Vol. 274, pp. 788-803, ISSN 0022-5193

Lovisolo, C. & Schubert, A. (1998). Effects of water stress on vessel size and xylem hydraulic conductivity in *Vitis vinifera* L. *Journal of Experimental Botany,* Vol. 49, pp. 693-700, ISSN 0022-0957

Lygis, V.; Vasiliauskas, R. & Stenlid, J. (2005). Pathological evaluation of declining *Fraxinus excelsior* stands of northern Lithuania, with particular reference to population of *Armilliaria cepistipes. Scand. J. For. Res.,* Vol. 20, pp. 337-346, ISSN 0282-7581

Mohr, H. & Schopfer, P. (1995). *Plant physiology.* Springer-Verlag, ISBN 3-540-58016-6, Berlin

Neuman, S.P. (1977). Theoretical derivation of Darcy's law. *Acta Maechanica,* Vol. 25, pp. 153-170, ISSN 0001-5970

O'Kelly, B.C. (2009). Development of a large consolidometer apparatus for testing peat and other highly organic soils. *Suoseura,* Vol. 60, No. 1-2, pp. 23-36, ISSN 0039-5471

Reddi, N.R. (2003). *Seepage in Soils: Principles and Applications,* John Wiley & Sons, Inc., ISBN 0-471-35616-6, New Yersey

Shepherd, R.G. (1989). Correlations of Permeability and Grain Size. *Ground Water*, Vol. 27(5), pp. 633-638, ISSN 1745-6584

Scholander, P.F.; Ruud, B. & Leivestad, H. (1957). The rise of sap in tropical liana. *Plant Physiology*, Vol. 32, pp. 1-6, ISSN 0032-0889

Schultz, H.R. & Matthews, M.A. (1988). Resistance to water transport in shoots of *Vitis vinifera* L. *Plant Physiology*, Vol. 88, pp. 718-724, ISSN 0032-0889

Schultz, H.R. & Matthews, MA. (1993). Xylem development and hydraulic conductance in sun and shade shoots of grapevine (*Vitis vinifera* L.): evidence that low light uncouples water transport capacity from leaf area. *Planta*, Vol. 190, pp. 393-406 ISSN 0032-0935

Schweingruber FH. (1990). *Microscopic wood anatomy*. Swiss Federal Institute for Forest, Snow and Landscape Research, London

Sperry, J.S. (2003). Evolution of Water Transport and Xylem Structure. *International Journal of Plant Sciences* Vol. 164 (3), pp. 15–127, ISSN 1058-5893

Sperry, J.S.; Stiller, V.; Hacke, U.G.; Oren, R. & Comstock J.P. (2002). Water deficits and hydraulic limits to leaf water supply. *Plant, Cell and Environment*, Vol. 25, pp. 251–263, print ISSN 0140-779, online ISSN 1365-3040

Sperry, J.S.; Stiller, V. & Hacke, U.G. (2003). Xylem hydraulics and the Soil-Plant-Atmosphere Continuum: Opportunities and Unresolved Issues. *Agron. J.*, Vol. 95, pp. 1362-1370, ISSN 0002-1962

Tyree, M.T. & Ewers, F.W. (1991). The hydraulic architecture of trees and other woody plants. *New Phytologist*, Vol. 119, pp. 345-360, ISSN 0028-646X

Tyree, M.T.; Davis, S.D. & Cochard, H. (1994). Biophysical perspectives of xylem evolution: Is there a tradeoff of hydraulic efficiency for vulnerability to dysfunction? *IAWA J.* Vol. 15, No. 4, pp. 335-360, ISSN 0928-1541

Tyree, M. & Zimmermann, M.H. (2002). *Xylem structure and the ascent of sap*. T.E. Timell, (Ed.), Springer-Verlag, ISBN 3 - 540-43354-6, Berlin

Tyree, M.T, Sperry, JS. (1989). Vulnerability of xylem to cavitation and embolism. *Annual Review of Plant Physiology and Molecular Biology*, Vol. 40, 19-38, ISSN 1040-2519

Zimmermann, M.H. (1971). Dicotyledonous Wood Structure Made Apparent by Sequential Sections, Inst wiss Film, Göttingen

Zimmermann, M.H. (1983). *Xylem structure and the ascent of sap*. Springer, ISBN 3-540-12268-0, Berlin

Zimmerman, M.H & Milburn, J.A. (1982). Transport and storage of water. In: *Encyclopaedia of plant physiology*, O.L. Lane, P.S. Nobel, C.B. Osmond, H. Ziegler, (Eds), Vol. 12B, pp. 135-151, Springer Verlag, ISBN 038712103X ,New York

7

Impacts of Wildfire Severity on Hydraulic Conductivity in Forest, Woodland, and Grassland Soils

Daniel G. Neary
USDA Forest Service, Rocky Mountain Research Station
United States of America

1. Introduction

Forest, woodland, and grassland watersheds throughout the world are major sources of high quality water for human use because of the nature of these soils to infiltrate, store, and transmit most precipitation instead of quickly routing it to surface runoff. This characteristic of these wildland soils is due to normally high infiltration rates, porosities, and hydraulic conductivities generated by biological and physical processes (Neary et al. 2009). Many of these ecosystems are subject to prescribed fires and wildfires that affect not only above-ground natural resources but also the soil and hydrologic systems (Ice et al. 2004).

Watershed condition is a term that describes the ability of a watershed system to receive, route, store, and transport precipitation without ecosystem degradation. When a watershed is in good condition, rainfall infiltrates into the soil, and baseflows are sustained between storms. Well-vegetated watersheds in good condition generally do not produce damaging peakflows (flash floods) and large amounts of erosion. However, in some regions of the world, these destructive streamflows are common irrespective of watershed condition. Severe fires, poor harvesting practices, over-grazing, conversion to agriculture and urban uses, and other disturbances alter watershed condition, reducing it to a moderate or poor level (Ffolliott et al. 2003). With poor watershed condition, rainfall infiltration and hydraulic conductivities are reduced significantly. Rainfall then runs over the surface of the soil, and there is little or no baseflow between storms. Erosion is considerable during high stormflows. This process is referred to as desertification and is, unfortunately, all too common in ecosystems currently being subjected to excessive wildfire (Neary 2006).

The surface conditions that determine watershed condition include: 1) the presence or absence of an organic litter layer (<5 mm to > 20 cm) and coarse woody debris, 2) herbaceous, shrub, and woody vegetation (variable cover), and 3) the geologic material (soil and rock). Disturbances like wildfire that destroy, remove, redistribute, or increase plant litter and vegetation, and change soil physical properties, alter the infiltration and percolation capacity of soil (DeBano et al. 2005). When watershed conditions deteriorate, the result is increased flood flows and erosion as watershed condition deteriorates.

Plant litter is a key factor in determining watershed condition (Neary 2002, DeBano and Neary 2005). In a *forest*, the organic "floor" consists of the Oi, Oe, and Oa horizons (also known as the L, F, and H layers; or the O1, O2, and O3 in other nomenclatures; Buol et al.

2003). The Oi layer consists of freshly fallen tree litter (leaves, branches). The Oe layer is made up of partially decomposed litter, and the Oa layer consists of well-decomposed organic matter. The term *woodland* refers to less dense vegetation units with lower vegetative structure that is sometimes referred to as *shrubland* or *scrubland* in the literature. In these ecosystems the distinct Oi, Oe, and Oa layers may occur only under continuous woody vegetation. In grassland ecosystems, easily identifiable layers may not be present and will be much thinner. Mesic grasslands have a complete herbaceous plant cover and well-developed organic soil horizons, but those in semi-arid climates may have only bare soil between plants. Organic material on the soil surface moderates the impact of rain drops, allowing water to infiltrate rather than running off over the surface. Loss of organic material by severe burning, harvesting, respiration, oxidation, site preparation, or other disturbances could result in adverse changes in hydrologic conditions in some instances.

Wildfires affect many water cycle processes. The specific hydrologic processes effects are summarized in Table 1. Changes in baseflow and stormflow definitely affect the quantity of water delivered from forested catchments, and can ultimately alter water quality. The occurrence and magnitude of these effects is a function of the general climate, precipitation, aspect, latitude, severity of fire, and the percentage of a watershed affected. The first three hydrologic processes affected by wildfire (interception, litter storage, and transpiration) listed in Table 1 are due to combustion of tree and herbaceous plant cover. Litter storage is the main process that is linked directly to hydraulic conductivity and infiltration. Loss of the litter layer during combustion is a highly significant process in producing direct effects on infiltration and the resulting watershed responses of streamflow, baseflow, and stormflow (DeBano et al. 1998 Moody et al. 2008). The heat flux during wildfire affects soil structure and porosity and produces water repellency that degrades hydraulic conductivity.

This paper examines the range of hydraulic conductivities measured in forest, woodland, and grassland soils produced by different levels of fire severity. It then discusses reductions in saturated hydraulic conductivities (K_{sat}) produced by degrees of severity-linked water repellency and O horizon destruction.

2. Forest, woodland, and grassland soils

Forests, woodlands, and grassland ecosystems usually develop deep and extensive root networks (Neary et al. 2009). Deposits of leaf, needle, limb, and herbaceous plant litter on the soil surface result in a surface soil horizon with relatively high levels of organic matter. The resulting soil environment produces a diverse micro- and macro-fauna as evidenced by the many invertebrates, insects, and small vertebrates found in these soils. Root growth and decay, cracking due to freeze/thaw and wetting-drying processes, animal burrowing, windthrow of weak trees, subsurface erosion, and other natural processes all increase soil porosity (ratio of void space to total soil volume), the number and size of macropores (>0.06 mm in diameter), and the hydraulic conductivity of the soil. Leaf and herbaceous plant litter on the soil surface dissipates raindrop energy and facilitates rainfall infiltration into the soil. The relatively high organic matter content of wildland soils increases the stability of soil aggregates, thereby preventing soil crusting by reducing detachment of small soil particles. This helps maintain high surface infiltration and hydraulic conductivity rates. For these reasons, most rainfall reaching the organic matter surface horizon infiltrates, and classical Hortonian overland flow occurs only during very intense rainfall events. Surface runoff occurs mainly as variable source area runoff (Hewlett and Troendle 1975) from rock outctrops, shallow soils, or low lying areas such as floodplains, wetlands, and ephemeral stream channnels where the surface water table rises to the soil surface during rainfall. These areas comprise only 5–15% of most wildland landscapes. Most of the infiltrated water either is used for plant transpiration needs or reaches streams by subsurface pathways (Jackson, 2006).

Forest, woodland, and grassland watersheds throughout the world are used as sources of municipal water supplies because of the stability of water yield and quality of the water (Neary 2002). High infiltration rates due to high hydraulic conductivities support baseflow hydrologic regimes that provide adequate supplies for human use.

Hydrologic Process	Type of Change	Specific Effect
1. Interception	Canopy consumed by fire	Moisture storage smaller
		Greater runoff in small storms
		Increased short-term water yield
2. Litter Storage	Litter Consumed	Less water stored (0.5 mm cm^{-1} of litter)
		Mineral soil exposed to raindrop impact
	Litter Scorched	No change
3. Transpiration	Temporary Elimination	Baseflow increased
		Soil moisture increased
4. Infiltration	Reduced	Hydraulic conductivity decreased
		Overland flow increased
5. Streamflow	Changed	Increased in most ecosystems
		Decreased in snow systems
		Decreased in fog-drip systems
6. Baseflow	Changed	Decreased with less infiltration
		Increased with less transpiration
		Summer low flow changes (+ and -)
7. Stormflow	Increased	Volumes greater
		Peakflows larger
		Time of concentration to peakflow shorter
8. Snow accumulation	Changed	Fires <4 ha, increased snowpack
		Fires > 4 ha, decreased snowpack
		Snowmelt rate increased
		Evaporation/sublimation increased

Table 1. A summary of the changes in hydrologic processes after wildfires (Adapted from Neary 2002).

3. Fire effects on ecosystems

3.1 Fire regime

The general character of fire that occurs within a particular vegetation type or ecosystem across long succession time frames, typically centuries, is commonly defined as the characteristic fire regime (Neary et al. 2005). The fire regime describes the typical or modal fire severity that occurs. But it is recognized that, on occasion, fires of greater or lesser severity also occur within a vegetation type. For example, a stand-replacing crown fire is common in long fire-return-interval forests (Figure 1).

The fire regime concept is useful for comparing the relative role of fire between ecosystems and for describing the degree of departure from historical conditions (Hardy et al. 2001, Schmidt et al. 2002). Brown (2000) contains a discussion of the development of fire regime classifications based on fire characteristics and effects, combinations of factors including fire frequency, periodicity, intensity, severity, season, size, pattern, and depth of burn. There are four commonly used fire regime classifications that are aggregated into fire regime groups depending on frequency of fire occurrence (0 to 35 years, 35 to 100+ years, and greater than 200 years) (Neary et al. 2005). *Understory fire regimes* are characterized by fires that are generally nonlethal to the dominant vegetation. They do not substantially change the structure of the dominant vegetation, and have minimal soil hydraulic conductivity effects. *Stand replacement fire regimes* frequently have fires that are lethal to most of the dominant aboveground vegetation. Approximately 80% or more of the vegetation is either consumed or dies as a result of fire, substantially changing the aboveground vegetative structure. Soil properties that influence hydraulic conductivity are frequently affected by this regime. In *mixed fire regimes* the severity of fires varies between nonlethal understory and lethal stand replacement fires with the variation occurring in space or time. Spatial variability occurs within the same fire when fire severity varies, producing a spectrum from fire effects characteristic of understory fire regimes to those of a stand replacement regimes. Hydraulic conductivity is affected in a spatial pattern that reflects the severity. The last fire regime is the *non-fire regime* which occurs in vegetation types that are not prone to fire such as temperate or tropical rain forests. However, hydraulic conductivity can be affected when large accumulations of woody debris burn during periodic droughts.

Fig. 1. High severity, stand replacing wildfire, Apache Sitgreaves National Forest, Arizona. (Photo courtesy of the U.S. Forest Service).

At finer spatial and temporal scales, the effects of a specific fire can be described at the stand and community level (Wells et al. 1979, DeBano et al. 1998, Ryan 2002, Neary et al. 2005). However, the fire regime concept does not work well for describing the soil impacts that alter hydrologic properties such as hydraulic conductivity. The commonly accepted term for describing the ecological effects of a specific fire is *fire severity*. Fire severity describes the magnitude of the disturbance and, therefore, reflects the degree of change in ecosystem components. Thus severity integrates both the heat pulse above ground and the heat pulse transferred downward into the soil. It reflects the amount of energy (heat) that is released by a fire that ultimately affects resources and their functions. Fire severity can be used to describe the effects of fire on the soil and water system, ecosystem flora and fauna, the atmosphere, and society (Simard 1991). It reflects the amount of energy (heat) that is released by a fire that ultimately affects soil hydraulic conductivity.

3.2 Fire intensity and severity

Although the literature historically contains confusion between the terms *fire intensity* and *fire severity*, a fairly consistent distinction between the two terms has been emerging in recent years. Fire managers trained in the United States and Canada in fire behavior prediction systems use the term fire intensity in a strict thermodynamic sense to describe the rate of energy released (Deeming et al. 1977, Stocks, 1991). Fire intensity is concerned mainly with the rate of aboveground fuel consumption and, therefore the energy release rate (Albini 1976, Alexander 1982). The faster a given quantity of fuel burns, the greater the intensity and the shorter the duration (Rothermel and Deeming 1980). Because the rate at which energy can be transmitted through the soil is limited by the soil's thermal properties, the duration of burning is critically important to the effects on soils (Frandsen and Ryan 1986, Campbell et al. 1995). Fire intensity is not necessarily related to the total amount of energy produced during the burning process. Most energy released by flaming combustion of aboveground fuels is not transmitted downward (Packham and Pompe 1971). Only about 5% of the heat released by a surface fire is transmitted into the ground. Therefore, fire intensity is not necessarily a good measure of the amount of energy transmitted downward into the soil, or the associated changes that occur in physical, chemical, and biological properties of the soil. Because one can rarely measure the actual energy release of a fire, the term fire intensity has limited practical application when evaluating ecosystem and soil responses to fire. Fire severity is the preferred measure of the magnitude of negative fire impacts on natural ecosystems and their components (DeBano et al. 1998).

3.3 Fire severity classification

Ryan and Noste (1985) and Ryan (2002) combined fire intensity classes with depth of burn classes to develop a two-dimensional matrix approach to defining fire severity. Their system was based on two components of fire severity: (1) an aboveground heat pulse due to radiation and convection associated with flaming combustion, and (2) a belowground heat pulse. In the literature there is common usage of a one dimension rating of fire severity (Wells et al. 1979, Agee 1993, DeBano et al. 1998, and many others). The single-dimension rating describes the overall severity of the fire and usually focuses primarily on the effects on the soil resource. At the spatial scale of a soil mapping unit, a tree stand, or a plant community, fire severity needs to be based on a sample of the distribution of fire severity classes. All fires produce a matrix of fire severities that cover the range of severity from low to high. The commonly accepted classes, definitions, and visual indicators of fire severity were noted in DeBano et al. (1998) and Neary et al. (2005). The classes are described as:

- Low severity: This class is typically indicated by scorching of smaller trees and seedlings, partial or complete combustion of herbaceous plants, <50% of plant brush canopy consumed, and >50% of trees showing no fire damage. Litter (Oi horizon) is charred or consumed with a 10-15% reduction. The Oe horizon (duff layer) is mostly intact and woody debris just charred. Mineral soil properties are usually unchanged and ash is mostly black. <2% of the area is severely burned, <15% is moderately burned, and the remainder of the area is burned at a low severity or unburned.
- Moderate severity: At this level of severity brush canopies are 60-80% charred or burned and 20-50% of tree canopies exhibit no visible scorch. There is extensive scorching of sapling and small tree crowns. The Oi horizon (litter) is consumed with a 50% reduction of cover and mass. The Oe horizon (duff) is deeply charred or consumed. Woody debris is extensively charred and the mineral soil is mostly unaffected. The signature ash color is gray. <10% of the area is severely burned, but >15% is burned moderately, and the remainder is burned at low severity or unburned.
- High severity: At this level of severity, <90% of brush canopies are charred or burned. Fewer than 20% of tree canopies exhibit no visible scorch, and all saplings and small trees are consumed. The entire organic horizons (Oi, Oe, and Oa) are consumed and woody debris is reduced to ash and charcoal. The mineral soil is often visible and exhibits a reddish or orange color. White ash is commonly found as the signature color. >10 percent of the area has spots that are burned at high severity, >80 percent moderately or severely burned, and the remainder is burned at a low severity.

Fire severity classifications were once done by on-the-ground visual surveys using these general definitions. They are currently being done by remote sensing from aircraft or satellites (van Leeuwen et al. 2010) using a BARC (Burned Area Reflectance Classification) system (Robichaud et al. 2007).

4. Fire severity effects on soils

4.1 General effects

Fire and associated soil heating can destroy soil structure, affecting both total porosity and pore size distribution in the surface horizons of a soil (DeBano et al. 1998). These changes in organic matter decrease both total porosity and pore size, and ultimately infiltration and hydraulic conductivity. Loss of macropores in the surface soil reduces infiltration rates and produces overland flow. Alteration of organic matter can also lead to a water repellent soil condition that further decreases infiltration rates and greatly decreases hydraulic conductivity. The scenario occurring during the destruction of soil structure by fire is:

- The soil structure collapses and increases the density of the soil because the organic matter that served as a binding agent has been destroyed.
- The collapse in soil structure reduces soil porosity (mainly macropores) and hydraulic conductivity.
- The soil surface is further compacted by raindrops when surface soil particles and ash are displaced, and surface soil pores become partially or totally sealed.
- Finally, the impenetrable soil surface reduces infiltration rates into the soil and produces rapid runoff and hillslope erosion. Loss of surface soil horizons leads to

further declines in soil and catchment hydraulic conductivities, and ultimately degradation of water resources.

The energy generated during the ignition and combustion of fuels provides the driving force that is responsible for the changes that occur in the physical, chemical, and biological properties of soils during a fire (DeBano et al. 1998). Mechanisms responsible for heat transfer in soils include radiation, conduction, convection, mass transport, and vaporization and condensation. The heat that is generated by the combustion of surface and aboveground fuels is transferred to the mineral soil surface where it is transferred down into the underlying soil by a series of complex pathways. Quantifying these different pathways for heat flow requires the mathematical modeling of fire behavior, duff ignition and combustion, and the transfer of heat downward to and through moist and dry mineral soil (Dimitrakopoulos and others 1994).

The heat radiated downward during the combustion of aboveground fuels is transferred either to the surface of the forest floor, or directly to the surface of mineral soil if organic surface layers are absent. In most forest ecosystems, heat is usually transferred to an organic layer of litter and duff. When duff is ignited it can produce additional heat that is subsequently transferred to the underlying mineral soil. The depth that heat penetrates a moist soil depends on the water content of the soil, and on the magnitude and duration of the surface heating during the combustion of aboveground fuels, litter, and duff (Frandsen 1987). During long-duration heating, such as that occurring under a smoldering duff fire or when burning slash piles, substantial heating can occur 40 to 50 cm downward in the soil. (Figure 2). This prolonged heating produces temperatures that are lethal to soil organisms and plant roots, and create water repellency that greatly diminishes hydraulic conductivity.

Fig. 2. White and gray ash typical of high severity fire remaining after a spruce-fir stand that was burned at high temperatures for a long duration, Coon Creek Fire of 2000, Tonto National Forest, Arizona. (Photo by Daniel G. Neary).

4.2 Water repellency

The creation of water repellency in soils involves both physical and chemical processes. It is discussed within the context of physical properties because of its strong, influence on infiltration and heat transfer (both physical processes). Although hydrophobic soils were observed in the early 1900s (DeBano 2000a, 2000b), fire-induced water repellency was first identified on burned chaparral watersheds in southern California in the early 1960s. In southern California, both the production of fire-induced water repellency and the loss of protective vegetative cover play a major role in the post-fire runoff and erosion. Normally, dry soils have an affinity for adsorbing liquid and vapor water because there is strong attraction between the mineral soil particles and water. In water repellent soils, however, the water droplet "beads up" on the soil surface where it can remain for long periods and in some cases will evaporate before being absorbed by the soil. Water, however, will not penetrate some soils because the mineral particles are coated with hydrophobic substances that repel water.

Water repellency is produced by soil organic matter and can be found in both fire and non-fire environments (DeBano 2000a, 2000b). Water repellency can result from the following processes involving organic matter: 1) An irreversible drying of the organic matter, 2) The coating of mineral soil particles with leachates from organic materials, 3) The coating of soil particles with hydrophobic microbial byproducts (for example, fungal mycelium), 4) The intermixing of dry mineral soil particles and dry organic matter, and 5) The vaporization of organic matter and condensation of hydrophobic substances on mineral soil particles during fire.

The magnitude of fire-induced water repellency and the reduction in hydraulic conductivity depend upon several parameters, including: 1) The severity of the fire, 2) Type and amount of organic matter present, 3) Temperature gradients in the upper mineral soil, 4) Texture of the soil, and 5) The water content of the soil. The more severe the fire, the deeper the layer, unless the fire is so hot it destroys the surface organic matter. Most vegetation and fungal mycelium contain hydrophobic compounds that induce water repellency. Steep temperature gradients in dry soil enhance the downward movement of volatilized hydrophobic substances that produce water. Early studies in chaparral showed that sandy and coarse-textured soils were the most susceptible to fire-induced water repellency (DeBano 1981). However, more recent studies indicate that water repellency frequently occurs in soils other than coarse-textured ones and that high water repellency may exist prior to wildfires occurring (DeBano 2000a, 2000b; Doerr et al. 2000). Soil water affects the translocation of hydrophobic substances during a fire because it affects heat transfer and the development of steep temperature gradients.

5. Forest, woodland, and grassland soil hydraulic conductivity

5.1 Examples of hydraulic conductivity rates

Table 2 shows the range in K_{sat} for a variety of forest ecosystems across the world. The K_{sat} rates for most undisturbed forests range from 143 to 4990 mm hr^{-1}. The rates for associated pastures that are up to 70 times lower are included for reference purposes. The highest K_{sat} rates in Table 2 are associated with thick Oe or Oa horizons (Lal 1996, Godsey and Elsenbeer 2002, Sauer and Logsdon 2002, Giertz and Diekkrüger 2003, and Sheridan et al. 2007).

5.2 Role of organic horizons in soil hydraulic conductivity

Plant litter is a key factor in determining watershed condition because the relatively deep organic layers have extremely high infiltration rates that exceed all but the most intense rainfall rates (Neary 2002, DeBano and Neary 2005). Organic horizons can range from a few centimeters deep to over 60 cm. Mollisols of grasslands have deep organic rich horizons that function in much the same manner as the O horizons of forest and woodland soils. However their K_{sat} levels are generally lower. Fiedler al. (2002) documented a K_{sat} of 83 mm hr^{-1} in a lightly grazed northern Colorado grassland compared to 8 mm hr^{-1} on a heavily grazed and mostly bare soil. On a mixed species grassland in southern Alberta, Canada, Dormaar et al. (1989) measured a K_{sat} of 45 mm hr^{-1} compared to 33 mm hr^{-1} on a grassland that was used for only short duration grazing. Li and Shao (2006) reported similar K_{sat} levels for climax oak forests and *Acer* - *Carex* shrubland on the central loess plateau of China (33-34 mm hr^{-1}). Grassland soils were lower at about 14 mm hr^{-1}, but double levels in agricultural fields.

Deep organic horizons found in forests have a profound effect on site hydrology because of their high hydraulic conductivities. The higher K_{sat} rates in Table 2 (e.g. Lal 1996 - 4,990 mm hr^{-1}; Sheridan et al. 2007 - 1,000 mm hr^{-1}; Sauer and Logsdon 2002 – 444 mm hr^{-1}) reflect the presence of these organic horizons. For the most part, these K_{sat} rates are well in excess of peak rainfall intensities. Shallow soils with bedrock, clay-textured horizons, or saturated conditions are the most limiting factors for infiltration and hydraulic conductivity in forest soils. Grace et al. (2006) provided a good example in their study of an organic soil in a hardwood forest of the Tidewater Region of eastern North Carolina. The surface horizon (Oa) of a loam-textured, thermic Terric Medisaprist soil varied between 0 and 60 cm deep. The K_{sat} was measured at 3540 mm hr^{-1}, well in excess of the most intense rainfalls for the humid climate of eastern North Carolina. The A horizon below the organic Oa ranged in depth from 60 to 107 cm, but the K_{sat} drops to 140 mm hr^{-1}. In the B horizon below the A horizon, K_{sat} fell to 40 mm hr^{-1}. The C horizon at 2.1 to 2.5 m below the surface is still above sea level (4.5 m below the soil surface) but it showed diagnostic characteristics of being poorly drained and it often had a shallow water table. The K_{sat} of the C horizon fell to 10 mm hr^{-1}. The impact of a potential wildfire fire on the site described by Grace et al. (2006), especially a high severity fire that consumes the surface organic horizon, is the reduction in the K_{sat} from 3540 to 140 mm hr^{-1}. Without factoring in water repellency or pore sealing by ash, the K_{sat} could be significantly reduced just by the removal of the Oa horizon. The importance of surface organic horizons in forest soils can also be demonstrated by data from Luce (1997). The K_{sat} of lightly disturbed forest soils in a harvested northern Idaho forest stand was reduced from 60 to 80 mm hr^{-1} to 1 mm hr^{-1} by road construction. Deep ripping of the road raised the K_{sat} to 22 to 35 mm hr^{-1}. Adding an artificial organic horizon of mulch raised the K_{sat} up to 80 to 85 mm hr^{-1}.

Woodlands tend to have the lowest hydraulic conductivities because they are less productive ecosystems and their soils are often lithic, shallow, poorly structured, less permeated by roots and soil organisms that develop macroporosity, and lower in organic matter content. The O horizons are thinner and they tend to be discontinuous. Thus the relationships of K_{sat} in these wildland soils is forests > grasslands > woodlands. The importance of surface organic horizons in determining the levels of K_{sat} in forest, woodland, and grassland soils can't be overstated.

Continent/Country		Reference	Hydraulic Conductivity K_{sat}
			mm hr^{-1}
Asia			
Nepal:	Undisturbed Forest	Gilmour et al. 1987	370
	Pasture Converted to Pine		183
	Pasture		39
China:	Forest	Chen et al. 2009	480
	Agriculture		360
	Bare Soil		6
Turkey:	Forest	Gol 2009	83
	Grassland		8
Africa			
Nigeria:	Forest	Lal 1996	4990
	Deforested area		460
Uganda:	Native Forest	Majaliwa et al. 2010	219
	Eucalypt Forest		149
Benin:	Forest	Giertz andDiekkrüger 2003	750
	Pasture		240
Australia			
New South Wales:	Undisturbed Forest	Moore et al. 1986	263
	Logged		19
Capital Territory:	Eucalypt Forest	Talsma and Hallam 1980	926
	Pasture To Pine		147
Western Australia:	Jarrah Forest Sand	Carbon et al. 1980	120
Victoria:	Eucalypt Forest	Sheridan et al. 2007	1000
Europe			
Czech Republic:	Forest Spodosols	Jacka et al. 2011	152
Finland:	Forest Spodosols	Mecke et al. 2000	5
Sweden:	Forest Spodosols	Lind and Lundin 1990	20
North America			
Canada:	Jack Pine	Cuenca et al. 1997	80
USA North Carolina:	Forest	Price et al. 2010	63
	Pasture		8
USA Arkansas:	Forest	Sauer and Logsdon 2002	444
	Pasture		113
South America			
Brazil:	Forest	Godsey and Elsenbeer 2002	250
	Pasture		15
Columbia:	Forest	Martinez and Zink 2004	143
	Pasture		2
Peru:	Forest	Allegre and Cassel 1996	420
	Pasture		41

Table 2. World-wide examples of K_{sat} measurements reported in the literature in surface horizons in forest soils and pastures.

6. Fire effects on hydraulic conductivity

6.1 Observed changes

Fire impacts on watershed hydrology have been reported for many years (DeBano et al. 1998, Neary et al. 2005, Moody et al. 2008). Wildfires exert a tremendous influence on the hydrologic conditions of watersheds in many forest ecosystems in the world depending on a fire's severity, duration, and frequency. Fire in these forested areas is an important natural disturbance mechanism that plays a role of variable significance depending on climate, fire frequency, and geomorphic conditions. This is particularly true in regions where frequent fires, steep terrain, vegetation, and post-fire seasonal precipitation interact to produce dramatic soil impacts (DeBano et al. 1998, Neary et al. 1999). A number of components of the hydrologic cycle can be impacted (Table 1), but K_{sat} reductions are often implicated as a major factor affecting baseflow and stormflow responses of burned watersheds. One clear signature of K_{sat} parameters is that they are highly variable (Doerr et al. 1998, Doerr et al. 2006).

A number of recent studies reporting changes in K_{sat} listed in Table 3 demonstrate clear reductions in conductivity after fires (Parks & Cundy 1989, Greene et al. 1990, Robichaud 2000, Ekinci 2006, Fox et al. 2007, Ekinci et al. 2008, Blake et al. 2009, Novák et al. 2009, and Nyman et al. 2011). Fire severity plays a key role in some of these reductions, but other investigators have demonstrated a surprising lack of correlation with severity (Rab 1996, Valzano et al. 1997, Sheridan et al. 2007, and Blake et al 2009). In the latter case, severity-related reductions in K_{sat} were measured in coniferous forests but not in oak woodlands. K_{sat} reductions of 20 to 48% are commonly reported (Table 3). Blake et al. (2009) also noted K_{sat} reductions of 88 to 92% with high severity wildfire.

An interesting trend emerging out of some of the recent Australian research on the impacts of wildfires on soil hydrologic properties including K_{sat} is that the soil surface K_{sat} values can be similar regardless of severity (Rab 1996 and Nyman et al. 2011) and that natural water repellency may produce K_{sat} values less than those measured in burned soils (Sheridan et al.2007). There also appears to be seasonal effects where natural summer water repellency breaks down and K_{sat} values return to the expected relationship of unburned soils > burned soils in the winter. It is obvious from these results that the fire severity - water repellency - hydraulic conductivity relations are more complex than once believed (DeBano 2000b, Doerr et al. 2000).

6.2 Mechanisms

A number of mechanisms have been discussed as the causative factors in post-wildfire hydrologic changes and K_{sat} reductions (DeBano et al. 1998, Neary et al. 2005, Doerr et al. 2000). The development of water repellency has received the most attention and is a test that is frequently carried out as part of wildfire Burned Area Emergency Response assessments (Keeley 2009, Neary 2009). Other mechanisms that have been suggested as major causes of K_{sat} reduction include pore clogging with fine ash and organic horizon destruction.

6.2.1 Hydraulic conductivity and water repellency

DeBano (1981, 2000a, 2000b) synthesized much of the knowledge about the effects of water repellency after wildfire and prescribed fire on forest, woodland, and grassland soils. Doerr et al (2000) discussed the biological sources of hydrophobic substances, physical factors affecting the formation and persistence of water repellency, temporal variations, and spatial

Location	Reference	Burned Condition Saturated Hydraulic Conductivity K_{sat} Unburned	Burned (Rx/Wf)	Soil Depth
		……………………mm hr^{-1}……………		cm
Australia				
	Greene et al. 1990			
	Eucalypt Woodland	92	74	0 - 10
	Nyman et al. 2011			
	Eucalypt Forest			
	Non Repellent Soil	40	30	0 - 05
	Repellent Soil	45	35	2 - 05
	Non Repellent Soil	120	240	>5
	Rab 1996			
	Eucalypts Mod. Sev.	32	34	0 - 10
	Eucalypts High Sev.	32	35	0 - 10
	Sheridan et al. 2007			
	Eucalypts Summer	490	855	0 - 05
	Eucalypts Winter	1409	459	0 - 05
	Valzano et al. 1997			
	Grassland	16	34	0 - 40
France				
	Fox et al. 2007			
	Pine Forest	210	155	0 – 06
Greece				
	Blake et al. 2009			
	Fir (Mod/High Sev.)	505	321/62	0 - 10
	Fir (Mod/High Sev.)	812	510/69	0 - 10
	Oak (Mod/High Sev.)	263	263/289	
Slovakia				
	Novák et al. 2009			
	Pine Forest	432	360	0 - 05
	Grassland	612	972	0 - 05
Turkey				
	Ekinci 2006			
	Oak/Pine Woodland	105	55	0 - 05
	Ekinci et al. 2008			
	Oak/Pine Woodland	48	33	0 - 05
USA				
	Parks & Cundy 1989			
	Douglas Fir& Pine	789	170	0 - 03
	Robichaud 2000			
	Douglas Fir & Pine			
	Low Severity	77 - 81	60 - 89	0 - 05
	High Severity	77 – 81	30 - 84	0 - 05

Table 3. Examples of soil K_{sat} changes after wildfires (Wf) and prescribed (Rx) fires.

variation of water repellent regions both laterally and vertically. As shown in Table 3, fire severity and any resulting water repellency can have a large effect on hydrologic processes including K_{sat} (Blake et al. 2009) or none at all (Rab 1996). The author has personally witnessed these effects during rain events (Ice et al. 2004). Water repellency usually breaks down within 1-3 years due to physical and biological processes (DeBano et al. 1998, Neary et al. 2005). While water repellency is certainly a major factor, other mechanisms can also be important.

6.2.2 Pore clogging

After wildfires, landscapes are blanketed by varying depths of ash until rainstorms remove the ash material in runoff (DeBano et al. 1998, Neary et al. 2005). While there is general concurrence that ash does contribute to post-wildfire hydrologic response, research studies have produced some conflicting results. In some cases ash has resulted in runoff increases but in others the ash increased infiltration (Cerda and Doerr 2008, Woods and Balfour 2008). More recent research by Woods and Balfour (2010) demonstrated that the degree of ash clogging of soil pores is soil texture related. A 1 cm layer of ash clogged pores and reduced infiltration by on a sandy loam soil but not a silt loam soil. The K_{sat} for the sandy loam soil was about 102 mm hr^{-1} pre-fire and dropped to about 30 mm hr^{-1} post-fire. The K_{sat} for the silt loam soil was 6 mm hr^{-1} pre-fire and increased to 8 mm hr^{-1} post-fire. Woods and Balfour (2010) concluded that thin, fine ash layers (1 cm) on a coarse soil with many macropores will clog the pores and increase the site post-fire hydrologic response. The same ash layer on a fine-textured soil with few macropores will have no effect on surface runoff. Thicker ash layers have the potential to delay surface runoff responses unless overwhelmed by intense rainfalls (e.g. 25 mm in 10 or 15 minutes).

6.2.3 O horizon destruction

Loss of the O horizon by combustion in high severity wildfires may play a larger role in post-fire hydrologic responses than previously thought possible. This phenomenon appears to be linked strongly to the loss of the O horizon and not necessarily any reduction in mineral soil K_{sat} values. High severity fires consume the entire O horizon and can decompose soil structure by combusting organic material involved in soil structure development. The example discussed previously of the study by Grace et al. (2006) in eastern North Carolina demonstrates the sharp drop in K_{sat} with the combustion of the surface organic layer (3540 to 140 mm hr^{-1}). A good deal of the large change in the hydrologic response after the Schultz Fire of 2010 in Arizona was most likely due to a similar loss of a 30-60 cm O horizon on steeply sloping soils, not necessarily any significant reduction in K_{sat} (Neary et al. 2011.). Intense rainfall (24 mm in 10 minutes) overwhelmed the infiltration capacity of the severely burned, silty gravel soils.

7. Conclusion

High severity fires produce the largest impacts on the hydrologic functioning of forest, woodland, and grassland soils. Fire severity can have a significant effect on K_{sat} by several mechanisms. These include development of water repellency, sealing of macropores, and combustion of surface organic horizons. High water repellency causes water droplets to sit on the surface of mineral soil, thereby reducing K_{sat} tolow values or even zero (DeBano et al. 1998; Neary et al. 2005) . Fine ash can effectively seal large pores at soil surfaces. The net effect is a reduction in K_{sat} by blocking macropores that are the cause of normally rapid infiltration in wildland soils. Although micropores can still infiltrate water, the rates are significantly reduced. Surface organic horizons have a high degree of porosity and can store, infiltrate, and conduct water at rates that exceed most peak rainfall intensities. Loss of the organic horizon in high severity wildfire is a major cause of K_{sat} reductions. In addition, high severity fires transmit large amounts of heat into the soil that often produce breakdowns in soil structure, leading to macropore size decreases and concomitant K_{sat} rate declines (DeBano et al. 1998).

Although there is a fairly clear correlation between high severity fire and K_{sat} reductions in the literature, some anomalies exist. Researchers in Australia have reported development of water repellency similar to wildfires in completely unburned watersheds. In addition, some moderate and high severity wildfires have not reduced K_{sat} to any significant extent. This could be due to soil physical properties, quality of the vegetation, or fire dynamics peculiar to the specific sites and soils studied or the characteristics of the wildfire. K_{sat} reductions of 20 to 48% are commonly reported after wildfires. Some studies have documented K_{sat} reductions of 88 to 92% with high severity wildfire. Reductions of this magnitude can have significant impacts on post-fire hydrological responses such as stormflows and peakflows.

8. References

Agee, J.K. (1993). *Fire ecology of Pacific Northwest forests*. Island Press, ISBN 1-55963-229-1, Washington, DC. 493 p.

Albini, F.A. (1976). Estimating wildfire behavior and effects. *General Technical Report INT-30*. Ogden, UT: U.S. Department of Agriculture, Forest Service, Intermountain Forest and Range Experiment Station. 92 p.

Alexander, M.E. (1982). Calculating and interpreting forest fire intensities. *Canadian Journal of Botany*, Vol. 60, pp. 349–357.

Allegre, J.C.; Cassel, D.K. (1996). Dynamics of soil physical properties under alternative systems to slash-and-burn. *Agriculture, Ecosystems, and Environment*, Vol. 58, pp. 39-48.

Blake, W.H.; Theocharopoulos, S.P.; Skoulikidis, N.; Clark, P.; Tountas, P.; Hartley, R.; Amaxidis, Y. (2009). Wildfire impacts on hillslope sediment and phosphorus yields. *Journal of Soils and Sediments*, Vol.10, pp. 671-682.

Brown, J.K.; Smith, J.K. (2000). *Wildland fire in ecosystems: Effects of fire on flora. General Technical Report RMRS-GTR-42-Vol. 2*, Fort Collins, CO: U.S. Department of Agriculture, Forest Service, Rocky Mountain Research Station. 257 p.

Buol, S.A.; Southard, R.J.; Grahma, R.J.; McDaniel, P.A. (2003). *Soil Genesis and Classification*, Iowa State University Press, ISBN: 978-0-8138-2873-2, Ames, IA. 460 p.

Campbell, G.S.; Jungbauer, J.D., Jr.; Bristow, K.L.; Hungerford,R.D. (1995). Soil temperature and water content beneath a surfacefire. *Soil Science*, 159(6): 363–374.

Carbon, B.A.; Bartle, G.A.; Murray, A.M.; Macpherson. (1980). The distribution of root length, and the limits to flow of soil water to roots in a dry sclerophyll forest. *Forest Science*, Vol. 26, pp. 656-664.

Cerda, A.; Doerr, S.H. (2008). The effect of ash and needle cover on surface runoff and erosion in the immediate post-fire period. *Catena*, Vol. 74, pp. 256-263.

Chen, X.; Zhang, Z.; Chen, X. (2009). Impact of land use and land cover changes on soil moisture and hydraulic conductivity along the karst hills of southwest China. *Environmental Earth Sciences*, Vol. 59, pp. 811-820.

Cuenca, R.H.; Stangel, D.E.; Kelly, S.F. (1997). Soil water balance in a boreal forest. *Journal of Geophysical Research*, Vol. 102, pp. 29,355-29,365.

DeBano, L.F. (1981). Water repellent soils: a state-of-the-art. *General Technical Report PSW-46*, Berkeley, CA: U.S. Department of Agriculture, Forest Service, Pacific Southwest Forest and Range Experiment Station. 21 p.

DeBano, L.F. (2000a). Water repellency in soils: a historical overview. *Journal of Hydrology* Vol. 231–232, pp. 4–32.

DeBano, L.F. (2000b). The role of fire and soil heating on water repellency in wildland environments: a review. *Journal of Hydrology*, Vol. 231–232, pp. 195–206.

DeBano, L.F.; Neary, D.G. 2005. Part A: The soil resource: its importance, characteristics, and general responses to fire. P. 21-28. In: Neary, D.G.; Ryan, K.C.; DeBano, L.F. 2005. *Wildland fire in ecosystems: Fire effects on soil and water. General Technical Report RMRS-GTR-42, Volume 4*, Fort Collins, CO: Rocky Mountain Research Station. 250 p.

DeBano, L.F.; Neary, D.G.; Ffolliott, P.F. (1998). *Fire's Effects on Ecosystems*. John Wiley & Sons, Inc., ISBN 978-0-471-16356-5, New York. 333 p.

DeBano, L.F.; Neary, D.G.; Ffolliott, P.F. (2005). Chapter 2: Effects on soil physical properties. Pp. 29-52. In: Neary, D.G.; Ryan, K.C.; DeBano, L.F. (editors) 2005. *Wildland fire in ecosystems: Fire effects on soil and water. General Technical Report RMRS-GTR-42, Volume 4*, Fort Collins, CO: U.S. Department of Agriculture, Forest Service, Rocky Mountain Research Station. 250 p.

Deeming, J.E.; Burgan, R.E.; Cohen, J.D. (1977). The national fire danger rating system, 1978. *General Technical Report INT-39*, Ogden, UT: U.S.Department of Agriculture, Forest Service, Intermountain Forest and Range Experiment Station. 63 p.

Dimitrakopoulos, A.P.; Martin, R.E.; Papamichos, N.T. (1994). A simulation model of soil heating during wildland fires. In: Sala, M.; Rubio, J.L. (Eds.). Soil erosion as a consequence of forest fires. Logrono, Spain. *Geoforma Ediciones*, pp. 207–216.

Doerr, S.H.; Shakesby, R.A.; Walsh, R.P.D. (1998). Spatial variation of soil hydrophobicity in fire-prone *Eucalyptus* and pine forests, Portugal. *Soil Science*, Vol. 163, pp. 313-324.

Doerr, S.H.; Shakesby, R.A.; Walsh, R.P.D. (2000). Soil water repellency: its causes, characteristics and hydro-geomorphological signifigance. *Earth Science Reviews*, Vol. 51, pp. 33–65.

Doerr, S.H.; Shakesby, R.H.; Blake, W.H.; Chafer, C.J.; Humphreys, G.S.; Wallbrink, P.J. (2006). Effects of different wildfire severities on soil wettability and implications for hydrologic response. *Journal of Hydrology*, Vol. 319, pp. 295-311.

Dormaar, J.F.; Smoliak, S.; Willms, W.D. 1989. Vegetation and soil responses to short duration grazing on fescue grasslands. *Journal of Range Management*, Vol. 42, pp. 252-256.

Ekinci , H. (2006). Effect of forest fire on some physical, chemical, and biological properties of soil in Çanakkale, Turkey. *International Journal of Agriculture and Biology*, Vol. 8, pp. 102-106.

Ekinci, H.; Kavdir, Y.; Yuksel, O; Yiğni, Y.; Cetin, S.C.; Altay, H. (2008). Fire induced changes in soil characteristics in Keşan, Turkey. Proceedings of the International Meeting on Soil Fertility, Land Management, and Agroclimatology, pp. 73-82.

Ffolliott, P.F.; DeBano, L.F.; Baker, M.B. Jr.; Neary, D.G.; Brooks, K.N. (2003). Chapter 4: Hydrology and impacts of disturbances on hydrologic functioning. In: *Hydrology, Ecology and Management of Riparian Areas in the Southwestern United States,* Baker, M.B. Jr.; Ffolliott, P.F.; DeBano, L.F.; Neary, D.G. (Eds.), pp. 51-76, Lewis Publishers, ISBN 9781566706261, Boca Raton, FL. 408 p.

Fiedler, F.R.; Frasier, G.W.; Ramirez, J.A.; Lajpat, R.A. (2002). Hydrologic response of grasslands: effects of grazing, interactive infiltration, and scale. *Journal of Hydrologic Engineering*, Vol. 7, pp. 293-301.

Fox, D.M.; Darboux, F.; Carrega, P. (2007). Effects of fire-induced water repellency on soil aggragate stability, splash erosion, and saturated hydraulic conductivity for different size fractions. *Hydrological Processes*, Vol. 21, pp. 2377-2384.

Frandsen, W.H. (1987). The influence of moisture and mineral soil on the combustion of smoldering forest duff. *Canadian Journal of Forest Research*, Vol. 17, pp. 1540-1544.

Frandsen, W.H.; Ryan, K.C. (1986). Soil moisture reduces belowground heat flux and soil temperature under a burning fuel pile. *Canadian Journal of Forest Research*, Vol. 16, pp. 244–248.

Giertz, S.; Diekkruger, B. (2003). Analysis of the hydrological processes in a small headwater catchment in Benin (West Africa). *Physics and Chemistry of the Earth*, Vol. 28, pp. 1333-1341

Gilmour, D.A.; Bonell, M.; Cassells, D.S. (1987). The effects of forests on soil hydraulic properties in the Middle Hills of Nepal: a preliminary assessment. *Mountain Research and Development*, Vol. 7, pp. 239-249.

Godsey, S.; Elsenbeer, H. (2002). The soil hydrologic response to forest regrowth: a case study from southwestern Amazonia. *Hydrological Processes*, Vol. 16, pp. 1519-1522.

Gol , C. (2009). The effects of land use change on soil properties and organic carbon at Dagdami River catchment in Turkey. *Journal of Environmental Biology*, Vol. 30, pp. 825-830.

Grace, J.M.; Skaggs, R.W.; Cassel, D.K. (2006). Soil physical changes associated with forest harvesting operations on an organic soil. *Soil Science Society of America Journal*, Vol. 70, pp. 503-509.

Greene, R.S.B.; Chartres, C.J.; Hodgkinson, K.C. (1990). The effects of fire on the soil in a degraded semi-arid woodland. I. Cryptogram cover and physical and micromorphological properties. *Australian Journal of Soil Research*, Vol. 28, pp. 755-777.

Hardy, C.C.; Schmidt, K.M.; Menakis, J.P.; Sampson, R.N. (2001). Spatial data for national fire planning and fuel management. *International Journal of Wildland Fire*, Vol. 10, pp. 353–372.

Hewlett, J.D.; Troendle, C.A. (1975). Non-point and diffused water sources: a variable source area problem. In: *Proceedings of the Watershed Management Symposium*, ISBN 0-939970-88-0, Logan, UT, August 1975, pp. 21–46.

Ice, G.G.; Neary, D.G.; Adams, P.W. (2004). Effects of wildfire on soils and watershed processes. *Journal of Forestry*, Vol. 102, pp. 16-20.

Jacka, L.; Pavlasek, J.; Pech, P. (2011). Estimated values of the saturated hydraulic conductivity of podzolic soils in the central part of the Sumava National Park – comparison of methods. *Geophysical Research Abstracts*, Vol. 13, p. EGU 2011-11751.

Jackson, C.R. (2006). Wetland hydrology. In: Batzer, D.P.; Sharitz, R. (Eds.), Ecology of 859 Freshwater and Estuarine Wetlands, University of California Press, Berkeley, pp. 43-81.

Keeley, J.E. (2009). Fire intensity, fire severity and burn severity: a brief review and suggested usage. *International Journal of Wildland Fire*, Vol. 18, pp. 116–126.

Lal, R. (1996). Deforestation and land use effects on soil degradation and rehabilitation in western Nigeria. *Land Degradation and Development*, Vol. 7, pp. 19-45.

Li, Y.Y.; Shao, M.A. (2006). Change of soil physical properties under long-term natural vegetation restoration in the Loess Plateau of China. *Journal of Arid Environments*, Vol. 64, pp. 77-96.

Lind, B.B.; Lundin, L. (1990). Saturated hydraulic conductivity of Scandinavian tills. *Nordic Hydrology*, Vol. 21, pp. 107-118.

Luce, C.H. (1997). Effectiveness of road ripping in restoring infiltration capacity of forest roads. *Restoration Ecology*, Vol. 5, pp. 265-270.

Majaliwa, J.G.M.; Twongyirwe, R.; Nyenje, R.; Oluka, M.; Ongom, B.; Sirike, J.; Mfitumukiza, D.; Azanga, E.; Natumanya, R.; Mwerera, R.; Barsasa, B. (2010). The effect of land cover change on soil properties around Kibale National park in southwest Uganda. *Applied and Environmental Soil Science*, Vol. 2010, 7 p. doi: 10,1155/2010/185689.

Martinez, L.J.; Zink, J.A. (2004). Temporal variation of soil compaction and deterioration of soil quality in pasture areas of Columbian Amazonia. *Soil and Tillage Research*, Vol. 75, pp. 3-18.

Mecke, M.; Westman, C.J.; Llvesniemi, H. (2000). Prediction of near-saturated hydraulic conductivity in three podzolic boreal forest soils. *Soil Science Society of America Journal*, Vol. 64, pp. 485-492.

Moody, J.A.; Martin, D.A.; Haire, S.L.; Kinner, D.A. (2008). Linking runoff response to burn severity after a wildfire. *Hydrological Processes*, Vol. 22, pp. 2063-2074.

Moore, I.D.; Burch, G.J.; Wallbrink, P.J. (1986). Preferential flow and hydraulic conductivity of forest soils. *Soil Science Society of America Journal*, Vol. 50, pp. 876-881.

Neary, D.G. (2002). Chapter 6: Environmental sustainability of forest energy production, 6.3 Hydrologic values. In: *Bioenergy from Sustainable Forestry: Guiding Principles and Practices,* Richardson, J.; Smith, T.; Hakkila, P., pp. 36-67, Elsevier, ISBN I978-1-4020-0676-0, Amsterdam. 344 p.

Neary, D.G. (2006). A global view of forest fires and desertification: Environmental and socio-economic consequences. Conferência Internacional sobre Incêndios e Desertificaçã, Lisbon, Portugal, 30 October 2006, Comissão Nacional de Coordenação do PANCD (Programa de Acção Nacional de Combate à Desertificação), General Directorat of Forestry Resources, Portugal. Programa de Acção Nacional de Combate à Desertificação Proceedings. 4 p.

Neary, D.G. (2009). Post-wildfire desertification: Can BAER treatments make a difference? *Fire Ecology,* Vol. 5, pp. 129-144.

Neary, D.G.; Ice, G.G.; Jackson, C.R. (2009). Linkages between forest soils and water quantity and quality. *Forest Ecology and Management.* Vol. 258, pp. 2269-2281.

Neary, D.G.; Ryan, K.C.; DeBano, L.F. (Editors) (2005) (Revised 2008). *Wildland fire in ecosystems: Fire effects on soil and water.* General Technical Report RMRS-GTR-42, Volume 4, Fort Collins, CO: U.S. Department of Agriculture, Forest Service, Rocky Mountain Research Station. 250 p.

Neary, D.G.; Koestner, K.A.; Youberg, K.A.; Koestner, P.E. (2011). Post-fire rill and gully formation, Schultz Fire 2010, Arizona, USA. *Proceedings of the Third International Conference on Fire Effects on Soil Properties,* Guimares, Portugal, March 2011. 4 p.

Novák, V.; Lichner, L.; Zhang, B.; Kňava, K. (2009).The impact of heating on the hydraulic properties of soils sampled under different plant cover. *Biologia,* Vol. 64, pp. 483-486.

Nyman, P.; Sheridan, G.J.; Smith, H.G.; Lane, P.N.J. (2011). Evidence of debris flow occurrence after wildfire in upland catchments of south-east Australia. *Geomorphology,* Vol. 125, pp. 383-401.

Packham, D.; Pompe, A. (1971). The radiation temperatures of forest fires. *Australian Forest Research,* Vol. 5, No. 3, pp. 1-8.

Parks, D.S.; Cundy,T.W. (1989). Soil hydraulic characteristics of a small southwest Oregon watershed following high-intensity wildfires. *General Technical Report PSW-109,* Berkeley, CA: U.S. Department of Agriculture, Forest Service, Pacific Southwest Research Station.

Price, K.; Jackson, C.R.; Parker, A.J. (2010). Variation of surficial soil hydraulic properties across land uses in the southern Blue Ridge Mountains, North Carolina, USA. *Journal of Hydrology,* Vol. 383, pp. 256-268.

Rab, M.A. (1996). Soil physical and hydrological properties following logging and slash burning in the Eucalyptus regnans forest of southeastern Australia. *Forest Ecology and Management,* Vol. 84, pp. 159-176.

Robichaud, P.R. (2000). Fire effects on infiltration rates after prescribed fire in Northern Rocky Mountain forests, USA, *Journal of Hydrology,* Vol.231-232, pp. 220-229.

Robichaud, P.R.; Lewis, S.A.; Laes, D.Y.; Hudak, A.T.; Kokaly, R.F.; Zamudio, J.A. 2007. Postfire soil burn severity mapping with hyperspectral image unmixing. *Remote Sensing of Environment*, Vol. 108, pp. 467-480.

Rothermel, R.C.; Deeming, J.E. (1980). Measuring and interpreting fire behavior for correlation with fire effects. *General Technical Report INT-93*. Ogden, UT: U.S. Department of Agriculture, Forest Service, Intermountain Forest and Range Experiment Station. 4 p.

Ryan, K.C. (2002). Dynamic interactions between forest structure and fire behavior in boreal ecosystems. *Silva Fennica*, Vol. 36, pp. 13-39.

Ryan, K.C.; Noste, N.V. (1985). Evaluating prescribed fires. In: Lotan, J.E.; Kilgore, B.M.; Fischer, W.C.; Mutch, R.W. (Eds.). Proceedings—symposium and workshop on wilderness fire. *General Technical Report INT* -182. Ogden, UT: U.S. Department of Agriculture, Forest Service, Intermountain Forest and Range Experiment Station, pp. 230-238.

Sauer, T.J.; Logsdon, S.D. (2002). Hydraulic and physical properties of stony soils in a small watershed. *Soil Science Society of America Journal*, Vol. 66, pp. 1947-1956.

Schmidt, K.M.; Menakis, J.P.; Hardy, C.C.; Hann, W.J.; Bunnell, D.L. (2002). Development of coarse-scale spatial data for wildland fire and fuel management. *General Technical* Report RMRS-87. Fort Collins, CO: U.S. Department of Agriculture, Forest Service, Rocky Mountain Research Station. 41 p.

Sheridan, G.; Lane, P.N.J.; Noske, P.J. (2007). Quantification of hillslope runoff and erosion processes before and after wildfire in a wet *Eucalyptus* forest. *Journal of Hydrology*, Vol. 343, pp. 12-28.

Simard, A.J. (1991). Fire severity, changing scales, and how things hang together. *International Journal of Wildland Fire*, Vol. 1, pp. 23-34.

Skaggs, R.W.; Chescheir, Amatya, D.M.; Diggs, J.D. (2008). Effects of drainage and forest management practices on hydraulic conductivity of wetland soils. *Proceedings of the 13th International Peat Congress. After Wise Use - The Future of Peatlands Volume 1*, ISBN Tullamore, Ireland, June 2008, pp. 452-456.

Stocks, B.J. (1991). The extent and impact of forest fires in northern circumpolar countries. In: *Global biomass burning: atmospheric climate and biospheric implications*, Levine, J.S. (Ed.), pp. 197-202, Massachusetts Institute of Technology Press, ISBN 13: 978-0-262-12159-0, Cambridge, MA.

Talsma, T.; Hallam, P.M. (1980). Hydraulic conductivity measurement of forest catchments. *Australian Journal of Soil Research*, Vol. 30, pp. 139-148.

van Leeuwen, W., Casady, G., Neary, D.G., Bautista, S., Carmel, Y.; Wittenberg, L.; Malkinson, D.; Orr, B. (2010). Monitoring post wildfire vegetation recovery with remotely sensed time series data in Spain, USA, and Israel. *International Journal of Wildland Fire*, Vol. 19, pp. 75-93.

Valzano, F.P.; Greens, R.S.B.; Murphy, B.W. (1997) . Direct effects of stubble burning on soil hydraulic and physical properties in a direct drill tillage system. *Soil& Tillage Research*, Vol. 42, pp. 209-219.

Wells, C.G., Campbell, R.E.; DeBano, L.F.; Lewis, C.E.; Fredrickson, R.L.; Franklin, E.C.; Froelich, R.C.; Dunn, P.H. (1979). Effects of fire on soil: a state-of-the-knowledge review. Gen. Tech. Rep. WO-7. Washington, DC: U.S. Department of Agriculture, Forest Service. 34 p.

Woods, S.W.; Balfour, V. (2008).Effect of ash on runoff and erosion after a severe forest wildfire, Montana, USA. *International Journal of Wildland Fire*, Vol. 17, pp.1-14.

Woods, S.W.; Balfour, V. (2010). The effects of soil texture and ash thickness on the post-fire hydrological response from ash-covered soils. *Journal of Hydrology*, Vol. 393, pp. 274-286.

Part 3

Determination by Mathematical and Laboratory Methods

8

Determining Hydraulic Conductivity from Soil Characteristics with Applications for Modelling Stream Discharge in Forest Catchments

Marie-France Jutras and Paul A. Arp
Faculty of Forestry and Environmental Management
University of New Brunswick
Fredericton, NB
Canada

1. Introduction

Many applications in watershed management, forestry, agriculture, and horticulture require hydrologically feasible estimates for assessing the rate at which water infiltrates and percolates through the soil, and how much of that is either taken up by the vegetation or passes through the ground until entering flow channels and streams further below in the landscape. In the literature, there are many approaches to do this, ranging from direct field measurements to numerical and theoretical constructs (Di Frederico and Tartakosky 2000; Pachepsky and Rawls 2005; Sudicky et al. 2010). Field measurements focus on, e.g., (i) direct measurements regarding the rate of infiltration, (ii) hydraulic gradients and hydraulic conductivities along hillslopes and aquifers, and (iii) stream discharge. Theoretical means infer soil and subsoil water retention and hydraulic conductivities from basic soil properties such as soil texture, organic matter content, and density. In turn, these estimates can then be used to determine temporal changes in soil moisture and soil moisture flow within fields (or hydrological response units), along hill slopes and across catchments, by way of simple trickle-down models (e.g., Church 1997), or complex geographically distributed hydrology models (Kim et al. 2008). The most elaborate models generate atmosphere-vegetation-soil transference fluxes based on empirical Eddy correlation techniques (Kuchment et al. 2006), while the simpler models use weather records involving precipitation and air temperature to assess daily changes in soil moisture and water flow (Balland et al. 2006; Murphy et al. 2009). This chapter (i) presents a generalized framework for estimating soil hydraulic conductivities at saturation, i.e., Ksat, at the soil-layer level, and (ii) applies this framework for modelling water retention and stream discharge for six well-studied forest catchments across Canada, from east to west. Within this framework, special attention is given to ensure that

i. soil moisture content at field capacity (FC) is always smaller than soil moisture content at the saturation point (SP),
ii. the permanent witting point (PWP) is always smaller than FC,

iii. the bulk density of the soil (Db) is always smaller than the mean particle density (Dp) of the soil,
iv. Ksat is strongly affected by the pore space of the soil, and
v. estimates for Db, Dp, SP, FC, PWP and Ksat are functionally related to soil depth, texture, and organic matter content (OM) across a wide range of natural soil conditions, from organic to mineral, from loose to compact, and from shallow to deep.

2. Estimating soil moisture flow, retention, and Ksat

Equations that generate layer-specific estimates for Db, Dp, FC, PWP and Ksat from generally available soil data such as OM, texture (i.e., sand, silt and clay content) and soil depth are known as "point pedotransfer functions", or point PTFs (Bouma, 1989; Gijsman et al. 2003; Børgesen and Schaap 2005; Pachepsky and Rawls 2005; Schaap 2006; Børgesen et al. 2007). The following equations (Balland et al. 2008) were used to this effect:

$$Db = \frac{a_{Db} + (Dp - a_{Db} - b_{Db}\ SAND)\,[1\text{-}\exp(-c_{Db}\ DEPTH)]}{1 + d_{Db}\ OM} \tag{1}$$

$$FC = SP\left(1\text{-}\exp\left(\frac{-a_{FC}\ (1\text{-}SAND)\text{-}b_{FC}\ OM}{SP}\right)\right) \tag{2}$$

$$PWP = FC\left(1\text{-}\exp\left(\frac{-a_{PWP}\ CLAY\text{-}b_{PWP}\ OM}{FC}\right)\right) \tag{3}$$

$$\log_{10}Ksat = a_{Ksat} + b_{Ksat}\log_{10}(Dp - Db) + c_{Ksat}\ SAND \tag{4}$$

where a, b, c, and d are Db-, FC-, PWP- and Ksat-specific calibration coefficients, and

$$\frac{1}{Dp} = \frac{OM}{Dp_{OM}} + \frac{1\text{-}OM}{Dp_{min}} \tag{5}$$

determines the average value for Dp, with $Dp_{OM} = 1.3$ gcm^{-3} and $Dp_{min} = 2.65$gcm^{-3} referring to the particle density of soil organic matter and minerals, respectively. SAND, CLAY, OM, FC, PWP, and FC refer to f dry soil weight fractions (fine earth fraction only). DEPTH refers to the mid depth of each soil layer, in cm. Ksat is expressed in cm hr^{-1}. Calibrating these equations with data taken from New Brunswick (NB) and Nova Scotia (NS) soil survey reports (CANSIS, 2000) produced the following results:

$$Db = \frac{1.23 + (Dp - 1.23 - 0.75\ SAND)\,(1\text{-}\exp(-0.0106\ DEPTH))}{1 + 6.83\ OM} \tag{6}$$

$$FC = SP\left(1\text{-}\exp\left(\frac{-0.588\ (1\text{-}SAND)\text{-}1.73\ OM}{SP}\right)\right) \tag{7}$$

$$PWP = FC\left(1\text{-}\exp\left(\frac{-0.511\ CLAY\text{-}0.865\ OM}{FC}\right)\right) \tag{8}$$

$$\log_{10}\text{Ksat} = -0.98 + 7.94 \log_{10}(\text{Dp} - \text{Db}) + 1.96\,\text{SAND} \quad (9)$$

with the best-fitted values for R^2, MAE, RMSE and the corresponding a, b, c and d regression coefficients listed in Table 1. These results show that the precision so achieved varied in the following order: FC > PWP > Db > Ksat. This order likely reflects the extent by which changes in soil structure (or state of soil aggregation) affect the measurement of these variables. It appears that such changes have (i) only small if any effects on the pressure-plate determinations for FC and PWP, (ii) moderate effects on the in-situ Db determinations, but (iii) large effects on Ksat on account of disproportionate flow rates through fine to large pores, root channels and cracks. With organic soils, varying degrees of humification also matter, with well-humified matter being more compactable and less permeable than fibrous matter (Pepin et al., 1992; Paquet et al., 1993; Balland et al. 2008). The modelled variations of Ksat with changing OM, sand content, and Db are shown in Figure 1, together with plots of actual versus best-fitted NB and NS data.

Property	ax	bx	cx	dx	R^2	MAE	RMSE
Db, g cm^{-3}	1.17 ±0.05	0.83 ±0.08	0.022 ±0.004	6.1 ±0.8	0.83	0.14	0.18
FC, g g^{-1}	0.588 ±0.016	1.734 ±0.049			0.96	0.032	0. 048
PWP, g g^{-1}	0.511 ±0.025	0.865 ±0.057			0.65	0.026	0.035
$\log_{10}$Ksat	-0.98 ±0.11	7.94 ± 0.48	1.96 ±0.21		0.80	0.38	0.49

Subscript x for a, b, c d mean Db, FC, PWP, or Ksat, as pertinent by row

Table 1. Best-fitted results for Db, FC, PWP, and Ksat (cm hr^{-1}) including their respective a, b, c, and d coefficients, coefficient of determination (R^2), mean absolute error (MAE) and root mean square error (RMSE) for the New Brunswick and Nova Scotia soils data, based on Eqs. 1 to 4.

The extent of inter-parametric correlations among the regression coefficients is shown in Table 2. These correlations should, ideally, be as close to zero as possible to narrow the equifinal solution space for the best-fitted a, b, c and d coefficients. For example, the -0.89 correlation between a_{Ksat} and c_{Ksat} implies that an increase in c_{Ksat} will produce a corresponding decrease in a_{Ksat}. Hence, large non-zero inter-coefficient correlations numbers imply large uncertainties about the best-fitted coefficients for the parameter pair so identified.

Applying the Ksat formulation to the Universal Soil Database (UNSODA, Leij 1996), instead of NB and NS data yielded,

$$\log_{10}\text{Ksat} = (-1.05 \pm 0.08) + (6.1 \pm 0.4) \log_{10}(\text{Dp} - \text{Db}) + (2.2 \pm 0.1)\,\text{SAND} \quad (10)$$

$$(n = 481;\ R^2 = 0.52;\ \text{RMSE} = 0.68;\ \text{MAE} = 0.54).$$

Db	a_x	b_x	c_x	d_x	$log_{10}Ksat$	a_x	b_x	c_x
a_x	1				a_x	1		
b_x	-0.0067	1			b_x	-0.043	1	
c_x	-0.45	-0.75	1		c_x	-0.89	-0.15	1
d_x	0.77	0.062	-0.16	1				
FC	a_x	b_x			PWP	a_x	b_x	
a_x	1				a_x	1		
b_x	-58	1			b_x	-0.34	1	

Table 2. Correlation coefficients between a, b, c and d parameters of equations 1 to 4, New Brunswick and Nova Scotia soil data.

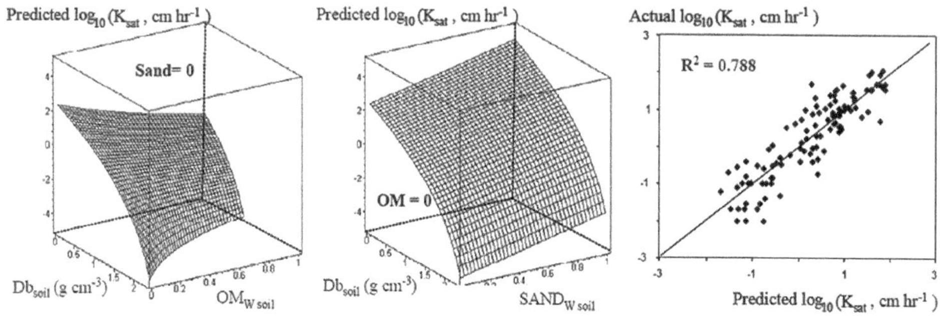

Fig. 1. Left and middle: 3-D visualisations of how $log_{10}Ksat$ varies with increasing soil bulk density (Db), sand fraction and organic matter (OM) fraction. Right: best-fitted $log_{10}Ksat$ versus actual data (right).

Hence, the $log_{10}Ksat$ formulation based on the NS and NB data alone remained valid in its general form, but the coefficient values changed slightly, with the largest change associated with the $log_{10}(Dp - Db)$ coefficient, i.e., dropping from 7.9 to 6.1. This change may relate to procedural differences, e.g., using estimated Dp values from known SP and Db values (NB and NS data) versus direct Dp measurements (UNSODA). The plot of actual versus best-fitted values in Figure 2 suggests a general agreement between the above Ksat formulation and the data from both sources.

3. Catchment hydrology

The Forest Hydrology Model (ForHyM; Balland et al. 2006; Fig 3) was used to simulate the the hydrothermal conditions within each of the discharge-monitored basins listed in Table 3. These simulations were driven by local weather records for daily precipitation (rain, snow) and air temperature. In this model, only gravitational water was allowed to flow, i.e., the amount of soil moisture above FC. The rate of this flow was set to be proportional to pore % of gravitational water multiplied by Ksat to estimate downward flow ("percolation", or "infiltration"), and adjusted for % slope of the

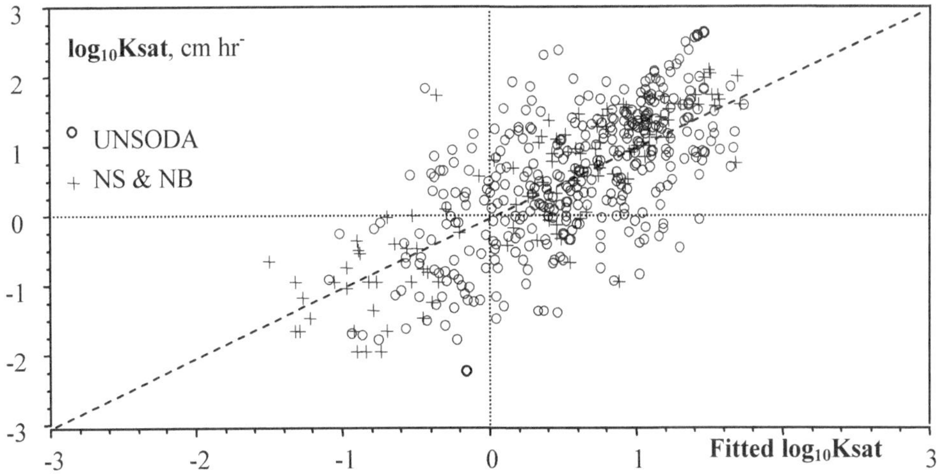

Fig. 2. Scatter plot of actual versus best-fitted Ksat values (Eq. 10) for two data sources (UNSODA (Leij et al. 1996; NB & NS soil survey data, Balland et al. 2008).

basin to estimate "lateral flow" or "interflow". At saturation, downward flow into the next mineral soil was typically slower than lateral flow on account of decreasing Ksat with increasing soil density at lower soil depth. Infiltration into unsaturated soils was determined (i) by directly filling the partially available pore space up to SP, and (ii) by accommodating downward as well as lateral flow as long as the soil moisture content would remain above FC. For simplicity, the soil at each location was represented by the forest floor (or LFH layer), the A and B layers combined, and the C layer. The substrate below the C layer was represented by 1 m intervals to a depth of 12 m, i.e., to the depth of isothermal conditions year-round. Values for Dp, Db, SP, FC, PWP, Ksat were generated for each layer from local soil surveys, using Eqs. 1 to 9 (Table 4). Estimates for the moisture- and frost-varying heat capacity and conductivity were also generated for each layer according to Balland and Arp (2005). The outcomes of these calculations are illustrated in Figure 4 for a basin at Turkey Lakes in Ontario near Sault St. Marie north of Lake Superior, and the Moosepit Brook basin east of Kejimkujik National Park, Nova Scotia. The general conformance between the modelled and actual stream discharge is documented in Table 5 by way of the best-fitted regression coefficient and the corresponding R^2 and RMSE values for each of the basins examined. The Ksat multiplier adjustments for the downward and lateral flow components are entered in Table 4. No adjustments were made to the layer-estimated values for Db, Dp, FC, and PWP (Eqs. 5 to 8). The following can be observed from Figs. 4 and 5 and Tables 1 to 5:

i. There is generally good agreement between the actual and basin calibrated snowpack depth, and stream discharge at the daily to annual time scaless (Fig. 4, Table 5).
ii. No adjustments were needed to match the monitored stream discharge with the incoming precipitation and model-assessed evapotranspiration rates.

iii. The Ksat adjustments for downward flow varied from 0.5 to 2 (Table 4), and were therefore within the generically determined precision for Ksat based on Eqs. 9 and 10 and the layer specification for soil texture, organic matter, soil depth and soil density (Table 1).

iv. The Ksat adjustments for lateral flow were more variable, thereby indicating that lateral flows through the basins were more complex and generally low thereby requiring downward Ksat adjustments for interflow, especially for the hummocky basins (Table 4). This was likely due to flow obstructions such as mounds and pits, empty or partially filled water pools above and below the regolith, erratic changes in soil depth, density, texture, organic matter, coarse fragment, and variations in the surface exposure of partially fractured bedrock especially along ridges, steep slopes and within crevices. Hayward Brook was particularly exceptional with its < Ksat multipliers for lateral flow. This undulating to rolling basin is underlain by calcareous shales, which generally have high flow variabilities, thereby enhancing downward flow (Schulze-Makuch et al. 1999).

v. For most soils, Db generally increases with soil depth (Eq. 1), and Ksat decreases accordingly (Eqs. 9 and 10). Within the A and B layers, Ksat values range from about 50 to 500 cm hr^{-1}. Within the subsoil, Ksat values are generally much lower by one to two orders of magnitude, especially on compacted tills.

vi. The Ksat value for the forest floor, as projected by way of Eq. 5, is rather low, but corresponds to Ksat values normally associated with organic soils. Due to the high porosity of this layer, infiltration occurs quickly. The water so received is, however, released rather slowly to the underlying forest soil and only so once the field capacity of the forest floor is exceeded. As a result, soil layers underneath the forest floor often remain quite dry during the later portion of the summer and during early fall. At this time, soils may also become hydrophobic. As a result, surface water would then flow laterally over short distances towards nearby pits and depressions, where the soil would be moister and permit gradual infiltration and downward percolation.

vii. The calibrated Ksat values for and depression lateral and downward flow generally fall within the Ksat uncertainty range associated with Eq. 9 and Eq. 10, with the best-fitted RSME values for Ksat varying from about ±0.4 to ±0.6. This range is similar to that obtained with (i) testing water recharge in wells receiving water from small depressions (about 50 m^2) to catchments up to about 1200 ha or more, and after taking care of the scaling-up effect that is associated with these measurements ($\log_{10}$Ksat RMSE = ±0.61; Schulze-Makuch et al. 1999), and (ii) using tension infiltrometers and Guelph permeameters to determine Ksat by soil depth at Turkey Lakes ($\log_{10}$Ksat, RMSE = ± 0.45; Murray and Buttle 2006) and at Lac Laflamme, as detailed in Table 4.

viii. Since the study locations represent a range of catchment size from about 70 to 1700 ha, there are no obvious trends with catchment size. Hence, the model-derived Ksat adjustments for the LFH, A&B and C layers are essentially independent of scale across this range. This is also in general agreement with Schulze-Makuch et al. (1999) who found that the up-scaling requirement for Ksat generally stops once the Ksat-determining flow fields offer no additional heterogeneity. However, Laudon et al. (2007) concluded that stream discharge is less dependent on scale than on wetland coverage per catchment, with discharge contributions of "event" water (or new water) amounting to 50% in wetland dominated catchment while limited to 10% to 30% in forest dominated catchments. Considering also

a. that forest catchments are generally permeated by many converging flow channels with varying and weather-dependent thresholds for flow initiation,
b. that forest catchments in glaciated landscapes such as the ones of this study are generally underlain by a layer of surface-fractured bedrock, and
c. that this layer provides additional space for water pooling and hydraulically activated flows towards the streams,

it is reasonable to suggest that the Ksat estimates and their multipliers in Tables 3 and 4 reflect similar flow heterogeneities within each of the many subcatchments for the catchments of this study.

ix. The above approach requires layer-specific estimates for Ksat. If these are not available, then Ksat can be derived from layer-representative values for sand content, Dp and Db. When estimates for Dp and Db are not available, one can derived these via Eqs. 5 and 6 for any soil depth and given values for sand and organic content. Generally, these values need to be representative of the LHF, A, B and C layers. The sensitivity of the resulting Ksat estimates to the natural variations of these quantities can be evaluated via Eq.s. 5, 6, and 9 or 10.

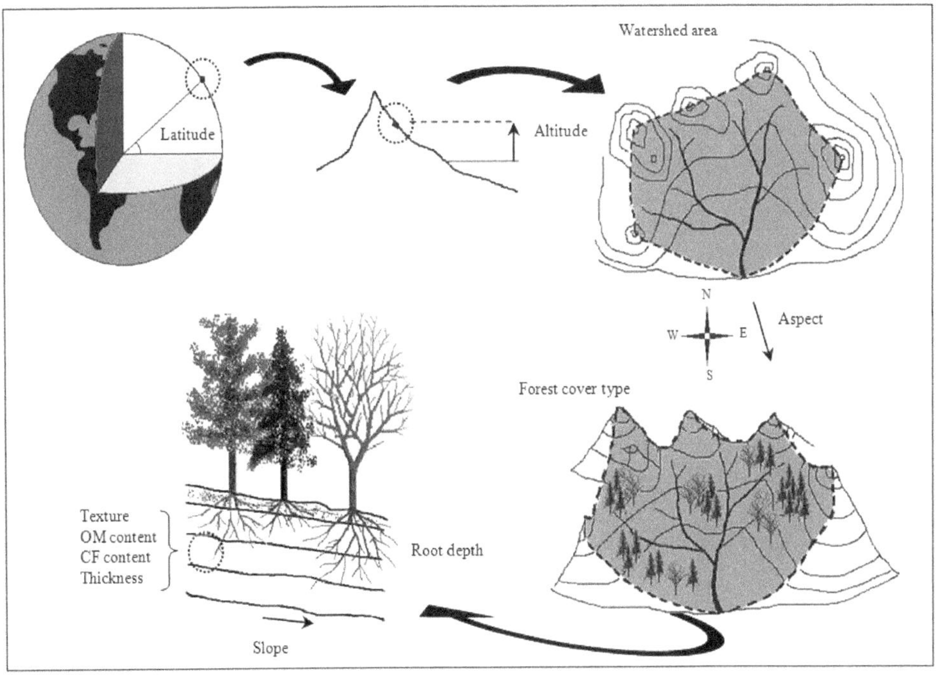

Fig. 3. Specification overview for the Forest Hydrology Model ForHyM (Balland 2002).

Site Parameters	Moosepit Brook Nova Scotia (NS)	Peggy Brook Nova Scotia (NS)	Hayward Brook New Brunswick (NB)	Turkey Lakes Ontario (ON)	Lac Laflamme Quebec (QB)	Rithet River Vancouver Island, British Columbia (BC)
References	Bobba et al. 1986 Jutras et al. 2011	Jutras et al. 2011	Chi 2008	Murphy et al. 2009; Beall et al. 2001 Murray and Buttle 2005	Barry et al. 1987; 1990 Stein et al. 1994	Zhu & Mazumder 2008
Abbreviation	MP	PB	HB	TL	LL	GB
Latitude (N)	44°28'	44°49'	45°52'	47°03'	47°17'	48°32'
Longitude (W)	65°03'	63°51'	65°11'	84°25'	71°14'	123°42'
Area (ha)	1670	196	508	1050	69	1630
Elevation (m)	100-150	172-227	96-162	350-400	777-884	375
Slope (%)	1	7	8	8	11	17
Deciduous:coniferous	50:50	75:25	50:50	100:0	10:90	0:100
Rooting habit	Medium	Shallow	Medium	Deep	Medium	Deep
Forest floor depth (cm)	5	5	5	7	8	15
Mineral soil: depth (cm); texture	42; SL	70; SL	40; S-LS	60; SilL	31; S	70; SL
Subsoil: depth (cm); texture	70; LS	100; SL	100; SL	100; LS	100; LS	100; SL
Adjustments for the snowpack calculations						
Snow-to-air temperature gradient multiplier	0.06	0.1	0.04	0.08	0.15	0.1
Density of fresh snow	0.15	0.15	0.12	0.16	0.3	0.12
Ksat multipliers						
Surface runoff	1	1	1	1	1	1
Forest floor infiltration	1	1	2	1	1	1
Forest floor interflow	2	2	0.05	1	1	2
A&B horizon infiltration	0.5	1	2	1	0.5	1
A&B horizon interflow	2	1.75	0.02	0.14	0.35	0.14
C horizon infiltration	0.5	1	2	1	0.5	1
C horizon interflow	2	2	0.05	0.5	2	2
Bedrock	Metamorphic greenschist	Granite	Sedimentary shale, sandstone	Metavolcanic Basalt	Metamorphic charnokitic gneiss	Volcaniclastic Rock; undivided Intrusive Rock
Landform	Glacial Till	Ablation Till	Ablation Till	Ablation till on basal till	Sandy Till	Colluvium
Topography	Undulating	Hummocky	Flat to rolling	Hummocky	Rolling to hummocky	Hummocky
Model Run	1999-2004	1999-2004	1994-1997	1995-2005	1990-1997	1994-2004

Table 3. Watershed description by site, with hydraulic conductivity adjustments.

Parameters	Moosepit Brook			Pockwock			Hayward Brook			Lac Laflamme			Turkey Lakes			Rithet River		
	FF	A& B	C	FF	A& B	C	FF	A& B	C	FF	A& B	C	FF	A& B	C	FF	A& B	C
Db, g cm^3	0.32	1.32	1.70	0.32	1.32	1.70	0.06	0.90	1.30	0.16	1.13	1.52	0.16	1.09	1.85	0.16	0.74	1.87
Dp, g cm3	1.33	2.48	2.60	1.33	2.48	2.60	1.33	2.48	2.58	1.33	2.58	2.48	1.33	2.48	2.58	1.33	2.27	2.55
FC, %	0.28	0.21	0.15	0.28	0.21	0.15	0.27	0.10	0.12	0.13	0.12	0.05	0.27	0.18	0.16	0.30	0.26	0.17
PWP, %	0.10	0.07	0.06	0.10	0.07	0.06	0.10	0.05	0.04	0.03	0.16	0.04	0.10	0.06	0.08	0.11	0.09	0.06
Thermal conductivity, W m^{-1} °C^{-1}	0.59	1.34	2.01	0.59	1.34	2.01	0.60	1.31	1.87	0.52	1.49	2.00	0.60	1.45	2.01	0.52	0.99	2.44
Heat Capacity, cal cm^{-3} °C^{-1}	0.42	0.46	0.57	0.42	0.46	0.57	0.35	0.36	0.48	0.46	0.35	0.43	0.49	0.47	0.56	0.41	0.44	0.54
OM, %	100	5	0.1	100	5	0.1	100	5	1	100	1	5	100	5	1	100	15	2
Coarse Fragments %	-	10	20	-	10	20	-	20	25	-	4	14	-	10	10	-	10	20
Sand, %	-	40	45	-	40	45	-	90	60	-	90	80	-	30	60	-	60	60
Clay, %	-	20	15	-	20	15	-	10	20	-	10	10	-	20	10	-	20	20
Ksat, cm hr^{-1} (Eq. 8)	0.30	41.50	6.65	0.30	41.50	6.65	0.30	150.00	21.70	0.30	105.60	20.53	0.30	39.80	0.10	0.30	45.67	0.58
Ksat, cm hr^{-1} (Eq. 8 x Table 3 multipliers)	0.30	20.75	3.33	0.30	41.50	6.65	0.60	300.00	43.40	0.30	52.80	10.27	0.30	39.80	0.10	0.30	45.67	0.58
Ksat, cm hr^{-1} (References)	-	-	-	-	-	-	-	-	-	-	56.63(1.8)[a]	17.07(3.1)[a]	-	-	-	-	-	-
Porosity	0.76	0.47	0.35	0.76	0.47	0.35	0.95	0.64	0.50	0.88	0.56	0.39	0.88	0.56	0.28	0.88	0.67	0.27
Porosity (References)	-	-	-	-	-	-	-	-	-	-	0.65(8.3)[a]	0.4(10.9)[a]	-	0.67[b]	0.30[b]	-	-	-

[a] Barry et al. (1990) [b] Murray and Buttle (2005)

Table 4. Hydrothermal soil profile, needed for the daily soil moisture, temperature and stream discharge calculations (Balland et al. 2008).

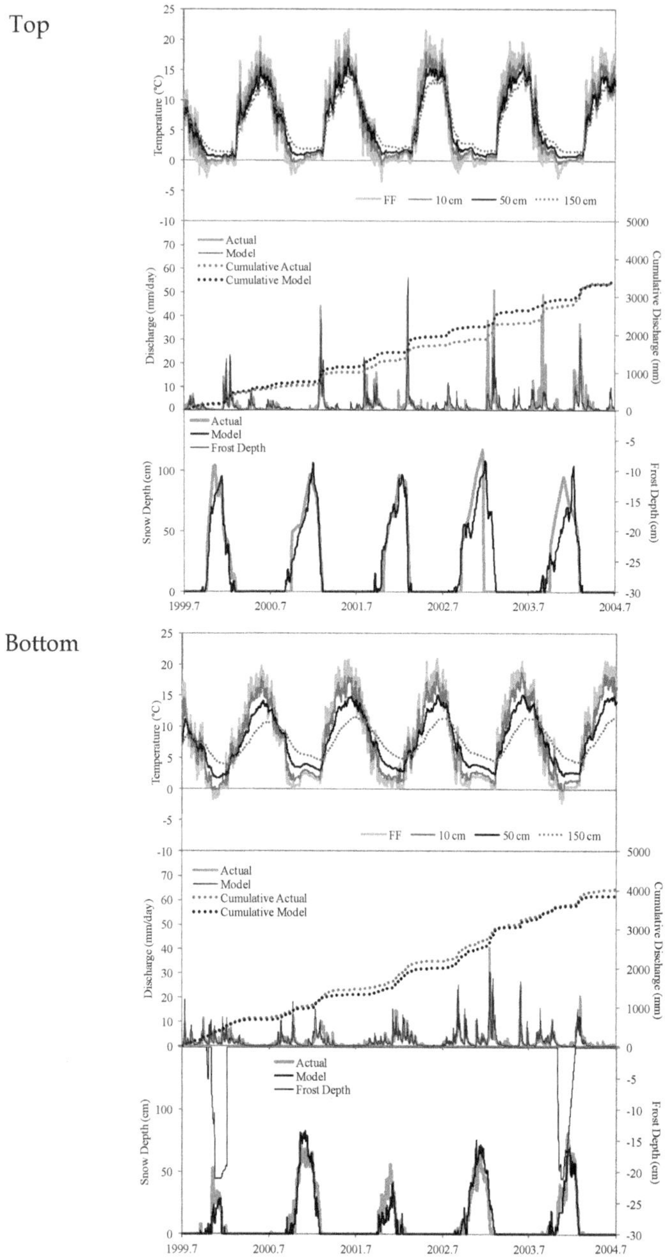

Fig. 4. ForHyM generated output for snow-on-the ground, frost depth, soil temperature, and stream discharge (daily as well as cumulative) within the forested basins at Turkey Lakes, Ontario (top) and for the Moosepit Brook basin in Nova Scotia (bottom). Basin details: Tables 3 and 4.

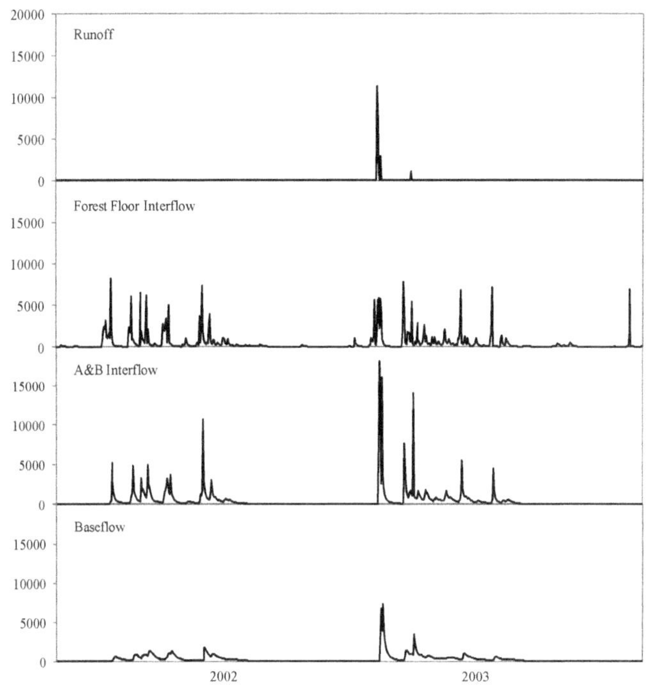

Fig. 5. ForHyM generated output for daily runoff, interflow (forest floor, A&B layers) and baseflow (in mm), for the Rithet River, BC. Basin details: Tables 3 and 4.

Site	Year		Month		Week		Day	
	R^2	β	R^2	β	R^2	β	R^2	β
Moosepit Brook	0.85	1.03	0.90	1.01	0.78	0.96	0.74	0.93
Peggy Brook	0.94	1.45	0.85	1.02	0.72	0.87	0.65	0.90
Hayward Brook	0.63	1.02	0.85	1.05	0.58	0.86	0.50	0.76
Lac Laflamme	0.81	0.95	0.55	0.85	0.55	0.74	0.48	0.67
Turkey Lakes	0.89	0.99	0.80	0.86	0.61	0.77	0.51	0.74
Rithet River	0.88	1.11	0.95	1.00	0.82	0.93	0.60	1.04

Table 5. Comparing ForHyM-modeled with measured daily, weekly, monthly and annual cumulative discharge: coefficient of determination (R^2) and linear regression coefficient (β ; intercept = 0).

4. Concluding remarks

The Ksat, SP, FC, PWP values generated from layer-specific values for Db, Dp, Sand, Clay, organic matter and soil depth produce reasonable results for the extent of water retention and flow rates for run-off, infiltration, interflow, percolation, baseflow and stream discharge across each of the six catchment areas of this study, with the layer-specific Ksat calibrations remaining within a factor of two of the generically derived Ksat estimates. For the well-drained watersheds at Turkey Lakes (Ontario), Lac Laflamme (Quebec), and Rithet River (Vancouver Island, British Columbia), however, extra downward adjustments for interflow were needed, likely due to the combined effects of (i) high slope heterogeneities, thereby leading to a slow-down of lateral flows, and (ii) the close proximity of glacially fractured bedrock beneath the regolith along ridges, thereby encouraging deep percolation instead of lateral flow. For the calcareous substrate of the Hayward Brook watershed in New Brunswick, the downward Ksat adjustments for interflow are likely due to the greater porosity of the calcareous shales, which - in turn - required upward Ksat adjustments for soil and subsoil percolation. Similar adjustments would have to be made for agricultural areas where the flow rates would be accelerated by drainage tiles and ditches. Additional Ksat adjustments would be needed where soil bulk density (Db) changes on account of surface and sub-surface compaction, weather-induced shrinking and swelling, and freezing and thawing. In conclusion, the process of:

i. estimating Ksat, FC, PWP and soil porosity from soil survey data for soil depth, texture and organic matter,
ii. using these estimates as initial values for modelling the daily changes in the hydrothermal conditions and flows through of forest catchments, and
iii. subsequently calibrating Ksat to improve the run-off, infiltration, percolation, interflow and base flow calculations

generated good agreements between modelled and monitored stream discharge for the six forest catchments of this study at the daily level, year-round. The Ksat adjustments required to do so generally remained within a factor of 2 for the downward flow components. Additional adjustments were required for the catchments on steep and calcareous terrains.

5. Acknowledgements

This research was supported by the NSERC Discovery Program, by an Environment Canada grant in support of the continuing development of the Forest Hydrology Model ForHyM (c/o T. Clair), and by NSERC's Sustainable Forest Management Network on modelling and mapping hydrologically sensitive areas.

6. References

Balland, V. 2002. Hydrogeologic watershed modelling with special focus on snow accumulation and snowmelt including retention and release of major ions. MScF thesis, University of New Brunswick, 175 p.

Balland, V. and Arp, P.A., 2005. Modelling soil thermal conductivities over a wide range of conditions. J. Eng. Env. Sci. 4, 549-558.

Balland, V, Bhatti, J. S., Errington,R., Castonguay, M., Arp, P.A., 2006. Modelling soil temperature and moisture regimes in a jack pine, black spruce and aspen forest stand in central Saskatchewan (BOREAS SSA). Can. J. Soil Sci. 86, 203–217.

Balland, V., Pollacco, J. A. P. and Arp, P. A. 2008. Modeling soil hydraulic properties for a wide range of soil conditions. Ecol. Model. 219, 300–313.

Barry R., Plamondon A.P., Stein J., 1987. Hydrologic soil properties and application of a soil moisture model in a balsam fir forest. Can. J. For. Res. 18, 427-434.

Barry R., Prévost M., Stein J., and Plamondon A., 1990. Simulation of snowmelt runoff pathways on the Lac Laflamme watershed. J. Hydrol. 113: 103-121

Beall, F. D., Semkin R. G., Jeffries D. S., 2001. Trends in the output of first-order basins at Turkey Lakes watershed, 1982-96. Ecosystems 4, 514-526.

Bobba A.G., Kam D.C.L., Jeffries D.S., Bottomley D., 1986. Modelling the hydrological regimes in acidified watershed. Water, Air and Soil Pollution. 31, 155-163.

Bouma, J., 1989. Using soil survey data for quantitative land evaluation. Adv. Soil Sci. 9, 177–213.

Børgesen, C. D., Schaap, M.G., 2005. Point and parameter pedotransfer functions for water retention predictions for Danish soils. Geoderma 127, 154– 167.

Børgesen, C. D., Iversen, B.V., Jacobsen, O.H., Schaap, M.G., 2007.Pedotransfer functions estimating soil hydraulic properties using different soil parameters. Hydrol. Process. DOI: 10.1002/hyp.6731.

CANSIS, 2000. Canadian Soil Information System. Available from http://sis.agr.gc.ca/cansis/

Chi X. 2008. Hydrogeological assessment of stream water in forested watersheds: temperature, dissolved oxygen, pH, and electrical conductivity. MSc thesis. University of New Brunswick, Federicton. 161p.

Church, M.R. 1997. Hydrochemistry of forested catchments. Annu. Rev. Earth Planet. Sci. 25, 23–59.

Di Frederico, V., Tartakosky, D.M., 2000. Effective hydraulic conductivity in multiscale random fields with truncated power variograms. Geol. Soc. Am. 348, 80-89.

Gijsman, A., Jagtap, S.S., Jones, J. W., 2003. Wading through a swamp of complete confusion: how to choose a method for estimating soil water retention parameters for crop models. Eur. J. Agronomy 18, 75-105.

Jutras, M.-F., Nasr, M., Castonguay, M., Pit, C., Pomeroy, J., Smith, T.P., Zhang, C.-F., Ritchie, C.D., Meng, F.-R., Clair, T.A., Arp, P.A. 2011. Dissolved organic carbon concentrations and fluxes in forest catchments and streams: DOC-3 model. Ecol. Model. in print.

Kim, K.W., Chung, I.M., Won, Y.S., Arnold, J.G. 2008. Development and application of the integrated SWAT-MODFLOW model. J. Hydrol. 356, 1-16.

Kuchment, L.S., Demidov, V.N., Startseva, Z.P. 2006. Coupled modelling of the hydrological and carbon cycles in the soil-vegetation-atmosphere system. J. Hydrol. 323, 4-21.

Laudon, H., Sjoblom, V., Buffam, I., Seibert, J., Morth, M. 2007. The role of catchment scale and landscape characteristics for runoff generation of boreal streams. J. Hydrol. 344, 198– 209.

Leij, F.J., Alves, W.J., van Genuchten, M.Th., Williams, J.R., 1996. The UNSODA Unsaturated Soil Hydraulic Database; User's Manual, Version 1.0. EPA/600/R-96/095, National Risk Management Laboratory, Office of Research and Development, U.S. Environmental Protection Agency, Cincinnati, OH. http://www.ussl.ars.usda.gov/models/unsoda.HTM

Murphy P.N.C., Castonguay M., Ogilvie J., Nasr M., Hazlett P., Bhatti J., Arp P.A., 2009. A geospatial and temporal framework for modelling gaseous N and other N losses from forest soils and basins, with application to the Turkey Lakes watershed project, in Ontario Canada. For. Ecol. Management 258, 2304-2317.

Murray, C.D., Buttle. J.M. 2005. Infiltration and soil water mixing on forested and harvested slopes during spring snowmelt, Turkey Lakes Watershed, central Ontario J. Hydrol. 306, 1–20.

Nemes, A., Schaap, M. Wösten, H., 2003. Functional evaluation of pedotransfer functions derived from different scales of data collection. Soil Sci. Soc. Am. J. 67, 1093-1102.

Pachepsky, Y., Rawls, W.L. 2005. Development of pedotransfer functions in soil hydrology. Developments in Soil Science, Vol. 30. Elsevier Science. 542 p.

Paquet, J.M., Caron, J., Banton, O., 1993. In-situ determination of the water desorption characteristics of peat substrates. Canadian J. Soil Sci. 73, 329–339.

Pepin, S., Plamondon, A., Stein, J., 1992. Peat water content measurement using time domain reflectometry. Canadian J. Forest Res. 22, 534–540.

Schaap, M.G., 2006. Models for indirect estimation of soil hydraulic properties. Encyclopedia of Hydrological Sciences. DOI: 10.1002/0470848944.hsa078

Schulze-Makuch D., Carlson D., Cherkauer D., Malik P., 1999. Scale dependency of hydraulic conductivity in heterogenous media. Ground Water 37, 904-919.

Sudicky, E.A., Illman, W.A. Goltz, I.K., Adams, J.J., McLaren, R.G. 2010. Heterogeneity in hydraulic conductivity and its role on the macroscale transport of a solute plume: From measurements to a practical application of stochastic flow and transport theory. Water Resources Research, 46, W01508, pp. 16; doi:10.1029/2008WR007558

Stein J., Proulx S., Lévesque D., 1994. Forest floor frost dynamics during spring snowmelt in a boreal forested basin. Water Resources Research 4, 995-1007.

Vega-Nieva, D. J., Castonguay, M., Ogilvie, J., Arp, P. A., 2009. Development of a modular terrain model to estimate daily variations in machine-specific forest soil trafficability, year-round. Can. J. Soil Sci. 89, 93-109.

Zhang, Y., Gable, C.W., Person, M. 2006. Equivalent hydraulic conductivity of an experimental stratigraphy: Implications for basin-scale flow simulations. Water Resources Research, 42, W05404, pp. 19; doi:10.1029/2005WR004720

9

Determination of Hydraulic Conductivity Based on (Soil) - Moisture Content of Fine Grained Soils

Rainer Schuhmann, Franz Königer, Katja Emmerich, Eduard Stefanescu and Markus Stacheder
Karlsruhe Institute of Technology (KIT)
Competence Centre for Material Moisture (CMM)
Karlsruhe
Germany

1. Introduction

The chapter will be divided into the subchapter material, processes and systems. The first one will focus on the physical, chemical and dynamic material properties and their measuring methods. The second specifies the dynamic of the surface moistening and fluid flow. The comprehensive characterization of materials is prerequisite to understand processes in large (geo)-technical systems and their manipulation. The transition from nano (material) via meso (processes) to macro scale (systems) will be illustrated with an example in the third chapter.

2. Materials

The properties of fine grained soils such as silt or clay considerably influence the migration of water. Especially their small pore sizes, their platy habit, and their high specific surface area generally lead to very low hydraulic conductivities. Therefore, it is indispensable to accurately determine these properties for a reliable assessment of the hydraulic conductivity. Fine-grained soils are soils with a grain size distribution ranging from 0,0002 to 0,2 mm, i.e. soil textures from clay and silt up to fine sand. The hydraulic conductivity, K_f, of these materials is generally smaller than 10^{-4} ms^{-1}. In the following the most important physical, chemical and dynamical properties of fine grained materials that affect the hydraulic conductivity will be explained in detail and their measurement methodologies will be illustrated.

2.1 Material properties

Generally the hydraulic conductivity depends on the soil matrix, the type of the soil fluid (density and viscosity), and the relative amount of soil fluid (saturation) present in the soil matrix. In this chapter we will focus on the important properties relevant to the solid matrix of fine grained soils which include the texture and fabric and its mineral phase content.

2.1.1 Physical material properties

2.1.1.1 Soil density

In soil science one can distinguish between the soil particle density ρ_p which is the mass of the soil particle m_p per volume of the particle V_p and the so called bulk density which is the ratio of the mass to the bulk volume V of a given amount of soil. For the latter two different definitions depending on the scientific discipline or the application are possible, i.e. the wet bulk density ρ_{wb} which is the soil mass m_s, plus the mass of the including water m_w, per unit volume V and its dry bulk density ρ_{db} which is the mass of oven-dry soil per unit volume of moist soil.

$$\rho_p = \frac{m_p}{V_p} \tag{1}$$

$$\rho_{wb} = \frac{m_s+m_w}{V} \tag{2}$$

$$\rho_{db} = \frac{m_s}{V} \tag{3}$$

2.1.1.2 Soil texture (grain size and grain size distribution)

Mineral soils are mostly classified according to grain size and grain-size distribution (grain-size fractions and grading) also generally summarized as the soil texture. A major factor that influences the rate of water flow through soils is the size of the particles. It can generally be stated that the smaller the particles, the smaller the voids and the stronger the flow resistance. This explains the very low hydraulic conductivity of fine-grained materials and their preferred use as sealing elements for example in buffers and backfills of underground storage facilities or landfills. The grain-size distribution gives a good picture of the content of clay minerals since they appear almost entirely in the clay fraction (< 2 μm). The compaction properties of the material can be estimated by the character of the entire grain size curve, from which one can conclude, if the material has very low hydraulic conductivities and can be used for water sealing purposes (Pusch, 2002).

Whereas the grain-size distribution of the coarse fractions (gravel and sand) can be determined by sieving, the fine fractions (silt and clay) must be determined by sedimentation of the dispersed soil.

2.1.1.3 Soil structure (porosity and pore size distribution, geometry and shape of the pores, tortuosity)

The term structure of soils generally refers to the pore geometry. The pore volume or porosity is a measure of the void spaces in the soil and is a fraction of the volume of voids over the total volume, between 0-1 or as a percentage between 0-100 %.

$$\Phi = \frac{V_V}{V} \tag{4}$$

where V_v is the volume of void-space and V is the total or bulk volume.

Porosity of surface soil typically decreases as particle size increases. This is due to soil aggregate formation in finer textured surface soils when subject to soil biological processes. Aggregation involves particulate adhesion and higher resistance to compaction. The transport of water, solutes, and gases in soil is not only influenced by the absolute size of the pore volume but also on the nature of how the pores are connected which is summarized under the term tortuosity. In soil the tortuosity is closely related to soil surface area and the pore-size distribution. Both porosity and tortuosity of fine-grained soils are considerably small due to the plat-like shape of the particles.

2.1.2 Chemical material properties

2.1.2.1 Composition of the mineral phases

Minerals in natural soils originate from degraded rock, most of them belonging to the silicates, sulphates, sulphides, and carbonates. Especially the clay minerals show a platy habit leading to a very high surface area to mass ratio and have a considerable influence on the hydraulic conductivity. The distinction of minerals is mostly based on crystal structure and chemistry. Their crystal structure is responsible for a number of their characteristic chemical properties such as cation exchange capacity or high sorption capacity (Pusch, 2002).

Hydraulic conductivity of swellable clayey and clay-enriched silty soils strongly depends on the density, the type of the adsorbed ions and the salinity of the percolating water (Scheffer, 1992). For example it is the high swelling properties that provide sodium bentonite's unique sealing qualities. As the clay hydrates and swells, the path for water to flow through becomes complex as the clay platelets intersperse. The large fraction of interlamellar, immobile water in smectites yields a much lower hydraulic conductivity than of soils with other minerals at any bulk conductivity. Thus, clays with micas, illites, and kaolinites as major minerals are about 100 to 100000 times more conductive than smectite in montmorillonite form at one and the same void ratio (Pusch, 2002). (See figure1)

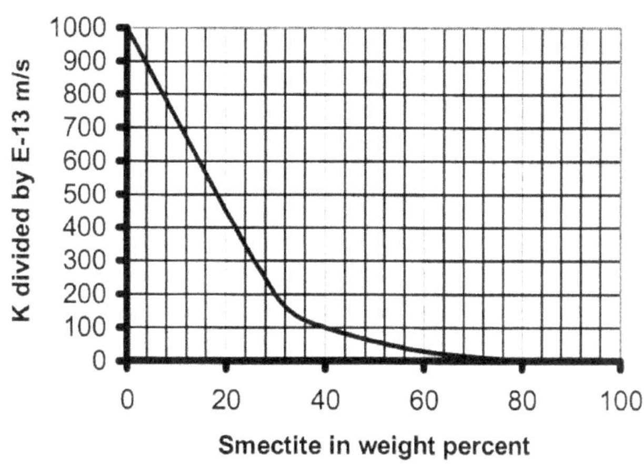

Fig. 1. Approximate relationship between smectite content and hydraulic conductivity (Pusch, 2002).

The type of absorbed cations of the clay is also important with respect to the hydraulic conductivity since bi- and polyvalent cations cause growth in the stack thickness and size, which means that the voids between the stacks of lamellae and thus the hydraulic conductivity are bigger in the Ca than in the Na montmorillonite (Pusch, 2001). Also the electrolyte concentration has a substantial influence on the interparticle distance at low and moderate densities and thus on the hydraulic conductivity because the stacks of lamellae that form a network with rather much space will coagulate at high electrolyte concentrations.

2.1.2.2 Specific soil surface

The magnitude of the specific surface area of a soil depends largely on the amount of clay and type of clay minerals in the soil. The specific surface area differs largely between types of clay minerals (Table 1). Especially in fine-grained materials one can distinguish between an external and an internal surface, the latter being the interlamellar space of the minerals. The total specific surface area is a factor that can relate grain-scale properties to macro-scale physical and chemical properties of a porous medium. Large specific surface areas lead to much interaction of ions and water molecules with the soil particles. Therefore the total surface determines many physical and chemical properties of the soils (Petersen et al. 1996).
In porous media filtration theory, a nonlinear relationship between saturated hydraulic conductivity and surface area has been established for structureless, randomly packed, non-compressible particles (Kozeny, 1927; Carman, 1937; Grace, 1953), the so-called Kozeny-Carman relationship.

Mineral	Specific surface area, m²/g	
	External	Internal
Kaolinite	10–15	10–15
Illite	100–150	100–150
Smectites	100–150	800–1000

Table 1. Typical specific soil surface area data of clay minerals (Pusch, 2001).

2.1.3 Dynamical material properties

2.1.3.1 Matric potential

Matric potential refers to the tenacity with which water is held by the soil matrix and, in the absence of high concentrations of solutes, is the major factor that determines the availability of water to plants. Differences in the value of matric potential between different parts of the soil also provide the driving force for the unsaturated flow of soil water after any differences in elevation have been allowed for (Mullins, 1991).
The total potential ψ_t of soil water refers to the potential energy of water in the soil with respect to a defined reference state and can be divided into three components:

$$\psi_t = \psi_p + \psi_g + \psi_o \tag{5}$$

with ψ_p as pressure potential, ψ_g as gravitational potential and ψ_o as osmotic potential.
The sum of gravitational and pressure potential is called the hydraulic potential ψ_h. Differences between the hydraulic potential at different places in the soil provide the driving force for the movement of soil water. Matric potential ψ_m is a subcomponent of pressure potential and is defined as the value of ψ_p where there is no difference between the pressure of air or gas in the soil and the gas pressure on the water in the reference state (Mullins, 1991).

The relationship between water content and soil water potential (capillary tension) is described in the water retention curve. The curve is characteristic for different types of soils and is also called soil moisture characteristic. It also depends on the geometry and network of the pores. Fine-grained soils show high residual water contents and high changes in capillary tension are necessary that they release the water.
The retention curve shows a hysteresis which means that depending on the history of the soil with regard to watering and drainage, the shape of the curve is different, a behaviour which is explained by the 'ink-bottle model' (Schuhmann, 2002).
For the analytical description of the water retention curve there are different approaches. Brooks and Corey (1964) combine the mathematical θ/ψ–relationship with the conductivity model of Burdine (1953), but the most common approach is the one from van Genuchten (1980) combined with the conductivity model of Mualem (1976a) which allows a direct determination of the hydraulic conductivity by numerical simulation. Up to now it was state of the art to determine the capillary tension by tensiometers and to determine the moisture with the θ/ψ–curve, yet the influence of the hysteresis does not allow distinct results. Therefore we propose other soil moisture measurement methods to derive soil hydraulic conductivity.

2.1.3.2 Moisture content

As with density there exist several different definitions of water content or moisture. The gravimetric water content can, similarly to the bulk density, be expressed on a dry mass, θ_{db}, or wet mass basis, θ_{wb}, and gives the ratio between the mass of the porewater and the mass of the dry solid substances resp. the solid substance plus mass of water. It can be expressed in percent units

$$\theta_{db} = \frac{m_w}{m_s} * 100 \tag{6}$$

$$\theta_{wb} = \frac{m_w}{m_s + m_w} * 100 \tag{7}$$

The volumetric water content can also be expressed on a volume basis as the ratio between the volume of water and the total volume of the soil sample:

$$\theta_{vb} = \frac{V_w}{V_t} * 100 \tag{8}$$

Since the soil water characteristic from the θ/ψ–curve especially for fine-grained soils is ambiguous, its transformation into a ψ/k_f-curve for the determination of the hydraulic conductivity is ambiguous too. Therefore, instead of the capillary tension, in the following the moisture content will be used as the relevant parameter for a more precise determination of the hydraulic conductivity.
For fine-grained soil samples, the wide distribution of void size means that the various pore water components play different roles. Determination of their relative amounts requires heating to different temperatures. The water in the large voids is lost at about 100 °C, the water in the fine capillaries at 105 °C or slightly more, while the hydration shell of interlayer cations in swellable clay minerals is lost at temperatures appreciably higher than 105 °C.
Although the determination of the water content by thermal analysis is a very accurate method and mostly used as a reference, this method is destructive and non-operational. But fortunately the range of possible soil moisture determination methods has increased

considerably since the beginning of the eighties of the last century. Especially the electromagnetic moisture measurement methods are promising new techniques for a reliable and in-situ determination of the soil moisture and thus of the soil moisture/hydraulic conductivity-relationship in the field. One of these new techniques will be presented in the following chapter.

2.2 Measuring methods

2.2.1 Physical methods

2.2.1.1 Bulk density

The methods available for the measurement of soil bulk density fall into two groups. In the first are the long established direct methods, which involve measurement of the sample mass and volume (core sampling, rubber ballon, sand replacement, clod). In the second group the attenuation or scattering of nuclear radiation by soil is used in conjunction with a calibration relationship to give an indirect measurement of bulk density (Mullins, 1991).

2.2.1.2 Soil texture (particle size distribution, grading curve)

The particle size distribution analysis is one of the most principle determinations in soil science and its knowledge already allows relatively good estimations of soil hydraulic properties. Grain size distribution is very important for the bulk density.
Principally with particle size distribution analysis there exist two problems: (1) from the wide range of particle sizes it follows that the analysis cannot be carried out by one single method alone and (2) since the particles show different stabilities it is nearly impossible to exactly distinguish between primary particles and aggregates, the latter being valid especially for fine grained materials such as clay and silt. Therefore a chemical and/or physical pre-treatment of the soil sample is indispensable to minimize aggregation of particles. The separation of the different particle sizes is carried out be sieving with exactly defined mesh sizes. The fine grained fraction, i.e. particle ranging from <63 μm to 2 μm, are normally separated by so called sedimentation analysis, using the different sinking properties of the grains in liquids (Stokes' law). This law is applied in several methods like aerometer according to Casagrande, Andreasen Pipete, Koehn Pipete, Atterberg or Kopecky, that yield the so called grading curve. This method reaches its limitations when the clay breaks up into fine, medium, and coarse clay. For determination of the clay fractions sedimentation is speed up by centrifugation applying several times of earth gravitational force.
All these methods are time consuming and determine size distribution of spherical equivalent particles as Stokes' law is based on the assumption of spherical particles. Faster methods of particle size distribution apply laser light scattering, X-ray absorption, acoustic spectrometry or dynamic light scattering. A systematic comparison of advantages and disadvantages in relation to particle size and particle shape was given by Latief (2010). The shape of platy clay minerals has a strong influence on their sedimentation behaviour and thus influences the determined equivalent particle size. Therefore, some methods allow introduction of a shape factor (e.g. Konert & Vanderberge, 1997) for calculation of particle size distribution.

2.2.1.3 Soil structure (pore distribution)

Beside the calculation of the porosity Φ from the density values, it can also be measured directly by an air pycnometer which is based on the Boyle-Mariotte law ($p_1V_1=p_2V_2$).

The relationship between the decrease of pressure (from p_1 to p_2) and the pore volume in the sample must be taken from calibration curves. This yields the air-filled pore volume V_A. To extract the total volume V, an additional water content determination to determine the volume of water V_W in the sample is necessary.

$$\Phi = \frac{(V_A + V_W)}{V} \tag{9}$$

2.2.1.4 Specific surface area

The determination of the specific surface area helps to identify expandable minerals and to estimate the degree of cementation of expandable clays. The measurement is carried out by determining the external and internal surface areas. The external surface area is measured by the adsorption of non-polar gases like nitrogen which cannot enter the interlamellar space. The internal surface is determined by measuring the total surface using water and subtracting from it the separately determined external surface (Pusch, 2002).

Water absorption capacity according to Enslin/Neff describes the property of soil at 60 °C dried up to weight constancy to absorb water by capillary action and to retain it. It mainly depends on content and type of the clay minerals. The Enslin-Neff values are known to be determined by the amount of exchangeable Na^+ and to a lesser extent by the smectite content (Neff, 1959). Hence the Enslin-Neff method provides an indirect estimate of the Na^+ content. Evaporation during the test has to be restrained as it is known to have a serious affect on the water-uptake capacity values measured by the Enslin-Neff device (Kaufhold & Dohrmann, 2008). Nevertheless, the test provides an index value which gives some indication of the size of the specific surface and activity of the fine grain.

2.2.2 Chemical and mineralogical methods

2.2.2.1 Phase analysis

Identification and quantification of mineral phases is performed by X-ray diffraction analysis and analysis of X-ray diffraction patterns from powdered samples by Rietveld method or pattern summation methods supported by complementary analyses (e.g. CEC, XRF, STA) (Omotoso et al., 2006, Emmerich, 2011).

- **X-Ray Diffraction Analysis**

X-ray diffractometry (XRD) is the standard analysis in mineralogy providing rapid information on clays and non-clay minerals present in a powder sample. An X-ray beam of defined wavelength (e.g. Cu Kα) is diffracted at the lattice of each mineral. According to Bragg's law the diffracted beam is commonly recorded at angles between 2-80° 2θ. The intensity ratio of peaks for each mineral depends on many factors (like chemical composition, preferred orientation and others, see e.g. Moore & Reynolds, 1997). Identification of the phases is made according to the JCPDS register (International Centre for Diffraction Data; JCPDS 1983) incorporated in commercial software. Identification of clay minerals often requires additional XRD analysis of so-called textured samples. The sample is slurried by appropriate chemical and/or physical treatment and the >2 µm fraction separated in a centrifuge and discarded. The remaining suspension (< 2 µm) is placed as drops on a slide or sucked on top of a ceramic disc whereby the clay mineral orientates during the sedimentation and drying process more or less on a parallel basis (texture) (Moore & Reynolds, 1997). Subsequent solvation with ethylene glycol will identify the swellable clay mineral phases. Calcinating at 550°C destroys the existing kaolinite through dehydroxilation allowing chlorite to be identified (http://pubs.usgs.gov/of/2001/of01-041/htmldocs/flow/index.htm).

- **Cation Exchange Capacity (Methylene Blue and Cu-Trien methods)**

Determination of cation exchange capacity (CEC) is performed by homoionic saturation of exchangeable positions by an Index cation (Dohrmann & Kaufhold, 2010) and indicates the amount of smectites in fine grained materials. Common index cations are methylene blue (MB), Ba^{2+}, NH_4^+ or copper triethylenetetramine(Cu-Trien). Despite MB is widely used (VDG P 69 1988) the method suffers certain restrictions that are discussed in detail by Kahr and Madsen (1995). If the CEC is determined by MB of a sodium exchanged material at neutral pH it would result in similar values as determined with the ammonium (NH_4^+) acetate method. MB can even be used for determination of SA if montmorillonite surface area per charge corresponds with the area of the MB of 130 $Å^2$, i.e. the interlayer charge of the montmorillonites must amount to 0.28-0.33 charges per half unit cell.

Occupation of exchangeable positions in natural state of clay containing is of particular significance for water adsorption, swelling and resulting microstructure and thus permeability of clays especially bentonites. Exchanged cations can be determined by AAS or ICP-OES from supernatant of CEC determination if precaution is taken to prevent dissolution of soluble minerals (Dohrmann & Kaufhold, 2010). According to Müller-Vonmoos & Kahr (1982) the cations bound to the exterior and interior surfaces of montmorillonite can be exchanged for an alcoholic ammonium rhodanide solution in the presence of soluble salts, too. The ion distribution of exchangeable bound ions (Ca^{2+}, Mg^{2+}, Na^+, K^+) is of particular significance for water absorption and swell capacity, the microstructure and thus the water permeability of bentonite.

- **Swell Capacity Method**

The ASTM D 5890 method id used to determine the swell capacity of bentonite. Two grams of dried (105°C) pulverised sample is placed in 90 ml of de-ionised water in standard 100 ml glass cylinders. Of the sample, 0.1 g at a time is sprinkled into the water within 30 seconds until the whole amount has been used up. The cylinder is filled to 100 ml with de-ionised water and temperature is measured. After 24 h, the swell volume is determined in ml ignoring any flocculated material.

- **Fluid Loss Test**

The fluid loss test is a very reliable test regarding the permeability behaviour of bentonites. It enables the evaluation of fluid loss properties of a clay mineral film deposited on a filter paper from a 6% solids slurry of clay mineral at 100 psi (690 kPa) pressure as a measure of its usefulness for the permeability of hydraulic conductivity reduction. This method was adapted from American Petroleum Institute drilling fluid specifications for bentonite.

2.2.3 Dynamical methods

2.2.3.1 Matric potential

Namely the tensiometer and the pressure-membrane (or pressure-plate) apparatus, that either measure, or generate, the matric potential as defined above are used. These instruments measure the difference in pressure across a semipermeable 'membrane' in contact with the soil on one side and the soil solution (i.e. the equilibrium dialysate) on the other. The 'membrane' is permeable to solution but not to solids (Passioura 1980).

Principally the tensiometers are suitable for the determination of the hydraulic conductivity in the field and a measurement precision of ± 1.0 vol.-% is principally sufficient, yet there can occur big uncertainties when transforming the capillary tension into moisture or hydraulic conductivity due to described hysteresis effect. Also maintenance and calibration of the tensiometers especially in fine-grained soils are quite complex since they tend to run dry very quickly due to the high water suction of fine-grained soils. Also the measurements are rather punctual. However they can be used as a reference method for our purpose.

2.2.3.2 Moisture content

Since moisture content is a decisive criterion of many porous materials, there exists a broad variety of different methods. Generally one can distinguish between direct methods, where the moisture is determined directly by physical or chemical methods, or indirectly by determining a property that is mainly a function of the water content.

The most common direct method is the thermogravimetric method, where a sample of the soil is dried at 105°C to a constant weight. Other direct methods use e.g. calcium carbide, sulphuric acid, or phosphorus pentoxide, which react with the water present in the sample. Because all these methods require sampling and considerable laboratory equipment, their use for soil moisture measurement in the field is not very practicable. Here the indirect methods are more common and more differentiated using mostly physical parameters such as electrical, radiometric, acoustic or thermal soil properties. A good survey is given in Schmugge et al. (1980).

In this chapter we will focus especially on the electromagnetic methods that use the dielectric properties of the soil. The principle is based on a functional relationship between the dielectric permittivity of the soil and its volumetric water content. Different methods take advantage of the high relative permittivity of water (ε_r = 80) compared to that of dry soil (ε_r = 3-5). One of the most well known meanwhile is the Time Domain Reflectometry (TDR) (Topp et al. 1980), where the transit time t of an electromagnetic pulse on a wave guide of length l, which is buried in the soil, is measured. The relative permittivity ε_r is determined according to:

$$\varepsilon_r = \left(\frac{c_0 t}{2l}\right)^2 \tag{10}$$

with c_0 as the velocity of light in free space. Relating the measured ε_r of different soil samples to the volumetric water content determined by the thermogravimetric method, allows to establish a so called calibration function. One of the most well-known calibration functions is the Topp-polynomial (Topp et al. 1980) which yields the volumetric water content θ_v according to:

$$\theta_v = -5.3 * 10^{-2} + 2.92 * 10^{-2}\varepsilon_r - 5.5 * 10^{-4}\varepsilon_r^2 + 4.3 * 10^{-6}\varepsilon_r^3 \tag{11}$$

Conventional TDR-sensors are normally fork-like metallic wave guides of several tens of centimetres that penetrate the soil, giving a rather punctual measurement. Yet the determination of hydraulic conductivity on a field-scale basis based on soil moisture measurement requires more large-scale sensors why a flat-band-like TDR-cable sensor called TAUPE has been developed (Brandelik & Huebner, 1999). Due to a plastic coating of the copper wave guides this sensor is capable of sensing up to 30 m of the surrounding soil. Both an integral soil moisture value and a moisture profiling along the length of the buried cable according to a new TDR inversion technique can be accomplished (Schlaeger, 2005).

3. Processes

3.1 Dynamic of the surface moistening

The moistening of surfaces obeys certain natural laws which were established during the last 200 years. The development was started in the beginning of the 19th century by several scientists from the fields of physics and chemistry. In this chapter we will exemplify some important laws with respect to surface moistening.

3.1.1 Laws

3.1.1.1 Young-laplace equation

In 1805 Thomas Young and Pierre-Simon Laplace both described independently from one another a fundamental equation with respect to interface science. The Young-Laplace equation describes the correlation between surface tension, pressure and surface curvature of a system consisting of two phases. Such a system could be e.g. a liquid drop on a solid surface or a liquid in another immiscible liquid. Surface tension of a liquid results from attractive interaction of the liquid molecules. A molecule located within a liquid is surrounded by other molecules, so the resultant force is zero. This does not apply to a molecule at the surface, since a part of interaction is missing at this place. The molecule is bordered by air molecules on the upper side and these intermolecular forces are of weak nature. This leads to an inward looking force. The energy required to overcome this force is the surface tension, sometimes also called surface energy.

3.1.1.2 Lucas-Washburn & modified Lucas-Washburn

The predefined aim of the studies of Lucas (1918) and Washburn (1921) was to develop a theoretically established law, which determined the capillary head existing in an arbitrary capillary system, as a function of time. The first approach was to immerse a wettable cylindrical tube vertically into a solution. The surface tension of the liquid becomes noticeable as the length of the cross section (2*r*п) multiplied by the surface tension (σ), perpendicular to the direction of the tube. This force elevates the liquid to a height where it is equilibrated by the gravity.

$$2r\pi\sigma = r^2\pi m_s h_0 \tag{12}$$

where h_0 is the maximum pressure head and m_s the specific mass. Thus the maximum height entirely depends on the surface tension, on the radius of the tube and on the specific mass of the solution measured. The penetration speed of the liquid due to the pulling force diminishes with the height because the mass of the liquid increases. Moreover the rise of the liquid is slower the tougher the liquid is. After the viscosity of the solution has been taken into account (Poiseuille) and assuming that wetting isn't complete, the Lucas-Washburn equation for the capillary rise is

$$h^2 = \frac{r\sigma \cos\theta t}{2\eta} \tag{13}$$

Here θ is the contact angle, t is the time and η is the viscosity. The contact angle is between 0° and 90°, so cos(θ) lies between 0 and 1.

3.1.2 Methods

3.1.2.1 Contact angle measurement

The basis of the contact angle measurement goes back to Thomas Young (1805), who related the contact angle to the surface tension:

$$\cos\theta = \frac{\sigma_s - \sigma_{ls}}{\sigma_l} \tag{14}$$

θ is the contact angle, σ_s the surface free energy, σ_{ls} the solid-liquid surface energy and σ_l is the surface tension of the liquid. We can distinguish between three cases relating to the contact angle. If $\theta < 90°$, the liquid wets the solid surface, if $\theta > 90°$, the sample is hydrophobic and the liquid doesn't wet it or wets it only partially and if $\theta = 0°$, the solid surface is totally wettable, the liquid spreads over the surface.

In practice the liquid drop is put on the solid surface, which has to be as straight as possible, plane and also clean. A light source, which is positioned in the rear lets the drop appear dark. θ can be measured directly using a goniometer or with the help of an optical calculating system which employs the equation of Young-Laplace. The goniometer measuring leads to a relatively large error (± 2%) and is not applicable for small angles and irregular contact lines (Dimitrov et al., 1991). In case of small drops the hydrostatic effects can be neglected and the contact angle can be calculated from the height of the drop (Butt et al., 2006).

3.1.2.2 Dynamic contact angle measurement

The processes happening at the solid-liquid interface during wetting and dewetting are best described by the dynamic contact angle. The interface at the contact between liquid drop and solid surface doesn't appear suddenly, but it needs a certain time until a dynamic equilibrium is reached. In practice the measuring of the dynamic contact angle works in the way that a liquid drop is spread on the solid surface and then extended by means of a needle. The solid-liquid interface migrates outwards and the contact angle can be measured by defining certain degrees steps. Studebaker & Snow (1955) developed an equation for the determination of the dynamic contact angle.

$$\cos\theta_2 = \frac{\sigma_1 \eta_2 t_1}{\sigma_2 \eta_1 t_2} \tag{15}$$

The authors determined dynamic contact angles of powder samples by measuring the time required for a liquid to imbibe the powder bed. This time was then compared to a reference sample with $\cos\theta = 1$ (contact angle = 0°). This method assumes that the differences in the penetration rate are due only to differences in contact angle, after taking surface tension and viscosity into account (Yang & Zografi, 1986).

3.1.2.3 Capillary rise method

Jones & Ray (1937) investigated the determination of the surface tension of water and several salt solutions. They developed a differential method to determine this property of liquids. The experimental set-up of the capillary rise method consists of a tight cylindrical tube and a broad tube being connected with each other. The vertical level difference between the meniscus in the tight tube and the extended one has to be measured. The density of the liquid, which also needed, may be determined directly by the use of a hydrometer.

$$\sigma = \frac{rhg(D-\beta)}{2} \cos \theta \tag{16}$$

r is the radius of the tight tube measured at the height of the meniscus, h the capillary rise, g the acceleration of free fall, D the true density of the liquid, β the density of the gas phase (air plus water vapor) at the temperature and the barometric pressure when the experiment is made and θ is the contact angle. θ should be zero in glass and silica tubes if the tubes are clean (Jones & Ray, 1937). First the elevation between the lowest levels of the menisci must be read off to get the approximate value of the capillary rise. This value has to be corrected for the liquid by means of the Rayleigh formula. Jones & Frizzell (1940) for their part examined the influence of the concentration of the solution on the capillary rise. Therefore they used diluted salt solutions of different concentrations. The most noticeable feature of the results was that the penetration height of water was higher than those of the diluted solutions. This was interpreted as an evidence for a higher surface tension of water compared to the salt solutions. Measurements based on Washburn's equation do not only depend on the particle size but also on the pore size distribution. Addition of fine particles to the measured bed increases the penetration rate of liquid and improves precision of the measurement (Dang-Vu & Hupka, 2005).

3.1.2.4 Wilhelmy-Plate

The Wilhelmy-Plate method is utilized to determine the surface tension of a liquid. It can also be used in order to study the contact angle during capillary rise. In doing so a plate is contacted with the surface of the examined liquid where a meniscus forms at the contact point of the two phases. Due to this meniscus a force between the phases appears which originates from the wetting. By pulling the plate upwards a force (surface tension) manifests itself.

3.1.2.5 Sessile drop

The interface science makes use of different methods with regard to measure both, properties of liquids (e.g. surface tension) and properties of solids (e.g. static/dynamic contact angle or surface energy). The sessile drop method is an example for a measurement on a solid. For this purpose a drop of liquid (in most of the cases a reference solution is used) is spread on a solid surface and the static contact angle of the liquid is measured optically. Bachmann et al. (2000) developed a sessile drop method by modifying Young's equation on two points, since it is strictly applicable only to completely uniform and plain surfaces:

i. A correction factor was introduced, which is defined as the ratio between the actual and the apparent area. This leads to the equation of Wenzel

$$\cos(\theta_{obs}) = r \cos(\theta) \tag{17}$$

Eq. (17) was developed due to the fact that the observed contact angle is smaller than the ideal (intrinsic) angle as long as this is below 90° and larger if the intrinsic angle is above 90°. So, the precision of the contact angle measurement therefore depends on the magnitude of it.

ii. The Cassie-Equation is considered as an empirical approach describing the apparent contact angle on a chemically heterogeneous surface

$$\cos(\theta_{obs}) = f_1 \cos(\theta_1) + f_2 \cos(\theta_2) \tag{18}$$

3.2 Fluid flow

3.2.1 Hydrology

According to the physical law of conservation of mass the equation of water balance is valid. The input precipitation (N) is equal to the sum of the current total evaporation (Haude, *1958*) ET_a (evaporation of soil, transpiration of plants and evaporation of interception), the sum of runoff (surface runoff Q_O, lateral runoff Q_L and leaching Q_V) and change of water content in soil layers within the observation period ($\Delta\theta = W_A - W_E$).

$$N = ET_a + (Q_O + Q_L + Q_V) + (W_A - W_E) \tag{19}$$

The description of water balance in vertical soil profiles is the fundamental requirement for the description of water balance in areas. Numerous investigations on the regionalization of point methods of measurement nowadays provide well-founded transmission options. The following sections consider the water balance of a vertical profile.

3.2.2 Analytical basics

The major focus of the investigation lies on the flow of a solution (in this case water) through the soil matrix. The water flow not only underlies gravity, but is also influenced by the soil properties (Hillel, 1980). Driving forces like gravity (hydrostatic forces), adsorption, cohesion, osmotic forces due to dissolved salts etc. cause water movement through their resultant (Czurda, 1994). Concerning the water movement we distinguish between advective flow and diffusion. Diffusion is irrelevant in the case of materials owing high hydraulic conductivity. The mathematical expression of the potential can be used to describe the flow of water through soil. It should be noted that the theory of the potentials considers neither the geometry of the pore space nor the mechanisms of water binding. These are included in the matric potential. The mechanical energy is taken into account but not the thermal energy. The stationary flow within the unsaturated zone is described by Darcy/Buckingham (equation of motion). Darcy's law features flow by means of a unit volume in dependency of the hydraulic conductivity of soil and in dependency of a potential gradient. The flow velocity depends on the hydraulic conductivity (k_f-value) of the soil and on the total potential (Ψ). k_f depends on the water content θ.

$$v_x = -k_{fx}(\theta) \cdot \frac{\partial \Psi}{\partial x} \quad v_y = -k_{fy}(\theta) \cdot \frac{\partial \Psi}{\partial y} \quad v_z = -k_{fz}(\theta) \cdot \frac{\partial \Psi}{\partial z} \tag{20}$$

k_f does not vary linearly depending on water content, but it follows a relationship which is characteristic to each soil. The hydraulic conductivity decreases with the square of the capillary radius. The air in the soil is considered to be stationary. In case of transient conditions Darcy's law (equation of motion) is combined with the equation of continuity (validity of conservation of mass).

$$\frac{\partial v_x}{\partial x} + \frac{\partial v_y}{\partial y} + \frac{\partial v_z}{\partial z} = -\frac{\partial \theta}{\partial t} - S \tag{21}$$

S represents a term containing a sink or a source. Combining (20) with (21) considers the change in water content during water flow. This equation is known as the partial differential equation of unsaturated flow (unit s^{-1}) within the soil matrix.

$$\frac{\partial \Psi}{\partial t} \cdot C + S = \frac{\partial}{\partial x}\left[k_{fx}(\theta) \cdot \frac{\partial \Psi}{\partial x}\right] + \frac{\partial}{\partial y}\left[k_{fy}(\theta) \cdot \frac{\partial \Psi}{\partial y}\right] + \frac{\partial}{\partial z}\left[k_{fz}(\theta) \cdot \frac{\partial \Psi}{\partial z}\right] \quad (22)$$

C ($=d\theta/d\psi$) is defined as specific moisture capacity of the soil. Richard's equation contains the relationship between water content (θ) and soil water tension (Ψ, set equal to the potential) and also the relationship between hydraulic conductivity and water content. These relationships are extremely non-linear. Therefore, the solution of equation (24) (calculation of water and solute transport) requires numerical methods (Jentsch, 1992; Philip et al., 1974). It is foreseeable that we need to know two parameters in order to describe the water movement within the soil: total potential (Ψ, represented by the soil water tension under described boundary conditions) and water content.

3.2.3 Soil mechanics and soil hydraulics

Soils store and transport water within their pore system. This property is determined by the hydraulic conductivity and the texture of the pores. The hydraulic conductivity depends on the water content of the soil. Considering an unsaturated "ideal" soil with a uniform microstructure without macropores the following relationships are relevant:

- ratio volumetric water content to soil water tension also called the soil moisture characteristic or pF-curve
- ratio soil water retention to hydraulic conductivity
- ratio volumetric water content to hydraulic conductivity

The soil water tension equals to the sum of bonding forces performed by the soil matrix with regard to the water. The pF-curve shows a characteristic course for each different soil (Scheffer, 1992). It depends on the particle size distribution, the formation and cross-linking of the pores and also of their size. The proportion of organic matter and the chemical composition of the wetting phase also affect the pF-curve. The water holding capacity of sand is low (if pF>4.2) the residual saturation θ_r equals a water content less than 3 vol. %), so sand releases water situated within the pore volume at low soil water tension differences (low specific water capacity C, that means a slight gradient of the tangent with respect to the pF-water content-curve at $0 < pF < 2.5$). In contrast, clay shows a high residual saturation (if pF = 4.2, water content equals ~ 30 vol. %) and the release of water needs high changes of the soil water tension (high specific water capacity C, that means high gradient).

Depending on the history of the soil with respect to watering and dewatering the shapes of the pF-curve for a single soil are different, an effect called hysteresis. It is assumed that the capillaries which connect the pores show a smaller cross section than the pores themselves. If the soil water tension is high, the convex meniscus holds on to the upper border of the pore (by means of retention forces). This soil water tension is higher than necessary in order to move the concave meniscus (formed by wetting resistance) into the capillary in case of watering. The main influencing factors of the pF-curve are grain size, soil structure and hysteresis.

There are manifold approaches for the analytical description of the pF-curve. The approach of Brooks & Corey (1964) combines the mathematical ratio of θ to Ψ with the conductivity model of Burdine (1953). The definition of effective water content, also called relative saturation index or soil-water-retention, is

$$S_e = \frac{\theta - \theta_r}{\theta_s - \theta_r} \tag{23}$$

with θ as current water content, θ_r as residual water content (simplified = 0, easier to measure at pF = 4,2) and θ_s as water content at saturation (measured at pF = 0), conform to porosity.
Besides the approach of Campbell (1974) the common approach originates from Van Genuchten (1980) and Mualem (1976a), VGM in the following. The VGM approach prevailed in the literature and will be considered subsequently. Therefore we combine the (Ψ)-relationship with the conductivity model from Mualem. Unlike the model of Brooks & Corey-Burdine (Berger, 1998) this approach doesn't take any sharp air inlet into account. Thus this approach is solvable from the analytical point of view if we consider certain boundary conditions

$$S_e = \frac{\theta - \theta_r}{\theta_s - \theta_r} = \left[\frac{1}{1+(\alpha h)^n}\right]^m \tag{24}$$

α,m,n parameters of shape, dependent on grain distribution of the soil:
$\alpha=1/h_b$, ; h_b = soil water tension (ψ) at air inlet point,
$m=\lambda/(\lambda+1)$; λ=index of pore size,
$n=\lambda+1$; hence $m=1-1/n$

h soil water tension (ψ) at the water content θ

So the description of the hydraulic conductivity/soil moisture-relationship becomes possible from the analytical point of view based on the prediction model of Mualem:

$$\frac{k_f(\theta)}{k_s} = S_e{}^{\gamma} \cdot \left[1-\left(1-S_e^{\frac{1}{m}}\right)^m\right]^2 \tag{25}$$

$k_f(\theta)$ hydraulic conductivity at the current water content

k_s hydraulic conductivity at water saturation. k_s is often set constant, but also depends on the texture of the soil

γ takes the influence of tortuosity into account, usually $\gamma = 0.5$.

By means of inverse identification of parameters (Schultze et al., 1996) the approximation of the values from numerical simulations to those from field trials is possible. Thus the sought parameters are provided directly. The measuring of water tension by means of tensiometer represented until now the state-of-the-art. The water content of the soil was determined via pF-curve (with knowledge of the water tension) and furthermore the k_f -value. Until now no exact results could be achieved due to the influence of hysteresis.

3.2.4 Soil moisture

In order to solve the Richards equation (22), knowledge of properties of the soil is required. Until now these properties were taken from the fundamental pF-curve. There the pF-value represents the common logarithm taken from Buckingham's "capillary potential" expressed as cm of the water column. The shape and uniqueness of the pF-curve depends on the texture. The texture includes pore size distribution (structure) as well as storage and arrangement of the particles (texture). For analytical modeling further parameters are necessary. These parameters describe the water content of the soil at different boundary conditions. Alongside the natural saturation water content (θ_s) where pF = 0, the absolute saturation (θ_s^*) is also an important parameter. θ_s^* correlates to the porosity of the soil. Absolute saturation cannot be achieved by rewatering because -according to structure and texture- certain parts of the pores remain air-filled. According to extensive investigations the following equation is valid

$$\theta_s = 0.8 \div 0.95\,\theta_s^* \tag{26}$$

Thus soil is liable to be seen as three-phase system. A further parameter is the remaining water or residual saturation defined as θ_r, where the aqueous phase is not constant anymore. The associated pF-value is 4.2. The distribution of the three components water, air and soil is described by various "mixing models". The aqueous phase is distinguished between "free water" and "bound water" that is adsorbed on or within the particles. The electrostatic forces surrounding the solid (here: soil particle) act outwards. These forces result from molecules that are not compensated electrical all-round. The wetting property of a solid with regard to water depends on the strength of these forces. If the cohesion forces of the water molecules are less than the surface forces, the water molecules absorb on the surface. Bound water is able to absorb further water molecules by means of associate forces, however, this binding is not stable (Huebner, 1999).

Bound water prefers ionic bonds. The surfaces of fine-grained materials such as clays are saturated by ions. The sorption forces between ions and the surface of the clays are greater than the non-polar sorption forces between surface and water molecule. As a result the wetting property of clays increases and a hydrate envelope around the metal cations is established. Also crystal water, i.e. water bound within the lattice of the soil particles, is present and water can condensate within the capillaries. If two water films get into touch, the water molecules flow together and form carrying menisci within the soil pores and more water molecules are attracted by the surface tension. If the soil air is saturated with vapor, the water condenses above the concave meniscus. At this place the vapor pressure is smaller than above the convex or the flat meniscus. The molecular forces get saturated by steam or by liquid water.

The main part of the water in the soil is not influenced by molecular forces. It has zero potential and underlies gravity. The water contents θ, θ_s and θ_r have been determined in on lab-scale depending on soil water tension. Because of the hysteresis it was distinguished between the watering and dewatering of the sample. The reasons for the hysteresis of the pF-curve are that the advancing contact angle between soil matrix and soil water is greater than the retreating contact angle, the effects regarding the geometry of pores, water bound on clay mineral surfaces, and enclosed air. Statements leading to a reliable approximation of the hydraulic conductivity are therefore impossible. The extension of uncertainty by 70 times results from empiric measures (Ψ is applied logarithmic, k_f is applied exponential). This uncertainty provides the basis for the assessment of the water content (θ) as a relevant measurement parameter.

4. Systems

The comprehensive characterization of materials is prerequisite to understand processes in large (geo)-technical systems and their manipulation. This can be achieved best by the knowledge of material properties, measuring methods to determine water content and processes that describe the interaction of matter and water. Examples for technical systems in that sense are e.g. sealing systems for landfills and subsurface storage of waste, monitoring of soil water content over large areas using power lines, or monitoring system for groundwater recharge in the unsaturated zone.

4.1 Monitoring system for surface sealings

4.1.1 Configuration of sealing system and monitoring layer

For the monitoring of the volumetric water content θ_V, the TAUPE TDR-system described in chapter 1.2.3.2 was used. Specifications for a monitoring system for surface sealing systems on landfills (figure 1) defined from legislating body (BAM, Federal Institute for Materials Research and Testing) are

- detection of increase over more than an order of magnitude in permeability of a mineral sealing or capillary barrier,
- detection of local relative variations in volumetric water content of more than 5 %,
- positioning information of 100 m^2, that means a circle with radius around 5.5 m around true position.

The sealing system installed at the landfill is build up as a capillary barrier as shown in figure 2.

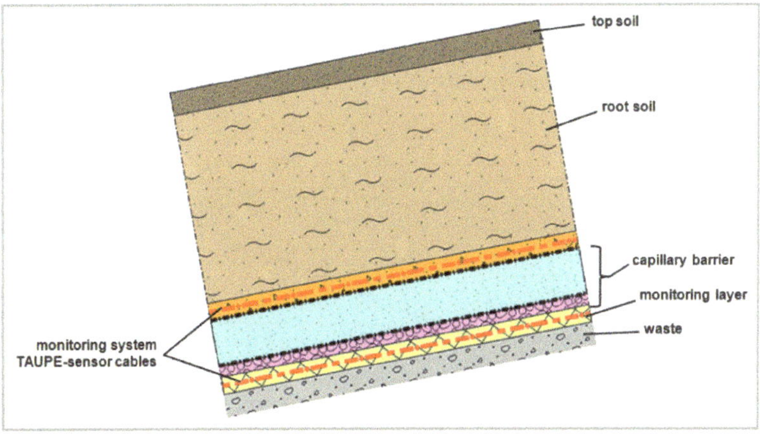

Fig. 2. Simplified schematic of mineral capillary barrier consisting of a fine-grained material capillary layer above a coarse-grained material capillary block in a surface sealing system of a landfill.

Sensors of the monitoring system were installed in monitoring layers above and below the capillary barrier to determine the quantity of permeated water through the barrier. Water content of a soil layer situated below the capillary barrier without appreciable hysteresis is directly related to its permeability, i.e. the unsaturated hydraulic conductivity. Landfill regulations determine saturated hydraulic conductivity as quality parameter of mineral sealing systems. A critical limit of $< 5 \times 10^{-9}$ m/sec can be deduced from experimental relation between volumetric water content and unsaturated hydraulic conductivity, which can reach values of a few orders of magnitude below that limit. Materials used in a monitoring layer must meet the following requirements

- its hydraulic properties must be defined unambiguously,
- inflow from layers above must lead to a significant change in volumetric water content,
- uniquely defined monitoring parameters for saturated/unsaturated hydraulic conductivity correspond to definite volumetric water contents. Percolation through monitoring layer must be equivalent to percolation in total sealing system.

Adequate for a monitoring layer are sandy to silty materials with a range of saturated hydraulic conductivity of 10^{-5} to 10^{-6} m/sec. A decrease in unsaturated hydraulic conductivity should be at least four to five orders of magnitude and cover the range of 10^{-10} m/sec to 10^{-5} m/sec. The monitoring layer then can discharge a break-through of the sealing system without building up backwater. Sensors are installed in monitoring layers at depth of around 2 m and 2.3 m, respectively.

4.1.2 Calibration of a material for the monitoring layer

An adequate material was tested in the laboratory using measurement equipment for the determination of complex dielectric permittivity. It consists of a vectorial network analyzer and a coaxial type probe cell (see inset in figure 2) in the same frequency range of 100 MHz to 1100 MHz as used with TDR method. The material has been exposed to different amounts of water, with a part of it being dried at 105 °C in oven for evaluation of gravimetric water content related to dry mass. A second part was inserted in the coaxial probe cylinder for determination of permittivity. Figure 3 shows the relation between volumetric water content and square root of measured permittivity. Results from TDR measurement can directly be converted via the indicated regression function.

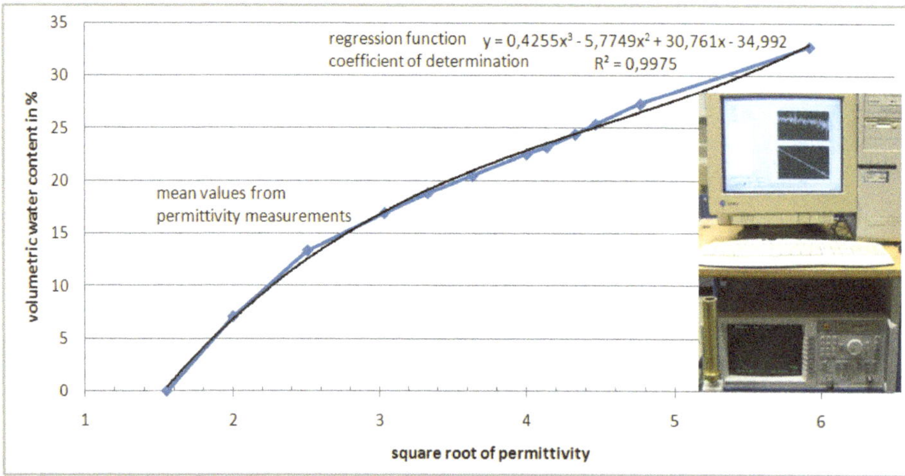

Fig. 3. Material calibration function for the monitoring layer. Inset shows the permittivity measurement system with coaxial probe cylinder.

4.1.3 Monitoring system

A test site for a monitoring system for surface sealing has been established from 2004 to 2005 at the landfill situated in Oberweier/Germany (Figure 4). In two monitoring layers (see figure 1) 230 sensors have been installed and connected via 34 multiplexer to a TDR system.

To keep the length of the connecting coaxial cables between TDR system and sensors below 150 m, two central units with separate TDR devices cover 120 and 110 sensors, respectively. Sensor length is 10 m and distances between adjacent sensors are between 8 and 10 m, depending on hill slope. Data collection takes place two times a day.

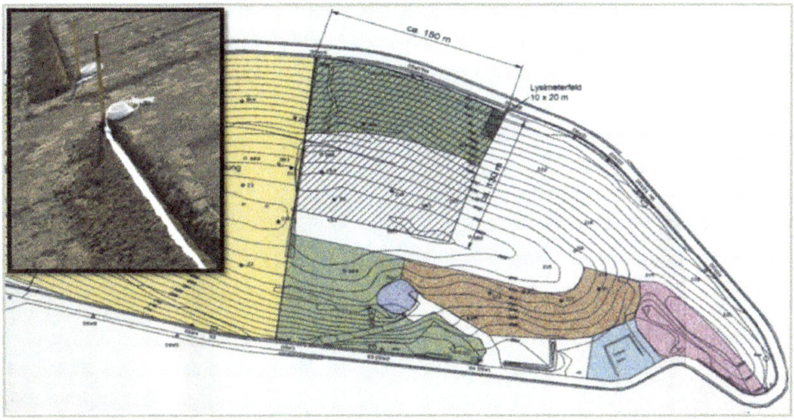

Fig. 4. Plan of landfill in Oberweier. Sensors and measurement equipment has been installed in the shaded area of approximately 150 m x 130 m. Inset shows TAUPE TDR-cable sensors during installation in lower monitoring layer.

4.2 Data evaluation

TDR signal data comprise only a part of the total signal length and is constraint to the transition between coaxial cable and start of sensor and shortly beyond end of sensor (see figure 5). First rise of the reflection signal occurs at start of sensor and second rise at end of sensor (Topp et al., 1980). Exact starting and ending points are defined by calculating the inflection points of the slopes to fit tangents to the curve and finding crossing points with horizontal lines. From time difference propagation time is calculated.

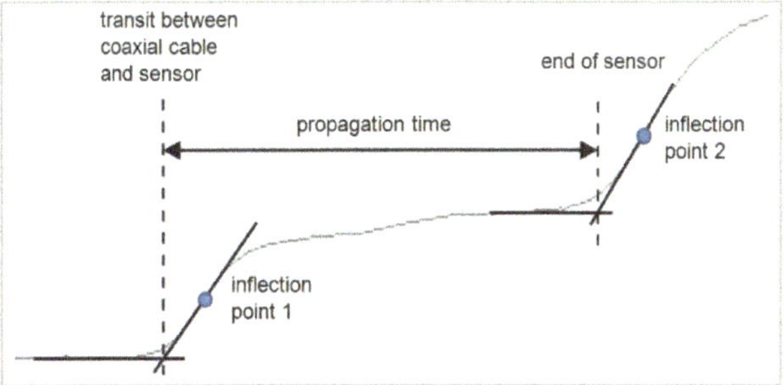

Fig. 5. Typical form of TDR reflection signal and determination of propagation time

Data from all sensors are automatically evaluated using an appropriate software system. Resulting propagation time defines an integral value for volumetric water content along a complete sensor according to the material calibration function in figure 2. Adding results for each sensor over time delivers variations in water content all over the landfill. To give an easier access to the hydraulic behavior at locations of different sensors the landfill is divided in vertical transects between top of the landfill and its base. This is shown for 2010 in figure 6 on eight sensors for both monitoring layers.

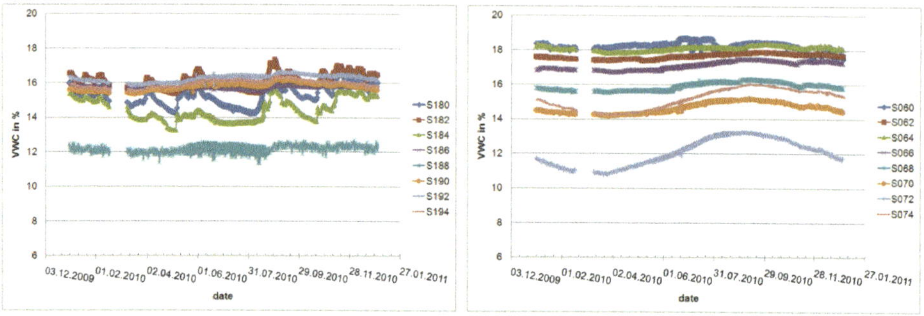

Fig. 6. Volumetric water content as a function of time in monitoring layer above (left: sensors 180, 182, 184, 186 188, 190, 192, 194) and below (right: sensors 60, 62, 64, 66, 68, 70, 72, 74) capillary barrier. Data gap in March 2010 was due to a failure in power supply.

In the monitoring layer above capillary barrier discharge of precipitation at the surface can be found directly, especially at the borders of the landfill due to a poor connection to the area outside of the test site. Preferential flow parallel to the slope in layers with local secondary capillary barriers, which have been built unintentionally during installation process, leads to a more steady value of volumetric water content. Normally short time reactions describe the hydraulic system above the sealing.

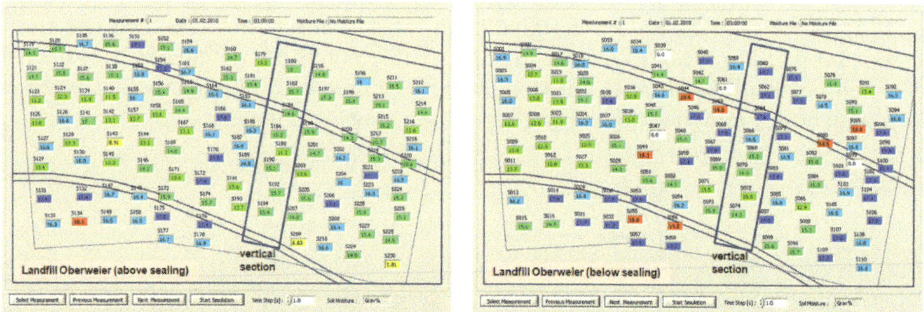

Fig. 7. Intrinsic situation of volumetric water content in monitoring layer above (left) and below sealing system (right). Vertical sections according to figure 6.

In contrast, situation below the sealing system shows little short time variations, what demonstrates the functionality of the capillary barrier. Except at the borders little influences of discharge from the surface occur. Depending on chemical reactions in waste and seasonal temperature changes the volumetric water content can vary locally in the order of up to 2 % due to temperature depending permittivity of water.

Graphs in figure 7 show the situation above and below the capillary barrier at a certain time as colored graphs. Rectangular blocks show the volumetric water content of each sensor and the red color indicates possible problems due to locally high water content above arbitrarily chosen 18 %. Situation below the sealing shows more dark blue and red spots than above, which is a result of the installation since the first section below the sealing has been constructed during heavy rain in October 2004 and the second section with the capillary barrier was built in spring 2005 during the dry season. Water exchange with atmosphere via evapotranspiration is low due to depth of monitoring layers.

4.2.1 Hydraulic conductivity

Volumetric water content of the layer below capillary sealing received from the monitoring system is the input parameter for the determination of hydraulic conductivity according to figure 8. In 2010 the sensors detected volumetric water contents between 12 and 18 %. Critical limit of 5x10^{-9} m/sec gives a monitoring value for volumetric water contents of 22 % and has not emerged during the observation period between 2005 and 2010.

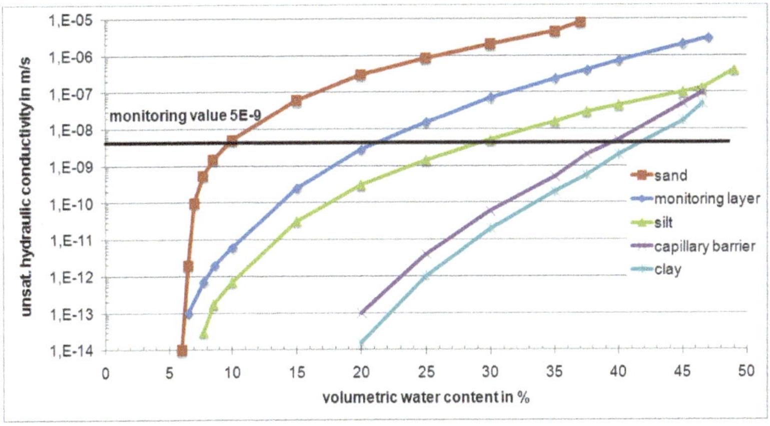

Fig. 8. Hydraulic conductivity of different materials as function of VWC; material of monitoring layers fulfills requirements of BAM.

5. Acknowledgement

We acknowledge support by Deutsche Forschungsgemeinschaft and Open Access Publishing Fund of Karlsruhe Institute of Technology.

6. References

Bachmann, J., Ellies, A., & Hartge, K. H. (2000). Development and application of a new sessile drop contact angle method to assess soil water repellency. *Journal of Hydrology* Vol. 231-232, pp. 66-75.

Berger, K. (1998). Validierung und Anpassung des Simulationsmodells HELP zur Berechnung des WHH. *BMBF-Schlussbericht*, Berlin.

Brandelik, A., & Hübner, C. (1999). Moisture monitoring with subsurface transmission line. *Proceedings of Conference on Subsurface Sensors and Applications*, Denver, June 1999.

Brooks, R.H. and Corey, A.T. (1964). Hydraulic properties of porous media. *Hydrological Papers*, No. 3, Colorado State University.

Brunauer, S., Emmet P.H., & Teller E. (1938). Adsorption of gases in multimolecular layers. *J. Am. Chem. Soc.*, Vol. 60, pp. 309-319.

Burdine, N.T. (1953). Relative permeability calculations from pore size distribution data. *Petroleum. Transactions of the American Institute, of Mining, Metallurgical and Petroleum Engineers*, Vol. 198, pp. 71-77.

Butt, H.-J., Graf, K., & Kappl, M. (2006). *Physics and Chemistry of Interfaces*. Wiley-VCH Verlag, Weinheim.

Campbell, G.S. (1974). A simple method for determining unsaturated conductivity from moisture retention data. *Soil Science*, Vol. 117, No. 6, pp. 311-314.

Carman, P. C. (1937). Fluid flow through granular beds. *Trans. Inst. Chem. Eng.* Vol. 15, pp. 150-166.

Cox, R. G., 1986a. The dynamics of the spreading of liquids on a solid surface. Part 1. Viscous flow. *Journal of Fluid Mechanics* Vol. 168, pp. 169-194.

Cox, R. G., 1986b. The dynamics of the spreading of liquids on a solid surface. Part 2. Surfactants. *Journal of Fluid Mechanics*, Vol. 168, pp. 195-220.

Czurda, A. (1994). Multimineralische Abdichtungen. *Schriftenreihe Angewandte Geologie Karlsruhe*, Vol. 30, pp. 1-21.

Dang-Vu, T. & Hupka, J. (2005). Characterization of porous materials by capillary rise method. *Physicochemical Problems of Mineral Processing*, Vol. 39, pp. 47-65.

Dimitrov, A. S., Kralchevsky, P. A., Nikolov, A. D., Noshi, H., & Matsumoto, M. (1991). Contact angle measurements with sessile drops and bubbles. *Journal of Colloid Interface Science*, Vol. 145, pp. 279-282.

Dohrmann, R. & Kaufhold, S. (2010). Determination of exchangeable calcium of calcareous and gypsiferous bentonites. *Clays and Clay Minerals*, Vol. 58 , pp. 79-88.

Emmerich, K., Kemper, G., Königer, F., Schlaeger, S., Gruner, M., Hofmann, M., Nüesch, R.,& Schuhmann, R. (2007). HTV-1: A semi technical scale testing of a multi-layer hydraulic shaft sealing system. *Proceedings of 3rd International Meeting on Clays in natural and engineered barriers for radioactive waste confinement*, Lille, Frankreich, 2007

Emmerich, K., Kemper, G., Königer, F., Buqezi-Ahmeti, D., Gruner, M., Gaßner, W., Hofmann, M., & Schuhmann, R. (2008). Sandwich - Sealing system with equipotential layers for underground storage of hazardous waste to ensure homogeneous wetting of sealing layers and to enhance long term stability. *Proceedings of Bodenkundliches Kolloquium des Instituts für Bodenkunde und Standortslehre der Uinversität Hohenheim*, Stuttgart, Deutschland, 2008

Emmerich, K. (2011): Thermal analysis for characterization and processing of industrial minerals. In: *EMU notes in mineralogy*, Christidis, G., Vol. 9: Industrial mineralogy, Mineralogical Society in press.

Genuchten, M.Th. van (1980). A closed-form equation for predicting the hydraulic conductivity of unsaturated soils. *Soil Sci. Soc. Am. J.*, Vol. 44, pp. 892-898.

Grace, H. P. (1953). Resistance and compressibility of filter cakes. Part I. *Chem. Eng. Prog.*, Vol. 49, pp. 303-318.

Haude, W. (1958). Über die Verwendung verschiedener Klimafaktoren zur Berechnung der potentiellen Evaporation und Evapotranspiration. *Meteorologische Rundschau*, Vol. 11, pp. 96-99.

Hillel, D. (1980). *Fundamentals of soil physics*, Academic press, Orlando.

Huebner, C. (1999). *Entwicklung hochfrequenter Messverfahren zur Boden- und Schneefeuchtebestimmung*, Wissenschaftliche Berichte FZKA 6329, Karlsruhe.

Jentsch, G. (1992). Bilanzierung des Stoff- und Schadstoffeintrags in das Grundwasser unter besonderer Berücksichtigung der ungesättigten Zone. *Schriftenreihe Angewandte Geologie*, Vol. 17.

Jones, G. & Frizzell, L. D. (1940). A theoretical and experimental analysis of the capillary rise method for measuring the surface tension of solutions of electrolytes. *Journal of Chemical Physics*, Vol. 8, pp. 986 - 997.

Jones, G. & Ray, W. A. (1937). The surface tension of solutions of electrolytes as a function of the concentration. I. A differential method for measuring relative surface tension. *Journal of the American Chemical Society*, Vol. 59, pp. 187-198.

Kahr, G. & Madsen, F.T. (1995). Determination of the cation exchange capacity and the surface area of bentonite, illite and kaolinite by methylene blue adsorption. *Applied Clay Science*, Vol. 9, pp. 327-336.

Kallioras, A., Piepenbrink, M., Schueth, C., Pfletschinger, H., Dietrich, P., Königer, F., & Rausch, R. (2010). Quantification of groundwater recharge through application of pilot techniques in the unsaturated zone. *Proceedings of EGU General Assembly*, Wien, Österreich, May, 2010.

Kaufhold, S. & Dohrmann, R. (2008). Comparison of the traditional Enslin-Neff method and the modified Dieng method for measuring the water uptake capacity. *Clays and Clay Minerals*, Vol. 56, pp. 686–692.

Kaufhold, S., Dohrmann, R., Klinkenberg, M. (2010). Water uptake capacity of bentonites. *Clays and Clay Minerals*, Vol.58, pp. 37-43 .

Koeniger, F., Emmerich, K., Kemper, G., Gruner, M., Gaßner, W., Nüesch, R., & Schuhmann, R. (2008). Moisture spreading in a multi-layer hydraulic sealing system (HTV-1). *Engineering Geology*, Vol. 98, pp. 41-49

Koeniger, F., Emmerich, K., Kemper, G., Gruner, M., Gaßner, W., Stacheder, M., & Schuhmann, R. (2009). Monitoring of moisture spreading in a multi-layer hydraulic sealing system during saturation with a rock salt brine by TDR sensors. *Proceedings of ISEMA*, Helsinki, June, 2009.

Koeniger, F., Schmitt, G., Schuhmann, R., & Kottmeier, C. (2010). Free Line Sensing', a new method for soil moisture measurements using high-voltage power lines. *Near Surface Geophysics*, Vol. 8, pp. 151-161

Konert, M., Vandenberge, J. (1997) Comparison of laser grain size analysis with pipette and sieve analysis: a solution for the underestimation of the clay fraction. *Sedimentology*, Vol. 44, pp. 523-535.

Kozeny, J. (1927). Soil permeability. *Sitzungsber. Oesterr. Akad. Wiss. Wien. Math. Naturwisss. Kl. Abt.* Vol. 136, pp. 271.

Latief, O. (2010): Korngrößenbestimmung an Tonmineralen Vergleich von Sedigraph-, Laserstreuung- und Zetasizermessungen, *Proceedings of Annual Meeting German Ceramic Society*, Hermersdorf, March 2010.

Lavi, B., Marmur, A., & Bachmann, J. (2008). Porous media characterization by the two-liquid method: Effect of dynamic contact angle and inertia. *Langmuir*, Vol. 24, pp. 1918-1923.

Madsen F.T., & Kahr G. (1992). Wasserdampfadsorption und spezifische Oberfläche von Tonen, *DTTG-Tagungsband*, Hannover, 1992.

Maubeuge, K. v. (2002). Investigation of bentonite requirements for geosynthetic clay barriers. *Proceedings of Clay Synthetic Barriers Symposium*, Nuremberg, April, 2002.

Maubeuge, K. v. & Egloffstein, T. (2004). Quality requirements for Bentonite in geosynthetic clay liners in the validity of test methods, In: *Advances in geosynthetic clay liner technology: 2nd Symposium*, Mackey, R. E. & Maubeuge K.v., pp. 11-30, ASTM International, ISBN 0-8031-3484-3, Mayfield.

Moore, D. & R.C. Reynolds, Jr. (1997). *X-Ray Diffraction and the Identification and Analysis of Clay Minerals*, 2nd ed.: Oxford University Press, New York

Mualem, Y., (1976a). A new model for predicting the hydraulic conductivity of unsaturated porous media. *Water Resour. Res.*, Vol. 12, No. 3, pp. 513-522.

Mullins, C.E. (1991). Matric potential, In: *Soil analysis - Physical methods*, Smith, K. A., & Mullins, C.E., pp. 75-110, Marcel Dekker Inc., 0-8247-8361-1, New York.

Neff, H. K. (1959). Über die Messung der Wasseraufnahme ungleichförmiger bindiger anorganischer Bodenarten in einer neuen Ausführung des Enslingerätes. *Bautechnik*, Vol. 39, No. 11, pp. 415-421.

Omotoso, O., McCarty, D., Hillier, S. & Kleeberg, R. (2006). .... *Clays and Clay Minerals*, Vol. 54, No. 6, pp. 748–760.

Passioura, J.B. (1980): The Meaning of Matric Potential. *Journal of Experimental Botany*, Vol. 31, No.123, pp. 1161-1169.

Petersen, L. W., Moldrup, P., Jacobsen, O. H., & Rolston, D. E. (1996). Relations between specific surface area and soil physical and chemical properties. *Soil Sci.*, Vol. 161, No. 1, pp. 9-21.

Philip, J.R., Knight, J.H. (1974). On solving the unsaturated flow equation: 3. new quasi-analytical technique, *Soil Science*, Vol. 117, No.1, pp. 1-13.

Pusch, R. (2001). The Buffer and Backfill Handbook, Part2: Materials and Techniques. *Technical Report TR-02-12*, Swedish Nuclear Fuel and Waste Management Co, ISSN 1404-0344, Stockholm.

Pusch, R. (2002). The Buffer and Backfill Handbook, Part1: Definitions, basic relationships, and laboratory methods. *Technical Report TR-02-20*, Swedish Nuclear Fuel and Waste Management Co, ISSN 1404-0344, Stockholm.

Ramirez-Flores, J. C., Bachmann, J., & Marmur, A. (2009). Direct determination of contact angles of model soils in comparison with wettability characterization by capillary rise. *Journal of Hydrology*, Vol. 382, pp. 10-19.

Scheffer, F. (1992). *Lehrbuch der Bodenkunde* (13. Ed.), Enke, ISBN 3-432-84773-4, Stuttgart

Schlaeger, S. (2005). A fast TDR-inversion technique for the reconstruction of spatial soilmoisture content . *Hydrology and Earth System Sciences*, Vol. 9, pp. 481–492.

Schmugge, T. J., Jackson, T. J., & McKim, H. L. (1980). Survey of methods for soil moisture determination. *Water Resour. Res.*, Vol. 16, No. 6, pp. 961-979.

Schultze, B., Zurmuehl, T., Durner, W. (1996). Untersuchung der Hysterese hydraulischer Funktionen von Böden mittels inverser Simulation, *Mitteilungen der deutschen bodenkundlichen Gesellschaft*, Vol. 80, pp. 319-322.

Schuhmann, R. (2002). Kontrolle von Barrieren: Bestimmung der hydraulischen Leitfähigkeit an Hand des Bodenwassergehaltes. *Mitteilungen des Instituts für Wasserwirtschaft und Kulturtechnik der Universität Karlsruhe*, Vol. 219.

Siebold, A., Walliser, A., Nardin, M., Oppliger, M., & Schultz, J. (1997). Capillary Rise for Thermodynamic Characterization of Solid Particle Surface. *Journal of Colloid and Interface Science*, Vol. 186, pp. 60-70.

Topp, G.C., Davis, J. L., & Annan, A.P. (1980). Electromagnetic Determination of Soil Water Content: Measurements in Coaxial Transmission Lines. *Water Resourc. Res.*, Vol. 16, No. 3, pp. 574-582.

Voinov, O. V. (1976). Hydrodynamics of wetting. *Fluid Mechanics*, Vol. 11, pp. 714-721.

Yang, Y. W. & Zografi, G. (1986). Use of the Washburn-Rideal equation for studying capillary flow in porous media. *Journal of Pharmaceutical Sciences*, Vol. 75, pp. 719-721.

10

Estimating Hydraulic Conductivity Using Pedotransfer Functions

Ali Rasoulzadeh
Water Engineering Dept., College of Agriculture
University of Mohaghegh Ardabili, Ardabil
Iran

1. Introduction

Hydraulic conductivity is an important soil physical property, especially for modeling water flow and solute transport in soil, irrigation and drainage design, groundwater modeling and other agricultural and engineering, and environmental processes. Due to the importance of hydraulic conductivity, many direct methods have been developed for its measurement in the field and laboratory (Libardi et al., 1980; Klute and Dirksen, 1986). Interestingly, comparative studies of the different methods have shown that their relative accuracy varies amongst different soil types and field conditions (Gupta et al., 1993; Paige and Hillel, 1993; Mallants et al., 1997). No single method has been developed which performs very well in a wide range of circumstances and for all soil types (Zhang et al., 2007). Direct measurement techniques of the hydraulic conductivity are costly and time consuming, with large spatial variability (Jabro, 1992; Schaap and Leij, 1998; Christiaens and Feyen, 2002; Islam et al., 2006). Alternatively, indirect methods may be used to estimate hydraulic conductivity from easy-to-measure soil properties. Many indirect methods have been used including prediction of hydraulic conductivity from more easily measured soil properties, such as texture classes, the geometric mean particle size, organic carbon content, bulk density and effective porosity (Wösten and van Genuchten, 1988) and inverse modeling techniques (Rasoulzadeh, 2010; Rasoulzadeh and Yaghoubi, 2011). In recent years, pedotransfer functions (PTFs) were widely used to estimate the difficult-to-measure soil properties such as hydraulic conductivity from easy-to-measure soil properties. The term PTFs were coined by Bouma (1989) as translating data we have into what we need. PTFs were intended to translate easily measured soil properties, such as bulk density, particle size distribution, and organic matter content, into soil hydraulic properties which determined laboriously and costly. PTFs fill the gap between the available soil data and the properties which are more useful or required for a particular model or quality assessment (McBratney et al., 2002). In the other hand PTFs can be defined as predictive functions of certain soil properties from other easily, routinely, or cheaply measured properties. PTFs can be categorized into three main groups namely class PTFs, continuous PTFs and neural networks. Class PTFs calculate hydraulic properties for a textural class (e.g. sand) by assuming that similar soils have similar hydraulic properties; continuous PTFs on the other hand, use measured percentages of clay, silt, sand and organic matter content to provide continuously varying hydraulic properties across the textural triangle (Wösten et al., 1995). In fact, continuous PTFs predict soil properties as a continuous function of one or more measured variables. Neural networks are an "attempt to build a mathematical model that supposedly works in an analogous way to the human brain" and were developed to improve the predictions of empirical PTFs. In brief, a neural network consists of an input, a hidden, and an output layer all containing "nodes". The number of nodes in input (soil bulk density, soil particle size data) and output (soil hydraulic properties) layers corresponds to the number of input and output variables of the model (Schaap and Bouten, 1996).

PTFs must not be used to predict something that is easier to measure than the predictor. For example, If we measure the water retention curve only to predict saturated hydraulic conductivity (K_s), this is not an efficient PTF, as the cost of measuring a water retention curve is greater than measuring Ks itself (McBratney et al., 2002).

2. Pedotransfer functions for estimating saturated hydraulic conductivity

PTFs have become a 'white-hot' topic in the area of soil science and environmental research. PTFs which used to estimate saturated hydraulic conductivity (K_s) were developed using texture classes, the geometric mean particle size, organic carbon content, bulk density and effective porosity as predictor variables. Many PTFs were presented to predict K_s. Here the following PTFs for K_s are considered. All PTFs give K_s in m.s^{-1}. Wösten et al. (1997) presented a function for determining K_s as follows:

$$K_s = 1.15741 \cdot 10^{-7} \exp(x) \tag{1}$$

where x for sandy soil is:

$$x = 9.5 - 1.471(BD^2) - 0.688(Om) + 0.0369(Om^2) - 0.332\ln(CS) \tag{2}$$

and x for loamy and clayey soils is:

$$\begin{aligned} x = & -43.1 + 64.8(BD) - 22.21(BD^2) + 7.02(Om) - 0.1562(Om^2) \\ & +0.985\ln(OM) - 0.01332(Clay)(Om) - 4.71(BD)(Om) \end{aligned} \tag{3}$$

where BD is bulk density in g.cm^3, $Clay$ is the percentage of clay, CS is the sum percentage of clay and silt, and Om is percent organic matter.

Wösten et al. (1999) represented another function as follows:

$$K_s = 1.15741 \cdot 10^{-7} \exp(x) \tag{4}$$

where x is:

$$\begin{aligned} x = & 7.755 + 0.0352(Silt) + 0.93(Topsoil) - 0.967(BD^2) - 0.000484(Clay^2) - 0.000322(Silt^2) \\ & +0.001/(Silt) - 0.0748/(Om) - 0.643\ln(Silt) - 0.01398(BD)(Clay) - 0.1673(BD)(Om) \\ & +0.02986(Topsoil)(Clay) - 0.03305(Topsoil)(Silt) \end{aligned} \tag{5}$$

where BD is bulk density in g.cm^3, $Clay$ and $Silt$ are the percentage of clay and silt, respectively, $Topsoil$ is a parameter that is set to 1 for topsoils and to 0 for subsoils, and Om is percent organic matter.

The Cosby's pedotransfer function (Cosby et al., 1984) was derived based on Sand and Clay contents as:

$$K_s = 7.05556 \cdot 10^{-6} \cdot \left(10^{[-0.6+0.0126(Sand)-0.0064(Clay)]}\right) \tag{6}$$

where *Clay* and *Silt* are the percentage of clay and silt, respectively.
Saxton et al. (1986) suggested a pedotransfer function to estimate K_s as follows:

$$K_s = 2.778 \cdot 10^{-7} \exp(x) \tag{7}$$

where

$$\begin{aligned} x &= 12.012 - 7.55 \cdot 10^{-2}(Sand) \\ &+ \left(-3.895 + 3.671 \cdot 10^{-2}(Sand) - 0.1103(Clay) + 8.7546 \cdot 10^{-4}(Clay^{2)}\right) / \theta_s \end{aligned} \tag{8}$$

where *Clay* and *Sand* are the percentage of clay and sand, respectively, and θ_s is the saturated water content.
Brakensiek et al. (1984) found a relationship between K_s and clay, sand and saturated water content as follows:

$$K_s = 2.778 \cdot 10^{-7} \exp(x) \tag{9}$$

where

$$\begin{aligned} x &= 19.52348(\theta_s) - 8.96847 - 0.028212(Clay) + 1.8107 \cdot 10^{-4}(Sand^2) - 9.4125 \cdot 10^{-3}(Clay^2) \\ &-8.395215(\theta_s^{\,2}) + 0.077718(Sand)(\theta_s) - 0.00298(Sand^2)(\theta_s^{\,2}) - 0.019492(Clay^2)(\theta_s^{\,2}) \\ &+1.73 \cdot 10^{-5}(Sand^2)(Clay) + 0.02733(Clay^2)(\theta_s) + 0.001434(Sand^2)(\theta_s) \\ &-3.5 \cdot 10^{-6}(Clay)(Sand) \end{aligned} \tag{10}$$

All parameters are defined before.
Campbell (1985) presented a pedotransfer function to estimate K_s based on empirical parameter of Campbell's soil water retention function as follows:

$$K_s = 4 \cdot 10^{-5} \left(\frac{1.3}{BD}\right)^{1.3b} \exp\left(-6.9(m_{clay}) - 3.7(m_{silt})\right) \tag{11}$$

where b is an empirical parameter of Campbell's soil water retention function. The coefficient b is derived from the geometric mean particle diameter (mm), d_g, and the standard deviation of mean particle diameter σ_g:

$$b = d_g^{\,-0.5} + 0.2\sigma_g \tag{12}$$

where d_g and σ_g are derived from soil main grain size fractions (m_{clay} , m_{silt} and m_{sand} are clay, silt and mass fractions, respectively) and geometric mean diameter of soil separates (d_{clay} , d_{silt} and d_{sand} are the geometric mean diameters of main grain size fractions in millimeters):

$$d_g = \exp \sum_{i=1}^{3} m_i \ln d_i \tag{13}$$

$$\sigma_g = \exp\left[\sum_{i=1}^{3} m_i(\ln d_i)^2 - \left(\sum_{i=1}^{3} m_i(\ln d_i)\right)^2\right] \quad (14)$$

where m_i is the mass fraction of textural class i, and d_i is the arithmetic mean diameter of class i. The assumption is taken over the three texture classes, sand, silt, and clay. For the three classes normally used in determining texture, d_{clay}=0.001 mm, d_{silt}=0.026 mm, and d_{sand}=1.025 mm. Vereecken et al. (1990) provided a equation for estimating K_s as follows:

$$K_s = 1.1574 \cdot 10^{-7} \exp\left(20.62 - 0.96\ln(Clay) - 0.66\ln(Sand) - 0.46\ln(Om) - 0.00843(BD)\right) \quad (15)$$

Ferrer-Julià et al. (2004) derived a relationship between K_s and sand content of soil as follows:

$$K_s = 2.556 \cdot 10^{-7}\, e^{(0.0491(Sand))} \quad (16)$$

All parameters in equations 15 and 16 are defined before.

3. Computer models for estimating saturated hydraulic conductivity

3.1 Rosetta

Some PTFs have been incorporated into standalone computer programs like Rosetta (Schaap et al., 2001). Rosetta uses a neural network and bootstrap approach for parameter prediction and uncertainty analysis respectively. Rosetta is able to estimate the van Genuchten water retention parameters (van Genuchten, 1980) and saturated hydraulic conductivity, as well as unsaturated hydraulic conductivity parameters, based on Mualem's (1976) pore-size model (Schaap et al., 2001). Here, Rosetta was used to estimate saturated hydraulic conductivity.

3.2 Soilpar 2

Soilpar 2 provides 15 PTF procedures to estimate soil parameters. The PTFs procedures are classified as point pedotransfer and function pedotransfer. Point PTFs estimate some specific points of interest of the water retention characteristic and/or saturated hydraulic conductivity. Two of these methods also estimate bulk density. Soilpar 2 uses PTFs of Jabro (1992), Jaynes and Tyler (1984), Puckett et al. (1985), and Campbell (1985) to estimate saturated hydraulic conductivity. Function PTFs, which estimate the parameters of retention functions are implemented: Rawls and Brakensiek (1989), to estimate the Brooks and Corey (1964) function parameters; Vereecken et al. (1989), to estimate the van Genuchten (1980) function parameters; Campbell (1985) to estimate the Campbell function parameters (Campbell, 1974)); Mayr and Jarvis (1999), to estimate the parameters of the Hutson and Cass (1987) modification of the Campbell function. All these methods require as input soil particle size distribution and bulk density. The Mayr and Jarvis, and Vereecken et al. methods also require organic carbon content (Acutis and Donatelli, 2003).

4. Pedotransfer functions for estimating unsaturated hydraulic conductivity

One of the most popular analytical functions for predicting unsaturated hydraulic conductivity ($K(\theta)$) is the van Genuchten-Mualem model which is Combination of soil water retention function of the van Genuchten (1980) and Mualem's (1976) pore-size model as follows:

$$K(\theta) = K_s S_e^{0.5}\left[1-\left(1-S_e^{n/(n-1)}\right)^{(1-1/n)}\right]^2 \tag{17}$$

and S_e, is

$$S_e = \frac{\theta(\psi)-\theta_r}{\theta_s-\theta_r} \tag{18}$$

where $\theta(\psi)$ is the measured volumetric water content ($cm^3.cm^{-3}$) at suction ψ (cm-water); θ_r and θ_s are residual and saturation water content ($cm^3.cm^{-3}$) respectively, the dimensionless n is the shape factor, and K_s is the saturated hydraulic conductivity.
Other popular function for predicting $K(\theta)$ is the Brooks and Corey (1964) model as follows:

$$K(\theta) = K_s S_e^{(3+2/\lambda)} \tag{19}$$

where λ is the pore size index.
Campbell (1985) proposed a function for determining $K(\theta)$ as:

$$K(\theta) = K_s\left(\frac{\theta}{\theta_s}\right)^{2b+3} \tag{20}$$

where K_s is saturated hydraulic conductivities, θ is measured volumetric water content($cm^3.cm^{-3}$), θ_s is the saturation water content ($cm^3.cm^{-3}$), and b is the slope of ln ψ vs ln θ in the soil water retention curve.
The unsaturated hydraulic conductivity function is particularly difficult and time-consuming to measure directly. So, in many model applications, reliance is often placed on predictions of unsaturated conductivity based on measurements of soil water retention and K_s. Direct measurements of soil water retention and K_s are time-consuming and costly, too. So here, PTFs are used to estimate unsaturated hydraulic conductivity.
Rawls and Brakensiek (1985) provided equations for the estimation of van Genuchten, Brooks - Corey and Campbell parameters as follows:

$$\begin{aligned} LAM = \exp[&-0.7842831 + 0.0177544 * ps - 1.062498 * por - 0.00005304 * ps^2 - 0.00273493 * \\ &pc^2 + 1.11134946 * por^2 - 0.03088295 * ps * por + 0.00026587 * ps^2 * por^2 - 0.00610522 * \\ &pc^2 * por^2 - 0.00000235 * ps^2 * pc + 0.00798746 * pc^2 * por - 0.00674491 * por^2 * pc] \end{aligned} \tag{21}$$

$$\begin{aligned} \theta_r = &-0.0182482 + 0.00087269 * ps + 0.00513488 * pc + 0.02939286 * por - 0.00015395 * \\ &pc^2 - 0.0010827 * ps * por - 0.00018233 * pc^2 * por^2 + 0.00030703 * pc^2 * \\ &por - 0.0023584 * por^2 * pc \end{aligned} \tag{22}$$

where LAM is pore size index, pc is percent clay, ps is percent sand, por is the porosity.
The unsaturated hydraulic conductivity of van Genuchten parameter (n) is then calculated from the above relations as follow:

$$n = LAM + 1 \tag{23}$$

Campbell's parameter (b) is estimated as follows:

$$b = -1 / LAM \tag{24}$$

In the Brooks and Corey function λ is equal to *LAM*.

5. Computer models for estimating unsaturated hydraulic conductivity

5.1 Rosetta

The van Genuchten parameters, θ_r, θ_s, and n were estimated from measured particle size and bulk density using Rosetta software (Schaap et al., 2001).

5.2 Soilpar 2

Using measured particle size and bulk density data, Campbell model parameter value (b) was estimated using the Soilpar 2 (Acutis and Donatelli, 2003).

Note that, for estimating unsaturated hydraulic conductivity, measured value of θ_s and K_s in the lab were used.

6. Statistical criteria for evaluation of PTFs

6.1 PTFs of saturated hydraulic conductivity

Two following statistical criteria were used for the evaluation of PTFs to estimate saturated hydraulic conductivity based on the approach presented by Tietje and Hennings (1996). Geometric mean error ratio (*GMER*) and geometric standard deviation of the error ratio (*GSDER*) were calculated from the error ratio ε of measured saturated hydraulic conductivity $(K_s)_m$ vs. predicted saturated hydraulic conductivity $(K_s)_p$ values:

$$\varepsilon = \frac{(K_s)_p}{(K_s)_m} \tag{25}$$

$$GMER = \exp\left(\frac{1}{n}\sum_{i=1}^{n}\ln(\varepsilon_i\right) \tag{26}$$

$$GSDER = \exp\left[\left(\frac{1}{n-1}\sum_{i=1}^{n}[\ln(\varepsilon_i) - \ln(GMER)]^2\right)^{0.5}\right] \tag{27}$$

The *GMER* equal to 1 corresponds to an exact matching between measured and predictive saturated hydraulic conductivity; the *GMER*<1 indicates that predicted values of saturated hydraulic conductivity are generally underestimated; *GMER*>1 points to a general over-prediction. The *GSDER* equal to 1 corresponds to a perfect matching and it grows with deviation from measured data. The best PTF will, therefore, give a *GMER* close to 1 and a small *GSDER*.

Also, other statistical criterion named deviation time (DT) was used to evaluate PTFs as follows:

$$\log DT = \left(\frac{1}{n} \sum_{i=1}^{n} (\log \varepsilon_i)^2 \right)^{0.5} \tag{28}$$

The *DT* equal to 1 shows an exact matching between measured and predictive saturated hydraulic conductivity.

6.2 PTFs of unsaturated hydraulic conductivity

Estimated unsaturated hydraulic conductivity using PTFs were compared by calculating modified index of agreement *d'* (Legates and McCabe, 1999):

$$d' = 1.0 - \frac{\sum_{i=1}^{n} |O_i - S_i|}{\sum_{i=1}^{n} \left(|S_i - O'| + |O_i - O'| \right)} \tag{29}$$

where O_i is the individual observed unsaturated hydraulic conductivity ($K(\theta)$) value at θ_i, S_i is the individual simulated value at θ_i, O' is the mean observed value and n is the number of paired observed-simulated values. The value of d' varies from 0.0 to 1.0, with higher values indicating better agreement with the observations. The interpretation of d' closely follows the interpretation of R^2 for the range of most values encountered (Legates and McCabe, 1999).

7. Estimation saturated and unsaturated hydraulic conductivity using Pedotransfer functions

7.1 Saturated hydraulic conductivity

Study area is located in northwest of Iran in Naghadeh county, Azarbaijanegharbi province, Iran (Fig. 1). The total area of the Naghadeh county is 52100 ha and is located at coordinates 36° 57′ N and 45° 22′ E.

Ten locations in the Naghadeh county was considered and undisturbed soil samples were taken by using a steel cylinder of 100 cm^3 volume (5 cm in diameter, and 5.1 cm in height) from 0-15 cm depth to measure saturated hydraulic conductivity and bulk density. The samples were transported carefully to avoid disturbance. Also, disturbed soil samples were taken using plastic bags to measure particle density, soil texture and organic matter. Saturated hydraulic conductivity was measured by the constant-head method (Israelsen and Hansen, 1962). Samples (steel cylinders) of soil were oven dried at 105°C and bulk density was calculated from cylinder volume and oven dry soil mass. Particle size distribution (sand, silt and clay percentages) was measured by the hydrometer method. Soil particle density was measured using a glass pycnometer; 10 g air-dried (<2 mm) soil sample was placed into the pycnometer and the displaced volume of distilled water was determined (Jacob and Clarke, 2002). Total porosity was calculated using bulk density (ρ_b) and particle density (ρ_p) according to the following equation:

$$Porosity = 1 - \frac{\rho_b}{\rho_p} \tag{30}$$

Fig. 1. Location of study area.

The organic matter was determined by Walkley and Black rapid titration method (Nelson and Sommers, 1996).
The measured soil properties are shown in Table 1.

Samples	Texture	Bulk density ($g.cm^{-3}$)	Particle density ($g.cm^{-3}$)	Organic matter	Porosity
1	Clay loam	1.37	2.57	2.07	0.46
2	Silty clay	1.21	2.57	3.03	0.52
3	Silty clay loam	1.07	2.58	1.34	0.58
4	Clay loam	1.23	2.61	1.68	0.52
5	Sandy loam	1.43	2.71	1.01	0.47
6	Silty clay loam	1.18	2.55	1.34	0.53
7	Silty clay	1.02	2.49	2.13	0.58
8	Silty clay loam	1.16	2.52	1.46	0.53
9	Silty clay	1.13	2.56	3.36	0.55
10	Silty clay	1.07	2.55	4.09	0.57

Table 1. Measured soil properties in the study area.

Saturated hydraulic conductivities were estimated according to the above mentioned PTFs (Eqs. 1 to 16) as well as Rosetta and Soilpar 2 software and compared to measured K_s of the 10 soils. Note that PTFs of Jabro, Jaynes and Tyler, Puckett et al. which are used in Soilpar 2 2, hereafter named Soilpar 2-Jabro, Soilpar 2- Jaynes - Tyler, and Soilpar 2- Puckett et al., respectively. Figures 2 and 3 show measured vs. estimated values for all models tested. With regard to Figures 2 and 3, it is clear that Soilpar 2-Jabro for estimating K_s was in excellent agreement with the measured value. After Soilpar 2-Jabro, Rosetta could estimate K_s with reasonable accuracy.

Three statistical criteria (Eqs. 25 to 28) were used for the evaluation of PTFs which estimate saturated hydraulic conductivity. Calculated values of *DT, GMER and GSDER* were shown in Table 2. Soilpar 2-Jabro resulted in lower DT (2.91), GMER and GMER equal to 1.13 and 3.06, respectively, performed better than the others PTFs (Table 2). The PTFs of Vereecken et al. tended to high overestimate saturated hydraulic conductivity, while the rest PTFs generally showed underestimate (Table 2).

It is expected that PTFs including organic matter such as Vereecken et al., Wösten et al., and etc could estimate K_s much better than the others PTFs. But the results showed (see Figures 2 and 3 as well as Table 2) these PTFs could not be able to estimate K_s with reasonable accuracy. The organic matter content is an important variable when infiltration rates are estimated in non-saturated soils, but it has less influence in saturated soils. The main explanation is that organic matter mainly affects retention forces (matric potential), the type of forces that almost do not work in saturated soils where forces are basically affected by gravity. For this reason when estimating water retention parameters in soils, organic matter is a valuable variable to use in PTF (Wösten et al., 1999), but the contribution of organic matter content in estimating K_s was very low and it was mainly limited to explain the relationship between soil structure and K_s.

PTF	DT	GMER	GSDER
Wösten et al., 1997	13.25	0.17	7.51
Wösten et al., 1999	6.90	0.21	3.36
Cosby et al.	22.82	0.06	4.74
Sxaton et al.	26.05	0.05	4.15
Brakensiek et al.	73.25	0.021	7.72
Campbell	11.78	0.12	3.61
Vereecken et al.	17925	16813.29	3.22
Ferrer Julia et al.	527	0.002	5.54
Rosetta	9.61	0.13	3.18
Soilpar 2- Jabro	2.91	1.13	3.06
Soilpar 2- Jynes-Tyler	302.6	0.005	11.16
Soilpar 2- Pukett et al.	291.86	0.007	21.42

Table 2. DT, GMER, and GSDER of the estimated K_s compared to measurement for 12 PTFs

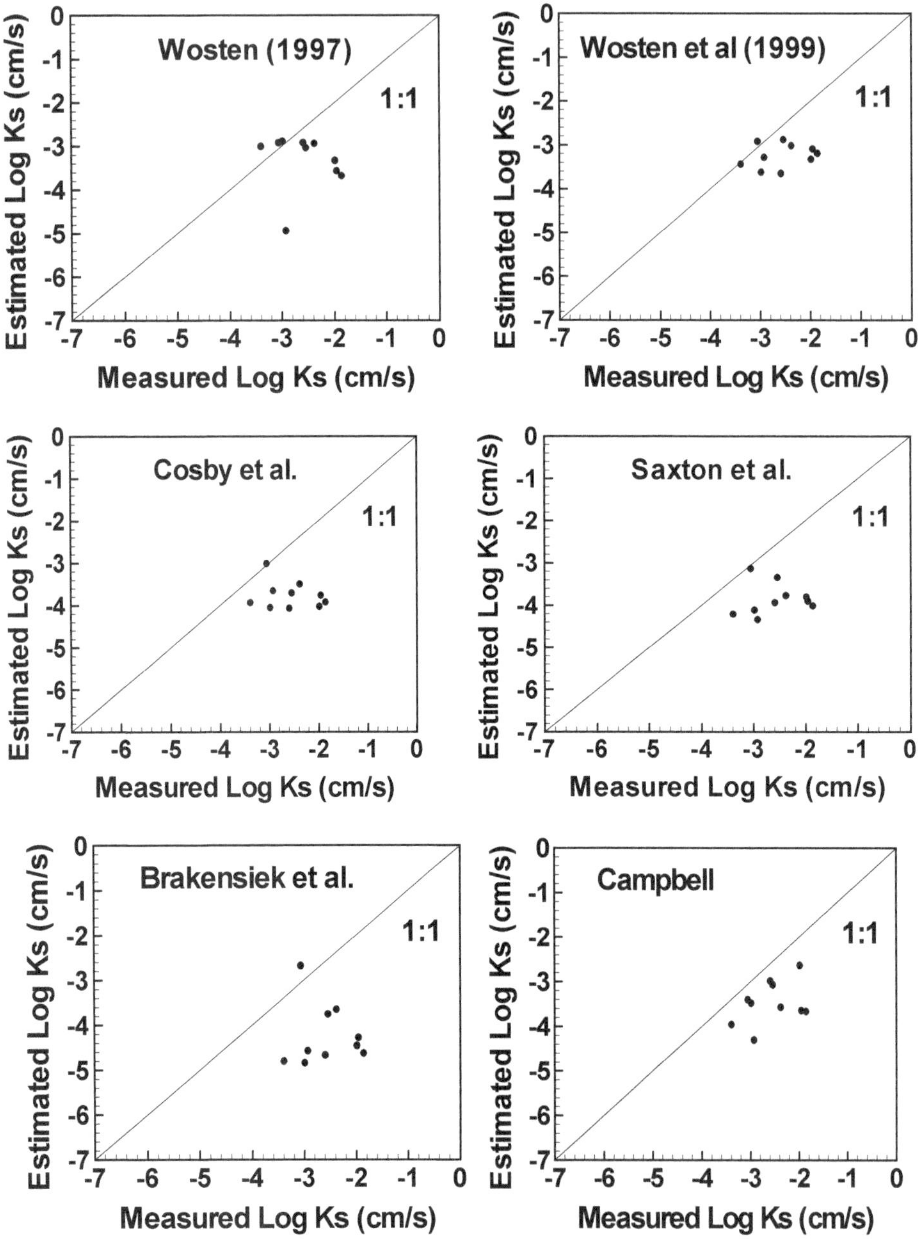

Fig. 2. Measured vs. estimated saturated hydraulic conductivities using PTFs of Wösten et al. (1997), Wösten et al. (1999), Cosby et al., sexton et al., Brakensiek et al. and Campbell for ten soils and 1:1 line

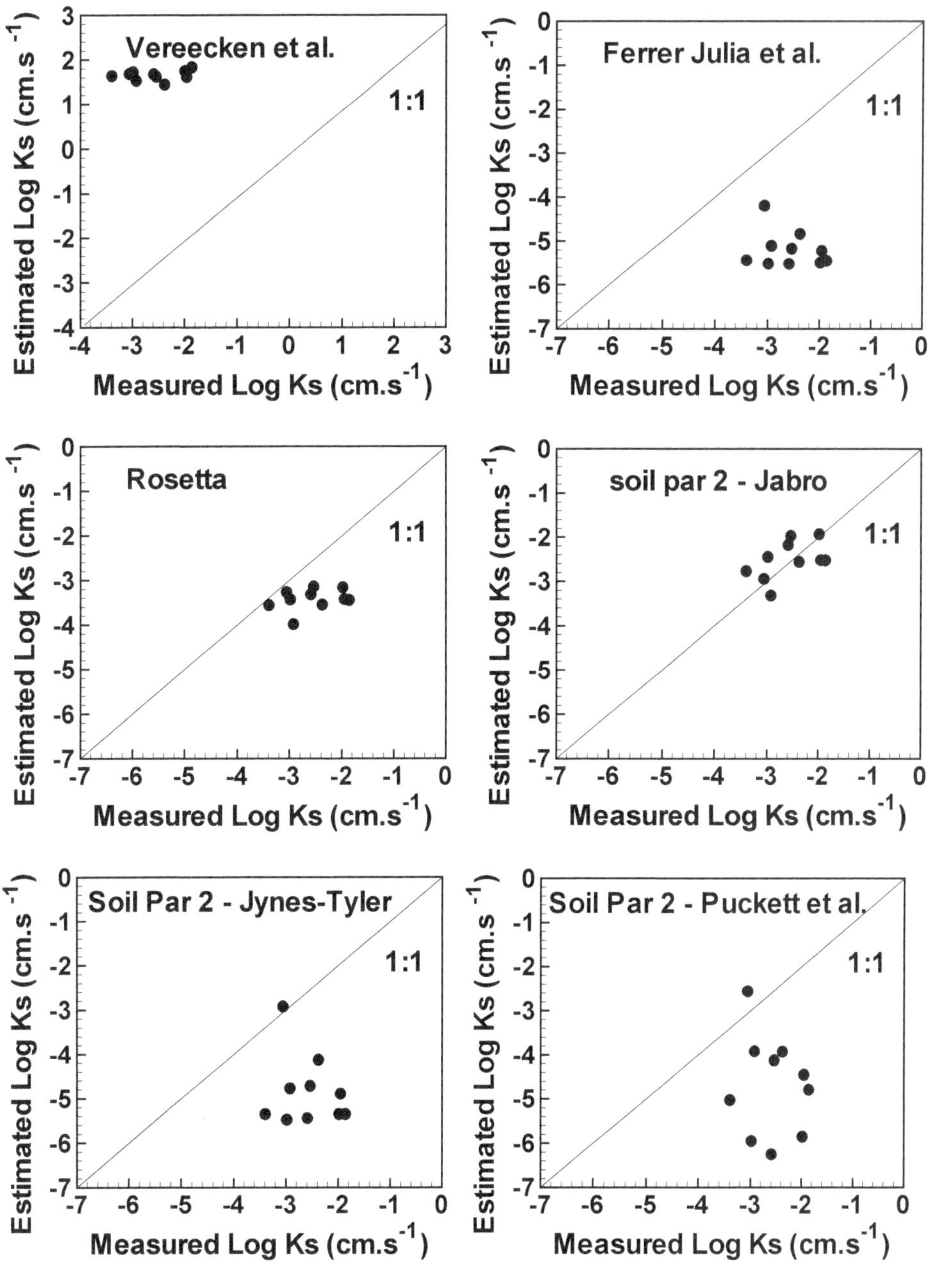

Fig. 3. Measured vs. estimated saturated hydraulic conductivities using PTFs of vereecken et al., Ferrer Julia et al., Rosetta, Soilpar 2-Jabro, Soilpar 2-Jynes-Tyler, and Soilpar 2- Puckett et al. for ten soils and 1:1 line

7.2 Unsaturated hydraulic conductivity

For comparison of the different models in predicting unsaturated hydraulic conductivity of soils, data sets, including data of K_s, measured data of unsaturated hydraulic conductivity, fractions of sand, silt and clay, bulk density of 27 soils were selected from the UNSODA hydraulic property database (Names et al., 1999), and used in the study. The measured soil properties from the UNSODA which used in this study, summarized in Table 3.

Properties	Number	Mean	Min	Max	SD
Bulk density (g. cm^{-3})	27	1.41	0.72	1.8	0.23
Sand (%)	27	49.76	4.30	95.00	30.98
Silt (%)	27	27.91	0.90	70.90	18.20
Clay (%)	27	22.33	1.00	62.00	18.86

Table 3. Mean, standard deviation (SD), Max and Min of soil samples parameters

By using fractions of sand, silt and clay and bulk density, unsaturated hydraulic conductivities ($K(\theta)$) were estimated according to the PTFs of Rawls and Brakensiek (1985) (Eqs. 21 to 24) as well as Rosetta and Soilpar 2 software and compared to measured $K(\theta)$ of the 27 soils. It is noted that PTFs of Rawls and Brakensiek (1985) which were used to estimate parameters of van Genuchten, Brooks - Corey and Campbell functions (Eqs. 17 to 20), hereafter named van Gen-R, B&C-R, and Cam-R, respectively. Also just Campbell model parameter value was estimated using the Soilpar 2, hereafter named Soilpar-Cam.

Figure 4 shows measured vs. estimated $K(\theta)$ by mentioned PTFs. To facilitate comparison of the PTFs, mean value of *modified index of agreement* (d') for the same soil texture classes was calculated (Table 4). With regard to Figure 4 and Table 4, one could conclude that for sand, loamy sand, sandy clay loam, and clay textures, the van Gen-R had the bigger d', indicating its higher accuracy in predicting $K(\theta)$ as compared to the other PTFs. The best PTF for loam, sandy loam, and silty loam textures is the Soilpar-Cam. Wagner et al. (2001) found that the performance of the Campbell model could be improved when the particle size distribution data used in the determining the Campbell parameters are as detailed as possible, while knowledge of only three fractions (clay, silt, and sand) may reduce the function performance considerably.

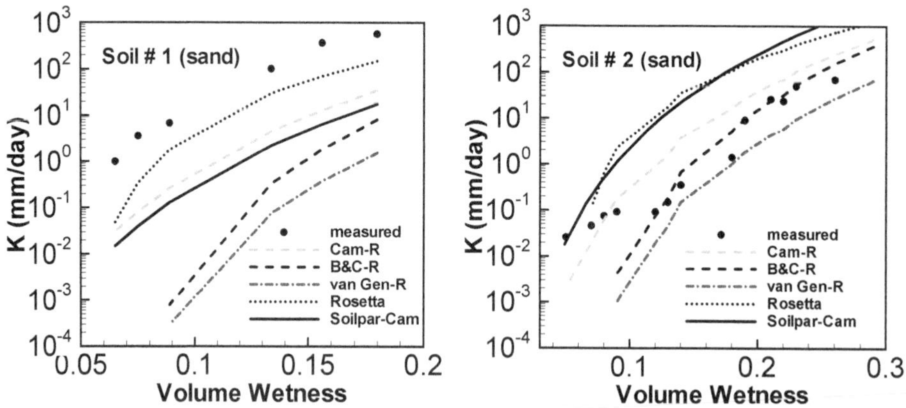

Fig. 4. Comparison of K(θ) measured and estimated by the five PTFs

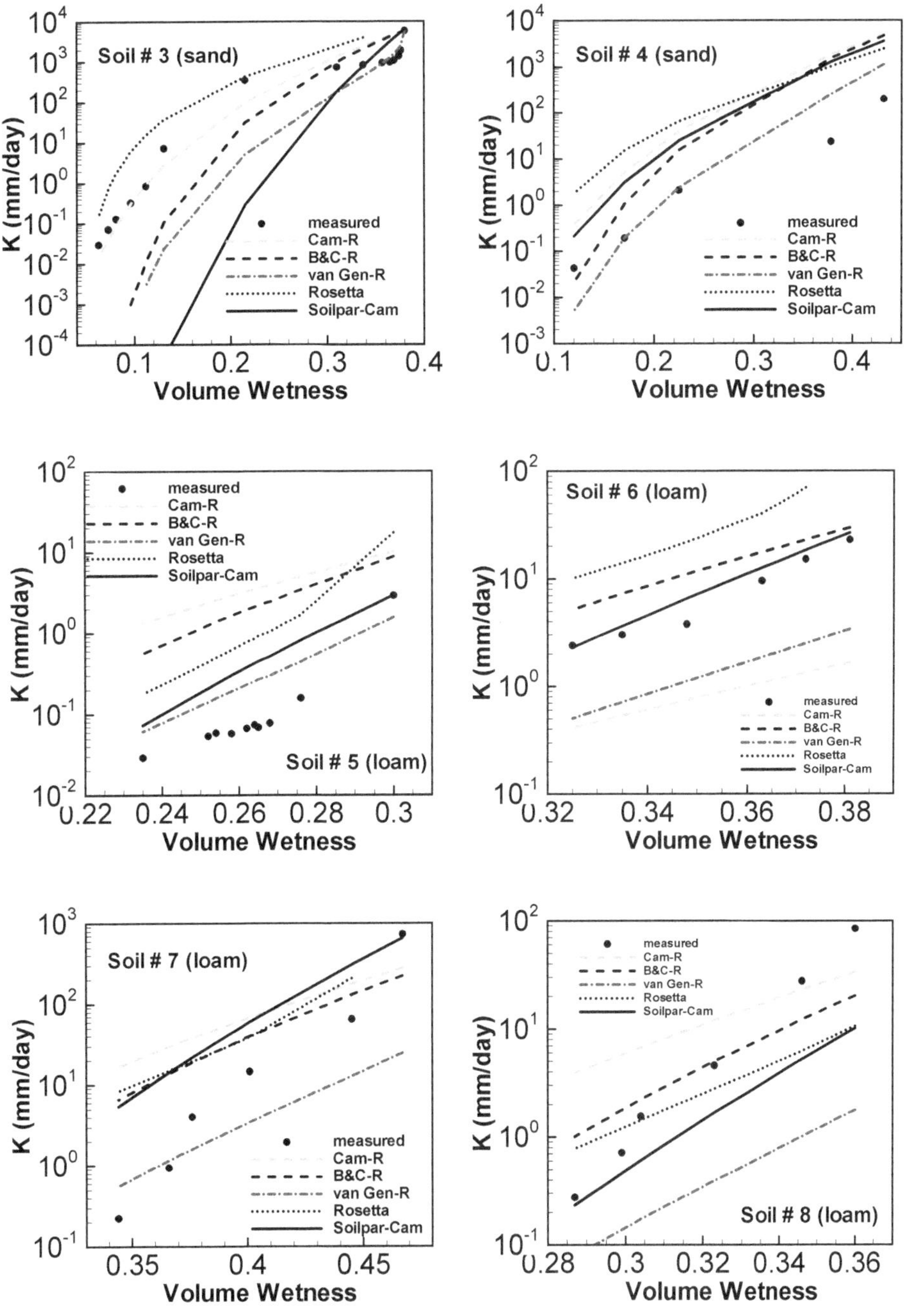

Fig. 4. Continued

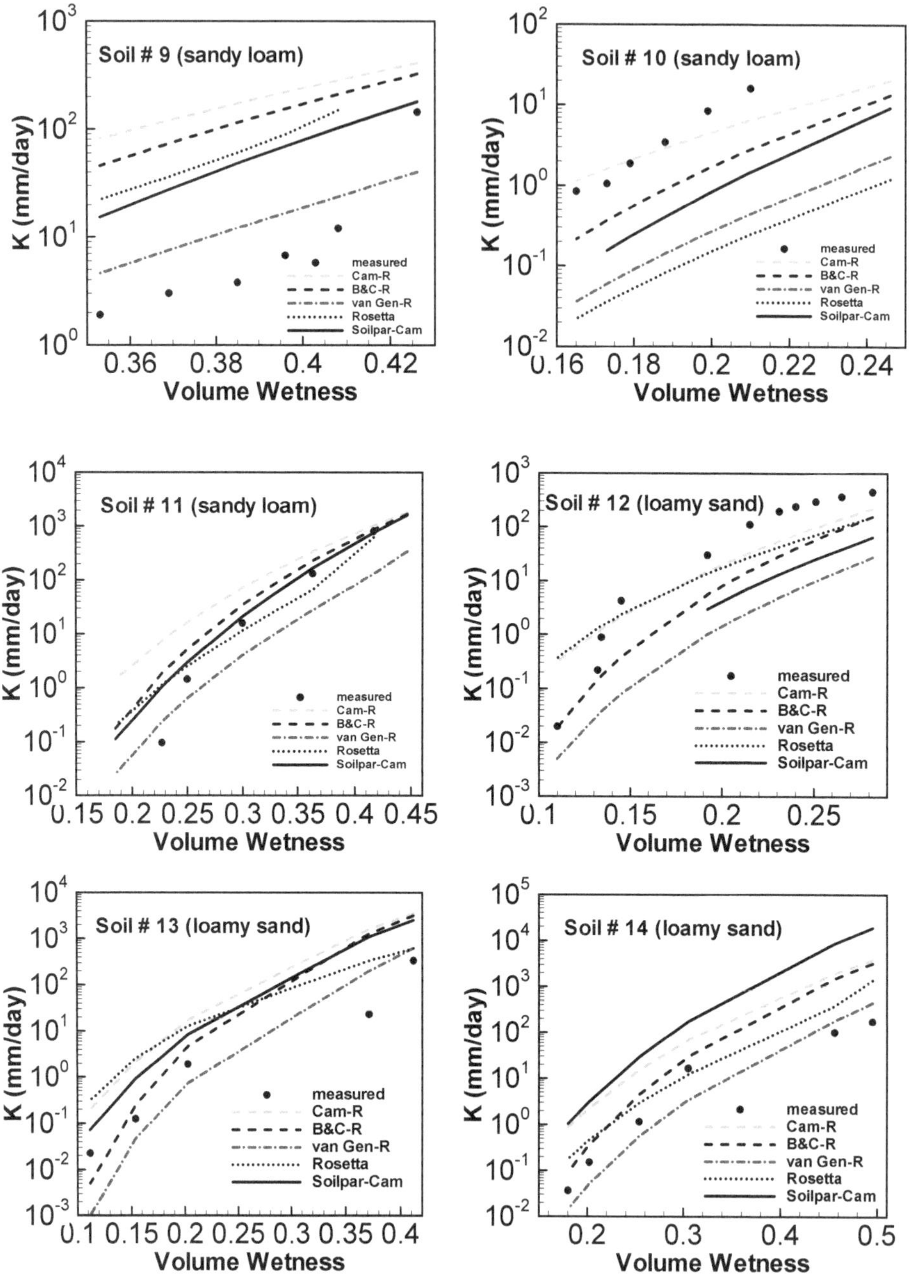

Fig. 4. Continued

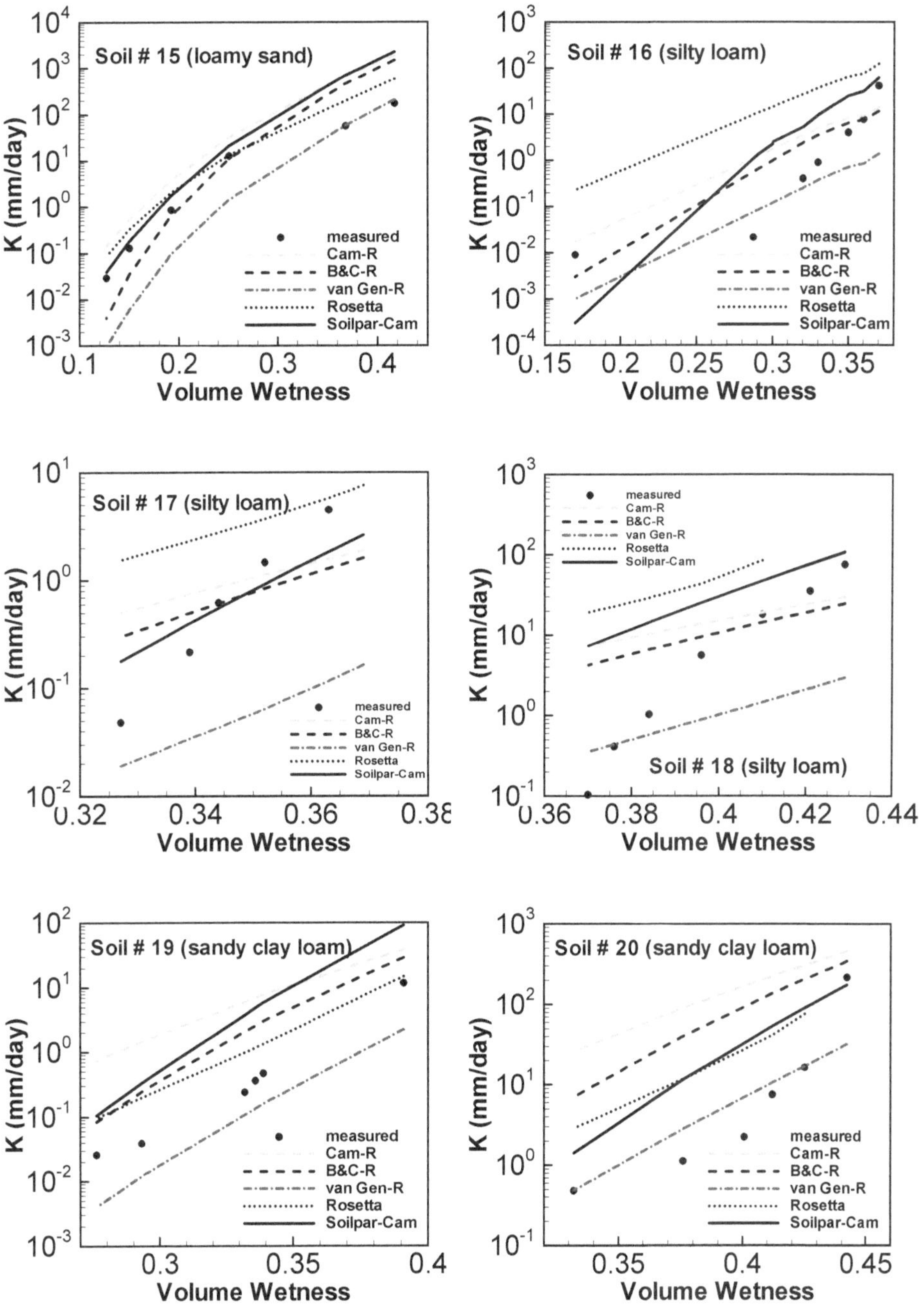

Fig. 4. Continued

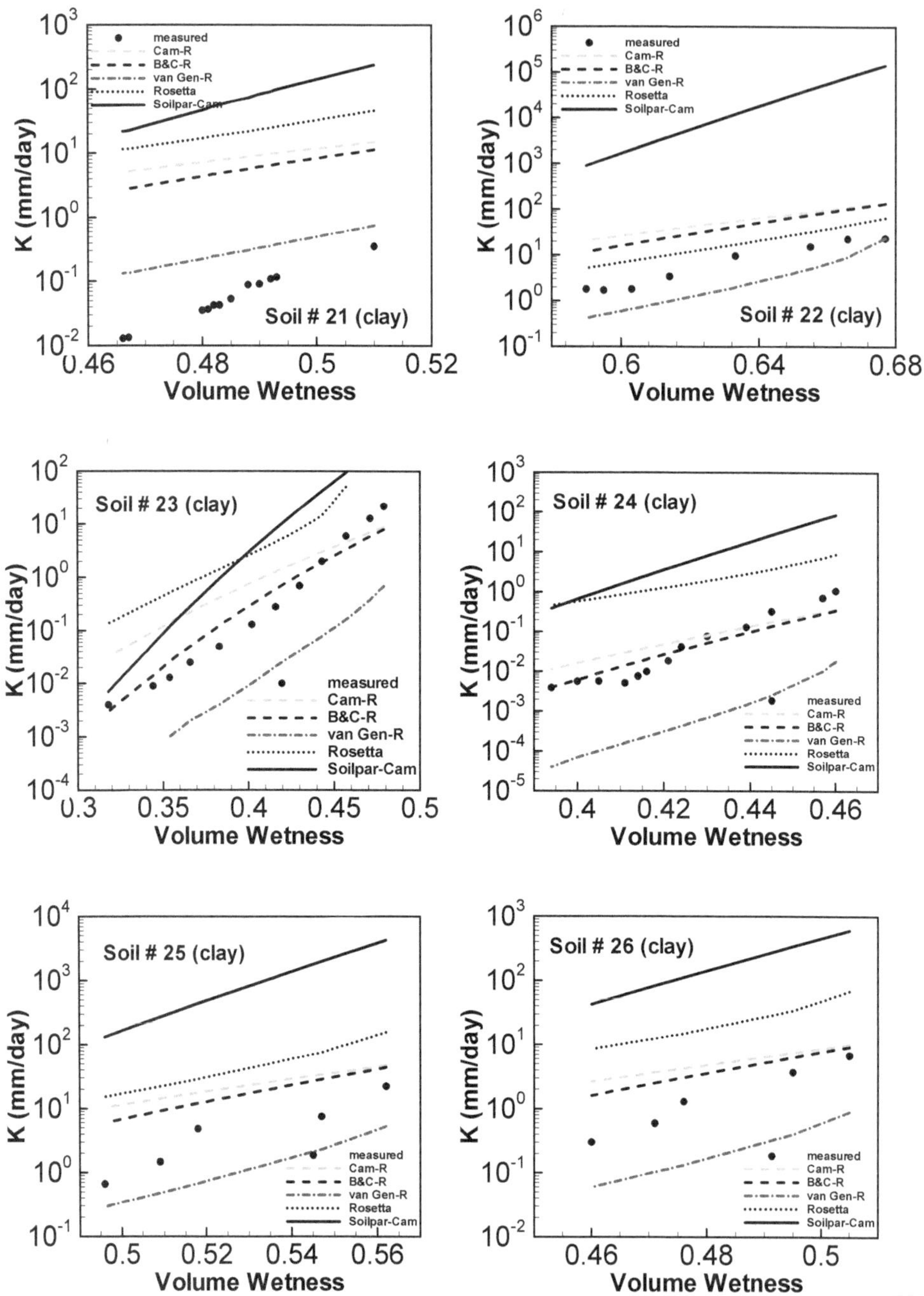

Fig. 4. Continued

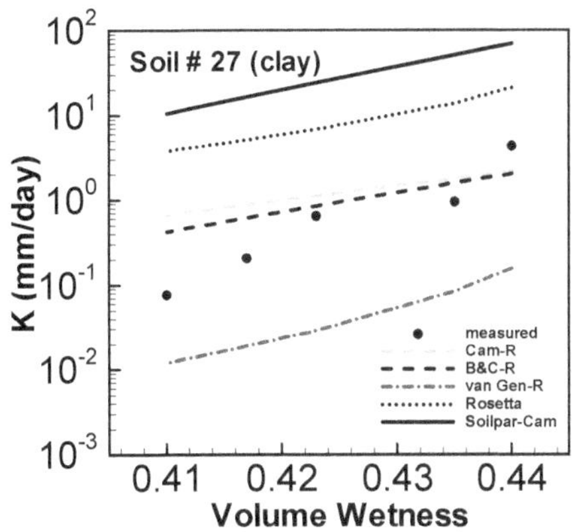

Fig. 4. Continued

Soil texture	Number	Cam-R	B&C-R	van Gen-R	Rosetta	Soilpar-cam
sand	4	0.474	0.485	0.595	0.467	0.425
loam	4	0.399	0.467	0.550	0.318	0.748
sandy loam	3	0.379	0.567	0.593	0.513	0.679
loamy sand	4	0.267	0.281	0.757	0.501	0.302
silty loam	3	0.546	0.578	0.539	0.318	0.590
Sandy clay loam	2	0.311	0.453	0.709	0.535	0.465
clay	7	0.429	0.490	0.513	0.229	0.125

Table 4. Mean value of *modified index of agreement* (d') for the same soil texture

8. Conclusions

Based on the results some of conclusions can be summarized as follows:

- PTFs are a powerful tool to estimate saturated and unsaturated hydraulic conductivity. Because PTFs estimate hydraulic conductivity from easy-to-measure soil properties so they have the clear advantage that they are relatively inexpensive and easy to use.
- The mean of error parameters DT, GMER and GSDER (Table 2) showed that Soilpar 2-Jabro for estimating K_s was in excellent agreement with the measured value in the study area. After Soilpar 2-Jabro, Rosetta could estimate K_s with reasonable accuracy. The PTFs of Vereecken et al. tended to high overestimate saturated hydraulic conductivity. Overestimated K_s by the PTFs of Vereecken et al. makes it a less likely candidate for estimating K_s at the study area or for similar soils. The rest PTFs generally showed underestimate (Table 2).
- The mean value of *modified index of agreement* (d') showed that for sand, loamy sand, sandy clay loam, and clay textures, the van Gen-R had the bigger d', indicating its higher accuracy in predicting $K(\theta)$ as compared to the other PTFs. One can be concluded that Gen-R was approximately good in describing the functional relationship between the soil moisture and unsaturated hydraulic conductivity for mentioned soils. The best

PTF to estimate unsaturated hydraulic conductivity for loam, sandy loam, and silty loam textures was the Soilpar-Cam.

9. References

Acutis M., and M. Donatelli. 2003. SOILPAR 2.00: software to estimate soil hydrological parameters and functions. European Journal of Agronomy 18: 373-377.

Bouma, J. 1989. Using soil survey data for quantitative land evaluation. Advances in Soil Science 9: 177–213.

Brakensiek D.L., W.J. Rawls, and G.R. Stephenson. 1984. Modifying SCS hydrologic soil groups and curve numbers for rangeland soils. ASAE Paper No. PNR-84-203, St. Joseph, MI.

Brooks, R.H., and A.T. Corey. 1964. Hydraulic properties of porous media. Colorado State University, Hydrological paper No. 3, p. 27.

Campbell, G.S. 1974. A simple method for determining unsaturated conductivity from moisture retention data. Soil Science 117: 311-314.

Campbell C.S. 1985. Soil Physics With Basic. Elsevier , New York.149pp.

Christiaens, K., and J. Feyen. 2001. Analysis of uncertainties associated with different methods to determine soil hydraulic properties and their propagation in the distributed hydrological MIKE SHE model. Journal of Hydrology 246: 63– 81.

Cosby B.J., G.M. Hornberger, R.B. Clapp, and T.R. Ginn. 1984. A statistical exploration of the relationship of soil moisture characteristics to the physical properties of soils. Water Resources Research 20 (6): 682-690.

Ferrer-Julià M., T. Estrela Monreal, A. Sánchez del Corral Jiménez and E. García Meléndez. 2004. Constructing a saturated hydraulic conductivity map of Spain using pedotransfer functions and spatial prediction. Geoderma 123: 275-277.

Gupta, R.K., R.P. Rudra, W.T. Dickinson, N.K. Patni, and G.J. Wall. 1993. Comparison of saturated hydraulic conductivity measured by various field methods. Transactions of American Society of Agricultural Engineers 36: 51–55.

Hutson, J.L., and A. Cass. 1987. A retentivity function for use in soil water simulation models. J. Soil Science 38, 105-113.

Islam, N., W. W. Wallender , J. P. Mitchell,S. Wicks, and R. E. Howitt. 2006. Performance evaluation of methods for the estimation of soil hydraulic parameters and their suitability in a hydrologic model. Geoderma, 134:135– 151.

Israelsen, O. W. and V.E. Hansen.1962. Irrigation principles and practices. 3nd ed. John Willey and Sins, Inc. 447 p.

Jabro, J.D., 1992. Estimation of saturated hydraulic conductivity of soils from particle size distribution and bulk density data. Transactions of American Society of Agricultural Engineers 35: 557– 560.

Jacob, H., and G. Clarke. 2002. Methods of Soil Analysis, Part 4, Physical Method. Soil Science Society of America, Inc. Madison, Wisconsin, USA. 1692 p.

Jynes D.B. and E.J. Tyler. 1984. Using soil physical properties to estimate hydraulic conductivity. Soil Science 138: 298-305.

Klute, A., C. Dirksen. 1986. Hydraulic conductivity of saturated soils. In: Klute, A. (Ed.), Methods of Soil Analysis. ASA & SSSA, Madison, Wisconsin, USA, pp. 694–700.

Legates, D.R., G. J. McCabe. 1999. Evaluating the use of "goodness-of-fit" measures in hydrologic and hydroclimatic model validation. Water Resources Research 35: 233–241.

Libardi, P.L., K. Reichardt, D.R. Nielsen, and J.W. Biggar. 1980. Simple field methods for estimating soil hydraulic conductivity. Soil Science Society of America Journal 44: 3–7.

Mallants, D., D. Jacques, P. Tseng, M.T. van Genuchten, and J. Feyen. 1997. Comparison of three hydraulic property measurement methods. Journal of Hydrology 199: 295–318.

Mayr, T., and N.J. Jarvis. 1999. Pedotransfer function to estimate soil water retention parameters for a modified Brooks -Corey type model. Geoderma 91, 1-9.

McBratney, A.B., B. Minasny, S.R. Cattle, and R. W. Vervoort. 2002. From pedotransfer functions to soil inference systems. Geoderma 109: 41-73.

Mualem, Y. 1976. A new model for predicting the hydraulic conductivity of unsaturated porous media. Water Resources Research 12: 513–522.

Nemes, A., M. Schaap, and F. Leij. 1999. the UNSODA unsaturated soil hydraulic database (Version 2.0).US Salinity Laboratory, Reverside,CA.

Nelson, D. W., and L. E. Sommers. 1996. Total carbon, organic carbon, and organic matter. In: D. L. Sparks (Ed.), Methods of Soil Analysis. Part 3: Chemical Methods. Soil Science Society of America, Madison, WI, 961-1010

Paige, G.B., and D. Hillel. 1993. Comparison of three methods for assessing soil hydraulic properties. Soil Science 155: 175–189.

Rawls, W. J. and D. L. Brakensiek.1985. Prediction of soil water properties for hydrologic modeling. In: E. Jones and T. J. Ward (eds.) Watershed Management in the Eighties, Proceedings of a symposium ASCE. 30 Apr.- 1 May. 293-299.

Rawls, W.J., and D.L. Brakensiek. 1989. Estimation of soil water retention and hydraulic properties. In: Morel, S. (Ed.), Unsatured Flow in Hydrologic Modeling. Theory and Pratice. Kluwer academic publishers, pp. 275-300.

Puckett W.E., J.H. Dane, and B.F. Hajek. 1985. Physical and mineralogical data to determine Soil hydraulic properties. Soil Science Society of America Journal 49: 831-836.

Rasoulzadeh, A. 2010. Evaluation of parameters estimation using inverse method in unsaturated porous media, 10th International Agricultural Engineering Conference, Bangkok, Thailand, December 7-10.

Rasoulzadeh, A. and A. Yaghoubi. 2011. Study of cattle manure effect on soil hydraulic properties using inverse method. 2nd International Conference on Environmental Science and Technology (ICEST), Singapore, February 26-28.

Saxton K.E., W.J. Rawls, J.S. Romberger, and R.I. Papendick. 1986. Estimating generalized soil water characteristics from texture. Soil Science Society of America Journal, 50: 1301-1036.

Schaap, M.G., and W. Bouten. 1996. Modeling water retention curves of sandy soils using neural networks. Water Resources Research 32: 3033–3040.

Schaap, M. G., and F. J. Leij. 1998. Using neural networks to predict soil water retention and soil hydraulic conductivity. Soil & Tillage Research 47: 37-42.

Schaap M.G., F.J. Leij, and M.T. van Genuchten. 2001. ROSETTA: a computer program for estimating soil hydraulic parameters with hierarchical pedotransfer functions. Journal of Hydrology 251: 163-176.

Tietje, O., and V. Hennings. 1996. Accuracy of the saturated hydraulic conductivity prediction by pedo-transfer functions compared to the variability within FAO textural classes. Geoderma 69, 71-84.

van Genuchten, M.T. 1980. A closed-form equation for predicting the hydraulic conductivity of unsaturated soils. Soil Science Society of America Journal 44: 892-898.

van Genuchten, M.T., and D.R. Nielsen. 1985. On describing and predicting the hydraulic properties of unsaturated soils. Ann. Geophys. 3: 615- 628.

Vereecken H., J. Maes, and J. Feyen. 1990. Estimating unsaturated hydraulic conductivity from easily measured soil properties. Soil Science 149: 1-12.

Vereecken, H., J. Maes, J., Feyen, and P. Darius. 1989. Estimating the soil moisture retention characteristics from texture, bulk density and carbon content. Soil Science 148: 389-403.

Wagner, B., V.R. Tarnawski, V. Hennings, U. Müller, G. Wessolek, and R. Plagge. 2001. Evaluation of pedo-transfer functions for unsaturated soil hydraulic conductivity using an independent data set. Geoderma 102: 275-297.

Wösten, J.H.M., P.A. Finke, and M.J.W. Jansen. 1995. Comparison of class and continuous pedotransfer functions to generate soil hydraulic characteristics. Geoderma 66: 227-237.

Wösten, J.H.M., and M.T. van Genuchten. 1988. Using texture and other soil properties to predict the unsaturated hydraulic conductivity. Soil Science Society of America Journal 52: 1762-1770.

Wösten J.H.M. 1997. Pedotransfer functions to evaluate soil quality. In:Gegorich, E.G., Carter, M.R. (Eds.), Soil Quality for Crop Production and Ecosystem Health. Developments in Soils Science, vol. 25, Elsesevier, Amesterdam 221-245.

Wösten J.H.M., A. Lilly, A. Nemes, and C. Le Bas. 1999. Development and use of a database of hydraulic properties of European soils. Geoderma 90: 169-185.

Zhang, S., L. Lövdahl, H. Grip, and Y. Tong. 2007. Soil hydraulic properties of two loess soils in China measured by various field-scale and laboratory methods. Catena 69: 264-273.

11

Analytical and Numerical Solutions of Richards' Equation with Discussions on Relative Hydraulic Conductivity

Fred T. Tracy
U.S. Army Engineer Research and Development Center
USA

1. Introduction

Hydraulic conductivity is of central importance in modelling both saturated and unsaturated flow in porous media. This is because it is central to Darcy's Law governing flow velocity and Richards' equation that is often used as the governing partial differential equation (PDE) for unsaturated flow. When doing numerical modelling of groundwater flow, two dominant challenges regarding hydraulic conductivity are heterogeneous media and unsaturated flow.

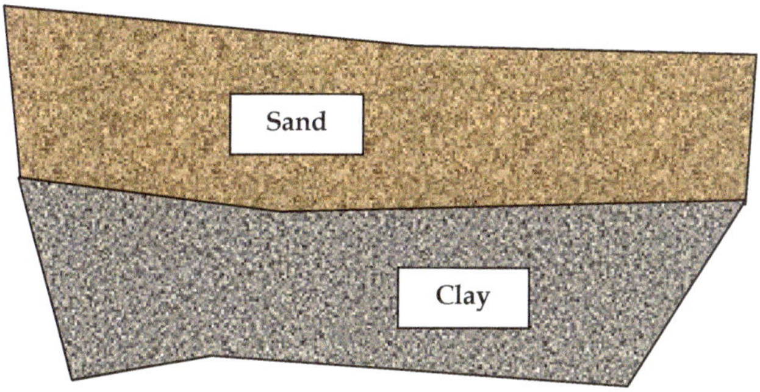

Fig. 1. Heterogeneous soil layers.

1.1 Heterogeneous media

Fig. 1 shows an example of soil layers full of heterogeneities that must be approximated in some way. Fig. 2 shows an idealization of a two-dimensional (2-D) cross section of a levee. Several layers representing different soil types are shown here. It is important to note that each layer is represented by a constant value of horizontal and vertical hydraulic conductivity rather than, for instance, a statistical variation. This is often done in numerical models and will be implemented in this work. The hydraulic conductivity values for sand and gravel are two to four times those of the silt and clay.

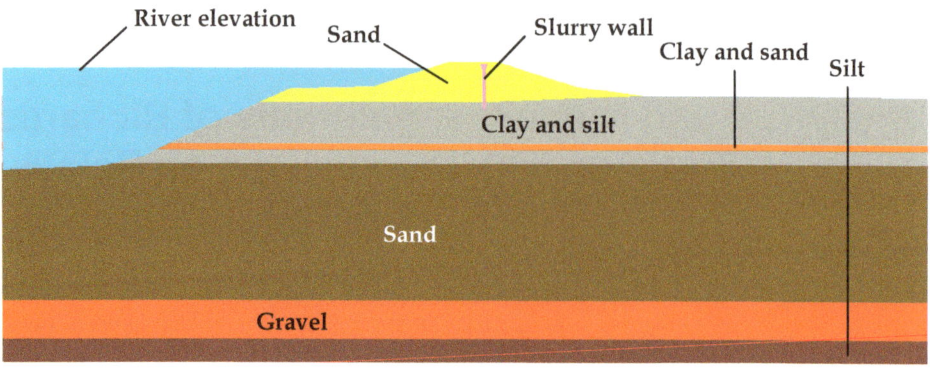

Fig. 2. Levee cross section with several soil types and a slurry wall.

An additional complexity is added in this problem by inserting the slurry wall. This type of wall is typically much less pervious than the surrounding soil, creating further stress on the computational model. This is because the numerical solution that is usually done requires a solution of a system of simultaneous, linear equations. The greater the span of orders of magnitude of hydraulic conductivity, the more challenging the solution of this system becomes.

1.2 Unsaturated flow

The last major concern and challenge discussed in this chapter regarding hydraulic conductivity with regard to computational and analytical solutions is unsaturated flow. Fig. 3 shows the location of the phreatic surface for steady-state conditions. The phreatic surface is where the ground goes from fully saturated when the soil voids are completely filled with water to partially saturated voids in the soil matrix. Above this phreatic surface, hydraulic conductivity is often modelled by

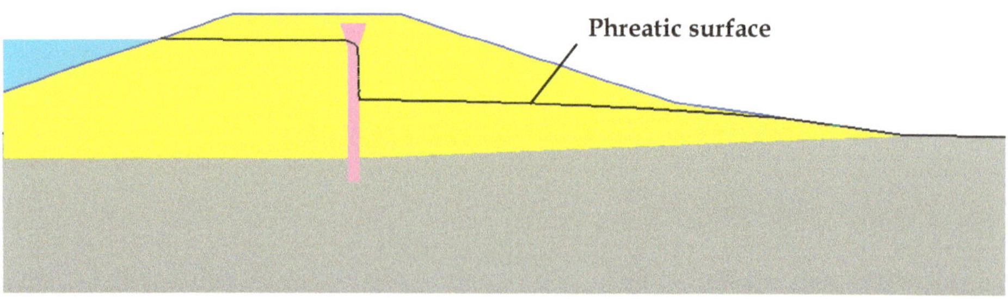

Fig. 3. Location of the phreatic surface at steady-state conditions.

$$k = k_r k_s \tag{1}$$

where

k = hydraulic conductivity of a given soil type

k_s = hydraulic conductivity for saturated soil

k_r = relative hydraulic conductivity for unsaturated soil

k_r is set to 1 in the saturated zone, but varies with the pressure head (h) in the unsaturated zone. There are many expressions for relative hydraulic conductivity in the literature and practice. Some of these will be discussed later in this chapter.

1.3 Obtaining computational results

A discretization of the flow region must be done to do the numerical analysis. Many techniques are available, but in this chapter, the finite element method (Cook, 1981) will be emphasized. Fig. 4 shows a zoom of the finite element mesh for the 2-D levee cross section given in Fig. 2 consisting of triangular elements. Define the total head as

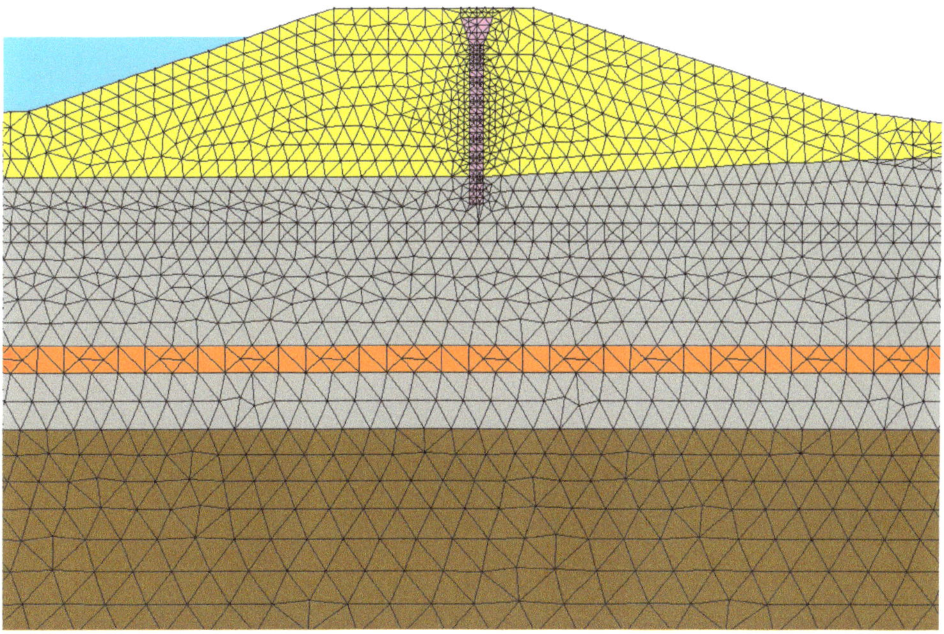

Fig. 4. Portion of the triangular mesh for the levee cross section.

$$\phi = h + z \tag{2}$$

where

h = pressure head

ϕ = total head

z = z coordinate or elevation

Then equipotentials or total head contours can be used as a good way to visualize the data computed at each node of the mesh. Fig. 5 shows this type of plot for the levee example.

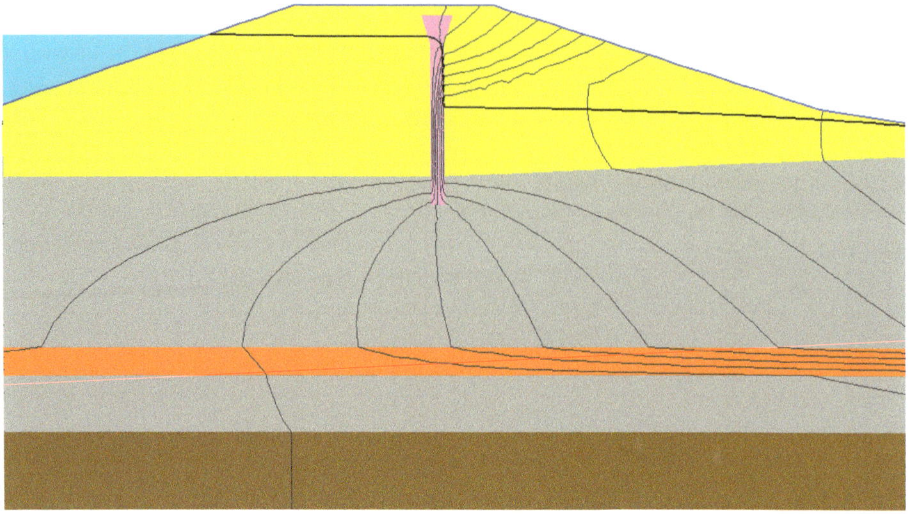

Fig. 5. Total head contours and phreatic surface.

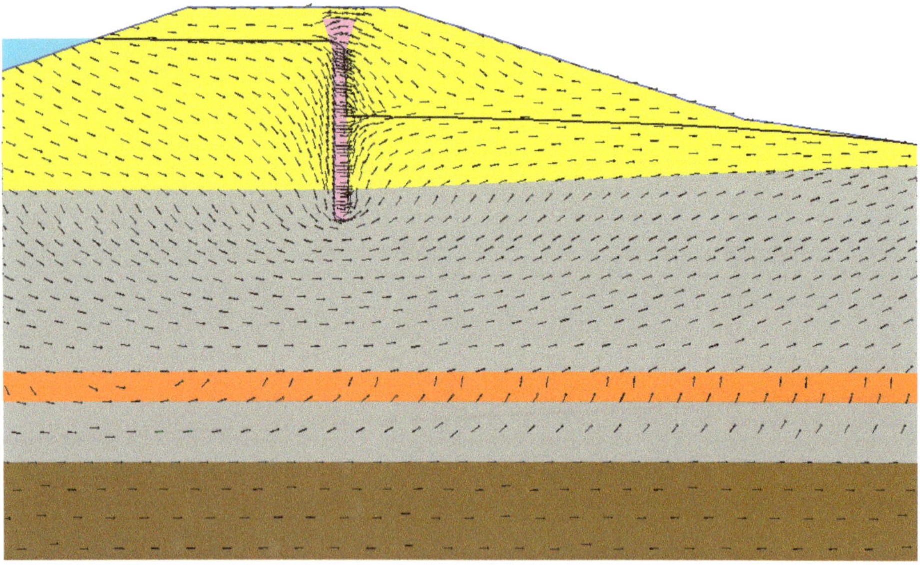

Fig. 6. Velocity vectors and phreatic surface.

Finally, using Darcy's Law for a homogeneous medium,

$$\mathbf{v} = -k\nabla\phi \tag{3}$$

where

$\mathbf{v}$ = flow velocity

A plot of velocity vectors for the levee cross section can be computed and plotted (see Fig. 6).

2. Relative hydraulic conductivity

One common way of representing relative hydraulic conductivity is using the van Genuchten expression (van Genuchten, 1980). First,

$$\begin{aligned} S_e &= \left[1+\zeta(-h)^n\right]^{-m}, \quad m = 1-\frac{1}{n}, \quad h \le 0 \\ &= 1, \quad h \ge 0 \end{aligned} \tag{4}$$

where

S_e = effective saturation

ζ = parameter based on soil type

n = parameter based on soil type

Then,

$$\begin{aligned} k_r &= \sqrt{S_e}\left[1-\left(1-S_e^{1/m}\right)^m\right]^2, \quad h \le 0 \\ &= 1, \quad h \ge 0 \end{aligned} \tag{5}$$

A simpler but less useful expression for relative hydraulic conductivity is the Gardner formulation (Gardner, 1958),

$$k_r = e^{\alpha h} \tag{6}$$

where

α = parameter based on soil type

Eq. 6 is shown here because this simpler equation is needed in the derivation of analytical solutions given later in this chapter. Regardless of the middle part of the curves, all relative hydraulic conductivity equations go from 1 at $h = 0$ to near 0 for negative values of h. In all these discussions, pressure head is greater than zero for saturated flow, equal to zero at the phreatic surface, and less than zero in the unsaturated zone.

3. Richards' equation

A common way of characterizing unsaturated flow is Richards' equation (Richards, 1931). A general version of this equation is

$$\nabla \cdot (\mathbf{K} \cdot \nabla \phi) = \frac{\partial \theta}{\partial t} \tag{7}$$

where

$\mathbf{K}$ = hydraulic conductivity tensor

θ = moisture content

t = time

For a homogeneous, isotrophic medium, **K** becomes k times the identity matrix, so after using Eqs. 1 and 2, Eq. 7 becomes,

$$\nabla \cdot (k_r \nabla h) + \frac{\partial k_r}{\partial z} = \frac{1}{k_s} \frac{\partial \theta}{\partial t} \tag{8}$$

Eq. 8 will be used for deriving the analytical solutions. The fact that k_r is a function of h creates significant difficulty both for solving this problem numerically and deriving analytical solutions since now Eq. 8 is often severely nonlinear.

4. Analytical solutions

Analytical solutions are an excellent tool for checking numerical programs for accuracy. In these derivations, hydraulic conductivity plays an important role. The challenge is finding a form of relative hydraulic conductivity such that the nonlinear Richards' equation can be converted from a nonlinear to a linear form. The derivations presented here are mirrored after those presented earlier (Tracy, 2006, 2007) because they lend themselves to one-dimensional (1-D), 2-D, and three-dimensional (3-D) solutions. First, 1-D and 2-D analytical solutions will be derived, and then numerical finite element solutions highlighting accuracy for different representations of relative hydraulic conductivity will be investigated.

4.1 1-D analytical solution of the Green-Ampt problem

Fig. 7 shows the 1-D problem that will be considered in detail. A column of soil of height, L, is initially dry until water begins to infiltrate the soil. A pool of water at the ground surface is then maintained holding the pressure head to zero. This is known as the 1-D Green-Ampt problem (Green & Ampt, 1911).

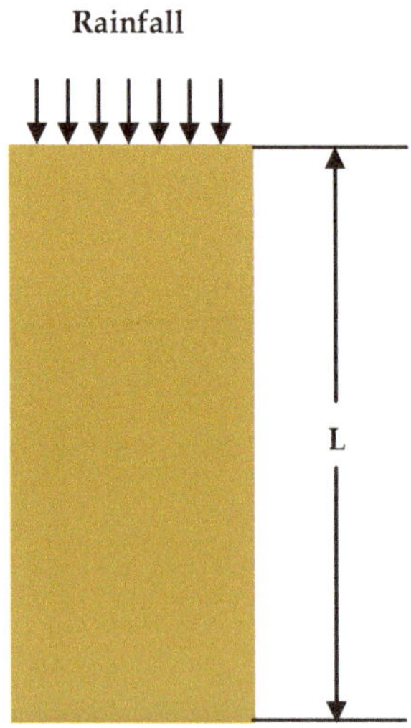

Fig. 7. A view of a 1-D column of soil that is initially dry until water is applied at the top of the ground surface from rainfall.

This problem is challenging numerically because the change in relative hydraulic conductivity is so dramatic, as it goes from small to one. There are several steps that are involved in the derivation for this problem, and they will now be summarized.

1. Provide a function of relative hydraulic conductivity and moisture content as a function of pressure head.
2. Establish initial and boundary conditions.
3. Perform a change of variables to linearize Richards' equation.
4. Solve this new PDE for the steady-state solution.
5. Obtain yet another PDE using a second change of variables.
6. Use separation of variables.
7. Use Fourier series to solve the current PDE.
8. Transform back to the original variables.

4.1.1 Relative hydraulic conductivity and moisture content

Gardner's equation (Eq. 6) is used for relative hydraulic conductivity, and moisture content is given by

$$\theta = \theta_d + (\theta_s - \theta_d) S_e \tag{9}$$

where

θ_d = moisture content when the soil is dry

θ_s = moisture content when the soil is saturated

Rather than use the van Genuchten expression for S_e, a simpler version is used (Warrick, 2003) as follows:

$$S_e = k_r \tag{10}$$

This equation is more limiting in actual practical application, but it allows easier derivation of the analytical solution. It is certainly good enough to test different computational strategies in computer programs.

4.1.2 Initial and boundary conditions

The initial conditions are that the soil is dry. Thus,

$$h(z,0) = h_d \tag{11}$$

where

h_d = the pressure head when the soil is dry

At $t > 0$, the boundary conditions at $z = 0$ and $z = L$ (top of the soil sample or at the ground surface) are

$$\begin{aligned} h(0,t) &= h_d \\ h(L,t) &= 0 \end{aligned} \tag{12}$$

4.1.3 Change of variables

The 1-D version of Eq. 8 is

$$\frac{\partial}{\partial z}\left(k_r\frac{\partial h}{\partial z}\right)+\frac{\partial k_r}{\partial z}=\frac{1}{k_s}\frac{\partial \theta}{\partial t} \tag{13}$$

Let the new variable, $\bar{h}$, be defined as

$$\bar{h}=e^{\alpha h}-\varepsilon, \quad \varepsilon=e^{\alpha h_d} \tag{14}$$

Then

$$\frac{\partial \bar{h}}{\partial z}=\alpha e^{\alpha h}\frac{\partial h}{\partial z} \tag{15}$$

and therefore,

$$k_r\frac{\partial h}{\partial z}=e^{\alpha h}\left(\frac{1}{\alpha}e^{-\alpha h}\frac{\partial \bar{h}}{\partial z}\right)=\frac{1}{\alpha}\frac{\partial \bar{h}}{\partial z} \tag{16}$$

In a similar manner,

$$\frac{\partial k_r}{\partial z}=\alpha e^{\alpha h}\frac{\partial h}{\partial z}=\frac{\partial \bar{h}}{\partial z} \tag{17}$$

and

$$\frac{\partial \theta}{\partial t}=(\theta_s-\theta_r)\frac{\partial S_e}{\partial t}=(\theta_s-\theta_r)\frac{\partial k_r}{\partial t}=(\theta_s-\theta_r)\frac{\partial \bar{h}}{\partial t} \tag{18}$$

Putting Eqs. 15-18 into Eq. 13 gives

$$\frac{\partial^2 \bar{h}}{\partial z^2}+\alpha\frac{\partial \bar{h}}{\partial z}=c\frac{\partial \bar{h}}{\partial t}, \quad c=\frac{\alpha(\theta_s-\theta_d)}{k_s} \tag{19}$$

with initial conditions,

$$\bar{h}(z,0)=0 \tag{20}$$

and boundary conditions for $t>0$ from Eq. 12,

$$\begin{aligned}\bar{h}(0,t)&=0\\ \bar{h}(L,t)&=1-\varepsilon\end{aligned} \tag{21}$$

4.1.4 Steady-state solution

The steady-state version of Eq. 19 will now be solved. It is important to note that this steady-state version now becomes an ordinary differential equation (ODE) as follows:

$$\frac{d^2\overline{h}_{ss}}{dz^2}+\alpha\frac{d\overline{h}_{ss}}{dz}=0 \tag{22}$$

where $\overline{h}_{ss}$ is the steady-state solution. The general solution to this equation is

$$\overline{h}_{ss}=A_1+A_2e^{-\alpha z} \tag{23}$$

where A_1 and A_2 are constants to be evaluated. When applying the boundary conditions of Eq. 21, the result is

$$\begin{aligned}0&=A_1+A_2\\1-\varepsilon&=A_1+A_2e^{-\alpha L}\\A_2&=-A_1\\A_1&=\frac{1-\varepsilon}{1-e^{-\alpha L}}\end{aligned} \tag{24}$$

The steady-state solution then becomes

$$\begin{aligned}\overline{h}_{ss}(z)&=(1-\varepsilon)\frac{1-e^{-\alpha z}}{1-e^{-\alpha L}}\\&=(1-\varepsilon)\frac{e^{-\left(\frac{\alpha}{2}z\right)}\left(\frac{e^{\left(\frac{\alpha}{2}z\right)}-e^{-\left(\frac{\alpha}{2}z\right)}}{2}\right)}{e^{-\left(\frac{\alpha}{2}L\right)}\left(\frac{e^{\left(\frac{\alpha}{2}L\right)}-e^{-\left(\frac{\alpha}{2}L\right)}}{2}\right)}\\&=(1-\varepsilon)e^{\frac{\alpha}{2}(L-z)}\frac{\sinh\left(\frac{\alpha}{2}z\right)}{\sinh\left(\frac{\alpha}{2}L\right)}\end{aligned} \tag{25}$$

4.1.5 Another transformation

Yet another transformation is now applied to Eq. 19. Define

$$\hat{h}=\overline{h}-\overline{h}_{ss} \tag{26}$$

Eq. 19 now becomes

$$\frac{\partial^2\left(\hat{h}+\overline{h}_{ss}\right)}{\partial z^2}+\alpha\frac{\partial\left(\hat{h}+\overline{h}_{ss}\right)}{\partial z}=c\frac{\partial\left(\hat{h}+\overline{h}_{ss}\right)}{\partial t} \tag{27}$$

Now since $\overline{h}_{ss}$ is the steady-state solution (Eq. 22), then

$$\frac{\partial^2 \hat{h}}{\partial z^2} + \alpha \frac{\partial \hat{h}}{\partial z} + \frac{\partial^2 \bar{h}_{ss}}{\partial z^2} + \alpha \frac{\partial \bar{h}_{ss}}{\partial z} = c \frac{\partial \hat{h}}{\partial t} + c \frac{\partial \bar{h}_{ss}}{\partial t}$$
$$\frac{\partial^2 \hat{h}}{\partial z^2} + \alpha \frac{\partial \hat{h}}{\partial z} = c \frac{\partial \hat{h}}{\partial t} \tag{28}$$

with initial and boundary conditions,

$$\hat{h}(z,0) = -\bar{h}_{ss}, \quad \hat{h}(0,t) = \hat{h}(L,t) = 0 \tag{29}$$

4.1.6 Separation of variables

Eq. 28 can be solved using separation of variables. $\hat{h}$ will be cast into the form,

$$\hat{h}(z,t) = \mathrm{Z}(z)\mathrm{T}(t) \tag{30}$$

where $\mathrm{Z}(z)$ is a function only of z, and $\mathrm{T}(t)$ is a function only of t. Substituting Eq. 30 into Eq. 28 and dividing by ZT gives

$$\mathrm{T}\frac{\partial^2 \mathrm{Z}}{\partial z^2} + \alpha \mathrm{T} \frac{\partial \mathrm{Z}}{\partial z} = c\mathrm{Z}\frac{\partial \mathrm{T}}{\partial t}$$
$$\frac{1}{\mathrm{Z}}\left(\frac{\partial^2 \mathrm{Z}}{\partial z^2} + \alpha \frac{\partial \mathrm{Z}}{\partial z}\right) = \frac{c}{\mathrm{T}}\frac{\partial \mathrm{T}}{\partial t} \tag{31}$$

The only nontrivial solution occurs when the left- and right-hand sides of Eq. 31 are set to the same arbitrary constant, η. Thus,

$$\frac{1}{\mathrm{Z}}\left(\frac{\partial^2 \mathrm{Z}}{\partial z^2} + \alpha \frac{\partial \mathrm{Z}}{\partial z}\right) = \eta = \frac{c}{\mathrm{T}}\frac{\partial \mathrm{T}}{\partial t}$$
$$\frac{\partial^2 \mathrm{Z}}{\partial z^2} + \alpha \frac{\partial \mathrm{Z}}{\partial z} - \eta \mathrm{Z} = 0, \quad c\frac{\partial \mathrm{T}}{\partial t} - \eta \mathrm{T} = 0 \tag{32}$$

This leads to the characteristic equations,

$$m_1^2 + \alpha m_1 - \eta = 0, \quad c m_2 - \eta = 0 \tag{33}$$

with solutions,

$$m_{1a} = \frac{-\alpha + \sqrt{\alpha^2 + 4\eta}}{2}, \quad m_{1b} = \frac{-\alpha - \sqrt{\alpha^2 + 4\eta}}{2}, \quad m_2 = \frac{\eta}{c} \tag{34}$$

The general solution for $\hat{h}$ now becomes

$$\mathrm{Z} = a_1 e^{m_{1a}z} + a_2 e^{m_{1b}z}, \quad \mathrm{T} = e^{m_2 t}$$
$$\hat{h} = \mathrm{ZT} = \left(a_1 e^{m_{1a}z} + a_2 e^{m_{2b}z}\right)e^{m_2 t} \tag{35}$$

where a_1 and a_2 are determined by initial and boundary conditions. For a physically realizable system, $\eta < 0$. To eliminate the radicals and to cast in a form that helps realize the general nature of the solution, the choice,

$$\eta = -\frac{\alpha^2}{4} - \lambda_k^2, \quad \lambda_k = \frac{\pi}{L}k, \quad k = 0,1,2,... \tag{36}$$

is made. This gives

$$\begin{aligned} \hat{h}_k &= \left(a_{1k}e^{m_{1a}z} + a_{2k}e^{m_{1b}z}\right)e^{m_2 t} \\ &= \left(a_{1k}e^{i\lambda_k z} + a_{2k}e^{-i\lambda_k z}\right)e^{-\frac{\alpha}{2}z-\mu_k t}, \quad \mu_k = \frac{1}{c}\left(\frac{\alpha^2}{4} + \lambda_k^2\right), \quad i = \sqrt{-1} \end{aligned} \tag{37}$$

It is best to rewrite Eq. 37 in terms of sine and cosine series and two other constants, A_k and B_k, to be evaluated. Thus, for all non-negative integers, k,

$$\hat{h} = \sum_{k=0}^{\infty}\left(A_k \sin\lambda_k z + B_0 + B_k \cos\lambda_k z\right)e^{-\frac{\alpha}{2}z-\mu_k t} \tag{38}$$

However, $\hat{h} = 0$ at $z = 0$, so $B_0 = B_k = 0$ and the final form is

$$\hat{h} = \sum_{k=1}^{\infty} A_k \sin\lambda_k z\, e^{-\frac{\alpha}{2}z-\mu_k t} \tag{39}$$

4.1.7 Fourier series solution

A_k in Eq. 39 can be evaluated by using Fourier series. Starting with

$$\hat{h}(z,0) = \sum_{k=1}^{\infty} A_k \sin\lambda_k z\, e^{-\frac{\alpha}{2}z} \tag{40}$$

the result from using Eqs. 25 and 29 is

$$\begin{aligned} A_k &= \frac{2}{L}\int_0^L e^{\frac{\alpha}{2}z}\hat{h}(z,0)\,dz \\ &= -\frac{2(1-\varepsilon)}{L\sinh\left(\frac{\alpha}{2}L\right)}e^{\frac{\alpha}{2}L}\int_0^L \sinh\left(\frac{\alpha}{2}z\right)\sin\lambda_k z\,dz \end{aligned} \tag{41}$$

The last item in determining $\hat{h}$ is to evaluate the integral,

$$
\begin{aligned}
I &= \int_0^L \sinh\left(\frac{\alpha}{2}z\right)\sin\lambda_k z\,dz \\
&= \frac{2}{\alpha}\cosh\left(\frac{\alpha}{2}z\right)\sin\lambda_k z\Big]_0^L - \frac{2\lambda_k}{\alpha}\int_0^L \cosh\left(\frac{\alpha}{2}z\right)\cos\lambda_k z\,dz \\
&= -\frac{2\lambda_k}{\alpha}\int_0^L \cosh\left(\frac{\alpha}{2}z\right)\cos\lambda_k z\,dz \\
&= -\frac{4\lambda_k}{\alpha^2}\sinh\left(\frac{\alpha}{2}z\right)\cos\lambda_k z\Big]_0^L - \frac{4\lambda_k^2}{\alpha^2}\int_0^L \sinh\left(\frac{\alpha}{2}z\right)\sin\lambda_k z\,dz \\
&= \frac{4\lambda_k}{\alpha^2}\sinh\left(\frac{\alpha}{2}L\right)(-1)^{k+1} - \frac{4\lambda_k^2}{\alpha^2}I
\end{aligned}
\tag{42}
$$

$$
\left(1+\frac{4\lambda_k^2}{\alpha^2}\right)I = \frac{4\lambda_k}{\alpha^2}\sinh\left(\frac{\alpha}{2}L\right)(-1)^{k+1}
$$

$$
\begin{aligned}
I &= \frac{\lambda_k}{\left(\frac{\alpha^2}{4}+\lambda_k^2\right)}\sinh\left(\frac{\alpha}{2}L\right)(-1)^{k+1} \\
&= \frac{\lambda_k}{c\mu_k}\sinh\left(\frac{\alpha}{2}L\right)(-1)^{k+1}
\end{aligned}
$$

The solution for $\hat{h}$ then becomes

$$
\hat{h} = \frac{2(1-\varepsilon)}{Lc}e^{\frac{\alpha}{2}(L-z)}\sum_{k=1}^{\infty}(-1)^k\frac{\lambda_k}{\mu_k}\sin\lambda_k z\,e^{-\mu_k t} \tag{43}
$$

4.1.8 Transform back

The last remaining task is to convert back to the original coordinates using Eqs. 14, 25, and 26. Therefore,

$$
\begin{aligned}
\bar{h} &= \hat{h} + \bar{h}_{ss} \\
&= (1-\varepsilon)e^{\frac{\alpha}{2}(L-z)}\left[\frac{\sinh\left(\frac{\alpha}{2}z\right)}{\sinh\left(\frac{\alpha}{2}L\right)} + \frac{2}{LC}\sum_{k=1}^{\infty}(-1)^k\frac{\lambda_k}{\mu_k}\sin\lambda_k z\,e^{-\mu_k t}\right]
\end{aligned}
\tag{44}
$$

$$
h = \frac{1}{\alpha}\ln\left(\bar{h}+\varepsilon\right) \tag{45}
$$

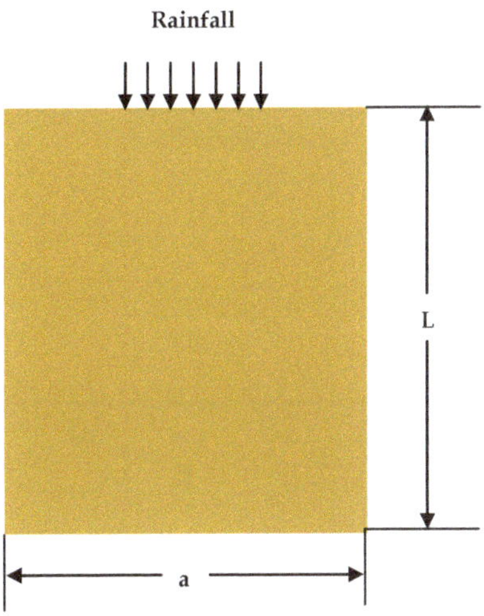

Fig. 8. A view of a 2-D cross section of soil that is initially dry until water is applied at the top

4.2 Analytical solution of a 2-D infiltration problem

The great thing about the above derivations is that they can be extended to two and three dimensions. Fig. 8 shows a 2-D cross section of a region of soil of dimensions, $a \times L$, where a 2-D Green-Ampt problem is presented. The soil is initially dry until water is supplied such that a specified pressure head is applied at the top with pressure head set to zero in the middle and tapering rapidly to h_d at $x=0$ and $x=a$. Fig. 9 shows the function selected to achieve this for h_d = -20 m, and a = 50 m. $h=h_d$ is maintained along the bottom and sides of the soil sample as well. The initial and boundary conditions are therefore

$$h(x,z,0)=h_d \tag{46}$$

$$\begin{aligned} &h(0,z,t)=h(a,z,t)=h(x,0,t)=h_d \\ &h(x,L,t)=\frac{1}{\alpha}\ln\left\{\varepsilon+(1-\varepsilon)\left[\frac{3}{4}\sin\left(\frac{\pi}{a}x\right)-\frac{1}{4}\sin\left(\frac{3\pi}{a}x\right)\right]\right\} \end{aligned} \tag{47}$$

The equation for $\bar{h}$ is now

$$\frac{\partial^2\bar{h}}{\partial x^2}+\frac{\partial^2\bar{h}}{\partial z^2}+\alpha\frac{\partial\bar{h}}{\partial z}=c\frac{\partial\bar{h}}{\partial t} \tag{48}$$

with

$$\begin{aligned} &\bar{h}(0,z,t)=\bar{h}(a,z,t)=\bar{h}(x,0,t)=0 \\ &\bar{h}(x,L,t)=(1-\varepsilon)\left[\frac{3}{4}\sin\left(\frac{\pi}{a}x\right)-\frac{1}{4}\sin\left(\frac{3\pi}{a}x\right)\right] \end{aligned} \tag{49}$$

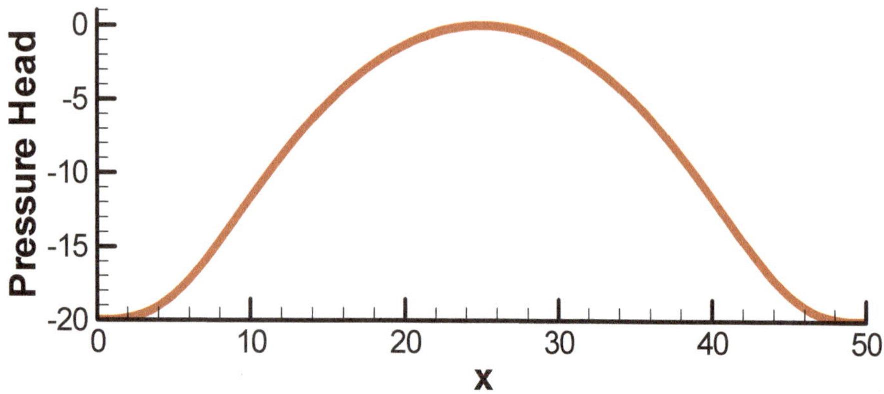

Fig. 9. Pressure head boundary condition applied at the top of the soil sample.

4.2.1 Steady-state solution for $\overline{h}$

The steady-state version of Eq. 48,

$$\frac{\partial^2 \overline{h}_{ss}}{\partial x^2} + \frac{\partial^2 \overline{h}_{ss}}{\partial z^2} + \alpha \frac{\partial \overline{h}_{ss}}{\partial z} = 0 \tag{50}$$

is now solved using separation of variables with $\overline{h}_{ss}$ taking the form,

$$\overline{h}_{ss} = \mathrm{X}(x)\mathrm{Z}(z) \tag{51}$$

This results in the equations,

$$\begin{gathered} \frac{1}{\mathrm{X}}\frac{\partial^2 \mathrm{X}}{\partial x^2} = -\lambda_i, \quad \frac{1}{\mathrm{Z}}\left(\frac{\partial^2 \mathrm{Z}}{\partial z^2} + \alpha\frac{\partial \mathrm{Z}}{\partial z}\right) = \lambda_i, \quad \lambda_i = \frac{\pi}{a}i, i = 0,1,2,... \\ \frac{\partial^2 \mathrm{X}}{\partial x^2} + \lambda_i \mathrm{X} = 0, \quad \frac{\partial^2 \mathrm{Z}}{\partial z^2} + \alpha\frac{\partial \mathrm{Z}}{\partial z} - \lambda_i Z = 0 \end{gathered} \tag{52}$$

with solutions,

$$\mathrm{X}_i = a_i \sin\lambda_i x + b_i \cos\lambda_i x, \quad \mathrm{Z}_i = \left(c_i \sinh\beta_i z + d_i \cosh\beta_i z\right)e^{-\frac{\alpha}{2}z}, \quad \beta_i = \sqrt{\frac{\alpha^2}{4} + \lambda_i^2} \tag{53}$$

where a_i, b_i, c_i, and d_i are constants to be evaluated. Applying boundary conditions on the sides and bottom yields the final form of the steady-state solution as

$$\overline{h}_{ss} = e^{-\frac{\alpha}{2}z}\sum_{i=1}^{\infty} A_i \sin\lambda_i x \sinh\beta_i z \tag{54}$$

where A_i is a constant to be evaluated. Applying the top boundary condition gives

$$\bar{h}_{ss}(x,L) = e^{-\frac{\alpha}{2}L} \sum_{i=1}^{\infty} A_i \sin \lambda_i x \sinh \beta_i L$$
$$A_i = \frac{2(1-\varepsilon)e^{\frac{\alpha}{2}L}}{a \sinh \beta_i L} \int_0^a \left[\frac{3}{4} \sin\left(\frac{\pi}{a}x\right) - \frac{1}{4} \sin\left(\frac{3\pi}{a}x\right) \right] \sin \lambda_i x \, dx \tag{55}$$

Only A_1 and A_3 are nonzero, so

$$A_1 = \frac{3(1-\varepsilon)e^{\frac{\alpha}{2}L}}{4 \sinh \beta_1 L}, \quad A_3 = -\frac{(1-\varepsilon)e^{\frac{\alpha}{2}L}}{4 \sinh \beta_3 L} \tag{56}$$

The steady-state solution for $\bar{h}$ thus becomes

$$\bar{h}_{ss} = (1-\varepsilon)e^{\frac{\alpha}{2}(L-z)} \left[\frac{3}{4} \sin\left(\frac{\pi}{a}x\right) \frac{\sinh \beta_1 z}{\sinh \beta_1 L} - \frac{1}{4} \sin\left(\frac{3\pi}{a}x\right) \frac{\sinh \beta_3 z}{\sinh \beta_3 L} \right] \tag{57}$$

4.2.2 Transient solution for $\hat{h}$

The equation for $\hat{h}$ is now

$$\frac{\partial^2 \hat{h}}{\partial x^2} + \frac{\partial^2 \hat{h}}{\partial z^2} + \alpha \frac{\partial \hat{h}}{\partial z} = c \frac{\partial \hat{h}}{\partial t} \tag{58}$$

with initial and boundary conditions, as before,

$$\hat{h}(x,z,0) = -\bar{h}_{ss}, \quad \hat{h}(0,z,t) = \hat{h}(a,z,t) = \hat{h}(x,0,t) = \hat{h}(x,L,t) = 0 \tag{59}$$

$\hat{h}$ now takes the form,

$$\hat{h} = \mathrm{X}(x)\mathrm{Z}(z)\mathrm{T}(t) \tag{60}$$

This yields

$$\frac{1}{\mathrm{X}} \frac{\partial^2 \mathrm{X}}{\partial x^2} + \frac{1}{\mathrm{Z}} \left(\frac{\partial^2 \mathrm{Z}}{\partial z^2} + \alpha \frac{\partial \mathrm{Z}}{\partial z} \right) = \frac{c}{\mathrm{T}} \frac{\partial \mathrm{T}}{\partial t} \tag{61}$$

and

$$\frac{1}{\mathrm{X}} \frac{\partial^2 \mathrm{X}}{\partial x^2} = -\lambda_i^2, \quad \frac{1}{\mathrm{Z}} \left(\frac{\partial^2 \mathrm{Z}}{\partial z^2} + \alpha \frac{\partial \mathrm{Z}}{\partial z} \right) = -\lambda_k^2 - \frac{\alpha^2}{4}, \quad \lambda_k = \frac{\pi}{L}k,$$
$$\frac{c}{\mathrm{T}} \frac{\partial \mathrm{T}}{\partial t} = -\left(\lambda_i^2 + \lambda_k^2 + \frac{\alpha^2}{4} \right), \quad i = 1,2,3,\ldots, \quad k = 1,2,3,\ldots \tag{62}$$

The general solutions are

$$\mathrm{X}_i = a_i \sin\lambda_i x + b_i \cos\lambda_i x, \quad \mathrm{Z} = \left(c_k \sin\lambda_k z + d_k \cos\lambda_k z\right) e^{-\frac{\alpha}{2}z}$$
$$\mathrm{T} = f_{ik} e^{-\gamma_{ik} t}, \quad \gamma_{ik} = \frac{1}{c}\left(\lambda_i^2 + \lambda_k^2 + \frac{\alpha^2}{4}\right) = \frac{1}{c}\left(\beta_i^2 + \lambda_k^2\right) \tag{63}$$

with the final form of $\hat{h}$ being

$$\hat{h} = e^{-\frac{\alpha}{2}z} \sum_{k=1}^{\infty} \sum_{i=1}^{\infty} A_{ik} \sin\lambda_i x \sin\lambda_k z e^{-\gamma_{ik} t} \tag{64}$$

Here, a_i, b_i, c_k, d_k, f_{ik}, and A_{ik} are constants to be evaluated. Evaluating the above equation at $t = 0$ using the double Fourier sine series gives

$$A_{ik} = -\frac{4}{aL} \int_0^L \int_0^a \overline{h}_{ss} e^{\frac{\alpha}{2}z} \sin\lambda_i x \sin\lambda_k z \, dx dz \tag{65}$$

with the two nonzero terms with respect to i being

$$\begin{aligned} A_{1k} &= -\left(\frac{2}{L}\right)\frac{3}{4}(1-\varepsilon) e^{\frac{\alpha}{2}L} \int_0^L \sin\lambda_k z \frac{\sinh\beta_1 z}{\sinh\beta_1 L} dz \\ &= \left(\frac{2}{Lc}\right)\frac{3}{4}(1-\varepsilon) e^{\frac{\alpha}{2}L} \left(\frac{\lambda_k}{\gamma_{1k}}\right)(-1)^k \\ A_{3k} &= \left(\frac{2}{L}\right)\frac{1}{4}(1-\varepsilon) e^{\frac{\alpha}{2}L} \int_0^L \sin\lambda_k z \frac{\sinh\beta_3 z}{\sinh\beta_3 L} dz \\ &= -\left(\frac{2}{Lc}\right)\frac{1}{4}(1-\varepsilon) e^{\frac{\alpha}{2}L} \left(\frac{\lambda_k}{\gamma_{3k}}\right)(-1)^k \end{aligned} \tag{66}$$

The solution for $\hat{h}$ now becomes

$$\hat{h} = \frac{2}{Lc}(1-\varepsilon) e^{\frac{\alpha}{2}(L-z)} \left\{ \begin{array}{l} \frac{3}{4}\sin\left(\frac{\pi}{a}x\right) \sum_{k=1}^{\infty} \left(\frac{\lambda_k}{\gamma_{1k}}\right)(-1)^k \sin\lambda_k z - \\ \frac{1}{4}\sin\left(\frac{3\pi}{a}x\right) \sum_{k=1}^{\infty} \left(\frac{\lambda_k}{\gamma_{3k}}\right)(-1)^k \sin\lambda_k z \end{array} \right\} \tag{67}$$

As done before, transforming back to the original coordinates gives

$$\overline{h} = \frac{2}{Lc}(1-\varepsilon) e^{\frac{\alpha}{2}(L-z)} \left\{ \begin{array}{l} \frac{3}{4}\sin\left(\frac{\pi}{a}x\right)\left[\frac{\sinh\beta_1 z}{\sinh\beta_1 L} + \frac{2}{Lc}\sum_{k=1}^{\infty}\left(\frac{\lambda_k}{\gamma_{1k}}\right)(-1)^k \sin\lambda_k z\right] - \\ \frac{1}{4}\sin\left(\frac{3\pi}{a}x\right)\left[\frac{\sinh\beta_3 z}{\sinh\beta_3 L} + \frac{2}{Lc}\sum_{k=1}^{\infty}\left(\frac{\lambda_k}{\gamma_{3k}}\right)(-1)^k \sin\lambda_k z\right] \end{array} \right\} \tag{68}$$

Also, as before, transforming back to the original h,

$$h = \frac{1}{\alpha}\ln\left(\bar{h} + \varepsilon\right) \tag{69}$$

A 3-D solution is done in a similar manner.

5. Numerical models

Hydraulic conductivity has an important role in numerical models. Many soil layers can be modelled by specifying hydraulic conductivity for the different layers. Because Richards' equation is nonlinear, the manner in which numerical models compute relative hydraulic conductivity is also important for both accuracy of the solution and the ability of the numerical algorithms to converge. When doing a 3-D Green-Ampt problem containing thousands of 3-D finite elements on a parallel high performance computing platform, the solution would not converge because of how relative hydraulic conductivity was computed inside each finite element. When the pressure head was averaged from the four nodes of each tetrahedral element and then used to compute a constant value for the relative hydraulic conductivity inside the element, the solution diverged. However, if relative hydraulic conductivity was considered to vary linearly inside each element, the solution converged quite well. Testing these different algorithms is greatly enhanced by the analytical solutions presented above. Some tests using the analytical solutions will now be illustrated.

5.1 1-D solution of the Green-Ampt problem

The 1-D version of Eq. 7 for a homogeneous, isotropic soil is

$$k_s \frac{\partial}{\partial z}\left(k_r \frac{\partial \phi}{\partial z}\right) = \frac{\partial \theta}{\partial t} \tag{70}$$

A finite element/finite difference/finite volume discretization of this equation (see Fig. 10) is

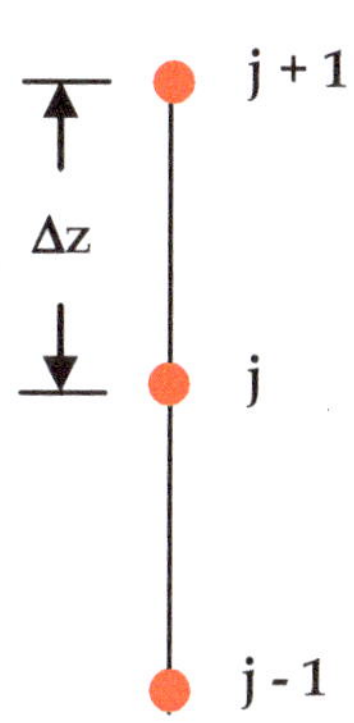

Fig. 10. Discretization of the 1-D soil sample showing two finite elements.

$$k_{r+}\left(\phi_j^{n+1}-\phi_{j+1}^{n+1}\right)+k_{r-}\left(\phi_j^{n+1}-\phi_{j-1}^{n+1}\right)+\frac{\Delta z^2}{k_s\Delta t}\left(\frac{d\theta}{dh}\right)_j^{n+1}\left(\phi_j^{n+1}-\phi_j^{n}\right)=0 \quad (71)$$

where

j = node number

k_{r-} = the relative hydraulic conductivity for the element between nodes j and $j-1$

k_{r+} = the relative hydraulic conductivity for the element between nodes j and $j+1$

Δt = time-step size

n = time-step number

The two ways of computing relative hydraulic conductivity inside each element will now be discussed.

5.1.1 Constant relative hydraulic conductivity inside each element

This way of computing relative hydraulic conductivity is to first compute the average pressure head (h_{av}) at the center of the element. For k_{r+}, this becomes

$$h_{av}=\frac{1}{2}\left(h_j+h_{j+1}\right) \quad (72)$$

Then compute relative hydraulic conductivity by

$$k_{r+}=e^{\alpha h_{av}} \quad (73)$$

k_{r-} is computed in the same way.

5.1.2 Linearly varying relative hydraulic conductivity inside each element

This way of computing relative hydraulic conductivity is the equivalent of first computing the relative hydraulic conductivity at the node points. For nodes j and $j+1$, designate relative hydraulic conductivity by

$$k_{r,j}=e^{\alpha h_j},\quad k_{r,j+1}=e^{\alpha h_{j+1}} \quad (74)$$

Averaging these values for the final result gives

$$k_{r+}=\frac{1}{2}\left(k_{r,j}+k_{r,j+1}\right) \quad (75)$$

k_{r-} is computed in the same way.

5.1.3 1-D numerical test results

The above equation was solved using L = 50 m; k_s = 0.1 m/day; h_d = -20 m; θ_d = 0.15; θ_s = 0.45; Δz = 0.25 m; α = 0.1 m^{-1}, 0.2 m^{-1}, and 0.3 m^{-1}; and Δt = 0.01 day for 100 time-steps with the two versions of computing relative hydraulic conductivity. The model used was a simple FORTRAN program written by the author. The largest in absolute value (worst) error in pressure head for each method is given in Table 1. It is important to note that the respective signs of these errors have been retained. From these results, it is seen that the linearly varying version gave the best results.

α (1/m)	0.1	0.2	0.3
Constant k_r (m/day)	-0.12	-0.28	-0.43
Linear k_r (m/day)	-0.09	-0.12	0.17

Table 1. Worst error in pressure head for different values of α for constant and linearly varying k_r.

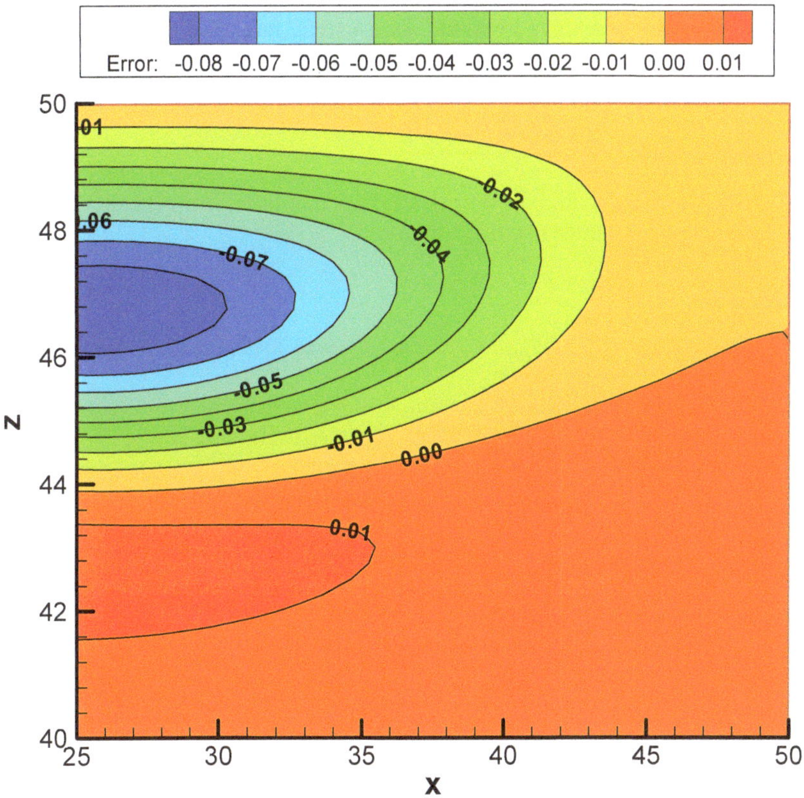

Fig. 11. Error plot for pressure head (h) for the upper, right-hand corner of the computational region.

5.2 2-D solution of the Green-Ampt problem

The 2-D version of Eq. 7 was solved for the problem given in Section 4.2 with the values of the parameters being the same as for the 1-D problem presented above but with the addition of a = 50 m. The model used for this computation was a transient version of Seep2D (Tracy, 1983, & Seep2D, 2011). A steady-state version of Seep2D is currently incorporated into the

Groundwater Modeling System (GMS) (Jones, 1999, & GMS, 2011). The transient version is not yet available.

Fig. 11 gives a color contour plot of the error for the linearly varying relative hydraulic conductivity option for α = 0.1 m^{-1} for the upper, right-hand region of 10 m × 25 m. Clearly, the results match well with the analytical solution.

6. Summary

This chapter has shown that hydraulic conductivity plays an important role in both deriving analytical solutions and doing numerical computations. Analytical solutions for both the 1-D and 2-D Green-Ampt problem were derived and computed numerically with the results compared. The derivations are presented in such detail that others can do additional solutions as well. Varying relative hydraulic conductivity linearly within each finite element not only makes the nonlinear convergence algorithm more robust, but it also produces more accurate answers than when it is considered constant inside each finite element.

7. Acknowledgment

This work was supported in part by a grant of computer time from the DoD High Performance Computing Modernization Program.

8. References

Cook, R. (1981). *Concepts and Applications of Finite Element Analysis* (2nd Edition), John Wiley & Sons, New York.

Gardner, W. (1958). Some steady-state solutions of the unsaturated moisture flow equation with application to evaporation from a water table. *Soil Science*, Vol. 85, pp. 228-232.

GMS. (2011). http://chl.erdc.usace.army.mil/gms.

Green, W., & Ampt, G. (1911). Studies on soil physics, part I, the flow of air and water through soils. *Journal of Agricultural Science*, Vol. 4, pp. 1-24.

Jones, N. (1999). Seep2D Primer. Groundwater Modeling System, Environmental Modeling Research Laboratory, Brigham Young University, Provo, Utah.

Richards, R. (1931). Capillary conduction of liquid through porous media. *Physics*, Vol. 1, pp. 318-333.

Seep2D. (2011). Wikipedia. http://en.wikipedia.org/wiki/SEEP2D.

Tracy, F. (1983). User's Guide for a Plane and Axisymmetric Finite Element Program for Steady-State Seepage Problems. Instruction Report No. IR K-83-4, Vicksburg, MS, U.S. Army Engineer Waterways Experiment Station.

Tracy, F. (2006). Clean two- and three-dimensional analytical solutions of Richards' equation for testing numerical solvers. *Water Resources Research*, Vol. 42, W08503.

Tracy, F. (2007). Three-dimensional analytical solutions of Richards' equation for a box-shaped soil sample with piecewise-constant head boundary conditions on the top. *Journal of Hydrology*, Vol. 336, pp. 391-400.

van Genuchten, M. (1980). A closed-form equation for producing the hydraulic conductivity of unsaturated soils. *Soil Science American Journal*, Vol. 44, pp. 892-898.

Warrick, A. (2003). *Soil Water Dynamics*, Oxford University Press, New York.

Part 4

Determination by Field Techniques

12

Field Measurement of Hydraulic Conductivity of Rocks

Maria Clementina Caputo and Lorenzo De Carlo
Water Research Institute (IRSA), National Research Council (CNR)
Via F. De Blasio, Bari
Italy

1. Introduction

Groundwater often represents the main and most precious source of drinking water supply for the population. In recent decades, overexploitation, uncontrolled anthropogenic actions and continuous reduction of rainfall, due to climate change, led to a depletion of the water resource by affecting both its quantity and quality. The scientific community pays attention to this particular environmental issue in order to implement effective strategies for the safeguarding, protection, and remediation of the aquifers. Over the last decade, considerable efforts have been made to develop methodologies and techniques aimed at improving knowledge of the processes that are based in the unsaturated zone or vadose zone—the portion of the soil above the groundwater level—within which flow and transport processes occur. In this portion of subsoil, in fact, important physicochemical phenomena take place, which regulate the environmental balance of the hydrogeological system, such as the ability to store water and transport it into the ground below. It also has a natural protective function as a filter for any potential pollutants carried by fluids circulating in the solid matrix before reaching the groundwater. The knowledge and the understanding of the processes that take place in the unsaturated zone are, therefore, essential for groundwater management and protection, to evaluate the recharge rate and assess groundwater vulnerability. In particular, the measurement and monitoring of the unsaturated hydraulic properties are very important, even though it is difficult and expensive (Castiglione et al., 2005). The methods and techniques developed are designed to investigate the unsaturated flow process in the soils. When the vadose zone consists of rock, rather than soil, technical aspects increase the difficulties in several ways (Bogena et al., 2007; Kizito et al., 2008). Usually, different kinds of probes are used to monitor the water infiltration in soils: the Time Domain Reflectometry (TDR) (Jones et al., 2004; Robinson et al., 2003), the Frequency Domain Reflectometry (FDR) and multi-sensor capacitance probes (Baumhardt et al., 2000; Seyfried et al., 2004) are used to measure the water content in the subsurface, while tensiometers measure water pressure (Masbrucha & Ferré, 2003). These devices are hard to utilize in the rocks, mainly because the probes are brittle. Therefore it is difficult to install them, as there needs to be good contact between the rock and the sensor to reduce the uncertainty of the measurements due to the gap effects. Field studies, set up to measure hydraulic conductivity, have employed infiltrometer tests under different conditions, but they have rarely been performed directly on the outcropped rock, owing to the difficulty of installation. Reynolds et al. (2002) conducted infiltrometer tests under different conditions, Ledds-Harrison & Youngs (1994) used very small diameter rings (from 1.45 mm to 2.5 mm) for field measurements on individual soil aggregates, Youngs et al. (1996) used a 20 m diameter infiltrometer cylinder to measure highly structured and variable materials that could not be sampled adequately by a smaller cylinder. Castiglione et al. (2005) developed in the laboratory a tension infiltrometer ring, 4 cm in height and 27.5 cm in diameter, suitable for accurate measurements of infiltration into a big sample of fractured volcanic tuff, at very low flow rates over long equilibration times. Most field studies employ cylinder infiltrometers with diameters ranging commonly from 1 to 50 cm, which are poorly representative of the heterogeneous media, such as fractured rocks, in which hydraulically important fractures may, typically, be spaced further apart than the cylinder's diameter. Indeed, up to now infiltrometer tests have rarely been performed directly on-site on rock outcrops. This chapter describes a methodology to obtain the field-saturated hydraulic conductivity, K_{fs}, by using a ring infiltrometer, with a large (~2 m) adjustable diameter, developed for measuring quasi-steady infiltration rates on outcropped rock. K_{fs} is the hydraulic conductivity of the medium (soil or rock) when it has been brought to a near-saturated state by water applied abundantly at the land surface, typically by processes such as ponded infiltration, copious rainfall or irrigation. The proposed device is inexpensive and simple to implement, as well as very versatile, owing to its large adjustable diameter that can be fixed on-site. Moreover, certain practical problems, related to the installation of the cylindrical ring on the rock surface, were solved in order to achieve a continuous and impermeable joint surface between the rock and the ring wall. An issue of major concern is linked to the edge effects, related to the radial spreading of the infiltrating water; obviously smaller rings are more influenced by these effects. Swartzendruber & Olson (1961) and Lai & Ren (2007) found that the ring infiltrometer needs a diameter greater than 1.2 m and 0.8 m, respectively, to avoid the edge effects. For this reason, the proposed large ring infiltrometer is made of a strip of flexible material with which build the cylinder on-site, with a suitable diameter in relation to the lithological and topographical features of the field. The flexible material, such as plastic or glass resin, allows the minimization of the size of the ring and, therefore, its movement easily, in order to acquire a large number of independent K_{fs} measurements over a given area. In fact, because of the extreme spatial variability of K_{fs}, its value finds statistical consistency in multiple tests. Geophysical techniques were coupled with the infiltrometer tests in order to monitor, qualitatively, the water infiltration depth, to allow a rapid visualization of the change in water content in subsurface and to ensure that the decrease in the water level in the ring was caused mainly by vertical water infiltration, and not by the lateral diversion of water flow. Since the late 1980s, many geophysical applications have been aimed at hydrogeological studies. White (1988) conducted electrical prospecting to determine the direction and the flow rate of saline aquifers using a tracer. Daily et al. (1992) used borehole electrical resistivity surveys to obtain the distribution of electrical resistivity in the subsurface, and compare the results with infiltrometer tests. Recently, electrical resistivity techniques have been used to monitor hydrogeological processes. Cassiani et al. (2006) conducted a monitoring test of a salt tracer by means of the application of electrical resistivity tomography using a time-lapse technique. The movement of the tracer was monitored with geophysical images. The methodology described in this chapter was carried out on different lithotypes in order to verify the applicability of the experimental apparatus in very different geological conditions. In particular two cases are described: the Altamura test site that represents a case of hard sedimentary rock consisting of limestone, and the San Pancrazio test site as an example of a soft porous rock, specifically calcarenite. Both sites, located in the Puglia region of southern Italy, have been studied because they are both contaminated areas, for which the understanding of the flow rate in the subsurface is very important in order to design a remediation strategy. The experimental approach gave good results in both situations that mean it could be used successfully on each different lithology of outcropped rock.

2. Method

The objective of this chapter is to describe a methodology to obtain the field-saturated hydraulic conductivity, using a large ring infiltrometer installed directly on the outcropped rock. The method considers an integrated approach that combines the infiltrometer test with geophysical techniques, in order to visualize the change in water content in the subsurface during the experiment. Among geophysical techniques, we have used electrical resistivity images (ERI) because the electrical resistivity is extremely sensitive to subsurface water content. Moreover, electrical time-lapse resistivity measurements (Binley et al., 2002; Deiana et al., 2007), carried out simultaneously with the infiltrometer test, allow us to check the performance of the experimental apparatus and, particularly, that the decrease of water level in the ring is due to the infiltration and not to the losses along the edge of the ring.

2.1 Large ring infiltrometer

A ring infiltrometer for studying water flow in rock formations needs to be made from a tough material, suitable for installation on the rock surface. At the same time, the material must be both flexible, to allow a cylinder to be built that is adjustable to the field conditions, and light, to facilitate transport and setup. Where the ground surface is very irregular, small adjustments to the ring diameter ensure efficient sealing, thus facilitating the setting-up of the device. If electrical methods are to be used for the detection of water content or salinity, the ring must also be made of a non-metallic material. The large ring infiltrometer was, therefore, designed in a light and flexible material, such as plastic or glass resin, to allow easy installation on the rock surface, and the adaptation of its size to the field conditions. Strips of the flexible material 30 cm wide and 0.2 cm thick were used to build in-situ rings of internal diameter of about 2 m, by sealing the two ends of the strip with impermeable adhesive tape to form the cylinder. The large diameter provides infiltration data that are representative both of the anisotropy and the heterogeneity of the rock, characteristics which cannot be sampled adequately by using smaller rings. It thus allows an improved representation of the natural system's heterogeneity, while also taking into consideration irregularities in the soil/rock subsurface. In addition, the large diameter allows the avoidance of the edge effects that usually occur during infiltration, such as the radial spreading of the infiltration water. Among a variety of materials tested to seal the ring to the ground, both in the laboratory and in the field, gypsum was shown to be the most suitable because it is cheap, easy to prepare in situ and to apply on the ground in a previously hollowed furrow, ensuring good sealing. Clay, instead, needed to be worked for a long time in order to obtain a consistency that would allow the ring to be sealed to the ground. It nevertheless gave poor results, because the clay did not grip the ring wall firmly on the ground. Moreover, tests carried out with clay at the laboratory showed water losses from cracks in the clay due to its characteristics of swelling and shrinking. Silicon was also tested, although using it for sealing a large-diameter infiltrometer would be expensive. However, it was also discarded, as it did not seal well due to the dissimilarity of the soil, rock, and plastic materials being sealed. Polyurethane foam was easy to apply and proved impermeable, but was also discarded because it was expensive and did not grip well on the rock and/or soil.

2.2 Geophysical techniques

Geophysical techniques are used routinely for subsurface investigation, as they provide information about the physical properties related to geological and hydrogeological conditions. There are two main groups of geophysical methods: the active methods, which measure the subsurface response to electromagnetic, electrical, and seismic artificially generated signals; and the passive methods, which measure the earth's natural fields, such as the magnetic, electrical, and gravitational ones. Among these, the electrical resistivity method is a relatively new imaging tool in geophysics; it is an active method that determines the subsurface distribution of electrical resistivity from a large number of resistance measurements. The electrical resistivity, ρ, is an intrinsic property of rock and soil, and is a measure of how strongly a material opposes the flow of electric current. The direct current resistivity tomography method considers spatial variations in electrical resistivity among geologic materials for mapping the subsurface structures. It is used extensively in the search for suitable groundwater sources, and also to monitor types of groundwater pollution. In engineering it is used to locate subsurface cavities, faults and fissures, permafrost, mineshafts etc., and in archaeology, for mapping out the areal extent of the remnants of buried foundations of ancient buildings, and many other applications (Reynolds, 1998). The electrical resistivity method is based on the Ohm's laws; the first of which states that when the electrical current, I (ampere, symbol A), passes through an electrically uniform medium of side length, L (m), the material has a resistance, R (ohm, symbol Ω), resulting in a potential drop, V (volt, symbol V), between opposite faces of the medium. The first Ohm's law correlates the three physical parameters into the formula:

$$V = RI \tag{1}$$

The resistance, R, is proportional to the length, L, of the resistive material and inversely proportional to the cross-sectional area, S (m^2) (second Ohm's law)

$$\rho = \frac{RS}{L} \tag{2}$$

where ρ is the electrical resistivity (ohm x m, symbol Ωm); it is the inverse of the conductivity (siemens/m, symbol S/m).

Electrical current can flow in rocks and soils by three main mechanisms: 1) electrolytic conduction that occurs by means slow migration of ions in a fluid electrolyte, controlled by ions type, ion concentration, and ionic mobility; 2) electronic conduction that occurs in metals through rapid movement of electrons; 3) dielectric conduction that occurs in conducting materials, or insulators, in the presence of an external alternating current when atomic electrons are shifted slightly relative to the nucleus. In most rocks, electrical current flows by electrolytic conduction (Telford et al., 1990). The electrical resistivity of the soils and rocks correlates with other soil/rock properties which are of interest to the geologist, hydrogeologist, and geotechnical engineer. Many factors affect electrical resistivity, such as texture, porosity, water saturation, clay content, permeability and temperature. Thus, the resistivity measurements cannot be related directly to the type of soil or rock in the subsurface without direct sampling, or some other geophysical or geotechnical information. Porosity is the major controlling factor of the changing resistivity of the rock, because the electricity flows in the near surface by means of the passage of ions through the pore space of the subsurface materials. The porosity (amount of pore space), the permeability (connectivity of pores), the water (or other fluid) content in the pores, and the presence of salts all are contributing factors to changing resistivity. Archie (1942) developed the empirical formula of the effective resistivity of the rock, ρ_e:

$$\rho_e = \rho_w a \varphi^{-m} S^{-k} \tag{3}$$

where φ (%) is the porosity, S (adimensional) is the volume fraction of pores with water, ρ_w is resistivity of pore fluid, a, m and k are physical parameters. ρ_w is influenced by dissolved salts and can vary between 0.05 Ωm for saline groundwater, up to 1,000 Ωm , for glacial melt water. Archie's law ignores the effect of pore geometry, but it is a reasonable approximation for many sedimentary rocks.

2.2.1 Electrical resistivity imaging

The technique of electrical resistivity imaging utilizes measurements of electrical potential associated with the subsurface electrical current flow, generated by a direct current. In the electrical resistivity method, the spatial variation of resistivity in the field is determined using four electrode measurements, and is based on measuring the potential difference between one electrode pair while another electrode pair, used as the current source, transmits the current (Dahlin, 2001). Measurements of the potential difference between the two electrodes allow the determination of the apparent resistivity, ρ_a:

$$\rho_a = \frac{K \Delta V}{I} \tag{4}$$

where K (m) is a geometric factor depending on the used array. The measured quantity is called apparent resistivity because the resistivity values measured are averages over the total current path length, but are plotted at one point for each potential electrode pair. The data can be arranged in a 2-D plot, called pseudosection, which displays both horizontal and vertical variations in apparent resistivity. The conventional presentation places each measured value at the intersection of two 45-degree lines through the centres of the dipoles (Edwards, 1977). The interpretation of the resistivity data consists of two steps: a physical interpretation of the measured data, by means of an inversion process that results in a physical model; and a geological interpretation of the resulting physical parameters (Dahlin, 1996). Many configurations of array can be used for measuring the distribution of electrical resistivity of the subsurface. The arrays most commonly used for 2-D imaging surveys are: Wenner, dipole-dipole, Wenner-Schlumberger. The choice of the best array considered for the surveys, depends on the depth of investigation, the sensitivity of the array to vertical and horizontal changes in the subsurface resistivity, the horizontal data coverage, and the signal strength (Loke, 2001). In general, Wenner array is relatively sensitive to vertical changes in the subsurface resistivity, but it is less sensitive to its horizontal changes. It is useful for recognized horizontal structures (vertical resistivity changes), but relatively poor in detecting narrow vertical structures (horizontal resistivity changes). Dipole-dipole is good for mapping vertical structures, such as dykes and cavities, but relatively poor for mapping the horizontal structures such as sedimentary layers or water table. Wenner-Schlumberger is moderately sensitive to both horizontal and vertical structures. In areas where both types of geological structures are expected, this array might be a good compromise between the Wenner and the dipole-dipole array.

3. Field tests

The infiltration experiments described in the chapter refer to two sites that differ in their geological and hydrogeological conditions, in order to show how the methodology can be applied widely. The sites consist of outcrop of hard sedimentary rock, fractured limestone, in the case of Altamura, and soft sedimentary rock, in the case of San Pancrazio (Fig. 1). The different lithology implies different infiltration process and, consequently, different experimental approaches and interpretations of the experimental data. All these aspects are described in detail for each test site.

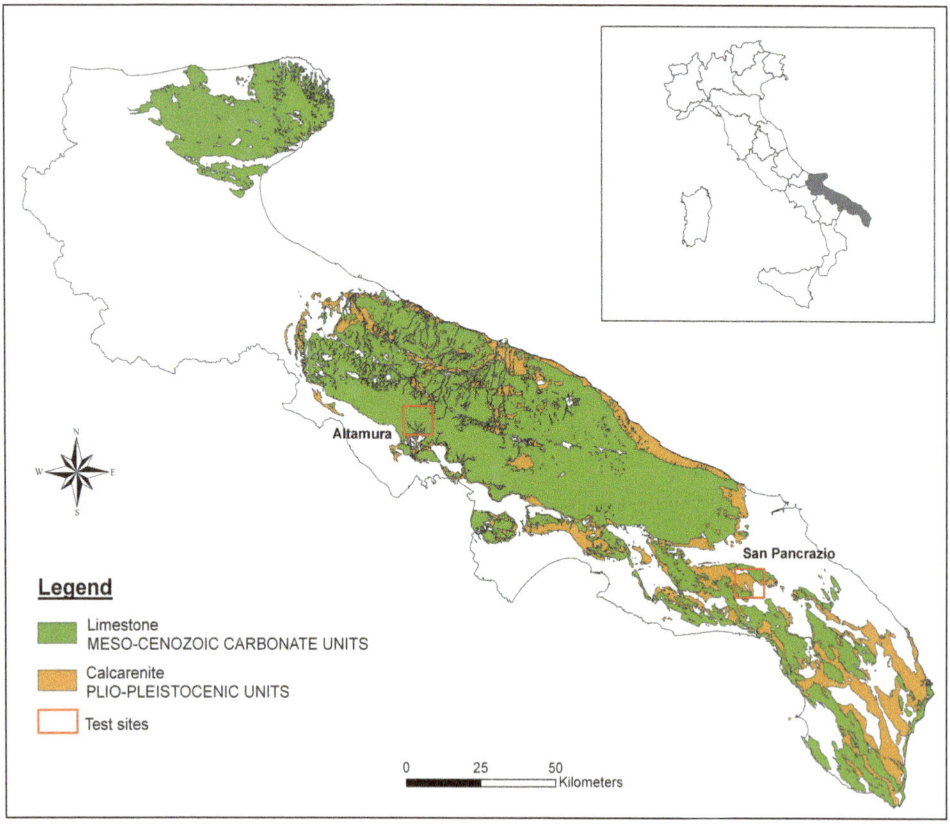

Fig. 1. Schematic geological map of the Puglia region.

3.1 Test on limestone

The infiltration experiments were performed in the study area belonging to Altamura territory, located in southern Italy (Fig. 1). The site is part of the Murge plateau, formed from Cretaceous limestone during the Paleogene and Neogene periods. The limestone is part of a sedimentary sequence of a carbonatic platform some thousands (up to 3,000) of meters thick. The main rock units consist of calcilutites, which in the study area constitute a unit known as *"Calcare di Altamura"* and *"Calcare di Bari"* (Ciaranfi et al. 1988). During a large part of the Neogene, the area was uplifted and affected by tectonic stress and karst morphogenesis. The experimental tests were conducted on the top of a karstic fractured limestone formation constituting the deep Murge aquifer, which has a water table ranging in depth from 400 to 500 m below the ground surface. The fractured limestone studied is characterized by low effective matrix porosity with respect to fracture porosity. Borgia et al. (2002) measured the effective matrix porosity of the limestone, at field scale, and found values up to 0.1%, together with a permeability lower than 24 mD (millidarcy), and a hydraulic conductivity of $2.1x10^{-2}$ m d^{-1} (for water at 20°C), on average. These values indicate a conductivity 1,000 times lower than that of the fracture network (8-77 m d^{-1}) of the same limestone formation (Masciopinto, 2005), which is characterized by an effective fracture porosity at field scale of up to 0.4% (Borgia et al., 2002). In light of these results, it is clear that the water flow rate component in the rock matrix can be considered negligible as compared to the water flow rate component in the network of fractures. This site was chosen for testing the newly developed large ring infiltrometers because it was affected by untreated sludge waste deposits that have caused soil and rock contamination by heavy metals (chromium, lead, zinc, nickel and cadmium), as well as hydrocarbons. Several soil chemical analyses carried out, showed chromium and nickel contamination, with concentrations higher than both Italian and European legal limits. The infiltrometer tests were carried out at two locations, 300 m apart, to evaluate the quasi-steady vertical flow rates in outcrops of limestone that showed different fracture frequencies and sizes on visual inspection. Specifically, the site of test #1 showed prominent visible fractures of about 5 cm spacing and about 1 mm aperture, whereas the site of test #2 showed no visible fractures (Fig. 2). A 2 m diameter ring, coupled with an external ring, was used in test #1. In this case, the infiltration area consisted of a central part with numerous visible fractures (Fig. 2a), while the rest was partially covered by soil, 5-10 cm thick, on average. For this reason, at this site, a second ring with a larger diameter (2.2 m) was positioned (Fig. 2b), in order to improve the hydraulic packing during the water infiltration test. Continuity between the ring infiltrometer wall and the rock-soil surface was obtained by filling the space between the external and internal rings with gypsum, up to a height of about 2 cm, to create a seal.
As a result, the second ring improved the gypsum sealing of the first ring to the ground, overcoming the challenge presented by the presence of different media (rock and soil) along the edge of the ring. In the second test (#2), located in an area with an outcrop of limestone characterized by no visible fractures, a 1.8 m diameter ring was inserted into a thin furrow 2 cm deep, previously hollowed into the rock (Fig. 3), which was then also sealed with gypsum. In both tests, the large diameter ring allowed the obtaining of data more representative of the anisotropy and heterogeneity of rocks than those obtained with a small ring. The difference in size of the rings used in the two tests was due simply to the local field conditions; in fact, the device is designed to allow the diameter to be adjusted easily when the irregularity of the surface at the test site so requires. In both cases the infiltration tests have been supported by ERI, carried out simultaneously by means "time-lapse" technique,

in order to monitor the infiltration and redistribution of water in the subsurface. The water used for the infiltration test had moderate salinity (electrical conductivity of 2.39 mS/cm) to enhance the subsurface electrical resistivity measurements.

Fig. 2. Test # 1 at the Altamura site: a) fractured limestone with visible fractures of about 5 cm of spacing (s); b) plastic ring infiltrometer resting directly on the ground and sealed with gypsum by using a second ring.

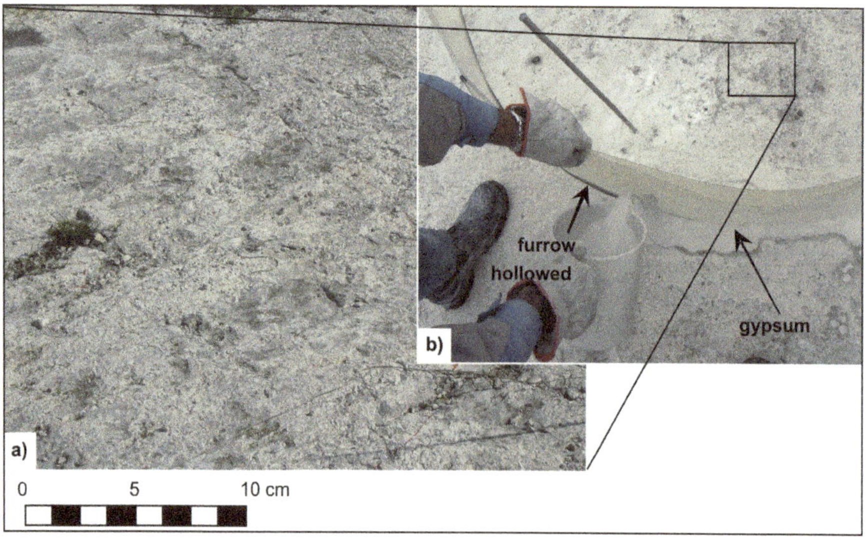

Fig. 3. Test # 2 at the Altamura site: a) limestone without visible fractures; b) plastic ring infiltrometer installed into a thin (2 cm) furrow hollowed in the limestone and sealed with gypsum.

In test #1, the resistivity data were collected using a 7.5 m long straight-line array of 16 steel electrodes with 0.5 m spacing, oriented in the parallel direction to the visible fractures. In test #2, two similar linear arrays were set up, perpendicular to each other, with the same length and electrode spacing, crossing at the center of the ring infiltrometer (Fig. 4.). The Wenner array was chosen over other array types because it provides a good signal-to-noise-ratio. Additionally, it is also highly sensitive to vertical changes in the subsurface resistivity. This makes the Wenner array a useful tool in studying the movement of the wetting front in time. The investigation depth is about 1.5 m, easily achievable with the chosen electrode configuration. Both tests used the IRIS-SYSCAL Pro Switch 48 instrument to acquire electrical resistivity measurements. To obtain 2-D resistivity models, the field data were inverted using Res2Dinv software (Griffiths & Barker, 1993; Loke & Barker, 1996). In the processing of resistivity data, an inversion routine based on the smoothness-constrained least-squares method was implemented (deGroot-Hedlin & Constable, 1990). The 2-D model used by the inversion program divides the subsurface into a number of rectangular blocks, whose arrangement is linked to the distribution of points in pseudosections. The distribution and size of the blocks is generated automatically by the program, using the distribution of the data points as a rough guide. The depth of the bottom row of blocks is set to be approximately equal to the equivalent depth of investigation (Edwards, 1977) of the data points. The optimization method basically tries to reduce the difference between the calculated and measured apparent resistivity values, by adjusting the resistivity of the model blocks. A forward modeling subroutine is used to calculate the apparent resistivity values, and a non-linear least-squares optimization technique is used for the inversion routine (Loke & Barker, 1996). A measure of this difference is given by the root-mean squared (RMS) error. However, the model with the lowest possible RMS error can sometimes show large and unrealistic variations in the model resistivity values, and might not always be the "best" model from a geological perspective. The water in the ring reached a maximum level of 0.13 m above the ground surface in both tests. During approximately 2 hours of the falling-head infiltrometer tests, the specific lateral water leakage fluxes were about 8% (0.1 m d^{-1}) and 3% (0.007 m d^{-1}) of the infiltrate water fluxes in the rings, for the first and second tests, respectively. The specific (i.e., per unit of ring area) leakage rates, due

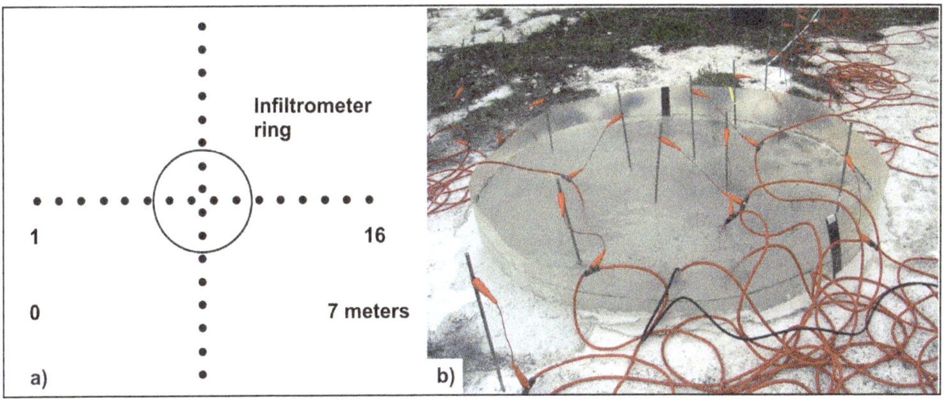

Fig. 4. Test # 2 at the Altamura site: a) scheme of electrode array; b) detail of field installation of electrodes.

not to the water flux from the edge of the ring (i.e., losses) but to the amount of infiltrated water which flowed laterally within the subsurface, was estimated by dividing the total water volume of the lateral diverted water flow by the duration of the infiltration (2 and 3.2 hours for tests #1 and #2, respectively). This estimation was performed on the basis of the areas of the wetted surfaces (1.6 and 1.5 m2 for tests 1# and 2#, respectively), which were observed around the rings at the end of each test, and multiplying them by the average thickness of the wetted layers (5 and 3 cm, respectively), and by their porosities (0.29 and 0.03) (Borgia et al. 2002). The thickness of the wetted layers was derived from measurements on core samples drilled at the test sites. During the experiments, water levels in the ring infiltrometer were monitored using a submersible pressure probe (PTX DRUCK LTD).

3.2 Test on calcarenite

The study area is an abandoned calcarenite open quarry, located near San Pancrazio city (Fig. 1). The quarry was used for about ten years, from 1980 to 1990, for the disposal of waste from a pharmaceutical company that produced antibiotics (in particular erythromycin), using fermentative processes and subsequent chemical transformations causing subsurface contamination. For this reason, field infiltrometer tests, coupled with electrical resistivity measurements, were carried out to evaluate the flow rate of potential contaminants in the porous aquifer. The geology of the study area consists of Cretaceous bedrock formed from dolomitic limestone and limestone, overlaid unconformably by Plio-Pleistocenic calcarenite. The oldest formations contain a deep and confined aquifer characterized by a potentiometric surface, ranging from 70 m to 80 m, in depth below ground surface. Electrical resistivity measurements have detected a shallow aquifer with a water table about 25 m below ground surface, and about 12 m below the bottom of the quarry in the Plio-Pleistocenic calcarenite. A large infiltrometer ring was installed in the field directly on the rock surface with an adaptable size to fit the condition in the field. A strip of 30 cm high flexible plastic material was used to build an in-situ infiltrometer ring, of about 2 m in diameter, sealing the two edges with impermeable tape. The ring was installed into a 5 cm deep thin furrow hollowed in the calcarenite at the bottom of the quarry, and sealed with gypsum (Fig. 5).

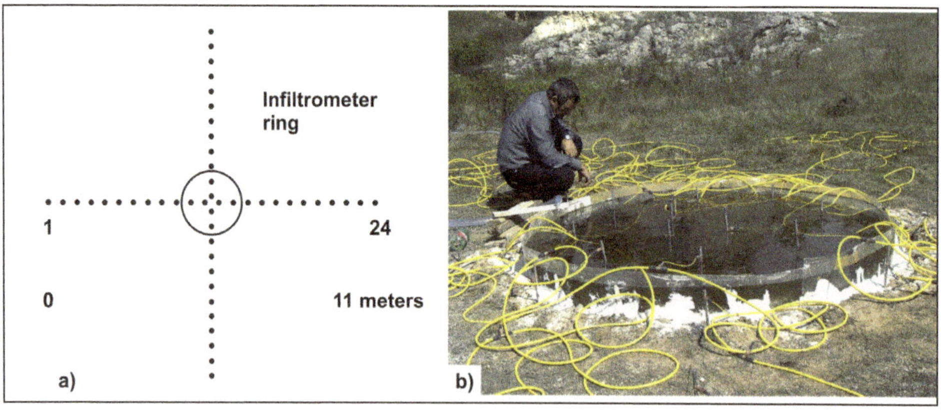

Fig. 5. Test at the San Pancrazio site: a) scheme of electrode array; b) detail of field installation of electrodes.

The infiltration tests were performed with constant and falling head conditions. About 0.27 m^3 of water was poured in the ring until the water level reached 8 cm from the bottom, and then another 0.2 m^3 to maintain that constant level in the ring for the duration of the test, which took 3 hours. No more water was added for the falling-head test because it started from the constant water level considered for the first test (under constant head condition). The second test took 2.5 hours. In order to study the infiltration/redistribution of water in the subsurface, electrical resistivity tomography was carried out simultaneously with the infiltrometer tests by collecting resistivity data along two electrical profiles, each of which was 11.5 m long, with 24 steel electrodes with 0.5 m spacing. The two electrical profiles were set up, perpendicular to each other, with the same length and the same electrode spacing for each direction (x, y), and in a symmetrical position with respect to the center of the ring infiltrometer (Fig. 5). The so-called Wenner array was chosen as at the Altamura test site. The same instrument resistivity meter, IRIS-SYSCAL Switch Pro 48, was used to measure the resistivity of the subsurface. During the infiltrometer tests, the monitoring of the infiltrated water was carried out by means of a time-lapse technique, collecting more than 2,000 measurements for each acquisition. The total acquisition time was about 3 hours for the constant head test, and about 2.5 hours for the falling head test.

4. Results

In the following paragraphs the results related to the two different test sites are described: the Altamura site consisting of limestone with and without visible fractures, such as an example of hard sedimentary rock, and the San Pancrazio site consisting of calcarenite, such as an example of soft sedimentary rock. Different approaches are used for the elaboration of the experimental data collected at the two geological outcrops. It is highlighted that in the case of fractured limestone the interpretation of the data collected experimentally, needs a numerical elaboration supported by mathematical models.

4.1 Limestone

The results of the infiltrometer tests are summarized in the graph in Figure 6, which shows a constant decrease in the water level in the ring during the tests, after the water level had reached its maximum depth of 0.13 m in both cases. The slope of the trend line of experimental sites gives an average infiltration rate equal to 1.33±0.0034 m d^{-1} for test #1, and 0.22±0.0027 m d^{-1} for test #2, where the standard deviations take into account the standard errors of the best fit procedures. In order to interpret the results of the falling head infiltrometer test, the simplified equation from Nimmo et al. (2009) was considered:

$$K_{fs} = \frac{L_G}{t} \ln\left(\frac{L_G + \lambda + D_0}{L_G + \lambda + D}\right) \qquad (5)$$

where K_{fs} (LT^{-1}) is the field-saturated hydraulic conductivity, t is the time, D_0 and D are the initial and final ponded depths, respectively, $L_G = C_1 d + C_2 b$ is the ring-installation scaling length, with 0.993 and 0.578 as recommended values for C_1 and C_2, while d and b are the ring insertion depth and ring radius, respectively, and λ is an index of how strongly water is driven by capillary forces in a particular medium. The macroscopic capillary length, λ, during downward water movement in a fracture plane, is related to the fracture aperture and the interfacial surface tension σ (M/T^2) between air and water (equal to 71.97 dyn/cm at 25°C) (de Gennes et al., 2002), using the following Young-Laplace equation (Pruess & Tsang, 1990):

$$\lambda = 2\frac{\sigma}{\rho g \times \bar{b}_c} \tag{6}$$

assuming that the water-air contact angle equals zero, and that the cutoff aperture, $\bar{b}_c$, equals the fracture aperture which delimits the region occupied only by the non-wetting phase, g (L/t^2) is the gravity acceleration and ρ (M/L^3) is the water density. In order to estimate the field-saturated hydraulic conductivity value, Kfs, by means of equation (5), the macroscopic capillary length, λ, is required. It is determined by inverse modeling, using an unsaturated fractured flow model in the vertical 2-D fracture (Masciopinto & Benedini, 1999). The model solutions were then calibrated on the basis of the comparison between the shapes of the wetting front obtained from the simulation outputs and those of the electrical resistivity images at the correspondent time (Figs. 7 - 8). After estimating the macroscopic capillary length (0.90 m and 0.95 m for tests #1 and #2, respectively) and the corresponding cutoff aperture (16.5 μm and 15.5 μm), equation (5) gave field-saturated conductivity values of 0.67±0.01 m d^{-1} and 0.054±0.001 m d^{-1} for tests #1 and #2, respectively.

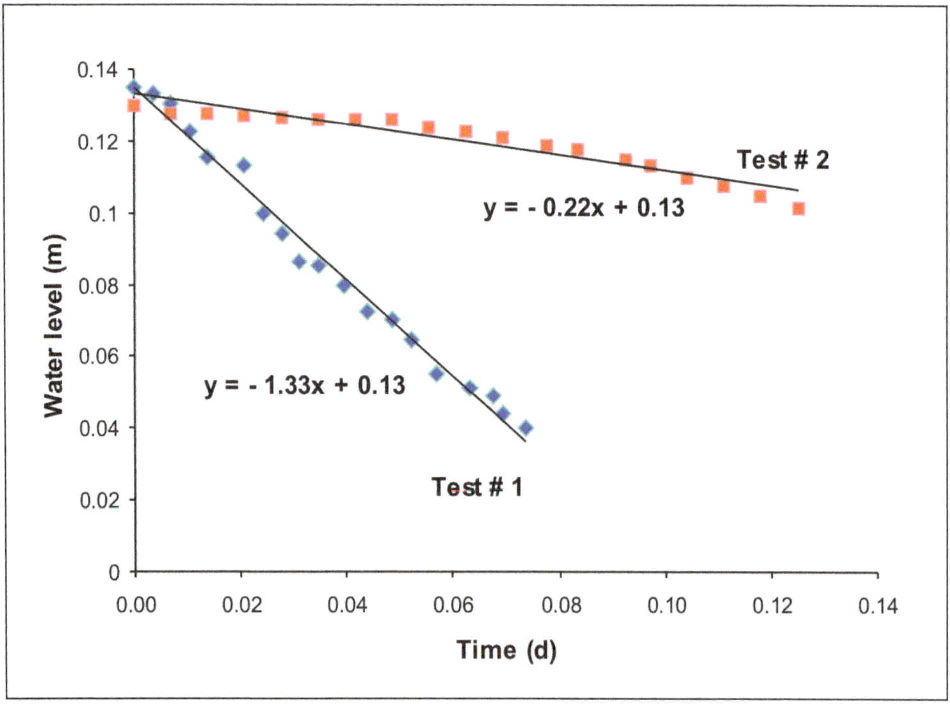

Fig. 6. Measured water level versus time during tests # 1 and #2 at the Altamura site.

The notable difference between the field-saturated hydraulic conductivity values of tests #1 and #2 highlights the great variability in hydraulic properties which characterizes the carbonate rocks studied. In fact, even though the two test sites were only 300 m apart, and the experiments were performed on the same geological formations, the infiltration rates observed in the areas tested were very different. For the Altamura fractured limestone, the modified Kozeny-Carman equation (Pape et al., 1999) leads to outcrop saturated conductivity values ranging from 0.6 m d^{-1} to 12 m d^{-1}, considering the rock outcrop porosity from 1% to 2%.

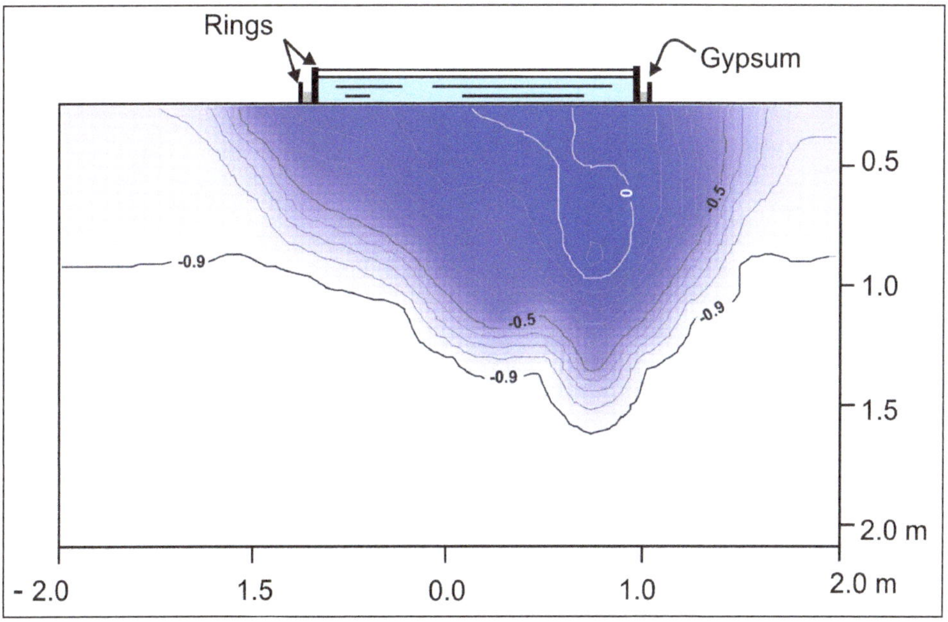

Fig. 7. Best selected model simulation output in order to fit the shape of the wetting front at t = 80 min. Contour lines represent model output expressed as matric potential (m) in the fracture.

It should be noted that the former conductivity value is very similar to the field-saturated conductivity obtained by the ring infiltrometer in test #1 (0.67 m d^{-1}). In contrast, the field-saturated conductivity (0.054 m d^{-1}) derived from test #2, where the limestone outcrop had no visible fractures, is close to the value (0.02 m d^{-1}) obtained by laboratory tests on the rock matrix of the Murge limestone (Borgia et al., 2002). The above comparison supports the field-saturated hydraulic conductivity values obtained using the large ring infiltrometers. Simultaneous electrical subsurface measurements, inverted by using the Res2Dinv software, also show a difference in behavior between the two tests as a consequence of the water infiltration during the tests (Fig. 8). Before the infiltrometer tests ($t \leq 0$ min), the subsurface resistivity was above 500 Ωm at the first site and more than 2,000 Ωm at the second site. In test #1, the low resistivity area, below 150 Ωm, was visualized after 80 minutes to about 1.5 m of depth; similar results were obtained for the electrode arrays in the perpendicular direction. The low resistivity zones can be related qualitatively to fractured rock with high water content and moderate salinity. The same resistivity value was visualized at shallower depth, at the same time, during test #2 than the previous test. In fact, in the last case, it is not possible to evaluate the depth of wetting front from a quantitative point of view, because the resolution of the resistivity images depends on the electrode spacing. As rule of thumb, a structure or object in the subsurface, having dimensions less than the electrode spacing, cannot be defined clearly. For both tests the electrical resistivity measurements confirmed the effectiveness of the ring sealing, showing that the water infiltrated from within the ring and not from outside (Fig. 8). Additionally, electrical resistivity imaging highlights not only that the water reached different depths in the two infiltrometer tests, but also that its variable redistribution, within the investigated vertical plane, results from the structural characteristics of the limestone outcrop. As expected, in test #1, the deepening of the level of the low resistivity anomaly associated with the infiltration of moderately saline water was much greater than in test #2, and was, in fact, consistent with the different number and size of the fractures observed visually in the limestone.

4.2 Calcarenite

The results of the field experiments carried out on the calcarenitic layer of the vadose zone outcropping at the bottom of the quarry, are shown in Figures 9 and 10. The Figures show the infiltration data and the electrical resistivity images, respectively. In order to interpret the results of constant head infiltrometer test we have plotted the infiltration rate, $q = dI / dt$, versus time, t, as shown it the graph of Figure 9a. The value of q, initially decreases rapidly with time and then approaches a constant value. Practically, the rate of infiltration falls, starting from very high values at the beginning of the test when the subsurface is in an unsaturated condition, up to a value of $9 \times 10^{-4} \pm 0.0001$ cm/s (i.e. between 0.86 md^{-1} to 0.69 md^{-1}), when the rock is in quasi-saturated condition. For the studied calcarenite, this value of the infiltration rate, obtained by considering the horizontal asymptote of the curve, was reached after about 2 hours from the start of the infiltration, depending upon the water content of the rock at the beginning of the test. This value should be approximately equal to the field-saturated hydraulic conductivity value, K_{fs}. Again for the falling head infiltrometer test, equation (5) was considered. In this case a λ value equal to 0.01 m was used for the calcarenite, taking into account that the sensitivity of conductivity calculations to the value of λ is small (Nimmo et al., 2009), and that Elrick et al. (1989) proposed a λ value of about 0.08 m suitable for most soils with structural development, a value of 0.03 m for gravelly soils, and 0.25 m for fine-textured soil without macropores. Figure 9b shows the plot of the effective infiltration length (right-hand side of equation (5), except the variable t) vs. time and the slope of this regression line is a convenient calculation of K_{fs}. The graph shows how the K_{fs} values decrease during the infiltration test, from the highest value of 1.22×10^{-3} cm/s (i.e. 1.05 md^{-1}), at the start of the test, until the lowest of 5.61×10^{-4} cm/s (i.e. 0.48 md^{-1}) at the end of the infiltration, with an average value of 8.13×10^{-4} cm/s (i.e. 0.77 md^{-1}). This last value is consistent with the quasi-steady infiltration rate value, obtained from the constant head test (Fig. 9a). The variability of the field-saturated hydraulic conductivity, K_{fs}, observed in Figure 9b depends on the change in water content with the time during the experiment: higher K_{fs} values are obtained in the initially dry rock than in that wet. In fact, the soil water content during the test has been considered an important factor that affects the K_{fs} value of the experimental area (Bagarello & Sgroi, 2007). In order to have a rough estimation of the wetting front depth, a porosity value of about 0.4 was considered. Using this value of porosity and the high infiltration rate of about 4 cm/s for the first 20 minutes of the constant head test, it was estimated that the wetting front should reach a depth of about 20 cm, while up to the end of the falling head test, by considering

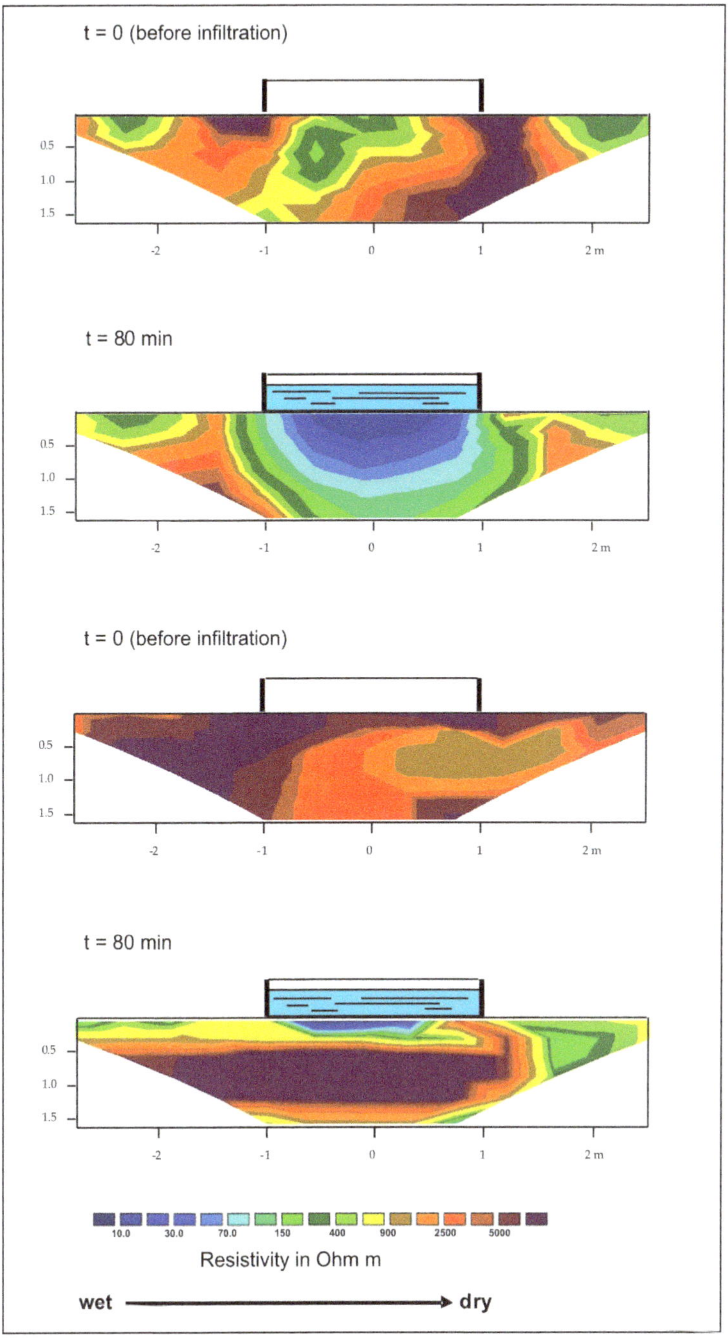

Fig. 8. Electrical resistity profiles during test # 1 and # 2, before (t = 0 min) and during (t = 80 min) water infiltration tests at the Altamura site.

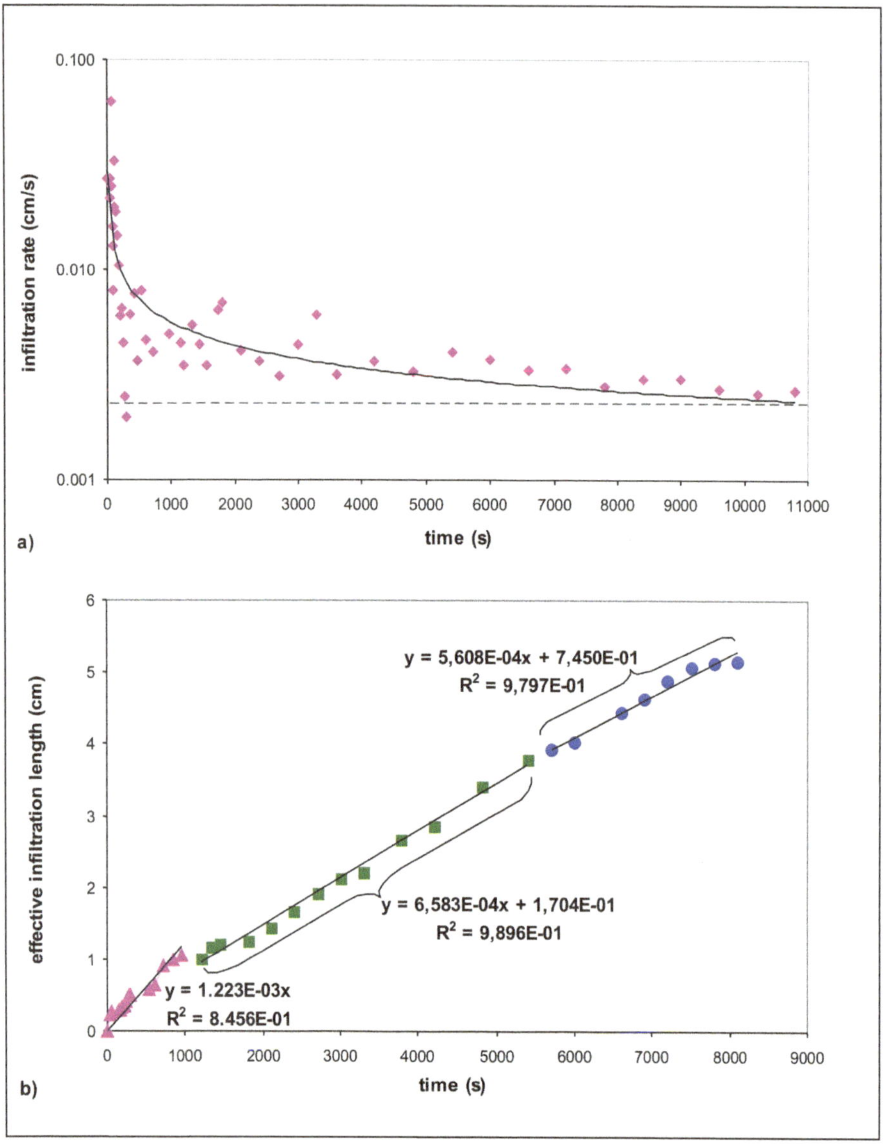

Fig. 9. Test at the San Pancrazio site: a) constant head infiltration test - infiltration rate versus time; b) falling head infiltration test - effective infiltration length versus time.

an average infiltration rate of about $9x10^{-4}$ cm/s, it should only deepen by about another 6 cm. Obviously, this so small a difference in depth that it is not perceptible in the chronological sequences of electrical resistivity images (Fig. 10). Knowing that the shallow aquifer is at 12 m below the bottom of the quarry where the waste was disposed, it was estimated that the time required for the pollutants to reach the shallow groundwater is just over 15 days. Concerning the results of the electrical resistivity surveys, using time-lapse

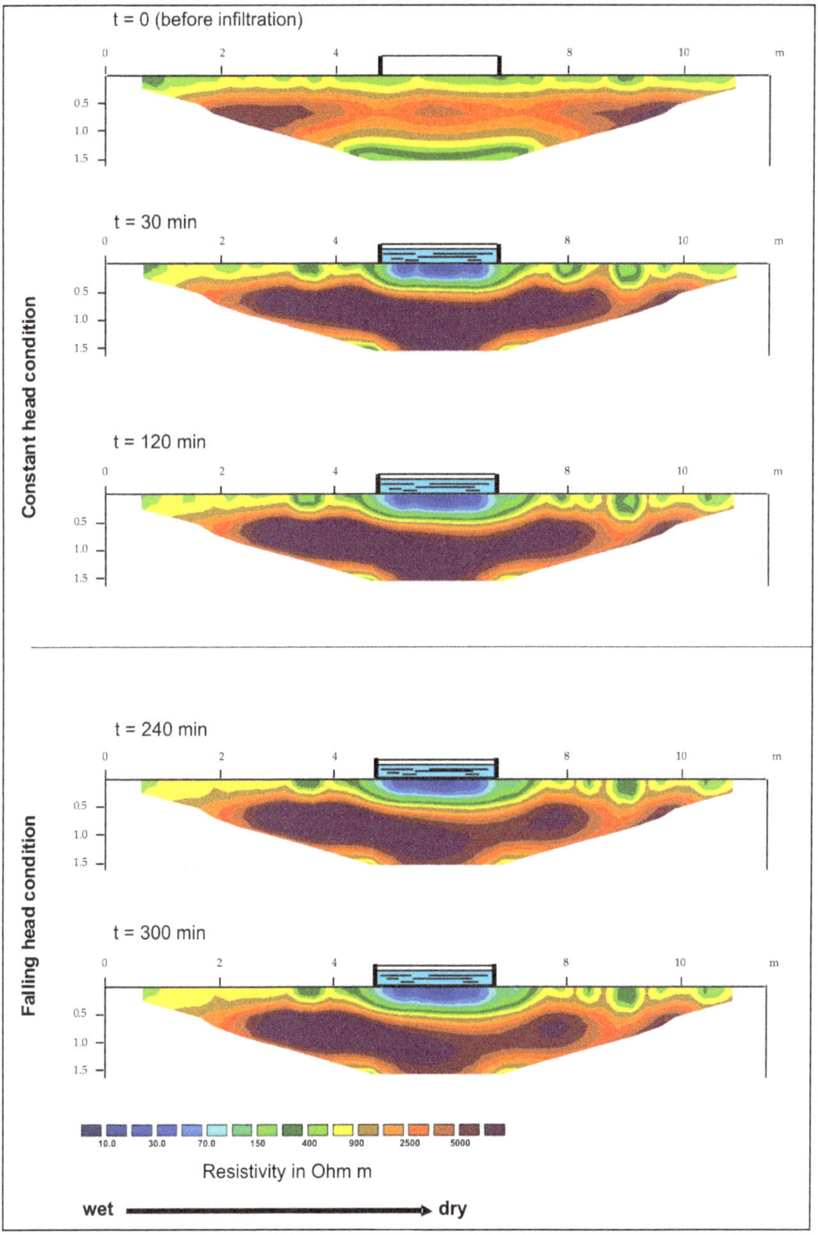

Fig. 10. Electrical resistivity profiles during infiltration test at the San Pancrazio site - profile 1.

techniques, 2-D resistivity images were obtained for different acquisition times. The electrical resistivity images showed no significant differences between the two perpendicular directions (Fig. 10). The first two profiles on the top of Figure 10, measured before starting water infiltration (t ≤ 0 min) and characterized by resistivity values above 1,000 Ωm, represent the profiles with which to compare the following profiles, in order to monitor the water flux in the subsurface. The images from the constant head test show that the wetting front (resistivity values around 60 Ωm) deepens in the first 20 minutes of the experiment, much more than in the following 4.5 hours, up to the end of the test. These results are consistent with the high values of infiltration rate, correspondent to the initial part of the constant head test. The images related to the sequences of the electrical resistivity profiles, acquired during the falling head test (Fig. 10), are similar to the last profile of the constant head condition, such as we expected by considering that the field-saturated hydraulic conductivity values, obtained from the tests under the different conditions, are similar.

5. Conclusion

The large ring infiltrometer, coupled with subsurface electrical resistivity measurements, described in this chapter, has proved to be a simple and inexpensive field tool, capable of evaluating the field-saturated hydraulic conductivity of rock formations, even if it requires numerical elaboration supported by mathematical models, in certain conditions. It is designed to be installed directly on an outcrop of both hard and soft rock, and easily constructed on site with inexpensive, lightweight materials, and has an adjustable diameter. These characteristics improve the versatility of the infiltrometer method and its adaptability to various geological conditions. Thus it expands the potential for exploring field-hydraulic conductivity of rocks, which until now has been investigated more frequently in laboratories, mainly because of the practical difficulties involved in field investigations. Simultaneous electrical resistivity measurements are used to monitor the subsurface water infiltration instead of humidity sensors or tensiometers, thus sidestepping the difficulty of inserting probes into the rock. Further technical difficulties, related to the installation in the field of the large ring infiltrometer, were solved using non-commercially available equipment during the on-site installation procedure. Specifically, the setup of the experimental apparatus requires the hollowing out of a furrow in which to install the ring and seal it to the rock surface with gypsum. Otherwise, if the field conditions of the infiltration area are very heterogeneous, consisting of different media, e.g. rock and soil, an infiltrometer made of two concentric rings is suggested in order to improve the hydraulic packing during the water infiltration test. In this case the continuity between the ring infiltrometer wall and the rock-soil surface is obtained by filling the space between the external and internal rings with gypsum, up to a height of about 2.0 cm,to create a seal. Thus the second ring improves the gypsum seal of the first ring with the ground, overcoming the challenge presented by the presence of different media along the edge of the ring. The efficacy of the seal of ring infiltrometer with soil/rock surfaces was confirmed by the simultaneous electrical resistivity measurements that show the deepening and spreading of the water during the infiltration tests. The time required for the installation was about two hours and the water volume used for each infiltration test was about 0.5 m^3 for the limestone, and about 0.3 m^3 for calcarenite, depending on the local rock permeability. On the whole, the field- hydraulic conductivity data obtained from the infiltrometer tests (0.67 m d^{-1} and 0.054 m d^{-1}, for tests #1 and #2 on limestone, respectively, and 0.77 m d^{-1} for calcarenite) are consistent with the nature of the rocks tested and are corroborated by laboratory measurements carried out by other authors (Quarto & Schiavone, 1994; Borgia et al., 2002). The difference between the field-saturated hydraulic conductivity values obtained by the infiltration tests, highlights the difference of the outcrops studied, owing to the different geological formations characterized from variable heterogeneity degree (number and size of fractures). The possibility of building a ring with a large adjustable diameter on site, has the advantage of including fractures and other features of the rock in order to obtain field-saturated hydraulic conductivity data at a more representative scale of measurements, with negligible border effects. The proposed installation procedure of a large ring infiltrometer on rocky outcrops extends the possibility of performing field infiltrometer tests on rocks that would, previously, have been very difficult to test. Simultaneous subsurface electrical resistivity imaging, using time-lapse techniques, is a useful indicator of water infiltration/redistribution. Moreover, the electrical resistivity survey provides evidence that the conductivity structure is confined under the ring infiltrometer, confirming that the apparatus works well in order to minimize the lateral lack of water, without having losses along the edge of the ring, allowing the acquisition of accurate and representative experimental data. On the other hand, it is important to highlight that the geophysical inversion model used, Res2Dinv, has not been able to represent the heterogeneities and anisotropies of the heterogeneous system, such as fractured or karstic rocks. This consideration encourages further research and study in order to define new algorithms able to derive real images closer to the real subsurface features than now, starting from the electrical resistivity measurements of heterogeneous systems. Using the simplified equation for determining the field-saturated hydraulic conductivity from the experimental data of falling head infiltrometer tests, it is demonstrated that the proposed method works well, in different cases, proving it is a good tool to know how quickly the water moves through the unsaturated zone in real field conditions.

6. Acknowledgment

The work was partially funded by the Italian Government (Regional Autority), which is gratefully acknowledged. We thank Costantino Masciopinto for his collaboration in modelling on fractured systems and for his fruitful cooperation during the field experiments. We also thank Rita Masciale and Francesco De Benedictis for their support for the field work and for their comments during the preparation and revision of this document.

7. References

Archie, G.E. (1942). The Electrical Resistivity Log as an Aid in Determining Some Reservoir Characteristics. *Petroleum Transactions of AIME*, Vol. 146, pp. 54-62.

Bagarello, V. & Sgroi, A. (2007). Using the Simplified Falling Head Technique to Detect Temporal Changes in Field-Saturated Hydraulic Conductivity at the Surface of a Sandy Loam Soil. *Soil & Tillage Research,* Vol. 94, pp. 283-294.

Baumhardt, R.L., Lascano, R.J. & Evett, S.R. (2000). Soil Material, Temperature, and Salinity Effects on Calibration of Multisensor Capacitance Probes. *Soil Science Society of America Journal*, Vol.64, pp. 1940-1946.

Binley, A.M., Cassiani, G., Middleton, R. & Winship, P. (2002). Vadose Zone Flow Model Parameterization Using Cross-Borehole Radar and Resistivity Imaging. *Journal of Hydrology*, Vol. 267 (3-4), pp. 147-159.

Bogena, H.R., Huisman, J.A., Oberdosrster, C.A. & Vereecken, H. (2007). Evaluation of a Low-Cost Soil Water Content Sensor for Wireless Network Application. *Journal of Hydrology*, Vol. 344, pp. 32-42.

Borgia, G.C., Bortolotti. V. & Masciopinto, C. (2002). Valutazione del Contributo della Porosità Effettiva alla Trasmissività di Acquiferi Fratturati con Tecniche di Laboratorio e di Campo (Evaluation of Effective Porosity Contribution to the Transmissivity of Fractured Aquifer Using Laboratory and Field Techniques). *IGEA, Groundwater Geoengineering*, Vol. 17, pp. 31-43.

Cassiani, G., Bruno, V., Villa, A., Fusi, N. & Binley, A.M. (2006). A Saline Tracer Test Monitored via Time-Lapse Surface Electrical Resistivity Tomography. *Journal of Applied Geophysics*, Vol. 59, pp. 244-259.

Castiglione, P., Shouse, P.J., Mohanty, B., Hudson, D. & van Genucthen, M.Th. (2005). Improved Tension Infiltrometer for Measuring Low Flow Rates in Unsaturated Fractured Rock. *Vadose Zone Journal*, Vol. 4, pp. 885-890.

Ciaranfi, N., Pieri, P. & Ricchetti, G. (1988). Note alla Carta Geologica delle Murge e del Salento (Puglia Centro-Meridionale) (Notes on the Geological Map of Murge and Salento (Central-Southern Puglia). *Memorie.della Società Geologica Italiana*, Vol. 41, pp. 449-460.

Dahlin, T. (2001). The Development of DC Resistivity Imaging Techniques. *Computers & Geosciences*, Vol. 27, pp. 1019-1029.

Dahlin, T. (1996). 2-D Resistivity Surveying for Environmental and Engineering Applications. *First Break*, Vol. 14 (7), pp. 275–283.

Daily, W., Ramirez, A., LaBrecque, D. & Nitao, J. (1992). Electrical Resistivity Tomography of Vadose Water Movement. *Water Resources Research*, Vol. 28(5), pp. 1429-1442.

de Gennes, P.G., Brochard-Wyart, F. & Quéré, D. (2002). Capillary and Wetting Phenomena: Drops, Bubbles, Pearls, Waves. Springer. ISBN 0-387-00592-7.

deGroot-Hedlin, C. & Constable, S. (1990). Occam's Inversion to Generate Smooth, Two-Dimensional Models from Magnetotelluric Data. *Geophysics*, Vol. 55, pp. 1613-1624.

Deiana, R., Cassiani, G., Kemna, A., Villa, A., Bruno, V. & Bagliani, A. (2007). An Experiment of Non-Invasive Characterization of the Vadose Zone via Water Injection and Cross-Hole Time-Lapse Geophysical Monitoring. *Near Surface Geophysics*, Vol. 5, pp. 183-194.

Edwards, L.S. (1977). A Modified Pseudosection for Resistivity and Induced-Polarization. *Geophysics*, Vol. 42, pp. 1020-1036.

Elrick, D.E, Reynolds, W.D. & Tan, K.A. (1989). Hydraulic Conductivity Measurements in the Unsaturated Zone Using Improved Well Analyses. *Ground Water Monitoring Review*, Vol. 9, pp. 184-193.

Griffiths, D.H. & Barker, R.D. (1993). Two-Dimensional Resistivity Imaging and Modelling in Areas of Complex Geology. *Journal of Applied Geophysics*, Vol.29, pp. 211-226.

Jones, S.B. & Or, D. (2004). Frequency Domain Analysis for Extending Time Domain Reflectometry Water Content Measurement in Highly Saline Soils. *Soil Science Society of America Journal*, Vol. 68, pp. 1568–1577, 2004.

Kizito, F., Campbell, C.G., Cobos, D.R., Teare, B.L., Carter, B. & Hopmans, J. W. (2008). Frequency, Electrical Conductivity and Temperature Analysis of a Low-Cost Capacitance Soil Moisture Sensor. *Journal of Hydrology*, Vol. 352, pp. 367-378.

Lai, J. & Ren, L. (2007). Assessing the Size Dependency of Measured Hydraulic Conductivity Using Double-Ring Infiltrometers and Numerical Simulation. *Soil Science Society of American Journal*, Vol. 71, pp. 1667-1675.

Leeds-Harrison, P.B., Youngs, E.G. & Uddin, B. (1994). A Device for Determining the Sorptivity of Soil Aggregates. *European Journal of Soil Science*, Vol. 45, pp. 269-272.

Loke, M.H. (2001). *A tutorial on 2-D and 3-D electrical imaging surveys*. Avalaible from www.geoelectrical.com.

Loke, M.H. & Barker, R.D. (1996). Rapid Least-Squares Inversion of Apparent Resistivity Pseudosections by a Quasi-Newton Method. *Geophysical Prospecting*, Vol. 44, pp. 131-152.

Masbrucha, K. & Ferré, T.P.A. (2003) A Time-Domaine Transmission Method for Determining the Dependence of the Dialectric Permittivity on Volumetric Water Content. An Application to Municipal Landifills. *Vadose Zone Journal*, Vol. 2, pp. 186-192.

Masciopinto, C. (2005). Pumping-Well Data for Conditioning the Realization of the Fracture Aperture Field in Groundwater Flow Models. *Journal of Hydrology*, Vol. 309 (1-4), pp. 210-228.

Masciopinto, C. & Benedini, M. (1999). Unsaturated Flow in Fractures with Anisotropic Variable Apertures, *Proceedings of XXVIII IAHR - AIRH Congress*, Graz, Austria, August 23–27, 1999.

Nimmo, J.R., Schmidt, K.S., Perkins, K.S. & Stock J.D. (2009) Rapid Measurement of Field-Saturated Hydraulic Conductivity for Areal Characterization. *Vadose Zone Journal*, Vol 8, pp.142-149.

Pape, H., Clauser, C. & Iffland, J. (1999). Permeability Prediction Based on Fractal Pore-Space Geometry. *Geophysics*, Vol. 64(5), pp. 1447-1460.

Pruess, K. & Tsang, Y.W. (1990). On Two-Phase Relative Permeability and Capillary Pressure on Rough-Walled Rock Fractures. *Water Resources Research*, Vol. 26(9), pp. 1915-1926.

Quarto, R. & Schiavone, D. (1994). Hydrogeological Implications of the Resistivity Distribution Inferred from Electrical Prospecting Data from the Apulian Carbonate Platform. *Journal of Hydrology*, Vol. 154, pp. 219-244.

Reynolds, J. M. (1998). *An Introduction to Applied and Environmental Geophysics*, Wiley, (Ed.) John Wiley and Sons, ISBN 0-471-9555-8, New York.

Reynolds, W.D., Elrick, D.E. & Young, E.G. (2002). Ring or Cylinder Infiltrometers (Vadose Zone), In: *Method of Soil Analysis*, J.K. Dane & G.C. Topp, (Eds.), pp. 818-826, Soil Science Society of America, Inc., ISBN 0-89118-841-X Madison, Wisconsin, USA.

Robinson, D.A., Jones, S.B., Wraith, J. M., Or, D. & Friedman, S. P. (2003). A Review of Advances in Dielectric and Electrical Conductivity Measurement in Soils Using Time Domain Reflectometry. *Vadose Zone Journal*, Vol. 2, pp. 444– 475.

Seyfried, M.S., & Murdock, M.D. (2004). Measurement of Soil Water Content with a 50-Mhz Soil Dielectric Sensor. *Soil Science Society of America Journal*, Vol. 68(2), pp. 394-403.

Swartzendruber, D., & Olson, T.C. (1961). Model Study of the Duble Ring Infiltrometer as Affected by Depth of Wetting and Particle Size. *Soil Science*, Vol. 92, pp. 219-225.

Telford, W.M., Geldart, L.P. & Sheriff, R.E. (1990). *Applied Geophysics* (2nd edition), Cambridge University Press, ISBN 0-521-33938-3, Cambridge.

Youngs, E.G., Spoor, G. & Goodall, G.R. (1996). Infiltration from Surface Ponds into Soils Overlying a Very Permeable Substratum. *Journal of Hydrology*, Vol. 186, pp. 327–334.

White, P.A. (1988). Measurement of Ground-Water Parameters Using Salt-Water Injection and Surface Resistivity. *Ground Water*, Vol. 26, pp. 179–186.

13

Contribution of Tracers for Understanding the Hydrodynamics of Karstic Aquifers Crossed by Allogenic Rivers, Spain

Rafael Segovia Rosales, Eugenio Sanz Pérez and Ignacio Menéndez Pidal
Laboratorio de Geología Aplicada, Escuela Técnica, Superior de Ingenieros de Caminos Canales y Puertos, Universidad Politécnica de Madrid
Spain

1. Introduction

Karstic aquifers provide a source of water supply over large parts of the Mediterranean region. Spain has many large karstic springs under natural regime that contribute to the base flow and quality of river water. Other aquifers are intensively exploited for human water supply, and their vulnerability to pollution needs to be defined and protection zones instated.

A large proportion of the karst systems are scarcely exploited and are under a near-natural regime. They are connected to rivers to a greater or lesser extent. Some of these large springs, such as La Galiana (Iberian Cordillera, Soria, Spain), represent the only drainage point from aquifers and contain significant high-quality water resources. They are responsible for the majority of the base flow of the rivers they give rise to, since the Mediterranean climate means that precipitation falls irregularly. These watercourses can also support a well-conserved aquatic fauna, whose survival depends largely on this base flow. These types of aquifer also require meaningful hydrogeological study and, specially, the knowledge of the hydrodynamics of the aquifer in general and their hydraulic conductivity in particular.

The relative influence of the geomorphological, geological and process factors determines the distribution of voids in the karst rock, and through that, the physical characteristics of effective porosity, hydraulic conductivity and specific storage.

Karstic aquifers are particularly vulnerable to pollution due, amongst other factors, to the presence of swallow holes through which polluted surface water can penetrate. In addition, polluted water circulates rapidly through conduits that possess little capacity for self-purification, transporting the polluted water over large distances. (COST 65; Goldscheider et al. 2000; Zwahlen, 2004; Goldscheider, 2005; Panno, 2006; Fournier et al. 2007).

The study area dealt with in this paper lies in the northwest of the Iberian Cordillera (Central Spain) and extends over approximately 400 km^2. Its relief is determined by the presence of a meseta of limestone layers that have been dissected by the River Lobos, giving rise to a canyon 26 kilometres long. In 1985, this was declared a Natural Park, due to its peculiar landscape and an important colony of Griffin Vultures, as well as colonies of other birds of prey like the Golden Eagle (C.M.A. - J.C.L., 1992). The River Ucero emerges from a spring at La Galiana situated at the end of the canyon, with a mean flow of some 2000 l/s. The Ucero is a right bank tributary of the river Duero. In turn, the river Lobos has a series of left-bank tributaries whose headwaters lie outside of the calcareous massif in the less permeable rocks around its perimeter.

Of the fish species present, the common trout is common and the river Ucero is one of the most productive in Spain, being famous as a trout stream ever since Medieval times. With respect to mammals, there are nearly thirty species represented, including otter, mountain cat, beech marten and roe deer. The presence of the otter is linked to the abundance and quality of the water. The majority of species depend on the deep pools of the Lobos to survive the summer. The Lobos itself, whose flow is not regulated, has a natural regime, ranging from the flood flows in winter and spring (which, in the narrowest reaches can cover the entire bed of the canyon) to a river bed that is dry except for the deep pools where the water table comes to the surface. The fauna are adapted to this dynamic - the yellow water lily, for example, which seems to be swept away by the winter flows, reappears each spring in the quiet backwaters. Even during the most unfavourable months of high summer, the largest deep pools still have water, whilst the rest of the watercourse is practically dry. As intimated above, it is essential to conserve the animal communities that depend on the deep pools.

Most of the Cretaceous aquifers and their recharge areas fall within the boundaries of the Natural Park. One of the main objectives of declaring the Natural Park was to conserve its waters and achieve compatible use of the aquifer as a groundwater reserve for supply high-quality water to the centres of population in the vicinity. Although the groundwater resources of the Weald Facies around the periphery of the karst would be sufficient in quantity to satisfy current demand, they are mostly degraded by the naturally poor quality of water, due to its high iron content. For this reason, the karstic aquifer of the Lobos Canyon assumes greater importance as a water resource and as a significant water reserve suitable for human consumption, which needs to be protected from contamination.

The object of this study was to provide a sufficiently detailed knowledge of the aquifer in a short time, bearing in mind the scarcity of data available. The methodology applied gave good results, yielding quite a clear model of its hydrogeological behaviour. The study is based on two preliminary studies of the geomorphology and hydrogeology, on the geological maps for the aquifer (only available on a scale of 1/50.000), and on data from a handful of drinking water supply wells. For this reason, it was necessary to devise a plan of work that could yield quantitative knowledge and include aspects important to the future management and conservation of the aquifer and its rivers, such as understanding the flow and hydrodynamics of the aquifer.

These tests of tracers can be the basis for estimating the hydraulic conductivity, because there is a direct relationship between the groundwater velocity and hydraulic conductivity. This research shows how hydraulic conductivity presents spatial variation with regards to variations of the karstification.

Existing wells are few and insufficient to draw an isopiezometric map. However, the aquifer is suitable for tracer studies because of its numerous sinkholes. In fact, there have been several qualitative antecedents with positive results, which have endorsed the viability of tracer experiments (Hernanz y Navarro, 1972; Sanz, 1992, 1996). It was hoped that tracer tests would determine how groundwater velocity varies under different rainfall regimes

and with distance from the discharge point, and throw light on the dispersion of contaminants within the aquifer.
Another important aspect was de determine the response of the aquifer to pollution. The autogenous recharge area of the aquifer is mostly covered by woodland and there are no pollution foci. Rather, the risk of pollution hails from discharges to allogenous rivers as they flow through towns and villages before reaching the karst. There have been cases of illegal discharges of purines from pig farms during the 1970s and 1980s, which reappeared through La Galiana spring, as well as sewage spills from towns and villages upstream.
Thus, it is important to have a detailed understanding of the relationship between the river Lobos and the karstic aquifer under different hydrological situations, since this river flows from one side of the aquifer to the other and is the collector stream for all the surface watercourses. It was not known if the river behaved solely as a losing stream along its entire length, nor what the infiltration capacity of the riverbed was. The riverbeds hydraulic conductivity of these streams determines the recharge of these allogenic streams. Tracer studies were needed to elucidate the hydraulic connection between the rivers and the aquifer and two temporary gauging stations were installed in the river Lobo, as well as three on its tributaries and one at La Galiana spring.

2. Hydrogeology of the karst of the river Lobos canyon

2.1 Geology

Sedimentary terrains in the Natural Park date from the Jurassic to the Present Day. The Jurassic rocks are essentially marine carbonate rocks. Overlying these, concordant with them and in transition towards the Cretaceous, are detritic sediments of the Purbeck-Weald Facies. Over these, lie sands of the Utrillas Facies (Albian), then Coniacian-Santonian-Campanian marls, and the Turonian limestones, which project to form the surface relief, and the cliff faces of the canyons in this region (including the canyons of the rivers Lobos, Espeja, Abión en Burgo de Osma and Boos). Above these strata comes the Garumnian Facies in transition with the Tertiary and, discordantly, the postorogenic sediments of the Miocene, Plioquaternary and Quaternary (IGME, 1982).

2.2 Definition of the karstic aquifer

The Coniacian–Santonian–Campanian limestones are 240 m thick and comprise a highly permeable hydrostratigraphic unit that forms the main aquifer. Its impermeable base consists of a marly series with calcareous intercalations towards the top, some 70-100m thick.
Due to the presence of the marly intercalations, the lower part of the aquifer can be locally confined, while the rest is classified as a free aquifer, forming a meseta, with a syncline structure oriented east-west. Its lateral borders are defined by impermeable Cretaceous marly outcrops that also define the base of the aquifer, except in the south, where there is a discordant contact between the clayey Miocene and the calcareous Upper Cretaceous. The anticlinal axis of Santa María de las Hoyas–Ucero–Ailagas raises the marly base and, whilst it does not always outcrop, it forms the hidden lateral barrier. The hinge and axis of the anticline form the underground watershed between the Lobos aquifer in the north, and the one to the south that feeds the springs at Rejas, Fuencaliente and Ucero. This behaviour was deduced from elevations, as indicated in the block diagram in figure 1.

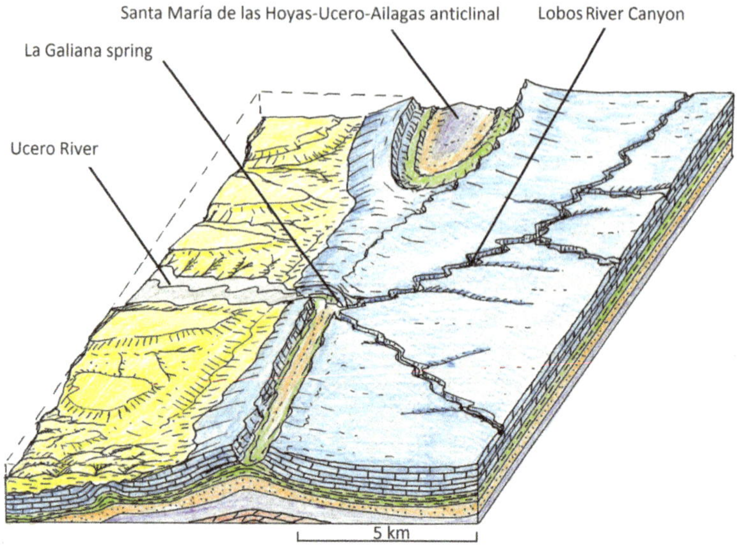

Fig. 1. Block diagram showing the southern part of Lobos River Canyon aquifer (E. Sanz). Green: Cretaceous marls. Blue: Cretaceous limestone (aquifer). Yellow: Tertiary clay.

The eastern border may be open towards Torreblancos, where a number of springs emerge carrying flows of 200 l/s. For this reason, it was decided to define the limit of the system there, where the Cretaceous is still covered by the Tertiary. The possibility of a hydraulic connection with the Tertiary of the Almazán Basin cannot be ruled out.

The marls mentioned above mark the base of the karst, and surround the synform on almost all sides. La Galiana spring emerges at a lower elevation (approximately 900 m a.s.l., at the entrance to the Lobos Canyon, upstream of Ucero), where the karst meets these marls.

The carbonate rocks outcrop over some 160 km^2 and represent the erosion surface of the meseta (SEF). As the SEF dates from the end of the Miocene, it means that these rocks have been exposed to karstification since this time. There are neither poljes nor uvalas, and dolines are infrequent. More abundant are karren fields. With respect to the endokarst, there is an overwhelming dominance of sinkholes over caves, and the only caves are restricted to former or present-day discharge points of the aquifer, all of which are concentrated at La Galiana, at the end of the canyon. The homogeneous distribution of the abundant sinkholes throughout the massif suggests recharge is predominantly diffuse, rather than concentrated.

Terra rossa soil is concentrated at the base of the slopes, in the stream beds, dolines and infilled fissures. Rendzina soils are dominant and have developed over the limestone. They contain more than 40% carbonate, and have a significant humus content, which has encouraged a shrubby vegetation cover of pine and juniper woodland.

2.3 Characterization of natural recharge and water balance of the aquifer

The Lobo catchment can be divided into two zones with distinct hydrogeological characteristics. The first contains Wealden deposits and Jurassic carbonates (197 km^2). These are located in the north and northeast and comprise mostly detritic deposits (Wealden), with a narrow band of Jurassic carbonates. Its outflows, both surface runoff and groundwater (or hypodermic) base flow, flow into the karst, since the headwaters of the rivers Lobos, Navaleno, Chico and others all flow into the Lobos karstic zone.

The second zone is the karst (164 km^2) in the south and southeast of the catchment and defined by low permeability deposits. This zone receives slightly less rainfall than the first. It discharges mainly towards La Galiana spring. The zone does not produce its own surface runoff, except after very intense rainfall. In addition to recharge from rain falling over its outcrops, it receives all of the runoff from the Wealden/Jurassic zone. When the flow from the Wealden/Jurassic zone exceeds the infiltration capacity of the riverbeds (of the river Chico and, especially, the river Lobos), the excess surface water leaves the karstic system via the river Ucero.

The water balance calculated for the period (during which the aquifer was under natural regime) was reported by Segovia (2008) and showed rainfall of 109.5 hm^3, natural recharge of 77 hm^3, evapotranspiration of 64.3 hm^3, and practically nil pumped water abstractions, zero surface runoff and zero lateral transfers to other aquifers. The water stored in the aquifer increased by 5.59 hm^3.

Allogenous recharge was differentiated from autogenous using data from the temporary gauging stations on the allogenous rivers. In this way, over the study period, autogenous recharge was calculated to be 42.2 hm^3 , or 54 % of the total recharge. Allogenous recharge was calculated to be 31.5 hm^3, or 41 % of the recharge. For the Lobos catchment as a whole during this period, R = 77 hm^3, which represents 32% of the precipitation. The recharge rate with respect to precipitation in the karstic zone (Rk) was R_k = 41% , and R_w = 24 % in the allogenous zone.

The sole discharges of the aquifer are via La Galiana spring, and through the river bed of the Lobos upstream of this spring, where discharge takes the form of permanent pools, interconnected by only a small stream of water under dry weather conditions, or by an appreciable current in times of mid to high water (10 to 60 l/s).

These outflows were monitored at a gauging station downstream, to reconstruct the hydrogram for the hydrological year 1995 - 1996 (1 September 1995 to 31 August 1996). The mean flow was 2,423 l/s, the great majority of which flows from the Galiana spring, and the remainder from the river Lobos discharges 4-5 km upstream.

As seen from figure 2, La Galiana spring has an irregular flow that is highly sensitive to rainfall and snow melt, indicating quite a rapid emptying of the aquifer. The spring at La Galiana reacts to heavy rainfall events with a delay of 4 or 5 days. This time lag is logical in large karstic aquifers like this one, where there is a delayed reaction to recharge to allogenic streams situated on the periphery at moderate or long distance. In addition, the allogenic recharge has to penetrate the 100 m thickness of the unsaturated zone. (This phenomenon is confirmed, to a certain extent, by the dripping from stalactites in the Lower cave at La Galiana after rainfall events (Sanz, 2000).

2.4 Hydrodynamic functioning

2.4.1 Flow monitoring

The synclinal structure determines that groundwater flow converges and accumulates in its centre, where the Lobos Canyon is set. In this way, the lower reach of the canyon acts as the natural drain, creating the patent and visible discharge at La Galiana, described above.

Tracer studies were required to determine the hydrodynamics of the aquifer, since the aquifer is under a natural regime and there is insufficient piezometric data from boreholes (there are only three boreholes) to draw isopiezometric maps.

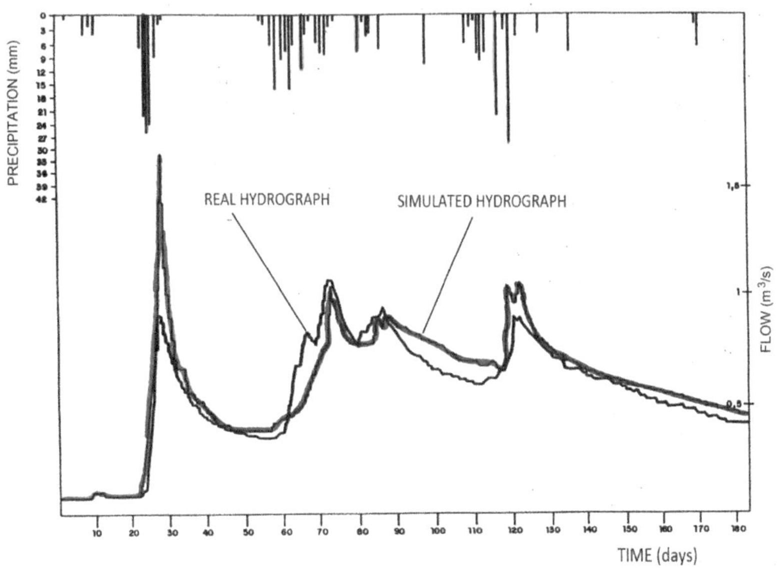

Fig. 2. Real and simulated hydrograms in 1989

2.4.2 Hydraulic gradient

Though no isopiestic map was available, the hydraulic gradient can be estimated using the elevation of the spring at La Galiana (900 m), in addition to the boreholes (at Puente de Siete Ojos, future supply wells for San Leonardo, Navaleno and Casarejos, and the well serving the bar within the canyon). Furthermore, swallow holes containing permanent water (like Valdecea) can be used because deep sinkholes that are always dry give us the maximum hydraulic gradient. They also give the longitudinal gradient of the course of the river Lobos, which provides a value for the maximum hydraulic gradient under high water conditions, where they act as collectors.

In this way, the following gradients were calculated: under high water conditions, the gradient could be around 2.5 per thousand (0.0025), whilst under low water (dry season), it could be 0.62 per thousand (0.00062).

2.4.3 Variation in phreatic level

It is well known that karstic aquifers can suffer sharp piezometric oscillations, since their porosity is relatively low but their recharge capacity is large - so that they can fill very quickly.

Variation in the phreatic level is also indirectly reflected by oscillations and sharp changes in flow seen in the hydrogram for La Galiana de Ucero.

Though no periodic piezometric measurements were available, the following observations have been made about variations in piezometric level (Sanz, 1992)

- The water level in a 25 m borehole on the talweg of the river Lobos at Puente de Siete Ojos, which is usually dry, rose quickly (in less than a week) following intense rain during December 1995/January 1996, reaching the same level as the river itself.
- Appearance of the resurgences at Las Raideras.

- Rapid rise of more than 20m in the cave at Hue Seca de Valdeavellano (a gallery that is normally dry to at least 20 m depth and 400 m length) which began to discharge after intense rain. Here, a spring bursts forth from the cave mouth, carrying up to 100 l/s that forms the headwater of a stream.
- The swallow hole at Hue Seca, in Santa María de las Hoyas, is 11 m deep and usually dry, but it produces a flow that can exceed 200 l/s as the phreatic level rises during exceptionally rainy spells. The flow usually persists only a few days and then the water level in the swallow hole recedes slowly, until the stream dries up.

3. Methodology

Ten tracer tests were carried out by dosing uranine in sinkholes between 5 and 14 km distance from each other. The tests were done in both low water and flood conditions during the hydrological year 1995 - 1996, a year that included both severe low water and a sudden winter recharge. A further test was done in the sporadic springs of Las Raideras via a connected sinkhole with the phreatic level at its base.

Monitoring and sample collection was done at the presumed emergence points of the tracer, i.e. at La Galiana spring, at a fish farm on the river Ucero, and at a point on the river Lobos. At the latter point, samples were taken upstream of the surgence (when there was any flow in the river). During the dry season, the lower reach of the river Lobos carries only a small volume of groundwater discharge. However, during the wet season, it was not known what connection the river flow has to the aquifer, nor its relevance. Samples were collected twice a day, giving a total of 700 samples, which was considered sufficient given the extended duration of the tests. In all cases, correct authorization had been obtained (though the tracer plumes were invisible to the human eye and so passed unnoticed by members of the public). Laboratory analysis employed precision fluoroscopy, with some samples being analysed for pH and conductivity as well. Fluocaptors were also installed in the springs at Fuencaliente, Rejas and Santervás, situated on the periphery of the karst, in order to check that there was no communication between these and the sinkholes used in the tracer test.

In karstic aquifers where there are few boreholes that can be used to draw isopiestic maps and understand groundwater flow, tracer studies can be used to understand the hydrodynamics of the system and other aspects, such as vulnerability to pollution.

In hydrogeology as well, quantitative experiments have been carried out to obtain the response curves (for example, Smart, (1988), Hauns et al.(1991), Meus and Bakalowicz, M. (1997), Käss (1998), Birk et al. (2005), Massei et al. (2006), Geyer et al. (2007), Morales et al. (2007), Goldscheider et al.(2008), Göppert and Goldscheider(2008), Perrin and Luetscher (2008)). G-Yélamos (1999) presents a clear summary of these kinds of experiments in Spain.

In the present case, the choice of this method is clearly justified, since we are dealing with a practically unexploited aquifer, with numerous permanent and temporary swallow holes distributed at various distances from the principal discharge point at La Galiana.

The methodology followed for the tracer tests during 1995 and 1996 involved the injection of a total of 5,100 g of uranine through nine different injection sites, selected according to the location of sinkholes and the hydrogeological knowledge available.

Injection sites were selected in order to elucidate the influence of orientation with respect to the spring at La Galiana. In addition, sinkholes were selected at different distances from the spring, to investigate the velocity and dispersion of the tracer. In addition to these two aspects of spatial distribution, tests were also distributed temporally, to determine variation in flow velocity under different hydrological situations (low and high water conditions, for example). The points selected are indicated in the following table 1 and on Figure 3 We consider that this range of distances and orientations is representative, given the possibilities offered by the aquifer.

Injection Site	Distance to spring at La Galiana (km)	Date	Mass of tracer injected (gr)	Mean time to appear (days)	Mean flow at La Galiana (l/s)	Time tracer was within aquifer (days)	Velocity (km/day)
River Navaleno sinkhole	9,5	5/10/1995	200	12	2.600	18	0,8
River Chico sinkhole	6,5	11/11/1995	300	14	2.300	30	0,46
River Navaleno sinkhole	9,5	17/12/1995	500	8,5	1.400	18	1,1
Sinkhole in Arroyo de Valderrueda (de Casarejos)	4,5	28/12/1996	600	3	3.000	10	1,5
	1 (distance to river Lobos)	28/12/1996	600	1 (approx)	3.000	hours	4
Sinkhole in Arroyo del Chorrón	10	4/1/1996	1.400	4,5	4.600	21	2,2
Sima de Valdecea	2,5	17/2/1996	500	< 1	3.0?00	4	3
sinkhole in the river Lobos (Apretadero)	12	13/3/1996	1000	21	2.200	15	0,6

Table 1. Some value obtained by tracers testing

3.1 Tracer injections made

Prior to the current investigation, two other tests had been done in the river Navaleno (16 August 1967) and in the river Chico (1992). In both cases, the uranine reappeared at La Galiana. The 1967 test formed part of research into karstic aquifers by the Centre for Hydrographic Studies, and consisted of a single dose of 3 kg uranine dissolved in alcohol and ammonia in a swallow hole in the bed of the river Navaleno, 9.5 km from the spring at La Galiana (Figure 3, measured as the crow flies). The dye re-emerged at La Galina after 6.5 days. The dye tracer test in the river Chico was done 1991. These preliminary tests provided orientation when planning the 1995-96 dye tracer survey.

3.2 Sampling

Water samples were taken periodically at La Galiana spring, and fluocaptors were also used at the springs of Rejas de Ucero, Valdelinares, Fuencaliente and Sentervás, in case the aquifer had any outflow, apart from the river Ucero, on its south-western edge. The tracer was recovered only from the spring at La Galiana, which indicates that most or all of the discharge from the karst emerges at this spring.

Samples were taken from the spring at La Galiana, from the river Lobos upstream of the springhead (since water also surcharges here) and, on occasions, from the fish farm downstream of the spring (which could be considered as the spring in its own right when the river Lobs does not carry any surface runoff).

Samples were taken in sterilized translucent plastic bottles (100 ml) twice a day (morning and afternoon), and more frequently for the injection sites close to La Galiana. Samples were stored in the dark to avoid photodegradation of the tracer and to reduce microbial activity, both of which could have reduced the concentration of the tracer in the samples.

Samples were analysed by measuring the sodium fluorescein spectra using a luminescence spectrometer. This was equipped with synchronized scanning, which allows conjunctive analysis of both the excitation and emission spectra (fluorimeter). Analyses were done within one month of the samples being taken, in the Centro de Estudios de Técnicas Aplicadas del CEDEX (Madrid) and in the ETSCCP Applied Geology Laboratory of the Universidad Politécnica de Madrid. A "zero" reference solution was used, corresponding to the concentration of fluourescein present in the spring water at La Galiana at the start of the dye tests. By this means, any background effect due to earlier tests is eliminated.

4. Results and discussions

Figure 3 shows the trajectories of groundwater confirmed by the tracer tests, while Figure 4 shows the response curves. From Figures 3 and 4 we can make the following observations:

1. Confirmation of the hydrogeological catchment

The tracer tests confirmed the recharge area, previously established from geological maps, and identified the boundaries of the aquifer. They verified the communication between all the swallow holes and the spring at La Galiana, and confirmed that there are no lateral transfers to the springs of Funcaliente, Santervás and Rejas de Ucero, situated to the south. The only zone of discharge is La Galiana spring - and the lower reach of the river Lobos connected to this spring, which, in effect, forms part of the same discharge.

2. Relationship of the river Lobos with the aquifer as observed using tracer tests.

Under low water conditions, groundwater flow is towards La Galiana spring only, since the river Lobos remains dry over its entire length, except for a number of deep pools where the phreatic level comes to the surface, from downstream of the San Bartolomé hermitage, over a stretch of approximately 4 km. The dye appeared only in La Galiana, in the form of a unimodal curve (Figure 4 and 5).

In times of intermediate flows, (neither dry season nor flood flows) which is the most common situation, the phreatic level rises and the lower reach of the river Lobos becomes a secondary surcharge for the aquifer, acting as a kind of overflow for La Galiana. This involves a flow between the pools of no more than 30 l/s, since the majority of the flow continues to emerge via La Galiana. The rest of the river Lobos watercourse, upstream of the hermitage, is a losing stream. In the intermediate flow situations, dye also emerged in the river Lobos. When sampling was done in the river Ucero, downstream of La Galiana and the river Lobos, the response curves were bimodal, a smaller peak presaging these two outflow plumes (Figure 5). When samples were taken in the river Lobos and at the Galiana spring, separately and together, it was observed that the tracer appeared first in the river Lobos and, later, at La Galiana. This makes sense, since the river Lobos sampling points are upstream of the spring. In the test where dye was injected into the Apretadero sinkhole, the delay at La Galiana with respect to the river Lobos was two days (Figure 6).

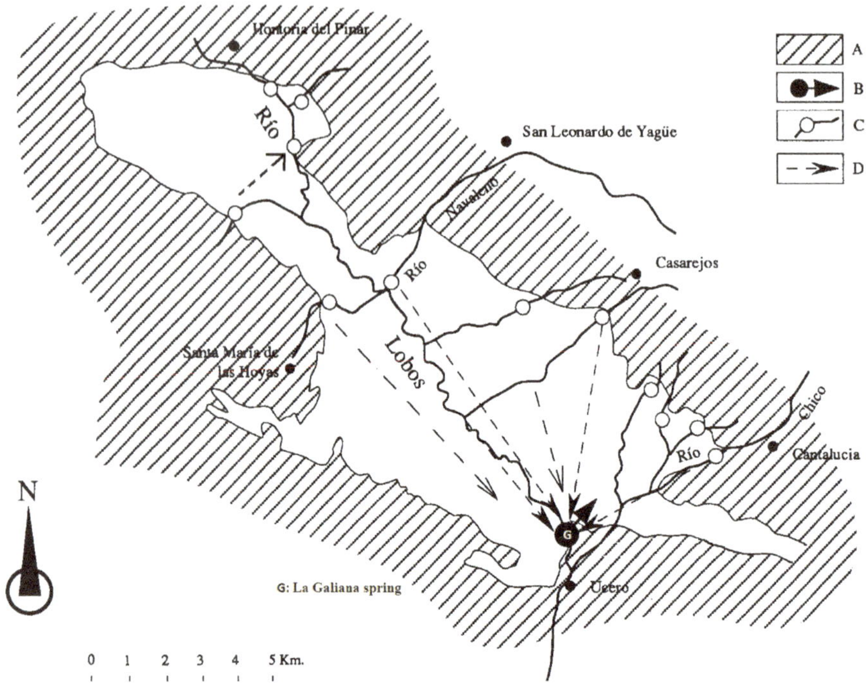

Fig. 3. Hydrogeological scheme at Lobos River Canyon and groundwater paths tested with tracers. A. Non-karst rocks. B. La Galiana spring. C. Connection proven by tracer. D. Streams with sinks

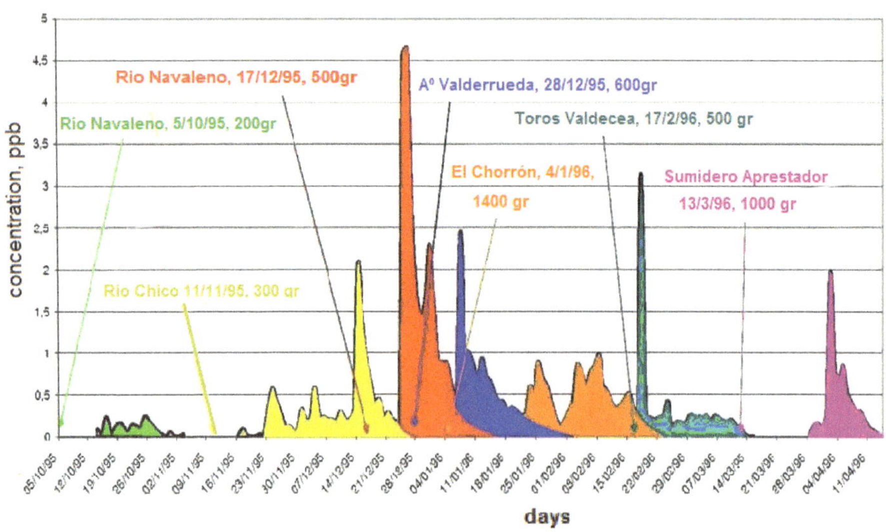

Fig. 4. Response curves concentration-time for different tracer tests.

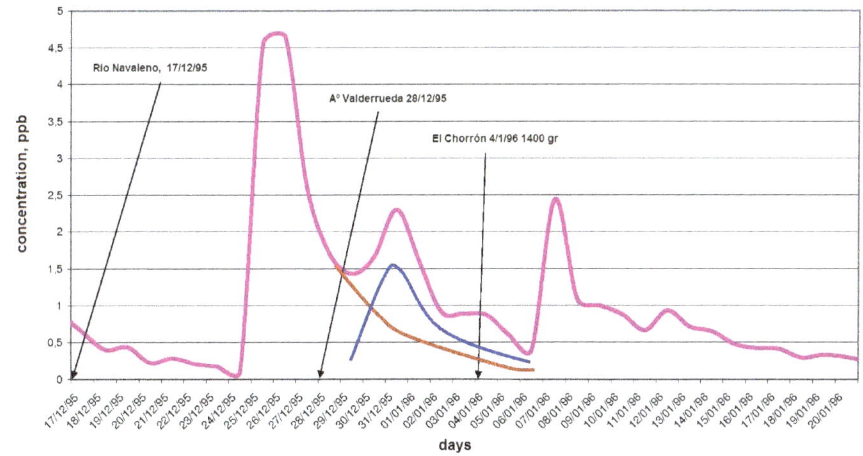

Fig. 5. Response curve detailed at La Galiana spring and Lobos River for testing Navaleno River, Arroyo de Valderrueda and Chorrones sink.

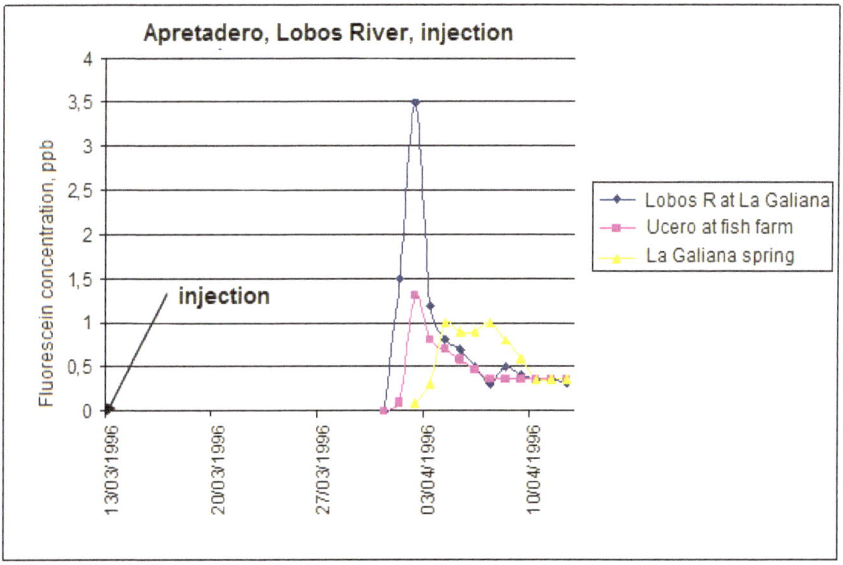

Fig. 6. Concentration-time in response curve for test in Lobos River, Apretadero sink place.

Under flood water conditions, the river Lobos contains water along the entire canyon. In its upstream 6 km, between the head of the canyon and the Puente de Siete Ojos, the river now acts as a gaining stream. This occurs as a consequence of producing copious ephemeral surcharges that drain a hanging syncline. In 2000, a tracer test confirmed the existence of a rapid, local flow between the groundwater flow from a cave called Sima del Portillo de

Hontoria to the principal of these springs (Las Raideras). In the reach between Puente de Siete Ojos and the San Bartolomé hermitage, more or less, the river is always a losing stream, as evidenced by differential gauging, and by the fact that the tracers did not appear here. The limit of the hermitage is approximate because, as the phreatic level rises again, it is the intersection between this and the watercourse that separates the losing and gaining parts of the river. As the river usually contains a great flow, it is difficult to appreciate this transition and the only way to check it is using tracers. Between the hermitage and La Galiana spring the river is a gaining stream, as shown by the appearance of the tracer substance.

4.1 Typology of the tracer curves

Although the curves overlap at times (when two tracer tests were run successively), in general it is possible to separate one from the other a *grosso modo,* as indicated tentatively in Figure 4.

Four of the curves (tracer injections in the river Navaleno on 17/12/95, and at Valderrueda, Torca de Valdecea and El Apretadero) follow the typical normal distribution, with an steeply-ascending arm and a steeply-descending arm. All the curves have the two peaks mentioned above: a principal wave followed by a much smaller one. It seems that this type of curve occurs mostly, though not exclusively, under flood and intermediate flow conditions.

However, three curves did not follow this pattern:

The first injection into the river Navaleno swallow hole (5/10/95) gave a serrated curve, which we attribute more to the poor ability of the fluorometer to detect low concentrations of tracer (only a small amount of uranine was used given the low water conditions).

The curve resulting from dye injection into the Chorrón sinkhole produced two clear peaks, with the second one, paradoxically, being larger than the first. It is difficult to interpret this feature, which once again, points to the complexity of karstic systems. It may have been due to a bifurcation in flow but, more probably, that the tracer plume was split into two plumes by the recharge caused by intense rainfall during the tracer test that caused a sharp change from low flow conditions to flood flow. The direction of groundwater flow coincides with the direction of flow in the river Lobos, so that the tracer plumes pass beneath the riverbed and, sometimes, beneath its tributaries. The recharge produced when these losing streams are carrying water can interfere and distort the geometry of the tracer plume.

The response curve for the tracer test in the river Chico under low flow conditions indicates low flow velocity and a long retention time of the plume within the aquifer. The curve has three peaks - two small ones followed by a larger one - all contrary to what is the norm. It is possible that, here, there were a number of bifurcations in flow, a phenomenon that is more common under a low water regime. The test cannot be repeated in this watercourse under flood water conditions because the river flow obliterates the swallow holes.

Two small response curves apparently have no correspondence to any of the tracer tests. These were recorded on 27/10/1996 and 8/02/1996, (Figure 5). The plume of 8/02/1996 can be explained by the rainfall and consequent spate in the river Lobos which washed out tracer remaining from the previous tests; this is a fair assumption as the plume coincides with an increase in the hydrogram recorded at La Galiana and in the river Lobos at Puente de Siete Ojos. The other plume (27/10/96) can be explained in the same way, though it occurred much later than the tracer tests and was not accompanied by any flood flow. It is relevant to note that the aquifer was "contaminated" by the fluorescein used in the second tracer test, which was translated as an increase in background concentration at both La Galiana and river Lobos springs.

4.2 Velocity of groundwater flow

According to the tests undertaken, the actual velocity of groundwater oscillated between around 460 m/day to more than 3,000 m/day. This attests to the high water velocity close to the Galiana spring and in more distant zones, both under high and low water conditions. Figure 7 shows the variation in groundwater velocity as a function of spring flow at La Galiana. It shows how the velocity of groundwater flow varies from low to high water conditions as a function of the phreatic level and natural recharge. As expected in karstic aquifers, phreatic level and recharge shows wide oscillation and this is reflected in the hydrogram for La Galiana spring as well as in the water levels in the borehole at Puente de Siete Ojos.

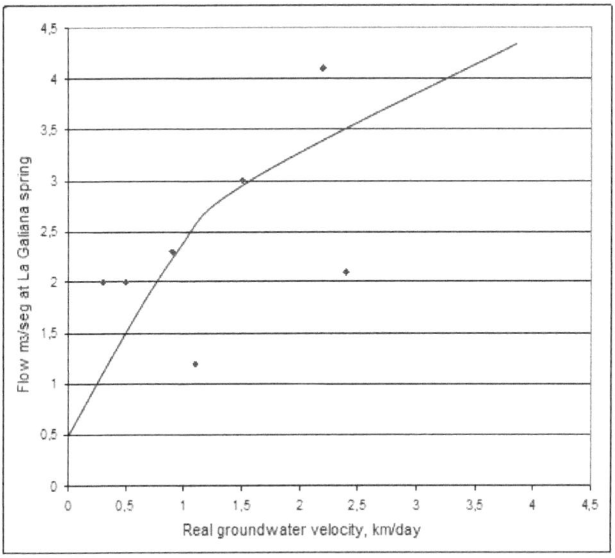

Fig. 7. Relationship of groundwater velocity and flow at La Galiana

Figure 8 shows the variation in groundwater velocity with respect to distance to the main discharge point at La Galiana. Two curves are shown - one for high water (rainy spells) and one for intermediate flow (drier spells). From these, a rough estimate can be made of the isochrones which, in the middle of the aquifer (which lies about 9 km from La Galiana), vary quite markedly - from 21 days for low water and 4 days under high water conditions. In other words, the hydrodynamic regime varies widely according to rainfall and recharge.

It also seems that flow velocity varies according to the direction of flow. The river Casarejos is the only point where flow is orthogonal to the general direction of flow and this supposes that groundwater flows sub-parallel to the axis of the canyon and at a lower velocity. This suggests certain anisotropy, since the velocity along the axis of the canyon (flowing north-west - south-west) is four times that flowing NE-SW. This feature is unsurprising because it makes sense that karstification has developed preferentially along the river, along the vertical plane that is hydrodynamically most active and where the greatest inflows and outflows to the aquifer occurred through the Quaternary.

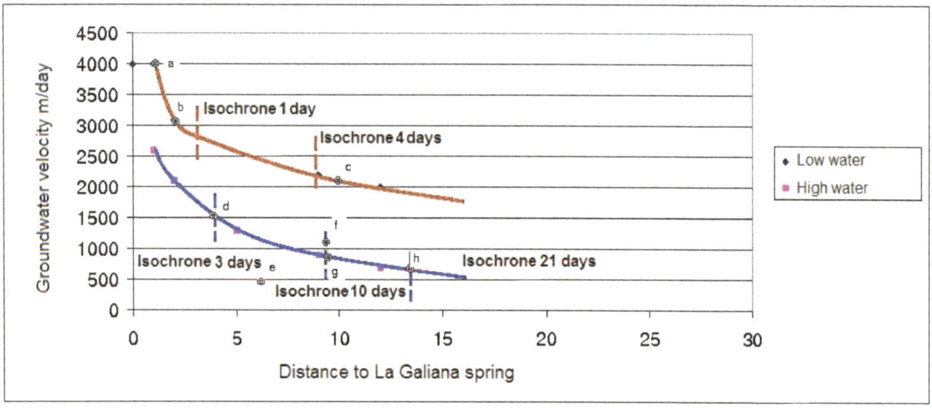

a. Arroyo de Valderrueda- Lobos River
b. Valdecea- La Galiana spring y Lobos River
c. Sumidero del Chorrón- La Galiana spring
d. Arroyo de Valderrueda- La Galiana spring
e. Chico River- La Galiana spring
f. Navaleno River- La Galiana spring
g. Navaleno River- La Galiana spring
h. Lobos River (Apretadero)- La Galiana spring

Fig. 8. Variation of groundwater velocity depending on the distance to La Galiana and isochrones calculation.

4.3 Turbidity as a natural tracer

The spring water emerging at La Galiana is normally clear but after very heavy rainfall it can become turbid. We observed two cases where the turbidity of this spring water could be used informally as a natural tracer, and allowed some conclusions to be drawn about the vulnerability of the aquifer to pollution:

During the torrential rainfall of December 1995 and January 1996 the river Lobos reached a peak flow of 30 m^3/s, flooding the entire canyon, whilst five or six days after the flood in the river, there was a rapid and significant increase in turbidity in the spring water issuing from La Galiana. This reflects a groundwater "flood" in the hypogeic drainage within the aquifer.

While the water in the river cleared a few days after the peak of the flood, the turbidity in the spring water persisted for one or two weeks more, until it too gradually cleared.

This interesting feature is illustrated in Figure 9 as a flow-turbidity response curve, alongside similar curves for conductivity and time. Measurements were made *ex profeso*, using the samples taken for the artificial dye tracer tests. The curves in figure enable the following observations:

- Turbidity follows a normal distribution, which fits the peak of the La Galiana hydrogram quite well.

- This turbidity curve is delayed by 5-6 days compared to the peak of the flood in the river Lobos.
- Remobilization of clays in the endokarst and from allogenous streams indicate erosion and a considerable capacity of hypogeic transport. During these periods, both the turbulent flow regime and the circulation through karstic conduits must be very significant.
- The conductivity curve is the inverse of the flood flow and turbidity, as is to be expected.

The turbidity during spate underlines the vulnerability of the aquifer to pollution under flood conditions, showing how a large quantity of contaminants, allogenous or autogenous, could be transported through the aquifer over a short period. It also indicates a marked capacity for self-purification and for natural contamination by the turbidity itself; in fact, the fish farm had to close its water intake for several days when the turbidity was greatest in order to protect the trout.

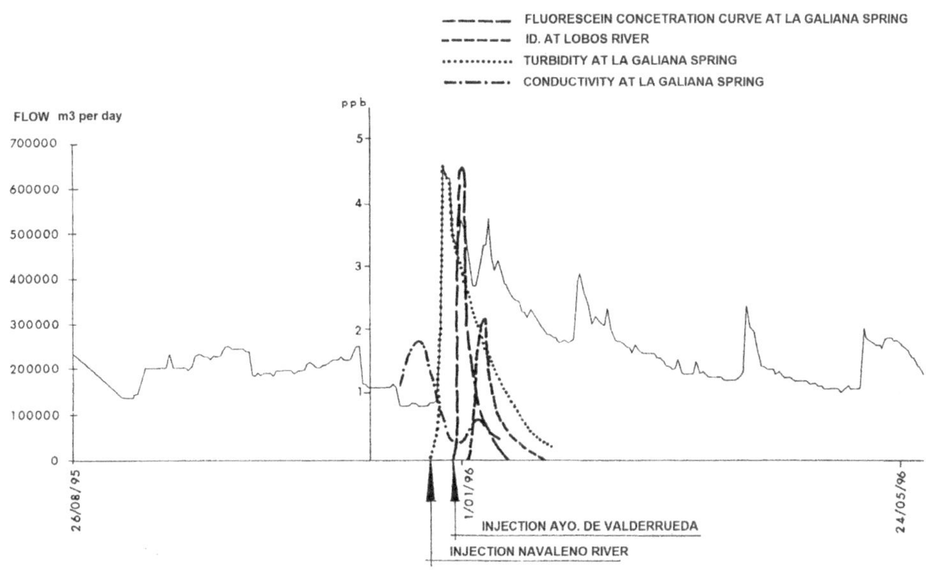

Fig. 9. Turbidity curves, conductivity and flow. La Galiana sping. (floods in december 95 to january 96)

Over the period 18 May to 7 June 2005, when practically no rain fell, the spring water at La Galiana was continually whitish in colour. This is very different from the red colour produced during natural floods. On this occasion, the cause of the turbidity was quite different, and due to the drilling and subsequent development of two wells to supply three villages (3580 permanent population and 8600 summer population), very close to each other. These were drilled very close to each other in April 2005 by rotary drilling, to depths of 275m and 280m at a distance of some 7 km from La Galiana spring. The boreholes penetrated thick deposits of white limestone and marly limestone and both the tailings and the water extracted during the drilling and development were drained directly into the watercourse (C.M.A. - J.C.L., 2006). From here they infiltrated the aquifer and were left hanging with respect to the position of the phreatic level. From 21 to 25 April, the boreholes were developed using pistons and compressed air, and it was this water, very turbid and milky, that was discharged into the watercourse. Without doubt, these four days of continuous development of the borehole are what caused the turbidity at La Galiana. What is interesting is that the plume took 27 days to appear at this spring, and lasted for about 20 days. All this took place under a low water regime; the "tracer" had to penetrate the unsaturated zone through sectors of matrix flow and where permeability was not necessarily very high at the beginning: this explains why the plume took so long to appear at La Galiana. The water velocity was calculated to have been 262 m/day. In contrast, the remaining tracer tests were carried out from injection points in river beds that lay very close to the saturated zone and in sectors that certainly contained developed conduits.

This is a good example of a tracer test from the surface of the meseta and it demonstrates the sensitivity of the aquifer to drilling and development of wells. It took the Natural Park Authorities by surprise, and, from 18 May to 7 June, the intake to the La Galiana fish farm had to be changed to take water from the river Lobos (which had not been visibly affected by the turbidity). Obviously, there was no option to call a halt to the borehole work, since this had been completed more than a month previously.

4.4 Mass of tracer recovered

Since some of the response curves from the tracer tests overlapped, it would have been difficult to separate the tracer recovery for each individual test. Instead, the cumulative recovery was calculated. This indicates that 3,300 g of the 4,500 g were accounted for, or 73.3%. It is clear that the remainder was retained in the ground. Some of this tracer is released during spells of intense rain, as occurred around the 8/02/96. This retention is not surprising since both the saturated and unsaturated zone of the aquifer, as well as the alluvium of the river where the injections were made, hold abundant clays (as confirmed by the turbidity increases described above).

The low concentrations recovered is also noteworthy, despite the volume of tracer used (sometimes 1.5 kg), and this indicates both that the aquifer stores considerable volumes of water and that significant dispersion must occur. Segovia (2008) estimated permanent reserves of 203 hm^3 and a refill rate of 2.55 years.

4.5 Hydrodynamic dispersion

It is recognized that the tracer mixture is displaced through a porous medium in three ways: carried along by the current to which it is added; molecular diffusion; and dispersion through the porous medium, whereby the tracer molecules take different paths through the conduits, dispersing it both longitudinally and transversely.

Taking the retention times of the tracer in the aquifer an approximate estimate was made of the parameters involved in longitudinal dispersion. Logically, these vary with distance. The graph in Figure 10 shows that there is a linear relationship between the distance of the injection point from the emergence point at La Galiana. This gives an approximate "longitudinal dispersion", which is measured from the leading edge of the plume to the end of its tail. The retention time of the dye in the aquifer and the average flow velocity in each tracer test is taken into account and the response curves deduced.

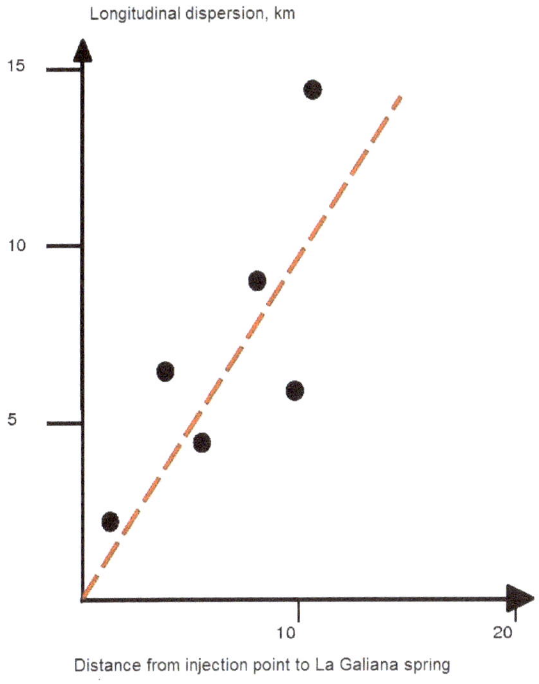

Fig. 10. Relationship between longitudinal dispersion and distance from La Galiana spring to each tracer injection point.

4.6 Observations on the vulnerability of the aquifer to pollution

In this section we outline some qualitative and semi-quantitative ideas. (We do not aim to calculate the transfer and evolution of pollution plumes to enable transit times to be calculated along with the predicted falls in pollutant concentration for point or diffuse pollution events.)

The first point to note is the very high velocity of the groundwater. Also, all of the preferential conduits (which is the network of conduits supposedly investigated in the tracer tests) would give residence times of less than one month, whatever the hydrological regime. Consequently, bacterial or viral pollutants could survive in the aquifer. This is an important point because there are several villages situated over the aquifer, and although these have wastewater treatment works, there could always be accidental or illegal spills of wastewater, or from farm slurry (pig and sheep purines).

The water flowing through the karst is extremely vulnerable. An accidental spill to the rivers that recharge the aquifer, or to any intermediate point within the calcareous outcrop, would contaminate the water that emerges from the aquifer at La Galiana and the lower reaches of the river Lobos, to some extent.

The dispersive effect of the aquifer is significant. It increases with distance from the spring under a low flow regime, with a NE-SW orientation. Closer to the discharge, and under high

water conditions, the curves are more unimodal in form. Where water levels are very high, the aquifer is vulnerable to acute turbidity "crises", and it is also very sensitive to turbidity arising from drilling of boreholes.
The aquifer has not demonstrated any self-purification capacity, except for the partial retention of tracer due, undoubtedly, to the absorption onto clay particles of alluvial deposits.

5. Conclusions

It is a relatively frequent occurrence, in hydrogeological practice, to search karst aquifers little known, little used, but with large resources. It is also quite common that these systems are connected, in some way, by allogenic surface rivers. For these cases we have proposed a methodology that can provide in a short period of time some key issues for future management, such as the hydrogeological conceptual model, knowledge about hydrodynamics and flow, water balance and preliminary assessment for risk of contamination.
The proposed method basically includes the following: hydrogeological interpretation of geological mapping, numerous tracer tests, control of allogenic river flows that recharge the aquifer and measures in the discharge capacity of the aquifer.
The methodology has been applied to the Lobos River Canyon karst system during one hydrological year and it was obtained the following more relevant results:
The tracer tests confirm the hydrogeological basin initially defined by geology. This is an area of Upper Cretaceous carbonate outcrops of 160 km^2 which forms an unconfined aquifer structured in a moderate syncline and is verified natural recharge of 42.2 hm^3. Allogenic rivers provide a recharge of 53.7 hm^3. The discharges are carried out mainly by the spring of La Galiana, and a little by Lobos river.
The groundwater velocity varies between 460 and 3.000 m/day depending on rainfall regime and increases with proximity to the La Galiana spring and as far as flow is subparallel to the axis of the syncline.
The tracer tests were used to find the connection Lobos river along 26 km through the aquifer in the canyon and in different hydrological situations. In case of low water table, the river Lobos and tributaries loose through simple sinks at headwater and only carries water on its final discharge path, where is also the source of La Galiana. In this case, sinks-underground discharges communication is showed. Hydraulic conductivity of sediment channel plays an important role in connection between Lobos river and aquifer.
Knowing spatial variation of groundwater velocity, is showed that hydraulic conductivity increases with proximity to aquifer outpoint, and also with direction. This one is greater in longitudinal direction to the flow, along the syncline, showing the heterogeneity of karst.
In high rainfall regime and in the header stretch of the canyon, the river becomes an effluent one, and its waters are traced. The intermediate stretch is a influent river and tracers do not show up, and at the final stretch, river is an effluent one. It emphasizes the high groundwater velocity. Also notorious is the fact that the whole preferential conduit net of the aquifer is located within the perimeter where residence time is less than a month in any hydrological regime. Therefore, any bacterial or viral contamination will survive in this aquifer. This is important because this aquifer system is located downstream of several villages. Although they have got wastewater treatment works, always could be accidental or illegal spills of wastewater or pig or beef cattle manure and slurries.

The dispersive effect in the aquifer is significant. It increases with distance from the discharge system in low rainfall regime and with NE-SW direction. Closer to the discharge, and under high water conditions, the response curves are more unimodal in form. It should be noted the existence of potential turbidity crises in water regime very high. The aquifer is also very sensitive to turbidity arising water from drilling of boreholes

6. Acknowledgements

We should like to thank the director of the Natural Park of the River Lobos Canyon, Don José Manuel Meneses Canalejo, for all the help given during field studies, from assistance obtaining necessary permissions, to help from the Natural Park rangers for water sampling during the tracer tests, taking readings at the gauging stations that they installed, etc. We should also like to thank the Confederación Hidrografica del Duero for giving permissions to undertake the tracer tests. Part of this investigation formed part of of Research Project AMB95-0154: "Hydrodynamic Function and Pollution Propagation in the karstic aquifer of the Natural Park of the River Lobos Canyon (Soria - Burgos)".

7. References

Birk S, Geyer T, R and Sauter M (2005) Process-based interpretation of tracer tests in carbonate aquifers. Ground Water 43(2): pp. 381-388.

C.M.A. - J.C.L. (Consejería de Medio Ambiente y Ordenación del Territorio - Junta de Castilla y León) (1992). Plan Rector de Uso y Gestión del Parque Natural "Cañón del río Lobos". Unpublished report. 457 pp.

C.M.A. - J.C.L. (Consejería de Medio Ambiente y Ordenación del Territorio - Junta de Castilla y León) (2006) Informe sobre los sondeos realizados en la obra: Navaleno, Casarejos, y San Leonardo de Yagüe. Abastecimiento Mancomunado. Clave: 21 - 50 - 202. Uxama Ingeniería y Arquitectura S. L.

COST 65 (1995) Hydrogeological aspects of groundwater protection in karstic areas, Final report (COST Action 65). European Commision, Directorate-General XII-Science, Research and Development, Report EUR 16547 EN, Brussels, 446.

Fournier, M., Massei, N., Bakalowicz, L. Rodet, S. and Dupont, J. P. (2007). Using turbidity dynamics and geochemical variability as a tool for understanding the behavior and vulnerability of a karst aquifer. Hydrogeology Journal, 15: 689 - 704.

Geyer, T., Birk, S., Licha, T.,R. and Sauter, M. (2007) Multitracer test approach to characterize reactive transport in karst aquifers. Ground Water. 45(1): pp. 36-45.

Goldscheider N, Klute M, Sturm S, Hötzl H (2000) The PI method: a GIS-based approach to mapping groundwater vulnerability whit special consideration of karst aquifers. Z. Angew Geol 463: pp. 157-166.

Goldscheider, N. (2005) Karst groundwater vulnerability mapping: application of a new method in the Swabian Alb, Germany. Hydrogeology Journal 13(4): pp. 555-564.

Goldscheider, N., Meiman, J., Pronk, M., and Smart, C. (2008) Tracer tests in karst hydrogeology and speleology. International Journal of Speleology 37(1): pp. 27-40.

González - Yélamos, J. (1999): Ensayos de trazadores en acuíferos kársticos: desarrollo histórico y anecdotario. In: Carrasco, F.; Duran, J.J. and Andreo, B. (Eds). Karst and Enviroment, 75 - 83.

Göppert, N., Goldscheider, N. (2008) Solute and Colloid Transport in karst conduits under low and high flow conditions. Ground Water, 46(1): pp. 61-68.

Hauns, M., Jeannin, P-Y, and Atteia, O. (1991) Dispersion, retardation and scale effect in tracer breakthrough curves in karst conduits. Hydrogeology Journal. 24(3-4): pp. 177-193.

Hernanz, A. and Navarro, J. M. (1973). La Cueva de La Galiana, Ucero (Soria). Revista Celtiberia n° 19, pp. 87 - 92.

IGME (Instituto Geológico y Minero de España)(1982). Mapa Geológico de España 1/50.000. Hoja 349. Cabrejas del Pinar.

IGME (Instituto Geológico y Minero de España) (1982). Mapa Geológico de España 1:50.000 Hoja 348. San Leonardo de Yagüe

Käss, W. (1998) Tracing techniques in geohydrology. Balkema, Rotterdam, The Netherlands.

Massei N, Wang HQ, Field MS, Dupont JP, Bakalowicz M, and Rodet J (2006) Interpreting tracer breakthrough tailing in a conduit-dominated karstic aquifer. Hydrogeology Journal. 14(6): pp. 849-858.

Meus, P. and Bakalowicz, M. (1997). Les traçages artificiels, outils de reconnaissance et f´étudedes aquifières karstiques. Hidrogéologie n°3, 43 - 50.

Morales T, Valderrama IF, Uriarte JA, Antigüedad I, and Olazar M (2007) Predicting travel times and transport characterization in karst conduits by analysing tracer-breakthrough curves. Hydrogeology Journal. 15(1): pp. 183-198.

Panno SV (2006) Karst aquifers: can they be protected? Ground Water 44(4): p. 494.

Perrin J and Luetscher M (2008) Inference of the structure of karst conduits using quantitative tracer tests and geological information: example of the Swiss Jura. Hydrogeology Journal. 16(5): pp. 951-967.

Segovia, R (2008). El drenaje subterráneo en el acuífero kárstico del Caños del río Lobos (Soria-Burgos). Tesis Doctoral(Inédita).Universidad Politécnica de Madrid.

Sanz, E. (1992). Las aguas subterráneas en el Parque Natural del Cañón del río Lobos (Soria - Burgos). Boletín Geológico y Minero. Vol. 103 - 102, 309 - 329

Sanz, E. (1996). Le karst du canyon du Lobos et son fonctionnement hydrogéologique. Karstologia, n° 28, 45 - 56.

Sanz, E. (2000). Infiltration measured by the dripping of stalactites. Groundwater, volume 38, n°2, 247 - 253.

Smart CC (1988) Artificial Tracer techniques for the determination of structure of conduit aquifers. Ground Water 26(4): pp. 445-453.

Zwahlen F (ed) (2004) Vulnerability and risk mapping for the protection of carbonate (karst) aquifers, final report (COST action 620). European Commission, Directorate-General XII Science, Research and Development, Brussels.

14

Estimating Hydraulic Conductivity of Highly Disturbed Clastic Rocks in Taiwan

Cheng-Yu Ku[1] and Shih-Meng Hsu[2]
[1]*National Taiwan Ocean University*
[2]*Sinotech Engineering Consultants, Inc*
Taiwan

1. Introduction

Understanding groundwater flow in fractured consolidated media has long been important when undertaking engineering tasks such as dam construction, mine development, the abstraction of petroleum, slope stabilization, and the construction of foundations. To study groundwater flow in support of these tasks, the focus of most hydrogeological investigations has been on the characterization of the hydraulic properties of the higher-permeability fractures in the rock mass.

Taiwan is situated on the edge of the Eurasian and Philippine Sea plate. Plate tectonics have created numerous fault lines that crisscross the island. As a result of high density of faults, rock core data with fractures, soft and cohesive gouges, and various lithologies are extensive in boreholes. In general, the permeability of clay-rich gouges has extremely low values. On the contrary, the fractures often have higher permeability. The hydraulic properties of fractured rocks in Taiwan, therefore, vary with highly disturbed geological structures and lithology. To obtain hydraulic properties of fractured rocks in Taiwan, the investigation of vertical variation of the fractures in a borehole is of importance. This study utilized a high-resolution BoreHole acoustic TeleViewer (BHTV, Williams and Johnson, 2004) to scan images of the borehole. The information gathered from BHTV was used to characterize lithology and fractures for the borehole and was essential to conduct a proper measurement of rock mass hydraulic conductivity. The double packer systems were then used to determine the hydraulic conductivity in a portion of borehole using two inflatable packers. Although this type of test can directly measure the hydraulic parameter, costs of the testing are fairly high. Several studies (Black, 1987; Carlson and Olsson,1977; Louis, 1974; Burgess, 1977; Wei et al., 1995, Zhao, 1998) have proposed the estimation of rock mass hydraulic conductivity using different empirical equations. These empirical equations provide a great feature for characterizing rock mass hydraulic properties quickly and easily. However, the applicability of these equations in highly disturbed clastic sedimentary rocks in Taiwan is very limited.

This study proposed the establishment of an empirical HC model for estimating rock mass hydraulic conductivity of highly disturbed clastic sedimentary rocks in Taiwan using the BHTV and the double packer hydraulic tests. Four geological parameters including rock quality designation (RQD), depth index (DI), gouge content designation (GCD), and lithology permeability index (LPI) were adopted for establishing the empirical HC model. To verify rationality of the proposed HC model, 22 in-situ hydraulic tests were carried out to measure the hydraulic conductivity of the highly disturbed clastic sedimentary rocks in three boreholes at two different locations in Taiwan. Besides, the model verification using another borehole data with four additional in-situ hydraulic tests from similar clastic sedimentary rocks was also conducted to further verify the feasibility of the proposed empirical HC model. This paper presents the measured hydraulic conductivity results and the relationship among the hydraulic conductivity, RQD, DI, GCD, and LPI. The application of the proposed HC model was also addressed.

2. Description of study areas and boreholes

2.1 Description of study areas

Taiwan's strata are distributed in long and narrow strips, almost parallel to the island's axis. Metamorphic rock lies under the Central and Snow Mountain Ranges. Sedimentary rock forms part of the island-wide piedmonts and coastal plains as well as the Coastal Mountain Range. The island of Taiwan has three geological zones divided by longitudinal faults: the Central Range, Western Piedmont and Eastern Coastal Mountain Range zones (Fig. 1(a)).

About 26 hydraulic conductivity measurements were conducted in four boreholes in Western Piedmont, primarily at three sites: Da-Keng, Shang-Ming, and Caoling (Fig. 1(a)) in which borehole HB-94-01 is in the Da-Keng site, boreholes HB-95-01 and HB-95-02 are in the Shang-Ming site, and borehole CH-04 is in the Caoling site. Besides, the Da-Keng and Caoling sites are in central Taiwan and the Shang-Ming site is in southern Taiwan. The dominant rock strata of the Shang-Ming site include Miocene sedimentary rock with layers of sandstone or shale or their alternation. The major structures consist of a series of parallel easterly inclined thrust faults and folds, which often form local fractured zones, including geological structures such as the Pingshi fault, the Biauhu fault, and the Chin-Shan fault. Figure 1(b) presents the distribution of these geological strata and structures. In addition, borehole HB-94-01 in the Da-Keng site and borehole CH-04 in the Caoling site also have similar rock strata but without geological structures.

Based on the loggings and geological analysis, HB-95-01 and HB-95-02 are strongly influenced by the faults; nevertheless, HB-94-01 and CH-04 are not.

2.2 Boreholes

The depth of the borehole HB-94-01 is 110 m. The principal lithologic units for the borehole are sandstone and siltstone. The interval of 36 m to 44 m is a fractured zone compared to other depths in the borehole. A total of 8 hydraulic tests using a double packer system were carried out to determine hydraulic conductivity (Sinotech, 2006). The strategy of the test design from the drilling work was to determine hydraulic properties from different geological structures such as no fracture, a single fracture, or multiple fractures at different depths.

The drilling depths of HB-95-01 and HB-95-02 are 250 m and 350 m, respectively. The principal lithologic units for HB-95-01 are sandstone, argillaceous sandstone, and sandy mudstone. The principal lithologic units for HB-95-02 are sandstone, argillaceous sandstone, and sandstone mixed with some mudstone. HB-95-01 and HB-95-02 are close to the Biauhu fault and Pingshi fault, respectively (Fig. 1(b)). Rock core photos (Fig. 2(a)) indicated soft and cohesive gouges are extensive in both boreholes in which the hydraulic properties of fault-related rocks can be studied. The study completed 3 and 14 hydraulic tests in HB-95-01 and HB-95-02, respectively (Sinotech, 2006). The strategy of the test design was to determine hydraulic conductivity in more permeable zones and clay-rich gouge zones. Besides, the borehole CH-04 is not influenced by the faults (Fig. 2(b)) and used for the model verification and it is described in Section 5.7.

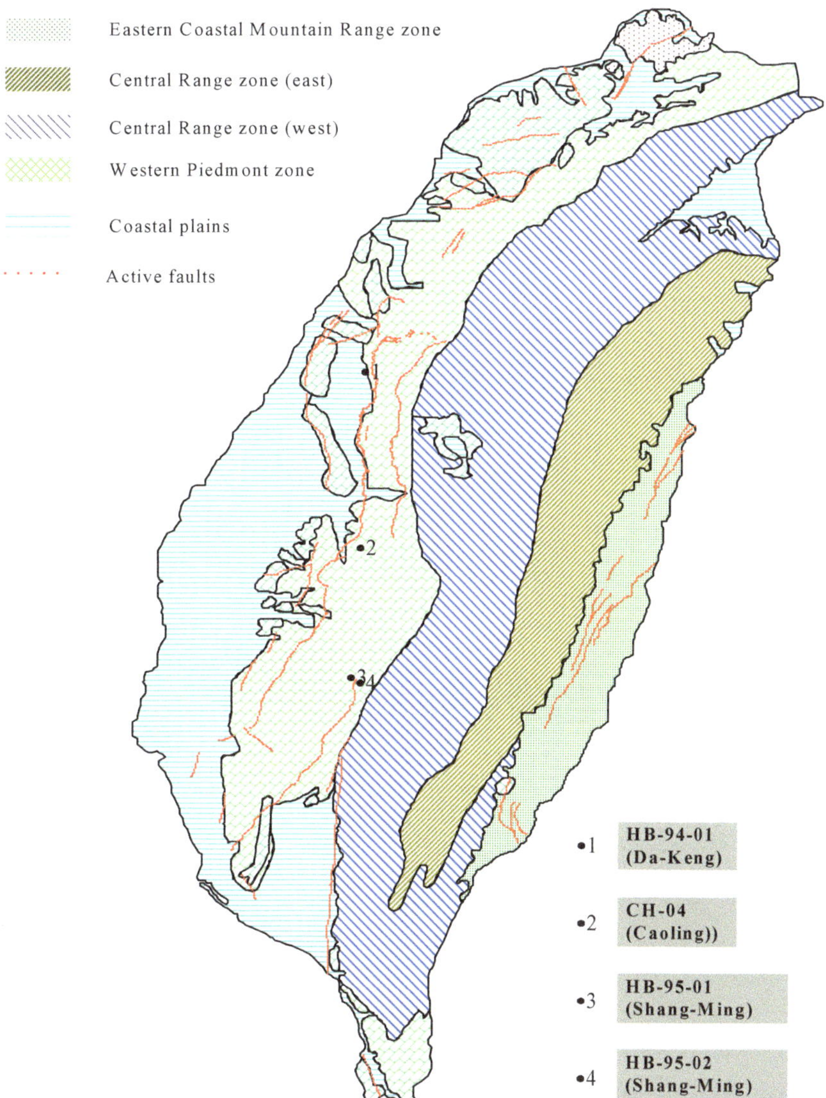

Fig. 1. (a) Location of major faults and four boreholes for this study in Taiwan.

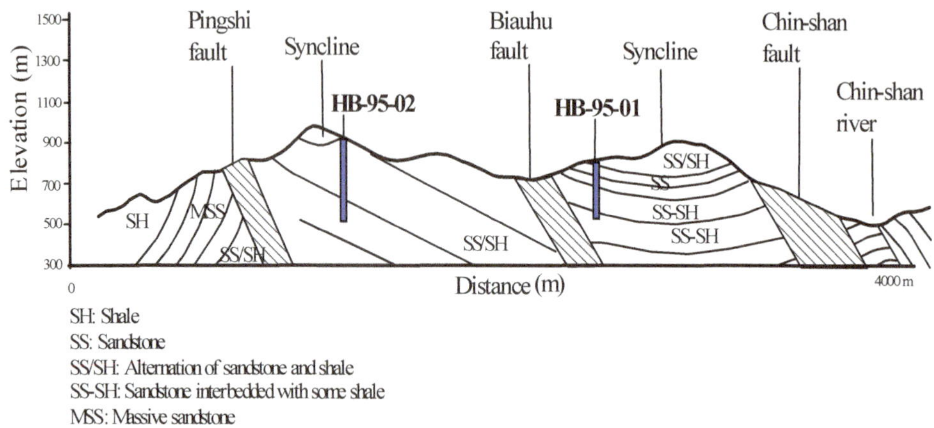

Fig. 1. (b) Detailed distribution of geological strata and structures of boreholes HB-95-01 and HB-95-02.

Fig. 2. (a) Rock core photos of borehole HB-95-2 with fault influence.

Fig. 2. (b) Rock core photos of borehole CH-04 without fault influence.

3. Hydraulic conductivity of fractured rock masses

It is widely recognized that fracturing plays a decisive role in rock hydraulics, especially in low permeability rocks such as crystalline, volcanic and carbonate rocks, and in some classic sedimentary formations, such as sandstones, shales, glacial tills and clays. In highly disturbed fractured rocks, hydraulic properties depend on density, size, infillings and interconnection of fractures. A distinction can be made between hydraulic conductivity of fracture and of intergranular (matrix) material. Previous study (Lee and Farmer, 1993) has proposed the hydraulic conductivity of a rock mass with three orthogonal joint sets with similar spacing ad constant aperture in all directions. The effect of stress on permeability is also of importance for estimating rock mass hydraulic conductivity (Snow, 1969).

Several studies, shown in Table 1, have also pointed out that rock mass permeability may decrease systematically with depth (Black, 1987; Carlson and Olsson,1977; Louis, 1974; Burgess, 1977; Wei et al., 1995, Zhao, 1998). The decrease in permeability with depth in fractured rocks is usually attributed to reduction in fracture aperture and fracture spacing. The reduction is due to the effect of geostatic stresses, and thereby the permeability of fractured rocks will be reduced. Accordingly, the depth may be considered as a factor in evaluating rock mass permeability.

4. Measurement of rock mass hydraulic conductivity

For decades, the determination of hydraulic properties in fracture rocks has been qualitatively estimated using the Lugeon test. It is now recognized that this approach is not suitable in highly disturbed fractured rocks. The type of test only gives an average value of hydraulic conductivity in a stratum and is not able to identify (1) aquifer's type in a required testing section; (2) storativity of an aquifer; and (3) relations between hydraulic properties and geological structures such as water-bearing fractured zones. Results from the test may be insufficient to characterize hydraulic properties for complex geological environments. They may be subject to hazards such as extensive water inflow during underground excavation.

To provide a better characterization of hydraulic properties of fractured rocks, a double packer technique can be adopted and is often utilized to overcome the shortcomings of the

Lugeon test. Packers can be used to isolate a portion of borehole for hydraulic testing. Hydraulic properties for a single of fracture, a group of fractures, or an entire rock formation can be easily identified by the technique.

Equation	Reference
$k = az^{-b}$	Black (1987) a and b are constants, z is the vertical depth below the groundwater surface.
$\log K = -8.9 - 1.671 \log Z$	Snow (1969) K (ft^2) is the permeability. z (ft) is the depth.
$K = 10^{-(1.6 \log z + 4)}$	Carlson and Olsson (1977) K (m/s) is the hydraulic conductivity. z (m) is the depth.
$K = K_s e^{(-Ah)}$	Louis (1974) K (m/s) is the hydraulic conductivity. K_s is the hydraulic conductivity near ground surface. h (m) is the depth. A is the hydraulic gradient.
$\log K = 5.57 + 0.352 \log Z$ $-0.978(\log Z)^2 + 0.167(\log Z)^3$	Burgess (1977) K (m/s) is the hydraulic conductivity. Z (m) is the depth.
$K = K_i[1 - Z / (58.0 + 1.02Z)]^3$	Wei et al. (1995) Z is the depth. K is the hydraulic conductivity. K_s (m/s) is the hydraulic conductivity near ground surface.

Table 1. Diverse approximations for estimating rock mass hydraulic conductivity.

4.1 Borehole acoustic televiewer(BHTV) investigation and hydraulic test design

Prior to hydraulic testing, the study utilized a high-resolution borehole acoustic televiewer (BHTV) to scan images of boreholes. The information (Fig. 3(a) and (b)) gathered from BHTV was used to characterize lithology and fractures for the borehole and was essential to the proper design of the hydrogeological program.

Test design is dependent on the characteristics of the zone tested and the desired information. Accordingly, the main testing strategy in this study was to detect water-bearing fractures. In addition, the study investigated the vertical variation of the hydraulic conductivity in a borehole and hydraulic property of fault-related rocks.

A water-bearing zone of subsurface commonly appears in the section with multiple fractures. According to BHTV logs from boreholes, the study selected locations with images that show multiple fractures as hydraulic test sections. Figure 3 shows that two test zones were selected by this strategy. Other testing zones for other study purposes can be selected by BHTV images.

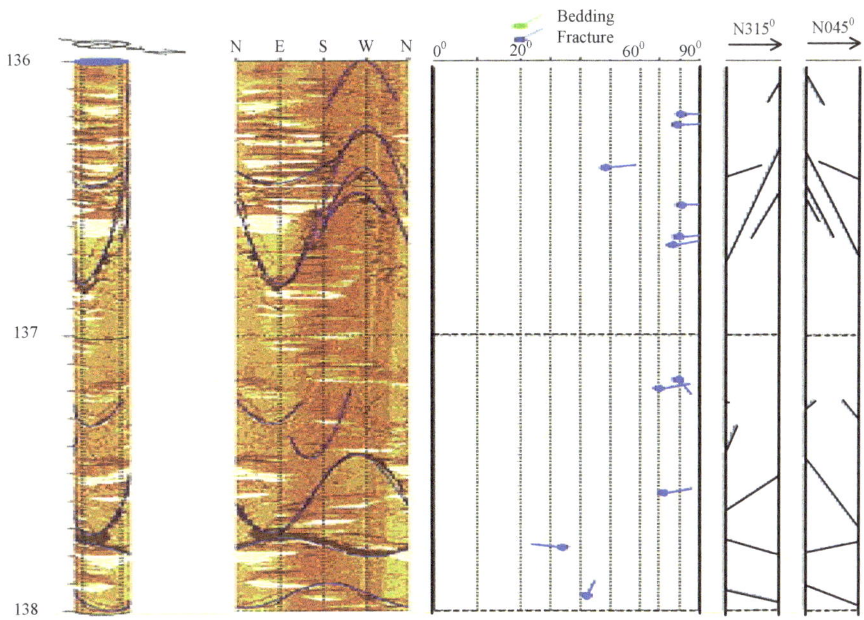

Fig. 3. (a) The pack-off zones and their corresponding BHTV images (depth 136m~138m, HB-95-01).

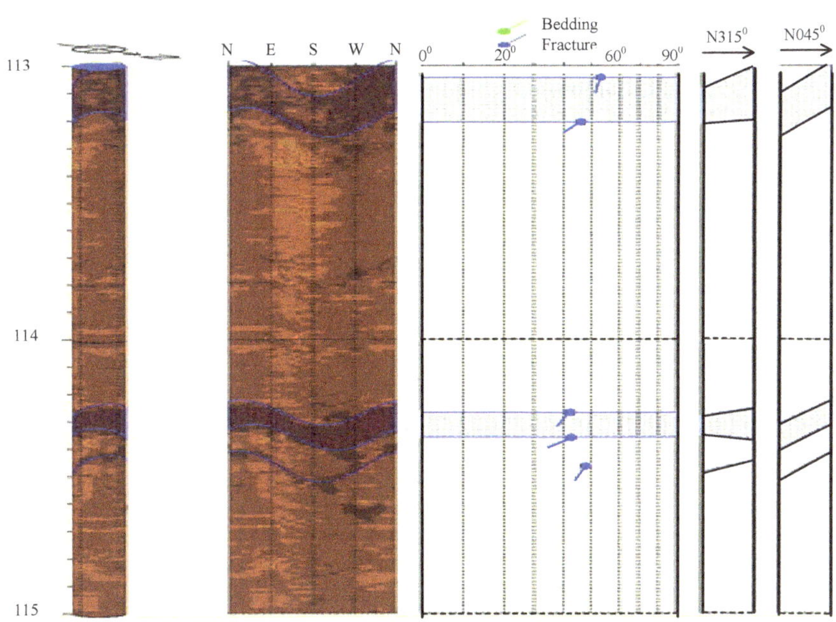

Fig. 3. (b) Identification of shear band from BHTV (depth 113m~115m, HB-95-02).

4.2 Double packer system

Double packer systems are the most commonly used tools for hydrogeological testing in boreholes. They can be used to determine the hydraulic property in a section of borehole based on two inflatable packers. It is now recognized that this approach is appropriate to investigate the variability of a borehole as it intersects various hydrogeological units.

The double packer system in this study (Fig. 4) consists of two inflatable rubber packers, a shut-in valve, a submersible pump, and pressure transducers for monitoring above and below the packers and the isolated interval. The shut-in valve can open or close the hydraulic connection between the pipe string and the test section. The rubber packers can be inflated using nitrogen delivered through a polyethylene air line. The pumping or injecting rate can be monitored at the land surface with a flow meter. To conduct each hydraulic test, the packers are inflated to isolate a section of borehole, and the rate of flow and/or pressure in the test interval over a period of time can be measured.

The BHTV images for different test intervals in Boreholes HB-95-01 and HB-95-02 are shown in Fig. 5(a) and 5(b). It is obvious that the fractures can be identified clearly using our high-resolution BHTV. The intervals in the depth from 118.5 m to 121.7 m and from 134.8 m to 138.0 m were sealed by double packers for conducting a pressure pulse test and a constant head injection test, respectively. Figures 5(a) and 5(b) show the results of hydraulic tests which were conducted by different type of hydraulic tests by means of AQTESOLVE. The type of hydraulic test chosen in this study for each test interval was decided by a hydraulic diagnosis test which mainly detects permeability of the test interval prior to a normal test. For the test interval of 118.5 m to 121.7 m, although three fractures and a fracture zone of approximately 7.25 cm thickness were seen on the borehole image, lack of interconnectivity of fractures and soft and cohesive gouges existing at the fractures may reduce the permeability of rock masses. In addition, four types of tests, including pumping tests, injection tests, slug tests, and pressure pulse tests can be applied to the double packer system.

Pumping tests involve pumping at a constant or variable rate and measuring changes in water levels during pumping. In injection tests, fluid is injected into a test interval while keeping the head of the test interval at a constant value. A slug test involves the abrupt removal, addition, or displacement of a known volume of water and the subsequent monitoring of changes in water level as equilibrium conditions return. In a pressure pulse test, an increment of pressure is applied to a packed zone. The pressure decay is monitored. Typically, the decision on which type of test to perform is based on the expected permeability of the test interval, the volume of rock to be sampled, and the availability of time and equipment (NRC, 1996). Hydraulic properties determined by slug tests or pressure pulse tests are representative only for the material in the immediate vicinity of the borehole.

To obtain hydraulic conductivity over a large area, the procedure of a single-hole hydraulic test is to perform a pumping test at a test interval first. If the pumping test cannot be performed due to low permeability of the test section, a constant head injection test will be conducted instead. Once the flow rate cannot be measured by limitation of the flow meter (less than 0.11 l/min) during the injection test, a slug test or pressure pulse test can be performed. The duration of a pressure pulse test is much shorter than that of a slug test. For this reason, the pressure pulse test is commonly applied to test intervals of very low permeability. However, the volume of rock tested by a pressure pulse test is significantly smaller when compared to a slug test.

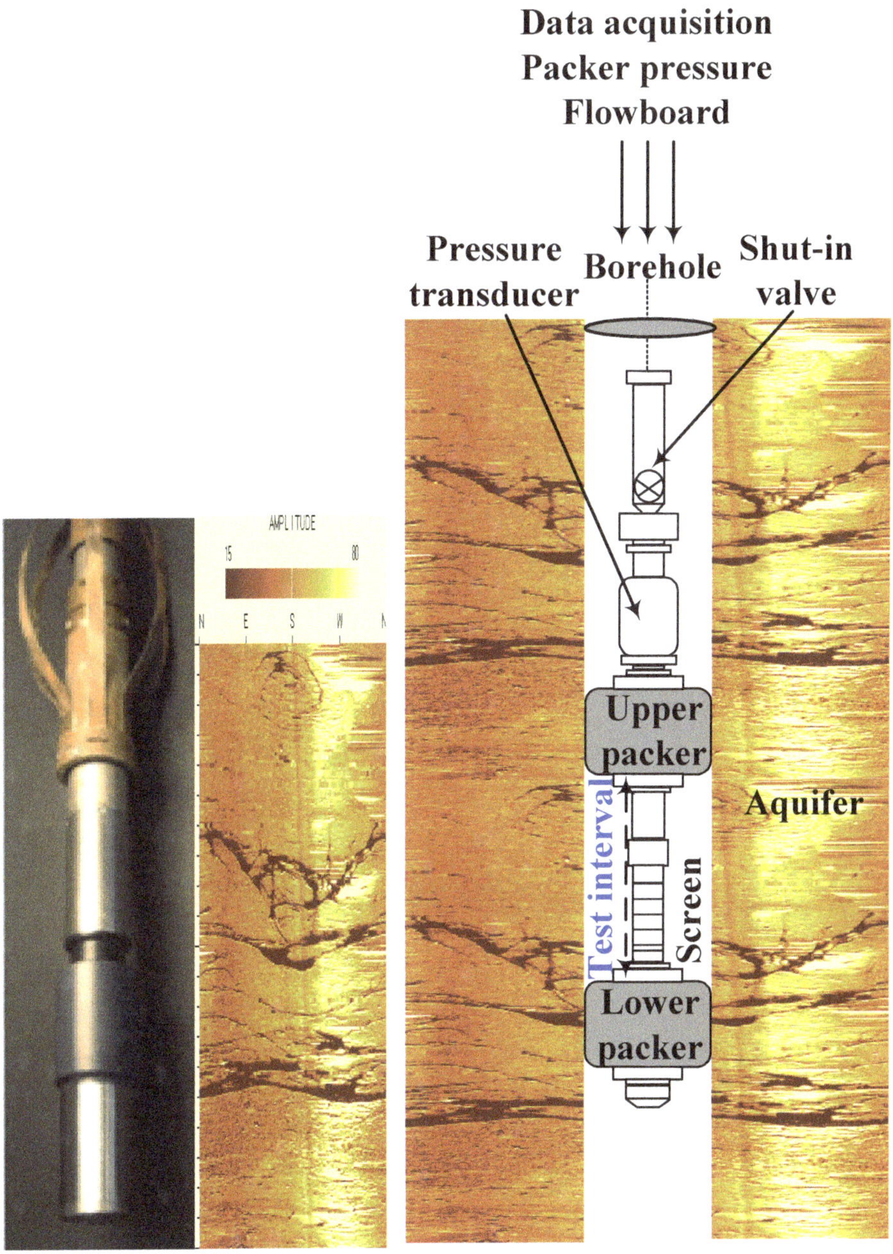

Fig. 4. Schematic drawing of BHTV, acoustic image of borehole, and the double packer system.

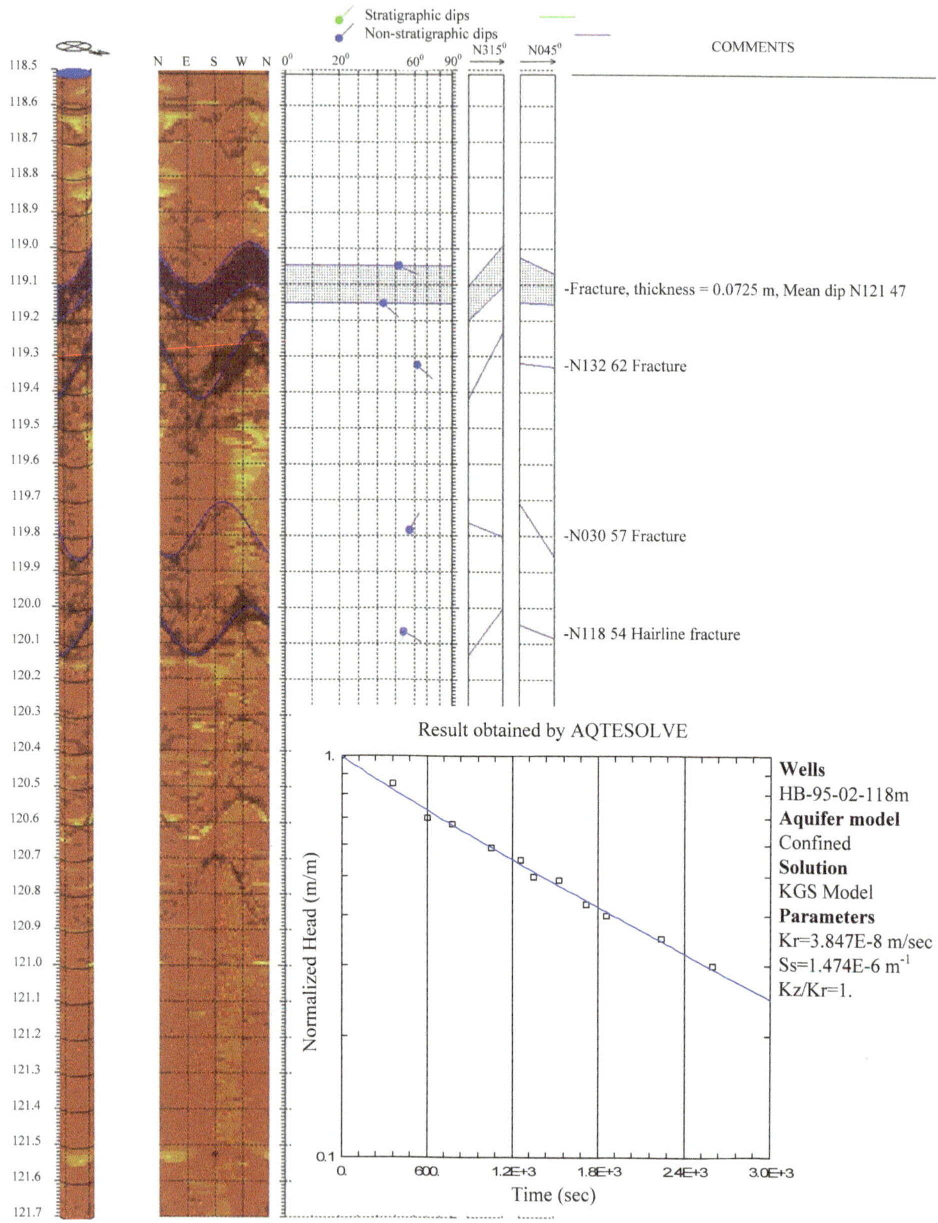

Fig. 5. (a) Evaluation of hydraulic parameters using AQTESOLVE (right lower figure) and BHTV images at pack-off zones 118.5 m to 121.7 m in Borehole HB-95-02.

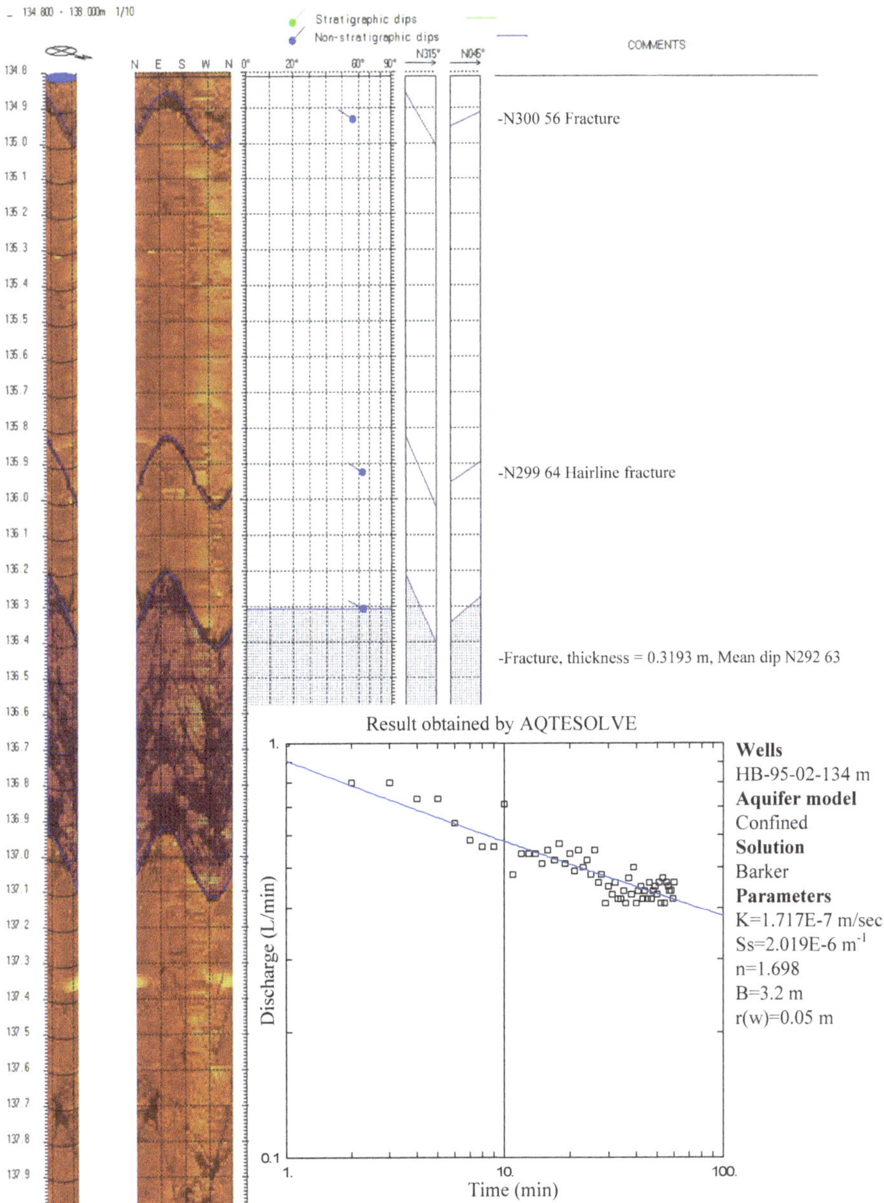

Fig. 5. (b) Evaluation of hydraulic parameters using AQTESOLVE (right lower figure) and BHTV images at pack-off zones 134.8 m to 138 m in Borehole HB-95-02.

A total of 26 hydraulic tests were designed to determine hydraulic conductivity in the boreholes. Data collected during hydraulic tests can be analyzed by analytical methods. Water pressure and discharge rate measurements with time for each hydraulic test were collected in this study. The data analysis was performed using a professional version of the AQTESOLVE test analysis software, which enables both virtual and automatic type curve matching (Duffield, 2004). The quantitative evaluation of hydraulic parameters was carried out as an iterative process of the best-fit theoretical response curves based on the measured data of the hydraulic test.

5. Empirical model of rock mass hydraulic conductivity

Prior to describing the empirical model of rock mass hydraulic conductivity, an attempt to find the decrease in permeability with depth was conducted. Figure 6 demonstrates that the testing data of HB-94-01 shows the tendency that the hydraulic conductivity decreases with depth. The form of the regression equation is close to the result obtained by Black, 1987. The coefficient of determination of the regression equation is 0.633. However, the testing data from HB-95-01 and HB-95-02 are very scattered. No relationship can be found between hydraulic conductivity and depth. Accordingly, potential factors, including rock quality designation (RQD), depth index (DI), gouge content designation (GCD), and lithology permeability index (LPI), that may affect the degree of permeability should be considered. The rating approach for each factor that represents the magnitude of permeability is also described as following.

5.1 Rock quality designation

To assess the influence of the fracture characteristic on permeability, the rock quality designation (RQD) index (Deere et al., 1967), can be adopted. The RQD index was introduced over 40 years ago as an indicator of rock mass conditions. The RQD value is defined as the cumulative length of core pieces longer than 100 mm in a run (RS) divided by the total length of the core run (RT) and can be obtained from the following equation.

$$\begin{aligned}\mathrm{RQD} &= \frac{\sum \text{Length of Intact and Sound Core Pieces} > 100\text{ mm}}{\text{Total Length of Core Run, mm}} \times 100\% \\ &= \frac{R_S}{R_T} \times 100\%\end{aligned} \tag{1}$$

In this study, a core run for calculating a RQD value is herein defined as a selected zone of a hydraulic test. Eq. 1. may be utilized to identify rock mass permeability.

5.2 Depth index

The decrease in permeability with depth in fractured rocks is usually attributed to reduction in fracture aperture and fracture spacing. The reduction is due to the effect of geostatic stresses, and thereby the permeability of fractured rocks will be reduced. The depth may be considered as a factor in evaluating rock mass permeability.

To assess the influence of the depth on permeability, a Depth Index, namely DI, was defined as the following equation.

$$\mathrm{DI} = 1 - \frac{L_c}{L_T} \tag{2}$$

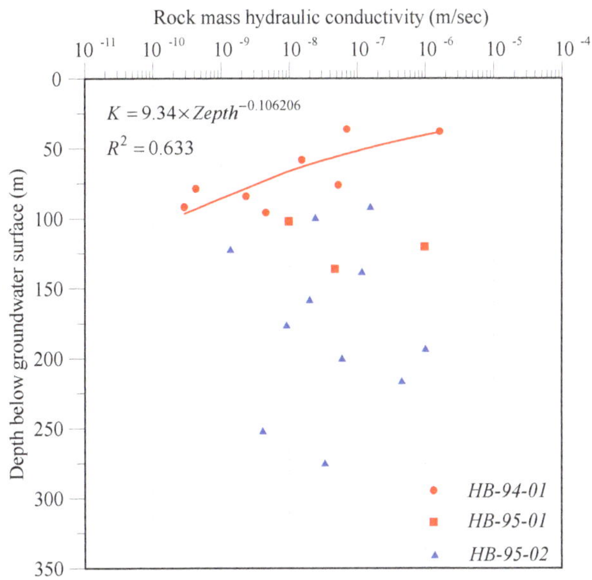

Fig. 6. Relationship between hydraulic conductivity and depth.

in which LT is the total length of a borehole; LC is a depth which is located at the middle of a double packer test interval in the borehole. The value of DI is always greater than zero and less than one. The greater the DI value, the higher the permeability.

5.3 Gouge content designation

The RQD value may decrease by an increase of fractures in a core run. If the fractures contain infillings such as gouges, permeability of the fractures will reduce. To assess the influence of the gouge materials on permeability, a Gouge Content Designation (GCD) index was defined as the following equation.

$$\mathrm{GCD} = \frac{R_G}{R_T - R_S}, \tag{3}$$

in which RG is the total length of gouge content. The value of GCD is always greater than zero and less than one. The greater GCD value stands for the more gouge content in a core run, and thereby it will reduce the permeability.

5.4 Lithology permeability index

Lithology is the individual character of a rock in terms of mineral composition, grain size, texture, color, and so forth. For an intact rock, the magnitude of permeability depends largely on the individual character of the rock. It may be affected by the average size of the pores, which in turn is related to the distribution of particle sizes and particle shape. In sedimentary formations grain-size characteristics are most important because coarse-grained and well-sorted material will have high permeability as compared to fine-grained sediments like silt and clay. Thus, the lithology may be regarded as a factor in evaluating rock mass permeability.

To assess the influence of lithology on permeability, a Lithology Permeability Index (LPI) was defined as Table 2.

Lithology	Hydraulic conductivity (m/s)				Range of rating	Suggested Rating
	Reference[1]	Reference[2]	Reference[3]	$K_{average}$		
Sandstone	10^{-6}~10^{-9}	10^{-7}~10^{-9}	10^{-7}~10^{-9}	$10^{-7.5}$	0.8-1.0	1.00
Silty Sandstone	—	—	—	—	0.9-1.0	0.95
Argillaceous Sandstone	—	—	—	—	0.8-0.9	0.85
S.S. interbedded with some Sh.	—	—	—	—	0.7-0.8	0.75
Alternations of S.S & Sh.	—	—	—	—	0.6-0.7	0.65
Sh. interbedded with some S.S.	—	—	—	—	0.5-0.7	0.60
Alternations of S.S & Mudstone	—	—	—	—	0.5-0.6	0.55
Dolomite	10^{-6}~$10^{-10.5}$	10^{-7}~$10^{-10.5}$	10^{-9}~10^{-10}	10^{-8}	0.6-0.8	0.70
Limestone	10^{-6}~$10^{-10.5}$	10^{-7}~10^{-9}	10^{-9}~10^{-10}	10^{-8}	0.6-0.8	0.70
Shale	10^{-10}~10^{-12}	10^{-10}~10^{-13}	—	$10^{-10.5}$	0.4-0.6	0.50
Sandy Shale	—	—	—	—	0.5-0.6	0.60
Siltstone	10^{-10}~10^{-12}	—	—	10^{-11}	0.2-0.4	0.30
Sandy Siltstone	—	—	—	—	0.3-0.4	0.40
Argillaceous Siltstone	—	—	—	—	0.2-0.3	0.20
Claystone	—	10^{-9}~10^{-13}	—	10^{-11}	0.2-0.4	0.30
Mudstone	—	—	—	—	0.2-0.4	0.20
Sandy Mudstone	—	—	—	—	0.3-0.4	0.40
Silty Mudstone	—	—	—	—	0.2-0.3	0.30
Granite	—	—	10^{-11}~10^{-12}	$10^{-11.5}$	0.1-0.2	0.15
Basalt	10^{-6}~$10^{-10.5}$	10^{-10}~10^{-13}	—	$10^{-11.5}$	0.1-0.2	0.15

[1]B.B.S. Singhal & R.P. Gupta (1999) ; [2]Karlheinz Spitz & Joanna Moreno (1996) ; [3]Bear(1972)

Table 2. Description and ratings for lithology permeability index.

5.5 Rock mass permeability index

As stated above, the rock mass permeability may be dependent on the following four parameters: RQD, DI, GCD, and LPI. However, the permeability is not simply affected by only one factor. It may account for the synthetic effect from the four parameters on permeability. Accordingly, Rock mass permeability index, called the HC index, was proposed herein.

$$HC = (1\text{-}RQD)(DI)(1\text{-}GCD)(LPI)\,., \tag{4}$$

The value of each parenthesis at the right hand side of Eq. 4. is always greater than zero and less than one depending on the values assigned to the four parameters. The greater the value of each parenthesis, the higher the permeability. Thus, the model performs a numerical assessment of rock mass permeability using the four parameters. Since it is rare to encounter the condition that RQD is 100% in highly disturbed clastic sedimentary rocks in Taiwan, the term of (1-RQD) is usually greater than zero. However, it should be noted that if (1-RQD) is zero, the value of 0.01 in the term of (1-RQD) is suggested to avoid the HC value to be zero. Currently, the study took the same weight for each factor in Eq. (4). In addition, Eq. (4) is limited in sedimentary rocks only and is only applied to vertical boreholes at present. With more testing data, a further study can be considered to assign a different weight for each factor to give a better correlation between the hydraulic conductivity and HC.

5.6 The empirical HC model

Regression analysis was performed to estimate the dependence of HC on hydraulic conductivity. A total of 22 hydraulic test data were applied to the study. HC-values for the hydraulic tests can be computed from borehole image data and rock core data, in which the values of RQD and GCD at each test interval can be calculated from borehole image data and rock core data with Eqs. 1. and 3., respectively. The value of DI can be calculated using Eq. 2. The value of LPI for each test zone can be obtained from rock core data and Table 3. Table 3 shows the calculated results for the HC model based on the verified data. The regression results indicated that a power law relationship exists between the hydraulic conductivity and HC with a coefficient of determination of 0.866 as shown in Fig. 7. The empirical HC model is obtained as shown in Eq. 5.

$$K = 2.93\times 10^{-6}\times (HC)^{1.380}, R^2 = 0.866 \quad (5)$$

If only HB-94-01 testing data were adopted, a better correlation with the coefficient of determination of 0.905 can be obtained as shown in Eq. 6.

$$K = 2.31\times 10^{-6}\times (HC)^{1.342}, R^2 = 0.905 \quad (6)$$

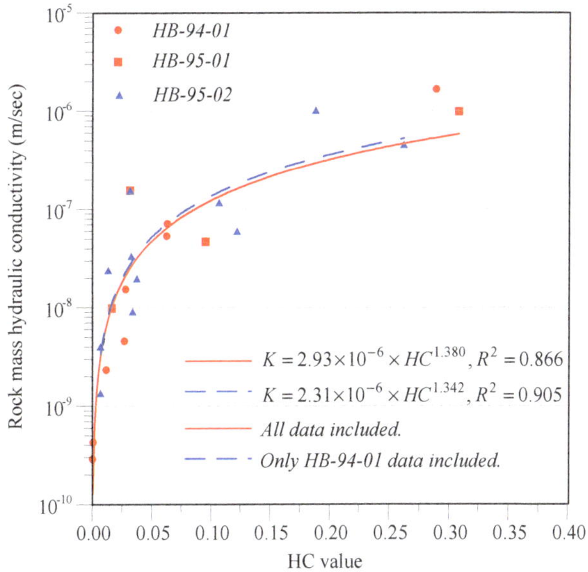

Fig. 7. Relationship between hydraulic conductivity and HC-values.

Boreholes	Test intervals (m)	1-RQD	DI	1-GCD	LPI	HC	K (m/s)
HB-94-01	34.7-36.3	0.094	0.677	1.000	1.000	0.0635	7.06E-08
	36.4-38.0	0.438	0.662	1.000	1.000	0.2895	1.64E-06
	56.7-58.3	0.063	0.477	1.000	0.950	0.0283	1.53E-08
	74.6-76.2	0.500	0.315	1.000	0.400	0.0629	5.3E-08
	77.2-78.8	0.010	0.291	1.000	0.400	0.0012	4.22E-10
	82.6-84.2	0.125	0.242	1.000	0.400	0.0121	2.31E-09
	90.2-91.8	0.010	0.173	1.000	0.400	0.0007	2.86E-10
	94.2-95.8	0.500	0.136	1.000	0.400	0.0273	4.53E-09
HB-95-01	99.0-101.9	0.345	0.598	0.200	0.400	0.0165	9.8E-09
	117.2-120.1	0.690	0.526	1.000	0.850	0.3081	9.76E-07
	133.2-136.1	0.724	0.461	0.286	1.000	0.0954	4.68E-08
HB-95-02	88.6-91.4	0.071	0.743	1.000	0.600	0.0318	1.56E-07
	96.0-99.2	0.031	0.721	1.000	0.600	0.0135	2.42E-08
	118.5-121.7	0.219	0.657	0.071	0.700	0.0072	1.36E-09
	134.8-138.0	0.344	0.610	0.727	0.700	0.1068	1.17E-07
	154.8-158.0	0.938	0.553	0.103	0.700	0.0376	1.99E-08
	173.0-176.2	0.938	0.501	0.103	0.700	0.0340	9.08E-09
	189.8-193.0	0.594	0.453	1.000	0.700	0.1883	1.01E-06
	196.6-199.8	0.563	0.434	0.500	1.000	0.1220	6.00E-08
	213.2-216.0	0.679	0.387	1.000	1.000	0.2625	4.54E-07
	249.0-251.8	0.393	0.285	0.091	0.700	0.0071	4.03E-09
	272.0-274.8	0.214	0.219	1.000	0.700	0.0328	3.36E-08

Table 3. The calculated results for HC-system based on 22 hydraulic test data.

It should be noted that the values of (1-GCD) in HB-94-01 borehole are all equal to 1. The results of Eq. 6 demonstrate that the empirical HC model may also be more accurate for the estimation of the rock mass hydraulic conductivity if the fractures do not contain infillings. There are a few limitations that need to be noted for the use of Eq. 5. The data used to develop the equation are limited in number and in the lithologies represented. From the definition of DI, DI cannot be determined for inclined boreholes because the data collected were from vertical boreholes.

5.7 Model verification

In order to further verify the feasibility of the proposed empirical HC model, the model verification is conducted. Another borehole data with the drilling depth of 120 m is adopted to verify the empirical HC model. The principal lithologic units of the borehole, namely CH-04, are mainly sandstone, shale, and sandstone with some thin shale. The depth from 24.5 m to 26.6 m, 32.5 m to 34.1 m, 65.7 m to 67.8 m, and 77.8 m to 79.9 m were sealed by double packers for conducting the hydraulic tests. The quantitative evaluation of hydraulic parameters was then performed using AQTESOLVE which uses an iterative process of the best-fit theoretical response curves based on the measured data of the hydraulic test. Figure

8 shows that the comparison of the rock mass hydraulic conductivity obtained by in-situ test and that from the estimation of the empirical HC model. Very good correlation can be found (Fig. 8). This verification example demonstrates that the empirical HC model is able to determine the rock mass hydraulic conductivity for different sites in which the lithologic conditions are similar.

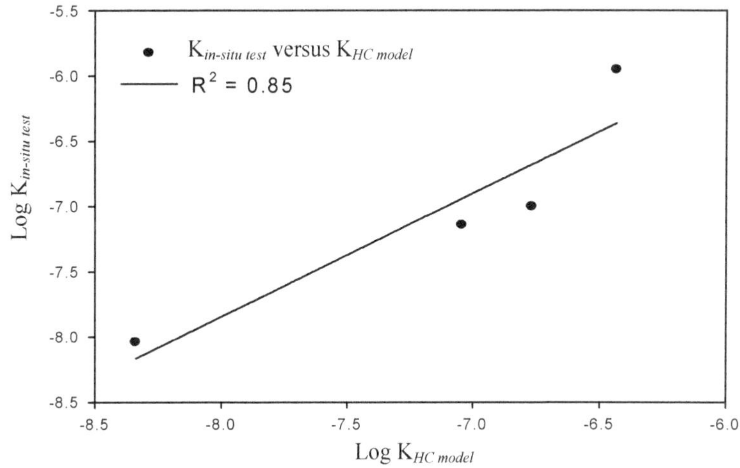

Fig. 8. Correlation between Kin-situ and KHC model.

Test intervals (m)	RQD(%)	DI	1-GCD	LPI	HC	$K_{HC\ model}$	$K_{in\text{-}situ}$
24.5~26.6	81.0	0.787	0.952	0.55	0.0785	9.06E-08	7.14E-08
32.5~34.1	43.8	0.723	0.975	0.55	0.2179	3.69E-07	1.11E-06
65.7~67.8	47.6	0.444	0.976	0.55	0.1248	1.71E-07	9.95E-08
77.8~79.9	95.2	0.343	1.000	0.55	0.0090	4.59E-09	9.09E-09
$K_{HC\ model}$ and $K_{in\text{-}situ}$ represent K obtained by Eq. 5 and the in-situ hydraulic test, respectively.							

Table 4. Four hydraulic test data for the model verification (Borehole CH-04).

6. Conclusions

The estimation of rock mass hydraulic conductivity of highly disturbed clastic sedimentary rocks in Taiwan was performed using the data of BHTV and double packer hydraulic tests. The field results indicated that the rock mass in the study area has the conductivity between the order 10-10 and 10-6 m/s at the depth between 34 m and 275 m below ground surface. The results demonstrate that the rock mass hydraulic conductivity of highly disturbed clastic sedimentary rocks in Taiwan mainly depends on the following four parameters: RQD, DI, GCD, and LPI.

This paper proposes an empirical HC model for estimating rock mass hydraulic using data collected for highly disturbed clastic sedimentary rocks in Taiwan. The HC-value can be calculated from borehole image data and rock core data. To verify rationality of the proposed HC model, the study collected data from the results of two hydrogeological investigation programs in three boreholes to determine a relationship between hydraulic conductivity and HC. Besides, good correlation is found from the model verification which demonstrates that the empirical HC model is able to determine the rock mass hydraulic conductivity for different sites in which the lithologic conditions are similar.

The regression results indicated that the relationship of a power law exists between the two variables with a coefficient of determination of 0.866. The empirical HC model may provide a useful tool to predict hydraulic conductivity of fractured rocks based on measured HC-values. By using this model, hydraulic conductivity data in a given site can be directly acquired, which removes the cost on hydraulic tests. For in-situ aquifer tests, the empirical HC model is valuable for preliminary assessment of the degree of permeability in a packed-off interval of a borehole.

7. References

Bear, J. (1972). Dynamics of Fluids in Porous Materials, American Elsevier.

Black, J. H. (1987). Flow and flow mechanisms in crystalline rock, in Fluid Flow in Sedimentary Basins and Aquifers. Geol. Soc. Special Publication No. 34, 186-200.

Burgess, A. (1977). Groundwater Movements Around a Repository—Regional Groundwater Analysis. Kaernbraenslesaekerhet, Stockholm, Sweden, 116.

Carlson, A. & Olsson, T. (1977). Hydraulic properties of Swedish crystalline rocks-hydraulic conductivity and its relation to depth. Bulletin of the Geological Institute, University of Uppsala 7, 71-84.

Deere, D. U.; Hendron, A. J.; Patton, F. D.& Cording, E. J. (1967). Design of surface and near surface construction in rock. Proceedings of 8th U.S. Symposium Rock Mechanics, AIME, New York 237-302.

Duffield, G. M. (2004). AQTESOLVE version 4 user's guide, Developer of AQTESOLV HydroDOLVE, Inc., Reston, VA, USA.

Lee, C. H. & Farmer, I.(1993). Fluid flow in discontinuous rocks. Chapman&Hall, London, UK.

Louis, C. (1974). Rock Hydraulics. in Rock Mechanics (ed. L. Muller), Springer Verlag, Vienna.

National Research Council. (1996). Rock fractures and fluid flow: contemporary understanding and applications. National Academy Press, Washington D. C., USA.

Singhal, B. B. S. & Gupta, R. P.(1999). Applied hydrogeology of fractured rocks. Kluwer Academic Publishers, The Netherlands, 400.

Spitz, K. & Morena, J. (1996). A Practical Guide to Groundwater and Solute Transport Modeling, Wiley.

Sinotech Engineering Consultants, LTD. (2006). Tseng-Wen transbasin diversion tunnel project-supplemental geology investigation, Southern Water Resources Office, Water Resources Agency, Ministry of Economic Affairs, Taiwan (in Chinese).

Snow, D. T. (1969). Anisotropic permeability of fractured media. Water Resources Research, 5(6), 1273-1289.

Wei, Z.Q., Egger, P., Descoeudres, F. (1995). Permeability predictions for jointed rock masses. International Journal of Rock Mechanics, Mineral Science and Geomechanics 32, 251-26l.

Williams, J. H. & Johnson, C. D. (2004). Acoustic and optical borehole-wall imaging for fractured-rock aquifer studies. Journal of Applied Geophysics, 55(1–2): 151–159.

Zhao, J. (1998). Rock mass hydraulic conductivity of the Bukit Timah granite, Singapor. Engineering Geology, V 50, 211-216.

15

Instrumentation for Measurement of Laboratory and *In-Situ* Soil Hydraulic Conductivity Properties

Jose Antonio Gutierrez Gnecchi et al.*
[1]*Instituto Tecnológico de Morelia, Departamento de Ingeniería Electrónica México*

1. Introduction

Measurement of soil hydraulic conductivity properties is very important for soil characterization, modelling of water transport and waste contaminant migration through soil, management of soil organic matter and management of water resources. Moreover, measurement of hydraulic properties is also important for developing strategies to increase crop productivity, and 3-D modelling of water migration properties to predict groundwater and aquifer recharge. Amongst the most common methods, used in laboratory and field test trials to determine the properties of water propagation through soil, are measurement of hydraulic conductivity and wetting front detection. However since the hydraulic conductivity properties vary considerably from region to region (and even for the same region and type of soil) numerous and diverse methods are continuously reported that fit particular needs. Despite the large number of methods and apparatus reported, and commercially available instruments for measuring the dynamics of water propagation through the soil, it is still necessary to continue developing new and improved instrumentation systems to increase the quality and quantity of reliable information and reduce systematic errors. In addition, commercial instruments may only be available from foreign distributors. Thus the use of imported technology, with little or no technical support locally, and the added import tax costs result prohibitive for the average producer and precludes the use of electronics instrumentation by producers without a technical background. Since 77% of the water in Mexico is used in agriculture, the availability is scarce in many wide areas, and the water usage efficiency is low, the situation becomes more critical due to the demand for increased productivity. Undoubtedly, research and development activities in higher education institutions should have scientific, technological, social and economical impact in the surroundings. This chapter presents the results of the cooperation between ITM-Electronics Engineering Department (*Spanish: Instituto Tecnológico de Morelia*), INIRENA-Research Centre for Natural Resources Studies (*Spanish: Instituto*

* Alberto Gómez-Tagle (Jr)[2], Philippe Lobit[3], Adriana Téllez Anguiano[1], Arturo Méndez Patiño[1], Gerardo Marx Chávez Campos[1] and Fernando Landeros Paramo[1]
[1]*Instituto Tecnológico de Morelia, Departamento de Ingeniería Electrónica, México*
[2] *Instituto de Investigaciones Sobre Los Recursos Naturales, Laboratorio de Suelos, Michoacán, México.*
[3] *Instituto De Investigaciones Agropecuarias y Forestales, México*

Nacional de Investigación Sobre Los Recursos Naturales) and IIAF (*Spanish: Instituto de Investigaciones Agropecuarias y Forestales*) to develop instrumentation for measuring some of the properties that govern the dynamics of water propagation through soil.

1.1 Water usage in Mexico

Water resources in Mexico are considered essential for national security. Urban, industrial and agricultural conservation of the environment, economic and social development depend on the rational management of water resources. In Mexico, the surface dedicated to agriculture is approximately 21 million hectare (*abbreviation ha*) (10.5% of the national territory) of which 6.46 million ha are irrigated zones and 14.5 million ha are rainfed zones. Most of the fresh water resources are dedicated to agriculture, where the 77% is allocated for consumptive use (Table 1).

USE	ORIGIN		TOTAL VOLUMEN	PERCENTAGE OF EXTRACTION
	SUPERFICIAL	SUBTERRANEAN		
Agriculture[1]	40.7	20.5	61.2	76.8
Public Water Supply[2]	4.2	7.0	11.2	14.0
Self sustained industry[3]	1.6	1.6	3.3	4.1
Thermoelectric	3.6	0.4	4.1	5.1
TOTAL	50.2	29.5	79.8	100.0

1 km^3 = 1 000 hm^3 = 1 thousand of millions of m^3.
Data correspond to volume allocated through to December 31 of 2008
[1] Includes agriculture, livestock, aquaculture, and other, according to the REPDA-CNA (Public Rights Register of Water- National Water Commission) classification. Includes 1.30 km^3 of water corresponding to irrigation districts pending registration.
[2] Includes urban public and domestic uses according to the REPDA-CNA classification.
[3] Includes industrial agro industrial, commerce and services according to the REPDA-CNA classification.
Source: National Water Commission (CONAGUA: http://www.cna.gob.mx).

Table 1. Consumptive use of water in Mexico according to the source of origin. (Thousands of millions of cubic metres, km^3)

The Free Trade Agreement of North America and the globalization of markets and the economy, impose more demands on Mexican producers to increase efficiency and quality of agricultural production, optimizing the use of resources in a sustainable manner. Now it is necessary to produce more, with better quality and lower costs to meet local demand, compete with imported agricultural products and eventually to produce products that meet the quality standards that exist in international markets (weight, size, color and texture). As part of Mexico's National Water Program 2007-2010 (Mexican National Water Commission, *Spanish: Comision Nacional del Agua* [CONAGUA], 2008) it is proposed that the use of technology for irrigation modernization would increase water productivity by 2.8% annually, measured in kilograms per cubic meter of water used in irrigation districts, going from 1.41 in 2006 to 1.66 in 2012, and will result in greater benefit to producers. At the same time, it is also proposed that the reduction of energy consumption will lead to achieve a more efficient use of water. However, it is common that the term "technification" generally corresponds to hydraulic infrastructure for drainage of surplus water. Although it has been reported (Mexican National Water Commission, *Spanish: Comision Nacional del Agua* [CONAGUA], 2010) that technification of agriculture has increased by 50% nationwide, compared to 2000, the reports do not specify what level or type of modernization is done, and generally consists of pumping equipment and/or hydraulic installations for the evacuation of excess water. To a lesser extent, the use of agro-meteorological stations is also included as part of efforts to introduce technology to the field. However, it is necessary to increase the level of modernization of the Mexican countryside in order to achieve precision agriculture practices at regional and national levels.

1.2 Soil hydraulic conductivity

Soil water infiltration is a process by which water propagates from the soil surface, inwards, through the porous media. One of the properties that govern the rate of propagation of water through the soil is hydraulic conductivity, which in turn, depends on a number of factors such as soil content and texture (Das Gupta et al., 2006), vegetation root hardness (Rachman et al., 2004; Seobi et al., 2005), soil preparation (Park & Smucker, 2005), chemical content (Schwartz & Evett, 2003), soil temperature and weather conditions (Prunty & Bell, 2005; Chunye et al., 2003), stability and continuity of the porous system (Soracco, 2003), including macro (Mbagwu, 1995), meso (Bodinayake et al., 2004) and microporosity (Eynard et al., 2004). Amognst the methods reported for studying the hydraulic properties of soils, the infiltrometer and permeameter are probably the most commonly used devices in field (Angulo-Jaramillo et al., 2000) and laboratory tests (Johnson et al. , 2005) respectively. Other methods used for characterizing soil hydraulic properties reported are heat-pulse soil water flux density measurements (Kluitenberg, 2001), electromagnetic measurements (Dudley et al., 2003; Seyfried & Murdock, 2004), radiation-based measurements (Simpson, 2006) image analysis (Gimmi & Ursino, 2004) and multimodal instruments (Pedro Vaz et al., 2001; Schwartz & Evett, 2003) that permit measurement of several variables simultaneously. The infiltrometer is a very popular instrument among researchers (Fig. 1A), because knowledge of soil hydraulic properties is a key factor in understanding their impact on hydrological processes such as infiltration (Esteves et al., 2005) the superficial flow and aquifer recharge. Basic infiltrometers are relatively simple devices, which essentially consist of a reservoir (fitted with a graduated scale), a metallic ring (single or double) partially inserted into the soil, and a stop valve. A test is conducted by allowing the liquid to exit the container, either directly or through a pipe into the ring, measuring the rate of water infiltration while maintaining a small positive pressure on the fluid. The infiltration process consists of two main parts: the transient and steady state (Fig. 1B). The transient state occurs from the beginning of the experiment up to the time when a constant rate of water infiltration is attained.

Once the soil sample is saturated with water, the constant pressure maintains a constant infiltration rate. Hydraulic conductivity can then be calculated using the entire data set (Wu1 method) (Wu & Pan, 1997) or the data corresponding to the steady state phase (Wu2 method) (Wu & Pan 1999) by measuring the slope of the resulting curve. However, recording the infiltration process data from direct, visual measurements is a highly demanding task, both, in time and economic resources; data has to be recorded in time intervals between 1 to 5 minutes in elapsed times ranging from 0.5 to 4 hours. Many authors

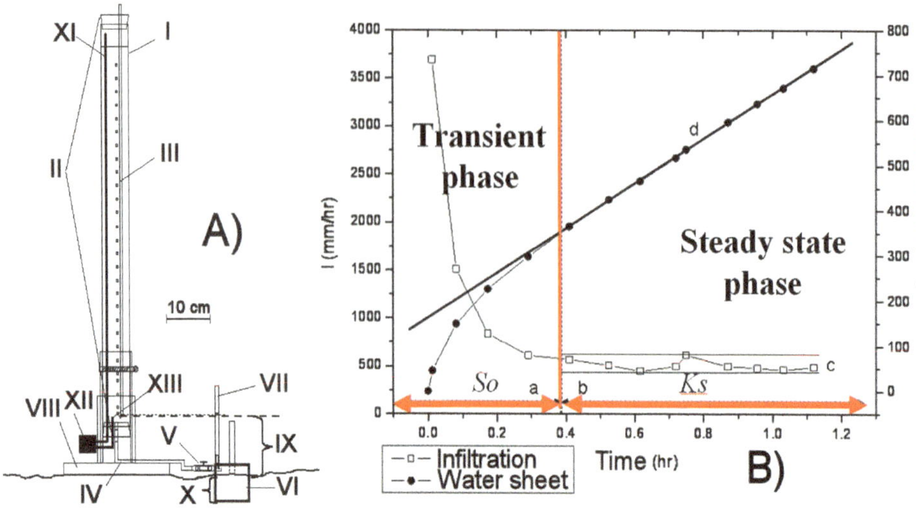

Fig. 1. A) Schematic diagram of an infiltrometer. I) Main reservoir (Mariotte), II) rubber stoppers, III) bubbling tubing, IV) water outlet, V) stop valve, VI) metallic ring, VII) purge and measurement of hydraulic load, VIII) stand base, IX) constant hydraulic load, X) insertion depth, XI) pressure sensor access to air chamber, XII) data logger, XII) pressure sensor access to the water column. B) Classical infiltration data results showing the transient and steady state phases of the infiltration process.

acknowledge the need for instrumentation and data recording devices (infiltrometer and permeameter) to automate the data acquisition, minimize human errors and reduce the time spent in taking measurements (Amezketa-Lizarraga et al., 2002; Johnson et al., 2005). Therefore many devices have been reported and patented since the 1940s (Bull, 1949) to try to automate the data acquisition process. Automated devices rely on data recording units (data loggers). However the main constraint is the local availability of such equipment, followed by cost. In many cases researchers implement their own devices without automation, such as a double ring infiltrometer (Carlon-Allende, 2006). Automating an infiltrometer requires accurate measurement of the change of height of the water column over time, as the water is allowed to exit the container. Some of the methods used for measuring the column height are the use of paired infrared sensors in a plastic cylinder (Wilson et al., 2000), float valve system with meter spool ring infiltrometer (Amezketa-Lizárraga et al., 2002), Time domain reflectometry (TDR) infrared detectors and float sensor or pressure sensors (Ankeny et al., 1988). The use of pressure transducers is probably the most common choice because of low-cost, simplicity, easy implementation and reliability. Overman et al. (1968) reported the application of pressure transducers since the mid 60s, to implement a variable load laboratory infiltrometer, designed specifically for low-permeability materials. Constanz & Murphy (1987) generated a system that could measure the height of a column of water from pressure changes in a Mariotte reservoir and thus infer the infiltration data. Their instrument used Transamerica CEC 4-312 pressure transducers, with pressure range ± 12.5 psi. The automated device allowed rapid data acquisition with minimal supervision. Ankeny et al. (1988) reported that the use of one transducer produced measurement errors due to bubbling inside the container and adapted the design of Constanz & Murphy (1987) to a tension infiltrometer (disc) with two PX-136 transducers with measurement range 0-5 PSI (Omega Engineering, Stanford, CT) and a 21X Campbell data logger (Campbell Scientific, Inc.). The resulting scheme using two transducers required precise timing, but minimized the variability generated by bubbling and reduced the standard deviation from 6.2 mm (single transducer) to 2.2 mm. Prieksat et al. (1992) used the two-transducer design in a single ring infiltrometer, to register data from multiple locations simultaneously, facilitating the characterization process. Casey and Derby (2002) used a differential pressure sensor (PX26-001DV, Omega Engineering, Stanford, CT) and evaluated the device in the field, achieving 0.05 mm standard deviation. The authors noted that the improvement in resolution might not change significantly the estimation of soil hydraulic properties, but could be useful when data are processed as exponential relations methods such as Ankeny (1992) or Reynolds and Elrick, (1991). Johnson et al. (2005) constructed six laboratory variable load permeameters, using pressure sensors PX236 (Omega Engineering) and perspex tubes, to work with undisturbed samples, using a data logger programmed to record readings at regular intervals. The comparison with the manual method showed no significant differences for texture analysis. Špongrová (2006), designed, built and tested a fully automated tension infiltrometer, that included both the measurement of water level and the control of the voltages applied, using a Honeywell differential pressure transducer with range 0 to 5 PSI (0 to 34.4 kPa), connected to a Campbell 21X datalogger (Campbell Scientific Inc.) and a laptop. The results showed that the equipment reduced the monitoring time, increasing the number of test trails per day. Although there are several commercial devices, such as manual or automated tension infiltrometers they generally depend on external data logger units.

2. Case study 1: Automated infiltrometer using a commercial data logger

One of the preferred methods for measurement the height changes of the water column involves the use of pressure transducers. Therefore, it is necessary to implement an instrumentation and data acquisition system that can be used to gather information of the infiltration process for off-line signal processing. Fig. 2 shows the classic data acquisition scheme used. The instrumentation scheme consists of a pressure transducer, a signal conditioning and amplifier stage, and a digitizing unit with data storage capabilities for transferring the measurements to a host computer. To allow some level of autonomy, and ease of use, it is necessary to use low-power, versatile analogue and digital devices.

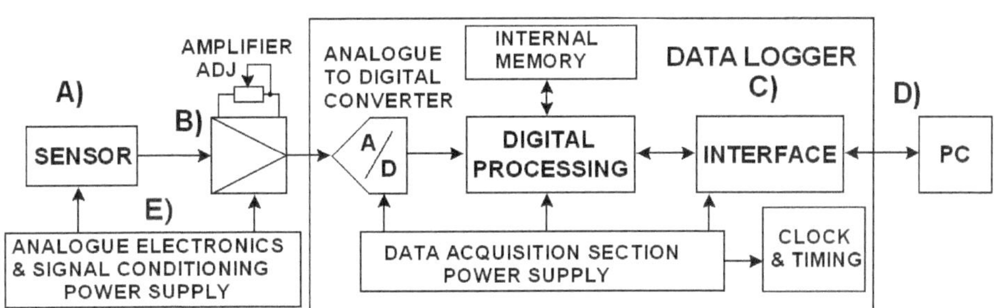

Fig. 2. Block diagram of the instrumentation and data acquisition system used for automating soil water infiltration measurements. A) Pressure sensor, B) instrumentation amplifier, C) data logger including analogue to digital converter, internal memory, interface circuitry and power supply for transferring data to D) a PC. E) Power supply circuit for the analogue electronics section.

Fortunately, the advances in electronics technology over the last two decades have resulted in a number of components that can be obtained from local and international distributors to build the analogue and signal conditioning circuitry. As to the digitizing section, a number of data logger units are commercially available with impressive operating characteristics. The case study presented in this section is based on the choice of a low-cost data acquisition unit.

2.1 Data logger

One particular low-cost, simple-to-use device is the EL-USB-3 data logger from Lascar Electronics (Fig. 3).

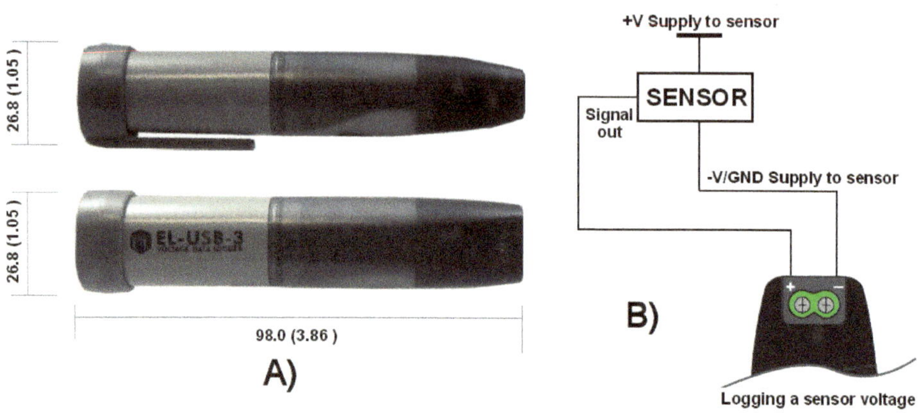

Fig. 3. EL-USB-3 Voltage data logger from Lascar Electronics. A) Physical dimensions and B) typical connection for logging sensor signals. Reproduced with permission. Copyright 2011 Lascar Electronics. All Rights Reserved from Lascar Electronics.

The EL-USB-3 is a stand-alone data logger powered internally by a 3.6 Volts battery, capable of taking 32,510 readings in the range of 0-30 Volts with 50 mV resolution. The signals are applied to the data logger through a detachable cap, so that it can be removed from the instrumentation electronics for programming and data transfer without disconnecting wires. The data logger includes a USB interface for setting the data acquisition sampling rate from 1 second to 12 hours (1 second, 10 seconds, 1 minute, 5 minutes, 30 minutes, 1 hour, 6 hours and 12 hours) and also for transferring the results to a host PC. The operation can be assessed by observing the activity of two LEDs, red and green, which are included (Fig. 4).
Once the test has concluded, the measured data can be transferred to a host PC for off-line analysis through the USB interface using the software included. Thus the EL-USB-3 includes all the necessary components shown in Fig. 2C corresponding to the digitizing section of the instrumentation scheme proposed. Using a commercial data logger reduces instrumentation development time. Nevertheless, the signal conditioning section must consider the operating characteristics of the data logger to maximize the measurement resolution. That is the voltage corresponding to the maximum height of the water column (100 cm) must be +30V.

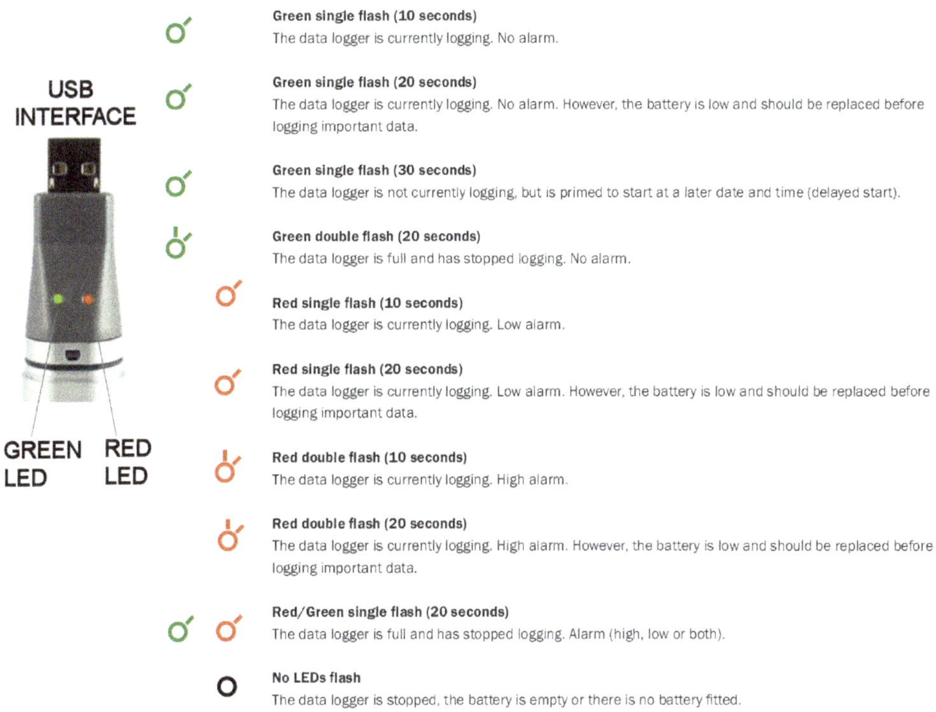

Fig. 4. Data logger USB connector and operation assessment depending on the LED activity. Reproduced with permission. Copyright 2011 Lascar Electronics. All Rights Reserved

2.2 Power supply

The first step in developing the instrumentation circuitry consists of obtaining a little over +30V from a +9V battery, because interfacing with the data logger requires that the analogue instrumentation operate with a voltage slightly over 30V. The circuit must be small and must consume very little current from the battery. Fig. 5 shows the block diagram of the power supply.

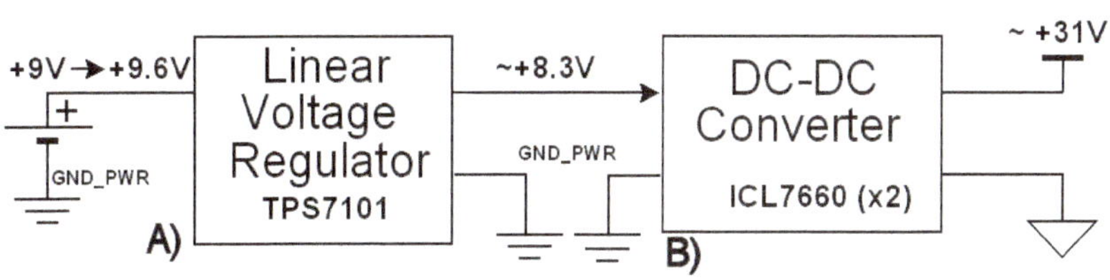

Fig. 5. Block diagram of the power supply. A) The battery feeds a linear voltage regulator. B) the output from the regulator is increased to (over) +30 V to power up the instrumentation amplifier.

The battery feeds a low dropout, adjustable linear regulator (TPS7101 from Texas Instruments©) which provides the regulated supply voltage to the DC/DC converter (Fig. 6).

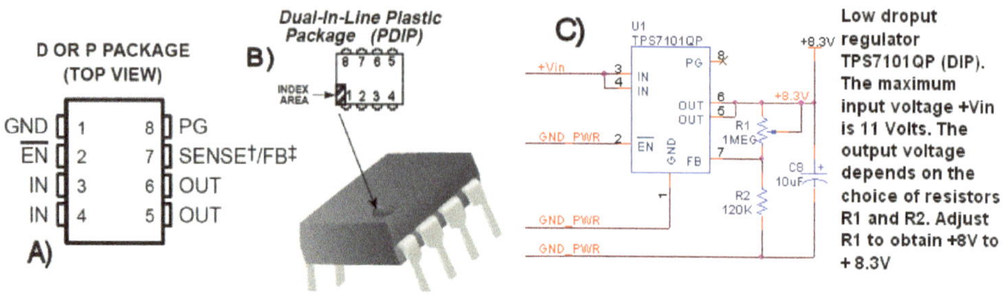

Fig. 6. A) Description of the TPS7101 (linear low dropout regulator) terminals. B) Identification of the index area (terminal 1). C) Regulator diagram. Copyright Texas Instruments©. All Rights Reserved.

The TPS7101 output voltage is adjusted to +8.3V by trimming R1 which is a 1 MegaOhm, 20 turns, trimming potentiometer. R2 is a fixed value ¼ Watt precision, metal film resistor. C8 is a tantalum capacitor, and is used for filtering the output and provides stability to the voltage regulator. The TPS7101's output feeds the DC/DC converter formed by the two ICL7660 integrated circuits from Intersil © (U2 and U3) (Fig. 7B). The ICL7660 is a low-power monolithic CMOS power supply that can be configured easily to double the input voltage and also to provide complimentary negative voltage, requiring a minimal amount of non-critical passive components. In this application, U2 is configured to perform two operations: U2 inverts the input voltage and also doubles the positive input. The circuit uses low forward-voltage-drop Schottky diodes to reduce the effect of the voltage drop across the circuit. Another feature of the ICL7660 is that it can be cascaded to increase the differential voltage. Thus, U3 is configured to double the negative voltage obtained from U2. In effect, the array of U2 and U3 increases the regulated voltage from the TPS7101 to approximately 31 Volts, enough to provide the energy for the instrumentation amplifier and pressure sensor.

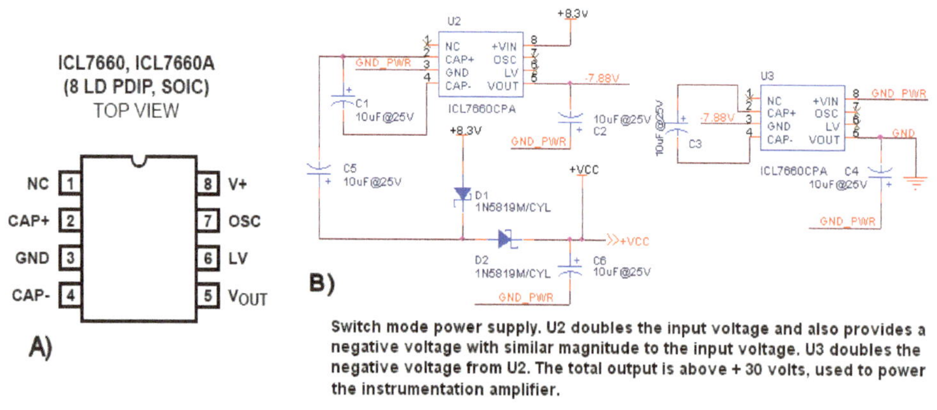

Fig. 7. A) Pinout of the ICL7660 CMOS converter. B) circuit diagram used to increase the TPS7101's output to ~+31V. Reproduced with permission. Copyright 2011 Intersil Americas Inc. All Rights Reserved

Typical current consumption values for the ICL 7660 are 80 μA (ICL7660A) which makes it suitable for this and other battery powered applications. In order to maintain the high efficiency of the CMOS voltage converters it is necessary to reduce the current consumption; therefore the analogue instrumentation must also be a low-power circuit.

2.3 Pressure transducer

The water reservoir is built using an 80 - 100 cm perspex pipe with rubber stoppers on each end. The pressure at the bottom, when the container is full (100 cm H_2O @ 4°C) is 9.806 kPa. Therefore, it is necessary to use a differential pressure transducer with 10 kPa measurement range (Fig. 8).

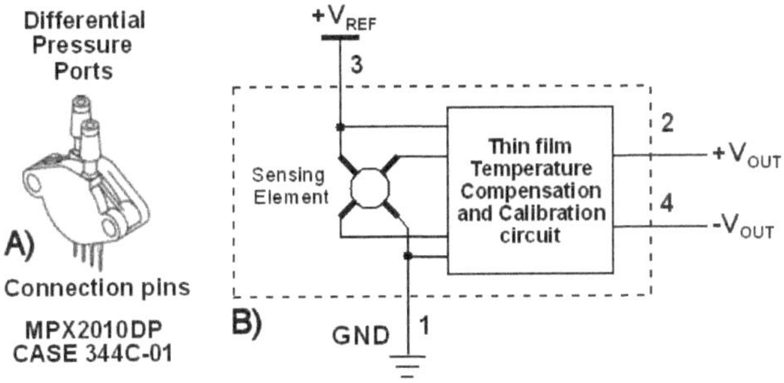

Fig. 8. MPX2010 DP differential pressure transducer A) physical model and B) connections. Reproduced with permission. Copyright Freescale Semiconductor, Inc. 2004 - 2011. All Rights Reserved.

The MPX2010DP differential pressure transducer from Freescale Semiconductor© is a temperature-compensated piezoresistive pressure sensor which provides a very accurate and linear output voltage proportional to the applied pressure in the range of 0 - 10 kPa. The recommended voltage supply is 10V, and the error and linearity figures are specified for 10V. However, the MPX2010DP is a ratiometric device; that is the maximum output voltage depends on the reference voltage which means that a different supply may be used so long as the reference voltage is very stable.

2.3.1 Pressure sensor reference voltage

Recalling that the MPX201DP pressure transducer is a ratiometric device, it is necessary to provide a highly-stable voltage reference signal to achieve correct operation, regardless of voltage and temperature variations.

It is common to find circuits that suggest the use of the voltage supply line to power up the pressure transducer (Fig. 9A). Unless the pressure transducer includes an internal voltage reference supply (i. e. it is a voltage compensated device), it is necessary to use a dedicated voltage reference circuit. (Fig. 9B). In terms of temperature variations, voltage reference circuits are specified in ppm/°C (parts per million per degree centigrade). Consider a reference circuit specified to change at a rate of 100ppm/°C. If the circuit output value is 10V @ 20 °C, exposing the integrated circuit to a 50 °C would change the output from 10V to 10.03 Volts ensuring the correct operation of the transducer. Some devices may also be specified to 10ppm/°C increasing the stability of the overall instrumentation circuit.

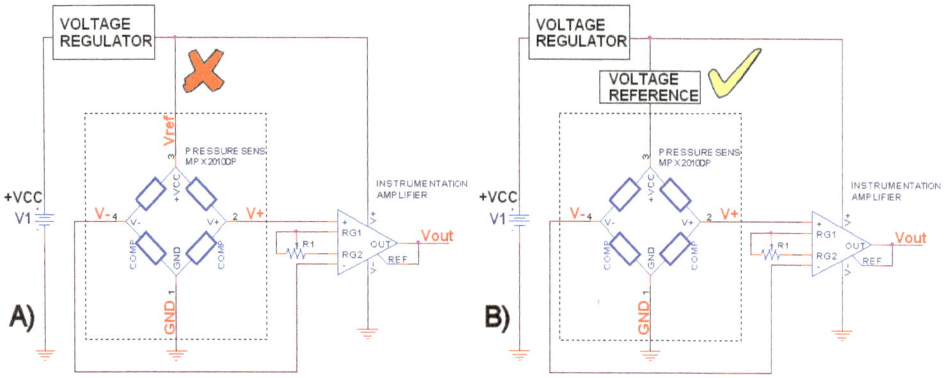

Fig. 9. A) Incorrect voltage supply for a bridge-type sensor. B) The correct use of bridge-type sensor involves the use of a reference voltage circuit.

2.4 Instrumentation amplifier

Instrumentation amplifiers are a type of differential amplifier with high input impedance and adjustable gain that constitute essential building blocks in analogue electronics. One of the classical configurations of instrumentation amplifiers uses three operational amplifiers to form a two-stage amplifying circuit (Figure 10).

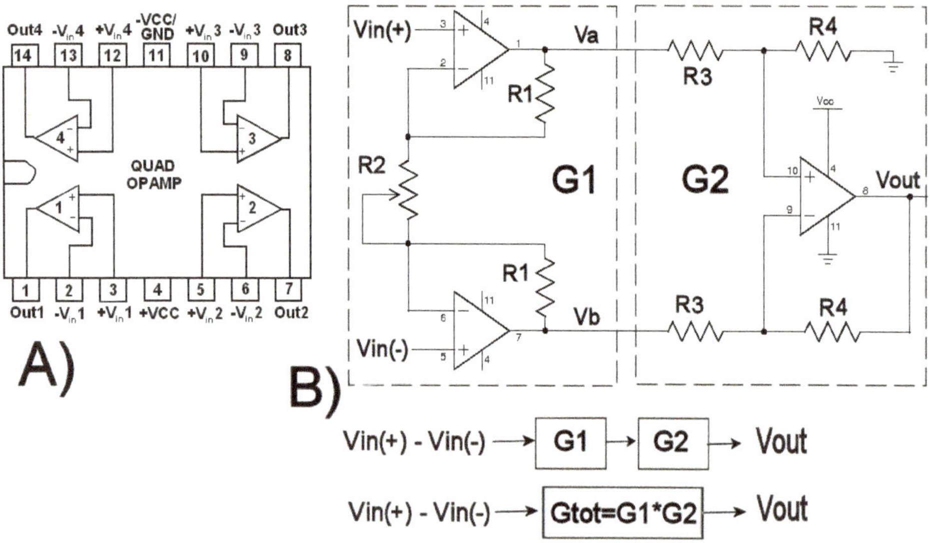

Fig. 10. A) A commercially available quad op-amp can be used to build B) a general-purpose instrumentation amplifier.

The differential input signal feeds the first amplifying stage formed by op-amp 1 and op-amp 2. The output between both op-amp 1 and op-amp 2 (Va and Vb respectively) is differential weighted version of the input voltage. The gain of stage one, G1 (1):

$$G1 = 1 + \frac{2R1}{R2} \tag{1}$$

can be adjusted using a single trimming potentiometer, R2. The differential output from stage one, enters a third operational amplifier (op-amp 3) configured as subtracting amplifier with gain, G2, (2):

$$G2 = \frac{R4}{R3} \tag{2}$$

The overall output is then a single-ended version of the differential input voltage. The overall gain of the instrumentation amplifier circuit is the multiplication of both amplifying stages given by (3):

$$G_{TOTAL} = G1\,G2 = \left(1 + \frac{2R1}{R2}\right)\left(\frac{R4}{R3}\right) \tag{3}$$

The importance and usefulness of instrumentation amplifiers have resulted in multiple versatile commercial integrated circuits, with impressive operating characteristics, that allow gain adjustment using a single variable resistor and/or digital signals. One particular integrated circuit that is suitable for portable applications is the INA125 from Texas Instruments© (Fig. 11).

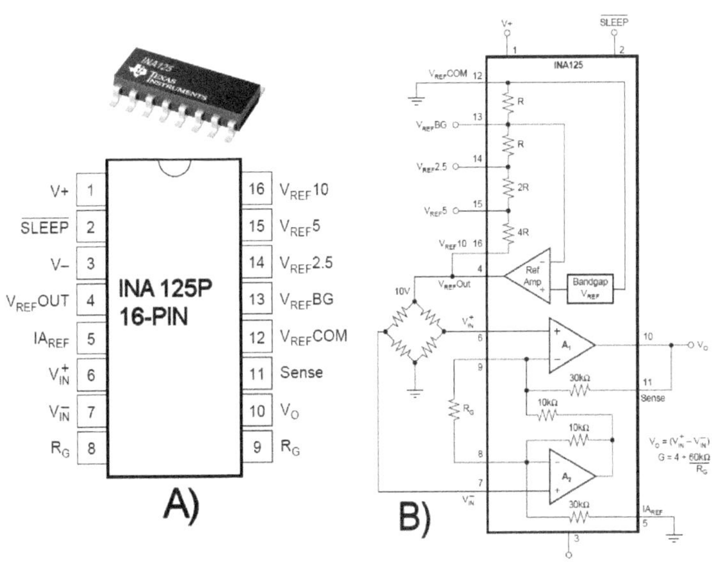

DESIRED GAIN (V/V)	R_G (Ω)	NEAREST 1% R_G VALUE (Ω)
4	NC	NC
5	60k	60.4k
10	10k	10k
20	3750	3740
50	1304	1300
100	625	619
200	306	309
500	121	121
1000	60	60.4
2000	30	30.1
10000	6	6.04

NC: No Connection.

C)

Fig. 11. A) Texas Instruments INA125 instrumentation amplifier pinout, B) typical application circuit and C) gain selection table. Copyright © 2009, Texas Instruments Incorporated. All Rights Reserved.

The INA125 also incorporates a selectable voltage reference circuit, which will be used to supply the reference voltage to the pressure transducer (Fig. 12). The INA125 is a low power device (quiescent current 460 μA) and can operate over a wide range of voltages from a single power supply (2.7V to 36V) or dual supply (±1.35V to ±18V), which makes it suitable for battery powered applications.

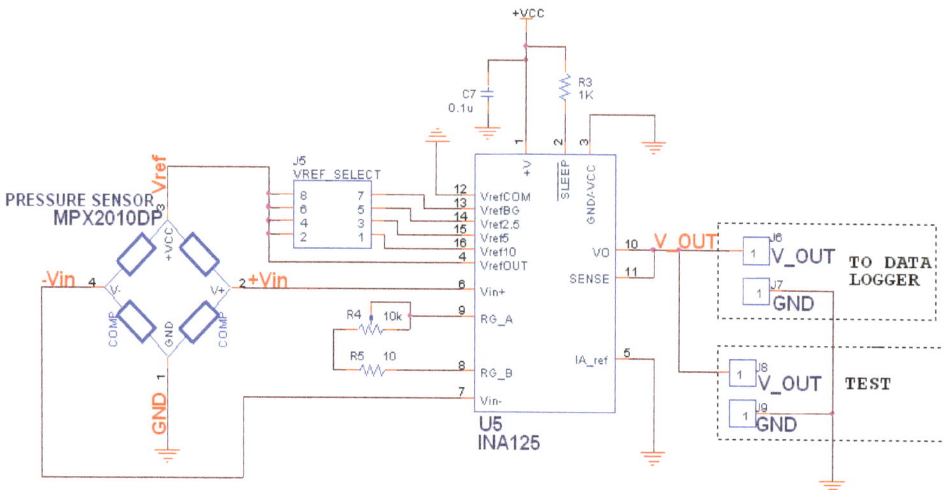

Fig. 12. Circuit diagram of the instrumentation amplifier and voltage reference circuit for the pressure sensor.

The voltage reference value can be adjusted using the jumper J5, to provide 10V, 5V, 2.5 V or 1.24V.

2.5 Complete instrumentation circuit

The complete circuit is shown in Figure 13.

The design includes an on-off switch and connectors to calibrate and monitor the output voltage using a multimeter.

2.6 Ring and reservoir

The size of both the reservoir and the ring may differ depending on the type of soil to be analysed. In addition, the hydraulic conductivity properties of soil vary throughout the test field, and thus a large metallic ring may be used to investigate a large area as much as possible. Analysis of sandy soils may require a larger reservoir compared to clay type soils, because coarse materials have higher hydraulic conductivity and require a larger amount of water to reach the saturated steady state, compared to fine particle soils. In laboratory test trials, it is of little concern the handling of a large reservoir, a heavy metallic ring, computers and electronics instrumentation, and there is tap water available nearby. However, carrying all the necessary materials in field tests may be a difficult task. Therefore the size of the reservoir and metallic ring is a compromise. The infiltrometer described here uses a 1 metre long, 6.35 cm diameter perpex pipe; the ring is made of an iron pipe (8.0 cm long and 8.8 cm diameter) .

2.7 Infiltrometer assembly

Fig. 14 shows the infiltrometer design and assembly, including instrumentation circuitry.

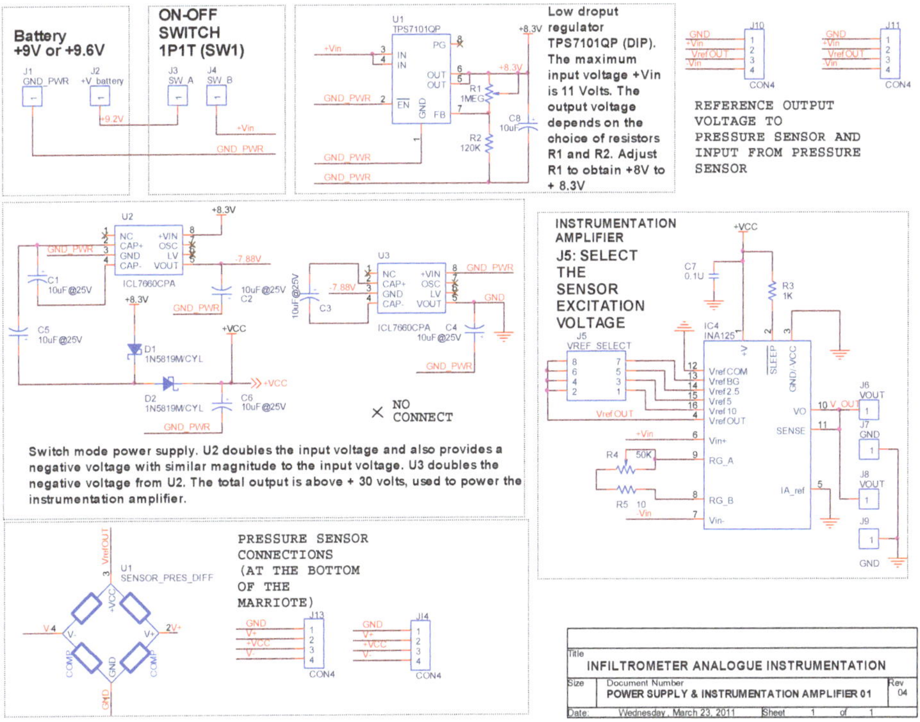

Fig. 13. Instrumentation circuit for measuring infiltration data using a pressure transducer.

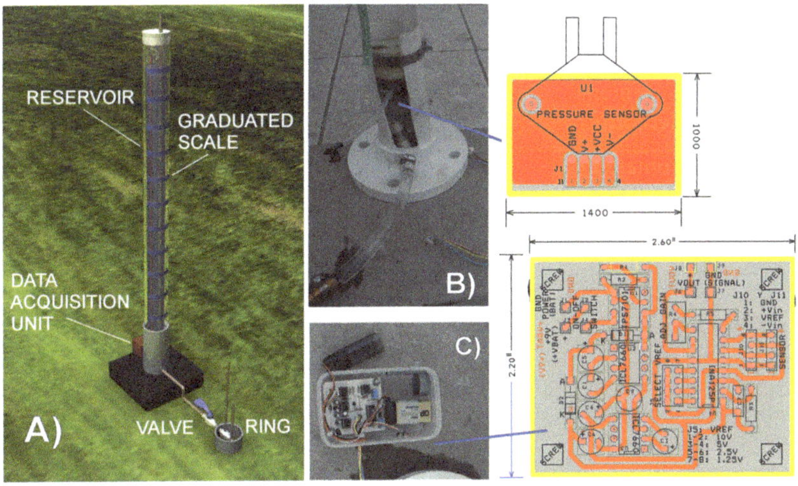

Fig. 14. A) Designed infiltrometer. B) close-up of the pressure sensor assembly. C) The single- sided circuit is fitted into a small (2.6" X 2.2") printed circuit board fits into the plastic enclosure.

The pressure sensor is installed at the bottom of the reservoir and soldered into a small printed circuit board to ensure the correct connectivity between the sensor and the instrumentation circuitry. The instrumentation circuit is installed, outside the infiltrometer, inside a small plastic enclosure, as well as the battery and data logger. Once the plastic enclosure is attached to the infiltrometer, the data logger can be removed from the electronics without removing any connections. The result is a compact versatile infiltrometer, which can easily be transported for field tests.

2.8 Test results

The infiltrometer was tested in two different test locations around the Cuitzeo Lake watershed (19°58' N, 101°08' W): sandy loam (Fig. 15A) and sandy soil (Fig. 15B).

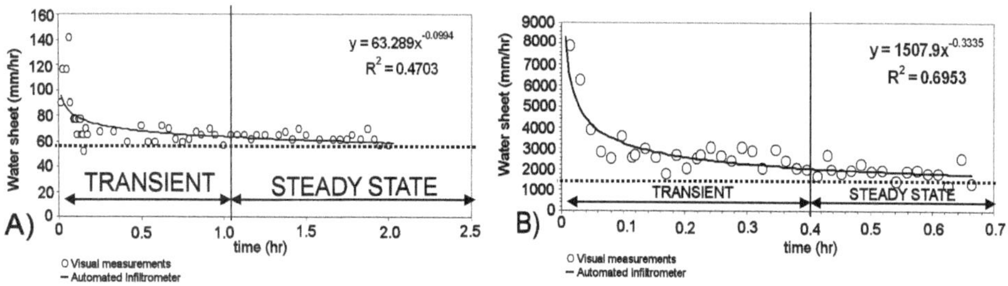

Fig. 15. Comparison of test results using the automated infiltrometer vs visual measurements for A) sandy loam and B) sandy soils.

The results from the automated measurement data acquisition system are much more consistent throughout the test, improving the quality of information compared to visual observations. Table 2 shows a summary of hydraulic conductivity results, using the Wu2 method and data from the steady state region.

Type of Soil	Measurement Method	Average Hydraulic conductivity K_{Fs} (mm/hr)	Standard Deviation	Number of test trials N
Sandy loam	Automated infiltrometer	49.65	21.72	7
	Visual Observations	68.31	42.85	7
Sandy soil	Automated infiltrometer	2282.2	429.98	5
	Visual Observations	1671.82	793.56	5

Table 2. Summary of the hydraulic conductivity results obtained from automated measurements and visual observations.

The automated infiltrometer can produce more reliable information than that obtained using visual measurements. For instance the standard deviation obtained from automatic data is smaller compared to visual observation. The resolution for a 1 metre water column is 1.66 mm approximately. The equipment presented in this case study is considerably easy to implement, low cost and can be used for laboratory and field test trials alike. Nevertheless, it still relies on the use of commercial data loggers. Alternatively, a dedicated data logger based on an ultra-low power microcontroller can be used to allow reviewing the data on-site and also to store the results of multiple test trials.

3. Case study 2: Automated infiltrometer using a dedicated data logger

Using a commercial data logger and relatively simple analogue electronics allows rapid development of test prototypes. On the other hand, since the determination of hydraulic conductivity infiltration requires performing multiple tests, it is desirable that all the results are stored in non-volatile memory without having to transfer the results to the host PC, immediately after each test has been concluded. Higher resolution may also be required for correct in-situ characterization of different types of soils. Moreover, since measurements are not taken continuously (i. e. the lowest sampling rate may be 1 second) it may be desirable to be able to shutdown the analogue circuitry in between measurements to extend battery life. Fig. 16 shows the schematic diagram of the proposed data acquisition system.

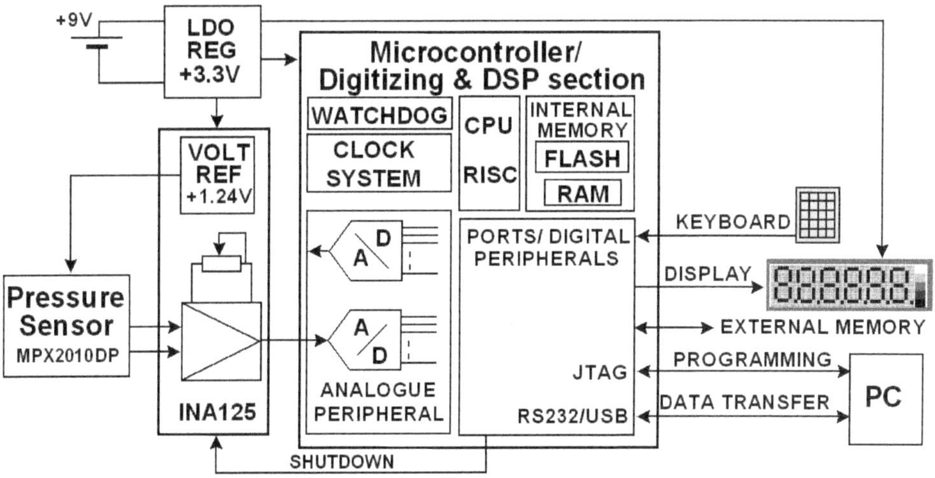

Fig. 16. Schematic diagram of a data acquisition system based on a low power microcontroller, specially designed for hydraulic infiltration measurements.

3.1 Dedicate data logger operation description

The equipment follows the same design philosophy for case study 1. The measurement system is based on the MPX2010DP pressure sensor, and the INA125 instrumentation amplifier is used to provide the reference voltage and measure the differential output from the transducer. However, the digitizing section is now based on a microcontroller.

3.1.1 Choosing the microcontroller

Several powerful microcontrollers are available from multiple companies that can be used to perform all the necessary data acquisition and signal processing operations. One particularly useful family of powerful microcontrollers suitable for low power operation is the MSP430 series from Texas Instruments© (Fig. 17). Case study 2 is based on MSP430F149IPAG microcontroller from Texas Instruments©. The MSP430 is a 16-bit RISC, ultra-low-power device with five power-saving modes, two built-in 16-bit timers, a fast 12-bit A/D converter, two universal serial synchronous/asynchronous communication interfaces (USART), 48 Input/Output pins, 60 kB of flash memory and 2 kB of RAM, which permits the implementation of all the functions required to build the data acquisition system. Initially, it was considered that basic signal processing algorithms (digital filter) are the main functions to be included. However, a JTAG interface implemented on the prototype allows in-system programming so that the equipment can be updated, and further signal processing algorithms can be included in the future, without changing the hardware.

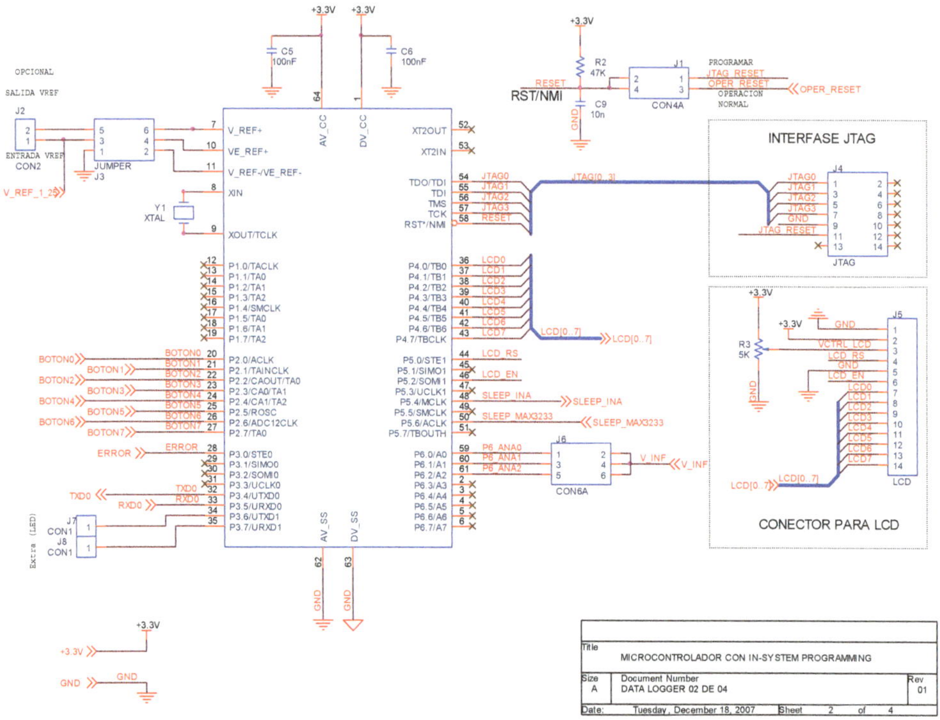

Fig. 17. Schematic diagram of the data logger based on the MSP430F148IPAG.

3.2 Operation of the data logger

The microcontroller interfaces with the user through a keyboard and LCD display, thus allowing the operation of the device in test fields, and reviewing the measured information in real time or right after the test has concluded. The microcontroller controls the data acquisition process, and stores each measurement in non-volatile flash memory. The microcontroller shuts down the analogue circuit in between samples to save battery power and enters a low-power mode. Prior to taking each sample, the microcontroller turns on the analogue circuit and waits 100 ms to allow the analogue output to settle and take a stable measurement. The data logger unit uses a 9V battery but it can also operate with a supply voltage as low as 4V. The MSP430 itself operates with 3.3V, so the same voltage is used throughout the circuit. The INA125 is very useful in this case, because it can operate with a voltage as low as 2.7V. The INA125's internal reference voltage circuitry requires that the power supply voltage is, at least, 1.25 V above the desired reference voltage and thus only the +1.24 V reference option can be used. The MSP430 internal voltage reference is adjusted to 2.5 volts, and the INA125 is adjusted to output 2.5V when the water column is full. In addition to accuracy, versatility, and compactness, it is necessary that the equipment can operate in low power mode to increase battery life. Therefore the MSP430 records data at fixed, programmable intervals, from 1 second, and then 10 seconds steps up to 60 minutes, selected by the user prior to each test. A real-time clock algorithm is implemented, using Timer A, so that the microcontroller can enter energy saving mode LMP3 consuming 2μA approximately in between samples. During the energy saving mode, the microcontroller also turns off the transducer voltage reference source, instrumentation amplifier and display. 100 miliseconds before each measurement is taken, the voltage reference source and instrumentation amplifier are activated, allowing the measurement to settle. The user can select the LCD to remain off while taking measurements. During operation, the LCD can also be activated temporarily to supervise the measured data, and then switched off again. The results of each test are stored in the flash memory, starting at memory block 0x3F. Before each test, the microcontroller detects which memory blocks are used and starts saving data in the next empty block. Thus, up to 90 tests can be conducted in-situ. The user can also select which memory block to erase, (i.e. which experiment) instead of erasing the entire memory, also contributing to saving battery life.

3.3 Electronics instrumentation assembly

Fig. 18A shows the double-sided printed circuit board. The board, keyboard and display and battery are fitted into a plastic enclosure (Fig 18B, 18C). In a similar manner to case study one, the pressure transducer is located below the reservoir and the wires carrying the voltage supply and signals are connected to the data logger.

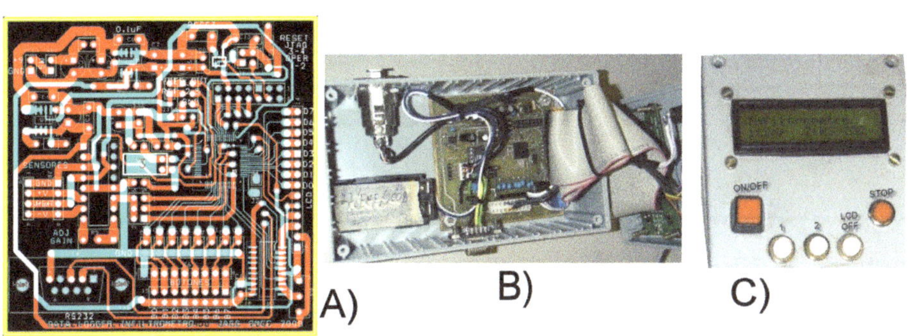

Fig. 18. A) Data Logger Printed Circuit Board (PCB). B) The PCB, C) keyboard and display and interface connections are also fitted in the plastic enclosure.

3.4 Transferring data for permanent storage and analysis

A C++ program interface was implemented to allow the user to transfer the data to a host PC for permanent storage, off-line results visualization and analysis (Fig. 19). Prior to each test the user can set the time, date and sampling rate for the experiment. The test information is stored at the beginning of each memory block, followed by the column height measurements. Once the experiment (or several experiments) has been completed, the user can review the measured data in-situ. Alternatively the user can transfer the results to a host PC, through the RS232 connection or with RS232-USB adaptors, to allow compatibility with current PC configuration ports.

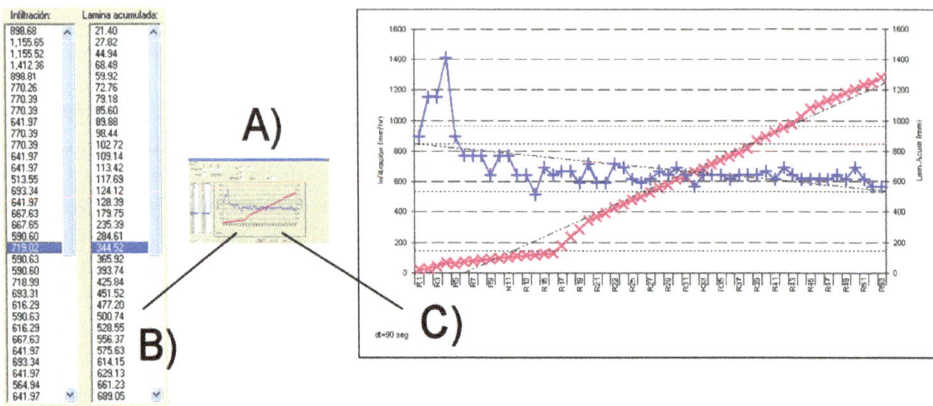

Fig. 19. A) The C++ software B) analyses the measurements using the Wu1 and Wu2 methods and C) plots the results.

The software processes the data and allows inspection of each value (Figure 18B). The program calculates hydraulic conductivity using the WU1 and WU2 methods, thus allowing result comparison.

3.5 Test results

The infiltrometer was tested in three different test locations around the Cuitzeo Lake watershed (19°58' N, 101°08' W): clay, loam and sand (Table 3).

Site		Wu1	Wu2	Guelph
	Average Kfs	5.497	2.231	2.782
Clay	Standard deviation	8.163	3.185	2.584
	Number of test trials, N	26	26	13
	Average Kfs	79.551	150.401	95
Loam	Standard deviation	63.58	82.86	97.05
	Number of test trials, N	36	36	3
	Average Kfs	708.30	963.41	----
Sand	Standard deviation	722.37	758.28	----
	Number of test trials, N	9	11	-----

Table 3. Summary of hydraulic conductivity tests using the automated infiltrometer.

Measurements where also obtained with a Guelph permeamenter, for comparison, except for sandy soil, because it was not possible to reach the required depth to introduce the 50 x

120 mm probe due to sample collapse. A commercial Guelph permeameter is a constant-head device, which also operates on the Mariotte siphon principle and allows simultaneous measurement of field saturated hydraulic conductivity, matric flux potential, and soil sorptivity in the field (Soilmoisture Equipment Corporation, Santa Barbara California, U. S.). In this work the Guelph permeameter operates as a "benchmark methodology" and is not to be considered it as the only valid acceptable method. Direct point-by-point comparison of results using different test methods is not valid since data is taken from different locations. In addition, the automated infiltrometer presented is limited to conduct tests at the surface, so Kfs variations at other soil depths is out of reach, in contrast with the Guelph permeameter, capable of measuring Kfs up to 80 cm depth without any special instruments. Nevertheless, the device described in this case study allowed the estimation of field saturated hydraulic conductivity in agreement with the Guelph permeameter in some cases.

4. Conclusion

The automated infiltrometers presented in this work, can produce reliable information about the infiltration process in-situ, with little supervision. The devices also allow the acquisition of a large number of measurements compared to visually obtained information, thus facilitating the calculation of Ks. Case study one shows the use of low-cost data loggers to automate the measurement process. A considerable simple instrumentation circuit is necessary to obtain the maximum resolution from the data logger. If a dedicated device is required, case study 2 shows the use of microcontroller technology to build the data logger unit. The DAQ units allow sample time adjustment on-site, which permits the investigation of different types of soils. The automated infiltrometer offers a ~0.25mm column height measurement resolution improving the quality of Ks calculations. Both cases present affordable and reliable instrumentation solutions, that can be built for about $ 100 US dollars without considering development time investment. Current and future work includes the development of a multi-channel simultaneous sampling system, so that the test field can be correctly characterized using multiple infiltrometers located around the test site.

5. Acknowledgment

The authors acknowledge the financial support from Public Education Secretariat-Mexico (Spanish: SEP-Secretaría de Educación Pública) under grants SEP-DGEST 4328.11-P and PROMEP ITMOR-CA1 103.5/11/1091 to carry out this work.

6. References

Amézketa Lizarraga, E.; Aragüés Lafarga, R. & Gazol Lostao R. (2002). Desarrollo de un infiltrómetro automático y su aplicación en campo. *Investigación agraria. Producción y protección vegetales*, Vol. 17, pp. 131-142, ISSN 0213-5000

Angulo-Jaramillo, R.; Vandervaere, J. P.; Roulier, S.; Thony, J. L.; Gaudet, J. P. & Vauclin M. (2000). Field measurement of soil surface hydraulic properties by disc and ring infiltrometers: A review and recent developments, *Soil and Tillage Research*, Vol. 55, pp. 1-29, ISSN 0167-1987

Ankeny, M. D.; Kaspar, T. C. & Horton. R. (1988). Design for an Automated Tension Infiltrometer. *Soil Sci. Soc. Am., J.*, Vol. 52, pp. 893-896, ISSN 0361-5995

Ankeny, M. D. (1992). Methods and Theory for Unconfined Infiltration Measurements. In G. C. Topp, W. D. Reynolds, and R. E. Green, (Eds). Advances in Measurement of Soil Physical Properties: Bringing Theory into Practice. *Soil Science Society of America, Inc.* pp. 123-141

Bodhinayake, W.; Si, B. C. & Noborio, K. (2004). Determination of Hydraulic Properties in Sloping Landscapes from Tension and Double-Ring Infiltrometers, *Vadose Zone J.* Vol 3, pp. 964-970, ISSN 1539-1663

Bull A. D. (1949). Automatic Recording Infiltrometer, US Patent 2,540,096, 1949

Carlón-Allende, T. (2006). Regionalización hidrológica en la cuenca del lago de Cuitzeo. MSc thesis. Universidad Michoacana de San Nicolás de Hidalgo, Morelia, Michoacán, México.

Casey, F. X. M. & Derby, N. E. (2002). Improved design for an automated tension infiltrometer, *Soil Sci. Soc. Am. J.*, Vol. 66, pp. 64-67, ISSN 0361-5995

Chunye, L.; Greenwald, D. & Banin, A. (2003). Temperature Dependence of Infiltration Rate during Large Scale Water Recharge into Soils, *Soil Sci. Soc. Am. J.*, Vol. 67, pp. 487-493, ISSN 0361-5995

CONAGUA. Mexican National Water Comission (Comisión Nacional del Agua) (2008). *Programa Nacional Hídrico 2007-2012,* ISBN 968-817-836-5, México, D. F., Mexico.

CONAGUA. Mexican National Water Comission (Comisión Nacional del Agua) (2010). *Estadísticas Agrícolas de los Distritos de Riego. Año agrícola 2008-2009*, Available from http://www.conagua.gob.mx, Last date visited: February 1st, 2011.

Constanz, J., & Murphy. F. (1987). An automated technique for flow measurement from Mariotte reservoirs. *Soil Sci. Soc. Am. J.*, Vol. 51, pp. 252-254, ISSN 0361-5995

Das Gupta, S.; Mohanty, B. P. & Köhne, J. M. (2006). Soil Hydraulic Conductivities and their Spatial and Temporal Variations in a Vertisol, *Soil Sci. Soc. Am. J.*, Vol. 70, pp. 1872-1881, ISSN 0361-5995

Dudley, L. M.; Bialkowski, S.; Or, D. & Junkermeier, C. (2003). Low Frequency Impedance Behavior of Montmorillonite Suspensions: Polarization Mechanisms in the Low Frequency Domain, *Soil Sci. Soc. Am. J.*, Vol. 67, pp. 518–526. ISSN 0361-5995

Esteves, M.; Descroix, L.; Mathys, N. & Lapetite. J. M. (2005). Soil hydraulic properties in a marly gully catchment (Draix, France), *CATENA*, Vol. 63, pp. 282-298. ISSN 0341-8162

Eynard, A.; Schumacher, T. E.; Lindstrom, M. J. & Malo D. D. (2004). Porosity and Pore-Size Distribution in Cultivated Ustolls and Usterts, *Soil Sci. Soc. Am. J.*, Vol. 68, pp. 1927-1934, ISSN 0361-5995

Gimmi, T. & Ursino, N. (2004). Mapping Material Distribution in a heterogeneous Sand Tank by Image Analysis, *Soil Sci. Soc. Am. J.*, Vol. 68, pp. 508–1514, ISSN 0361-5995

Johnson, D. O.; Arriaga, F. J. & Lowery. B. (2005). Automation of a falling head permeameter for rapid determination of hydraulic conductivity of multiple samples, *Soil Sci. Soc. Am. J.*, Vol. 69, pp. 828-833. ISSN 0361-5995

Kluitenberg, G. J. & Warrick, A. W. (2001). Improved Evaluation Procedure For Heat-Pulse Soil Water Flux Density Method. *Soil Sci. Soc. Am. J.*, Vol. 65, pp. 320-323, ISSN 0361-5995.

Mbagwu, J. S. C. (1995). Saturated hydraulic conductivity in relation to physical properties of soils in the Nsukka Plains, southeastern Nigeria, *Geoderma*, Vol, 68, pp. 51-66, ISSN 0016-7061

Overman, A. R.; Peverly, J. H. & Miller. R. J. (1968). Hydraulic conductivity measurements with a pressure transducer, *Soil Sci. Soc. Am. Proc.* Vol. 32, pp. 884-886.

Park, E. J. & Smucker, A. J. M. (2005). Saturated Hydraulic Conductivity and Porosity within Macroaggregates Modified by Tillage, *Soil Sci. Soc. Am. J.*, Vol. 69, pp. 38–45, ISSN 0361-5995.

Pedro Vaz, C. M. & Hopmans, Jan W. (2001). Simultaneous Measurement of Soil Penetration Resistance and Water Content with a Combined Penetrometer–TDR Moisture Probe. *Soil Sci. Soc. Am. J.*, Vol. 65, pp. 4–12, ISSN 0361-5995

Prieksat, M. A.; Ankeny, M. D. & Kaspar, T. C. (1992). Design for an automated, selfregulating, single-ring infiltrometer. *Soil Sci. Soc. Am. J.*, Vol. 56, pp. 1409-1411. ISSN 0361-5995

Prunty, L. & Bell, J. (2005). Soil Temperature Change over Time during Infiltration, *Soil Sci. Soc. Am. J.*, Vol. 69, pp. 766–775, ISSN 0361-5995

Rachman, A.; Anderson, S. H.; Gantzer, C. J. & Alberts, E. E. (2004). Soil Hydraulic Properties Influenced by Stiff-Stemmed Grass Hedge Systems, *Soil Sci. Soc. Am. J.*, Vol. 68, pp. 1386–1393, ISSN 0361-5995.

Reynolds, W. D., y D. Elrick. 1991. Determination of Hydraulic Conductivity Using a Tension Infiltrometer. *Soil Sci. Soc. Am. J.*, Vol. 55, pp. 633-639. ISSN 0361-5995

Schwartz, R, C. & Evett, S. R. (2003). Conjunctive Use of Tension Infiltrometry and Time-Domain Reflectometry for Inverse Estimation of Soil Hydraulic Properties, *Vadose Zone J.*, November 1; Vol. 2, Issue 4, pp. 530 - 538.

Seobi, T.; Anderson, S. H.; Udawatta, R. P. & Gantzer, C. J. (2005). Influence of Grass and Agroforestry Buffer Strips on Soil Hydraulic Properties for an Albaqualf, *Soil Sci. Soc. Am. J.*, Vol. 69, pp. 893–901, ISSN 0361-5995

Seyfried, M. S. & Murdock, M. D. (2004). Measurement of Soil Water Content with a 50-MHz Soil Dielectric Sensor. *Soil Sci. Soc. Am. J.*, Vol. 68, pp. 394–403. ISSN 0361-5995

Simpson, M. J. (2006). Nuclear Magnetic Resonance Based Investigations of Contaminant Interactions with Soil Organic Matter. *Soil Sci. Soc. Am. J.*, Vol. 70, pp. 995–1004, ISSN 0361-5995

Soracco, G. (2003). *Relación entre la conductividad hidráulica saturada y la densidad aparente en tres situaciones de manejo contrastantes,* in D. Lobo Luján, D. Gabriels, G. Soto, "Evaluación de Parámetros y Procesos Hidrológicos en el Suelo", pp. 35-38, UNESCO, Paris, France.

Špongrová, K. 2006. Design of an automated tension infiltrometer for unsaturated hydraulic conductivity measurement. M.Sc. by Research. Cranfield University.

Wilson, M. A.; Brown, R. J. E.; Hoff, W. D. & Carter, M. A. (2000). A falling head permeameter for the measurement of the hydraulic conductivity of granular solids. Review of Scientific Instruments, Vol. 71, pp. 3942–3946, ISSN 0034-6748.

Wu, L. & Pan, L. (1997). A Generalized Solution to Infiltration from Single-Ring Infiltrometers by Scaling, *Soil Sci. Soc. Am. J.*, Vol. 61, pp. 1318-1322, ISSN 0361-5995

Wu, L.; Pan, L; Mitchell, J. & Sanden, B. (1999). Measuring Saturated Hydraulic Conductivity using a Generalized Solution for Single-Ring Infiltrometers, *Soil Sci. Soc. Am. J.*, Vol. 63, pp. 788-792, ISSN 0361-5995

16

Contribution of Seismic and Acoustic Methods to Reservoir Model Building

Jean Luc Mari and Frederick Delay
IFP School and University of Strasbourg
France

1. Introduction

The seismic reflection method has the advantage of providing a picture of the subsurface in three dimensions (3D) with a regular grid. In high resolution seismic surveys, the size of the grid cell is of the order of tens of meters for horizontal distances, and of several meters for vertical distances. The initial success of the seismic reflection method was solely linked to structural interpretation: it only recognized geometric shapes, regardless of the content. Stratigraphic interpretation permitted a deeper understanding of seismic data. To be validated, seismic data must be tied to wells. The tying and calibration of seismic data are carried out with the use of geophysical measurements in wells. Well measurements are seismic data such as vertical seismic profiles (VSP) with a resolution comparable to that of seismic reflection and log data with a vertical resolution ranging from tens of centimetres to several meters. Acoustic logs establish the link between seismic data (surface and wells) and other logs. The acoustic log has the vertical resolution of the other logs (electric, nuclear, etc) but obeys the same propagation laws as the seismic methods, although it operates in a different frequency bandwidth. The purpose of this chapter is to show the contribution of seismic and acoustic methods to reservoir model building. After a description of the geophysical methods (acoustic logging and reflection seismics), we show, with field examples, how the geophysical data have been recorded and processed to estimate porosity and hydraulic conductivity of the studied geological formations.

2. Geophysical methods

In this section, we will provide the necessary background for the understanding of the following discussion. More information of geophysical methods and tools can be found in Mari et al. (1999) , Glangeaud & Mari (2000), Robein (2003) and Chaouch & Mari (2006).

2.1 Acoustic logging

The transmission of an acoustic wave through geological formations is used for formation characterisation. The tools are of the monopole type or the dipole type. Monopole-type tools are the most commonly used. Sources and receivers are multidirectional. Sources generate in the fluid a compression wave which creates in the formation a compression wave (P wave) and a shear wave (S wave) at the refraction limit angles. In a vertical well, these tools are used to record five propagation modes: the refracted compression wave, the refracted

shear wave (only in fast formations), the fluid wave, two dispersive guided modes: the pseudo Rayleigh waves (only in fast formations) and the Stoneley waves.
Acoustic logging allows the measurement of the propagation velocities and frequencies of the different waves which are recorded by an acoustic tool. Velocities and frequencies are computed from the picked arrival times of the waves (P-wave, S-wave and Stoneley wave). For a clean formation, if the matrix and fluid velocities are known, an acoustic porosity log can be computed from the acoustic P-wave velocity log. The analysis of the acoustic waves recorded simultaneously on both receivers of the acoustic tool is used to compute additional logs, defined as acoustic attributes, useful for the characterization of the formation, such as amplitude, shape index, wavelength and attenuation logs. Attempts to predict permeability from acoustic data have been made by several researchers. The historical focus has been on predicting permeability from P-wave velocity (*Vp*) and attenuation. The attenuation of a formation is dependent both on the clay content and the rock (clean part of the formation). Laboratory experiments (Morlier & Sarda, 1971) have shown that the attenuation of a clean formation can be expressed in terms of three structural parameters: porosity, permeability and specific surface. Both theoretical and experimental studies have identified the relation between acoustic attenuation and petrophysical parameters:

$$\delta=(C.S/\varphi).\left(2\pi.k.f.\rho_f/\mu\right)^{1/3} \tag{1}$$

with δ: attenuation (dB/cm) , f: frequency (Hz), ρ_f: fluid density, φ: porosity , μ: fluid viscosity (centipoise), S: Specific surface (cm2/cm3), C: calibration coefficient, k: permeability (mD).
To compute permeability from eq 1, it is necessary to measure the P-wave attenuation and the porosity of the formation and calculate the effective specific surface of the formation. Ida L. Fabricius et al. (2007) have found that specific surface with respect to grain volume (Sg) is apparently independent from porosity. In an attempt to remove the porosity effect on Vp/Vs and mimic a reflected φ vs log (Sg) trend, they propose to use the following relationship between porosity φ, Vp/Vs (Vs: S-wave velocity) and Sg:

$$\log(Sg.m)=a.\varphi+b.(Vp/Vs)+c \quad \text{with} \quad Sg=S/(1-\varphi) \tag{2}$$

where it should be observed that Sg is multiplied by m to make Sg dimensionless. The a, b, c coefficients have to be computed to minimize the fluctuations of the specific surface Sg. In practice we try to predict permeability from P-wave attenuation, acoustic porosity, Vp to Vs ratio, and P-wave frequency. The log thus obtained is proportional to permeability and must be calibrated on core measurements or pumping tests.

2.2 Seismic reflection method

Today, most surveys are three-dimensional (3-D) designs where source and receiver points are distributed on a region of the earth's surface. This distribution typically forms a grid pattern, composed of source lines and receiver lines. When source and receiver lines are orthogonal, the 3-D survey design is called an orthogonal design. Design considerations for an orthogonal design include the spacing of the source and receiver lines, and the spacing of the source and receiver locations along the lines. The signal sent from any given source point is monitored by multiple receivers on different lines. The set of receivers that monitor a single source signal are often referred to as the template or active spread.

The classical approach to seismic processing can be summarized in two main steps. The first step includes pre-processing of the data and the application of static corrections. The purpose of pre-processing is to extract reflected waves from individual shots, by filtering out the parasitic events created by direct and refracted arrivals, surface waves, converted waves, multiples and noise. It is intended to improve resolution, compensate for amplitude losses related to propagation, and harmonize records by taking into account source efficiency variations and eventual disparities between receivers. Static corrections, that are specific to land seismic acquisition, are intended to compensate for the effects of the weathering zone (Wz) and topography. Static corrections can be computed by conventional methods such as the Plus-Minus method (Hagedoorn, 1959) or the refraction tomography method. The refraction tomography method, which is used to solve the travel time tomography, considers the velocity model provided by the Plus-Minus method to be a good solution for the global search and applies a simultaneous iterative reconstruction technique (SIRT) for the local search to produce an even more accurate result. The optimization scheme uses a misfit function defined as the least-square error between the computed and picked first break times. The velocity model is ray traced and updated until the misfit function value reaches a value previously defined by the modeler. The inversion technique (Mendes, 2009) gives the velocity distribution in the weathering zone (Wz) and is an effective method to compute the static corrections in 2D or 3D geometries. Records are then sorted in common mid-point gathers or constant offset sections.

If the data are sorted in common mid-point gathers, the second processing step is the conversion of the seismic data into a migrated seismic section after stack. This second step includes the determination of the velocity model, with the use of velocity analyses, the application of normal move-out (NMO) corrections, stacking and migration. If dip values are sufficiently large, velocity analyses provide time-depth relationships that are affected by dips. To overcome this inconvenience, a correction for dip effects or dip move-out (DMO) correction prior to determining velocity relationships must be applied to the data. If the data are sorted in constant offset sections, a pre-stack time (PSTM) or depth (PSDM) migration procedure which simultaneously performs dip correction, NMO correction, common mid-point stack and migration after stack, is applied. It is indispensable to have a good velocity model to carry out the migration process. The role of migration is to place events in their proper location and increase lateral resolution, in particular by collapsing diffraction hyperbolas at their apex. Proper migration requires the definition of a coherent velocity field, which must be a field of actual geological velocities in migrated positions. Several methods can be used to generate velocity models such as tomography or stereo-tomography methods. The migrated section can then be depth converted and transformed in pseudo velocity or acoustic impedance sections if well data (such as acoustic logs) are available. The procedures used to obtain acoustic impedance sections are often referred to as Model-based seismic inversions which require an a priori impedance model (obtained from well data) which is iteratively refined so as to give a synthetic seismic section to match the seismic section to be inverted. The final impedance model can be converted in porosity by using an empirical relationship between porosity and acoustic impedance established at well locations.

3. First field example: Acoustic logging and 3D seismic surveying in a carbonate formation

The 3D seismic data were recorded in France at the boundary of the Meuse and Haute-Marne departments in the vicinity of the Andra Center (National radioactive waste management Agency). The acoustic data were recorded in well "EST 431" located on the Ribeaucourt township, in the national forest of Montiers-sur-Saulx, 8 km North-North-West of the Andra Center.

3.1 Geological context

One of the drilling platforms, located in the center of the studied zone, was used to study formations ranging from the Oxfordian to the Trias. The analysis presented here concerns borehole "EST 431" and covers the Oxfordian formation. The objective of this borehole is to complement the geological and hydrogeological knowledge of this formation. This formation consists essentially of limestone deposited in a vast sedimentary platform (Ferry et al., 2007). The limestone facies, which vary from one borehole to another, are generally bio-detritic with reef constructions. In this formation, porosity ranges between 5 and 20% and "porous horizons" of a kilometric extension have been identified. As far as hydrogeology is concerned the observed water inflows are usually located in high porosity zones (Delay et al., 2007). The base of the Kimmeridgian shales was observed at -258.3 m (100 m ASL) and the base of the Oxfordian limestones at - 544.3 m (-186 m ASL). During the drilling, water inflows were detected at - 368m and - 440 m. At the end of the drilling, the well was left in its natural water.

3.2 Acoustic logging

Figure 1 shows the 3 m constant offset section in the 333 - 510 m depth interval, opposite the geological description. On the acoustic section, the refracted P-waves appear in the 0.6 - 1.2 ms time interval, the converted refracted shear waves in the 1.2 - 2 ms time interval, and the Stoneley wave in the 2 - 2.4 ms time interval. On the acoustic section, we can differentiate: an event at 345 m showing a very strong attenuation of all the waves; an interval showing a very strong slowing down of the P and S waves (360 - 375 m); a relatively homogeneous mid-level interval (375 - 397 m); a level which stands out because of its strong variations in P, S and Stoneley velocities (397 - 462 m); a very homogeneous zone below 495 m with easily identifiable P and S waves, and an image of alteration between 501 and 507 m.

Figure 2 is a comparative display of acoustic logs (P-wave velocity, P-wave frequency, acoustic porosity), borehole diameter, gamma ray log and NMR porosity. A strong correlation between acoustic porosity and NMR porosity can be noted (correlation coefficient: 0.86).

The correlation coefficient between acoustic porosity and borehole diameter is high (0.84).The attenuation of the formation is computed from the amplitudes of the refracted P-wave acoustic signal in selected time windows. The energy ratio (between the two adjacent receivers of the acoustic tool) gives the attenuation (expressed in dB/m) of the formation. Two attenuation logs have been computed with two selected time windows. The first window is a short window centered on the first arch of the acoustic signal. The second is a large window centred on the first three arches of the acoustic signal. The correlation coefficient between the two attenuation logs is high (0.78). This allows the computation of an average attenuation log with its associated standard deviation which gives an indication of the uncertainty. The event at 345 m shows a strong attenuation, a high porosity value, a significant increase of the borehole diameter and a high gamma ray value. It has been interpreted as a shaly carbonate layer. P-wave attenuation and permeability are both strongly dependent on clay content. The gamma ray log was used to compute both shaliness and the shaliness corrected attenuation log in order to evaluate the formation permeability. It was also used to compute an effective porosity log (Figure 2, bottom right). Figure 3 (left) shows the attenuation logs with their associated standard deviation before (a) and after (b) shaliness correction.

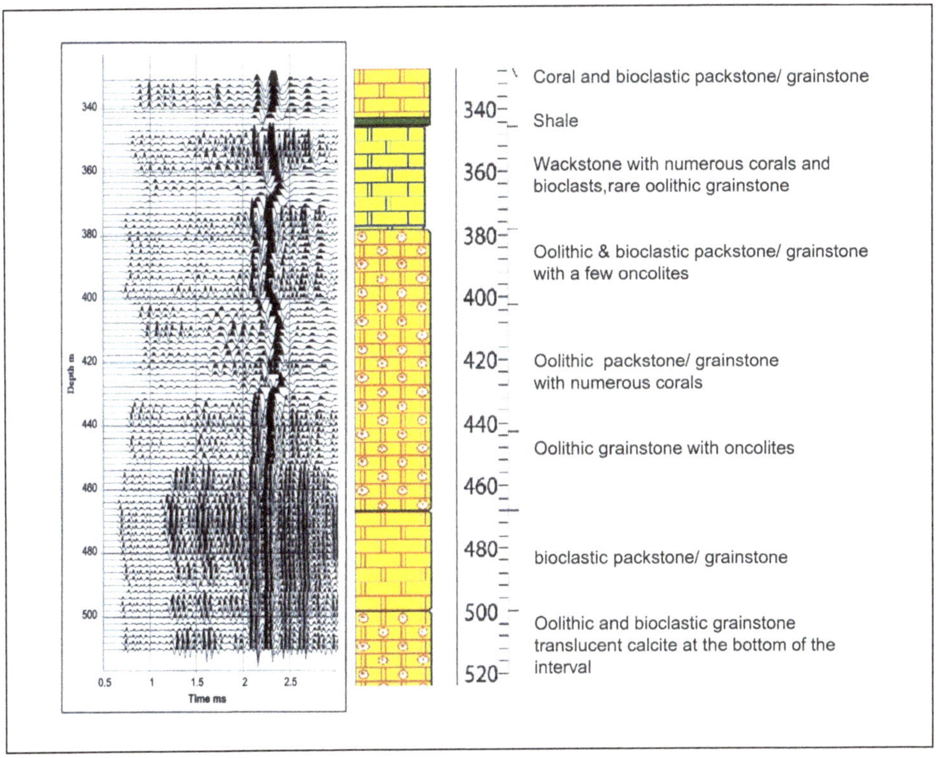

Fig. 1. Acoustic logging and lithological description (Courtesy of Andra)

The a, b, c coefficients of equation 2, computed in the 370 - 500 m depth interval (a = 0.02, b = 0.012 and c = 6.25), were used to calculate the specific surface Sg with respect to grain volume.

Figure 3 (right) shows the Sg specific surface log and the acoustic permeability log calculated from eq. 1. The fluid viscosity μ and density ρ_f have been assumed to be constant (μ = 1 centipoise, ρ_f = 1 g/cm^3). The acoustic permeability log detects three permeable zones at 368 m, between 400 and 440 m, and 506 m. The permeable zone located at 506 m corresponds to a high value of conductivity and is characterized by a low porosity (6 %), a 10 dB/m attenuation, but a significant decrease of the P-wave frequency and of the specific surface.

During the drilling, water inflows were detected at - 368 m and - 440 m. At the end of the drilling, the well was left in its natural water. The hydraulic tests and conductivity measurements conducted later on did not confirm the inflow at 368 m seen during the drilling, but they validated the 400 - 440 m and 506 m permeable zones detected by the acoustic logging. The absence of active hydraulic fracture in this borehole was confirmed by hydrogeological tests and acoustic logs.

The 3D seismic survey was performed to predict the porosity of a carbonate formation and to evaluate the hydraulic conductivity between the porous bodies of the aquifer.

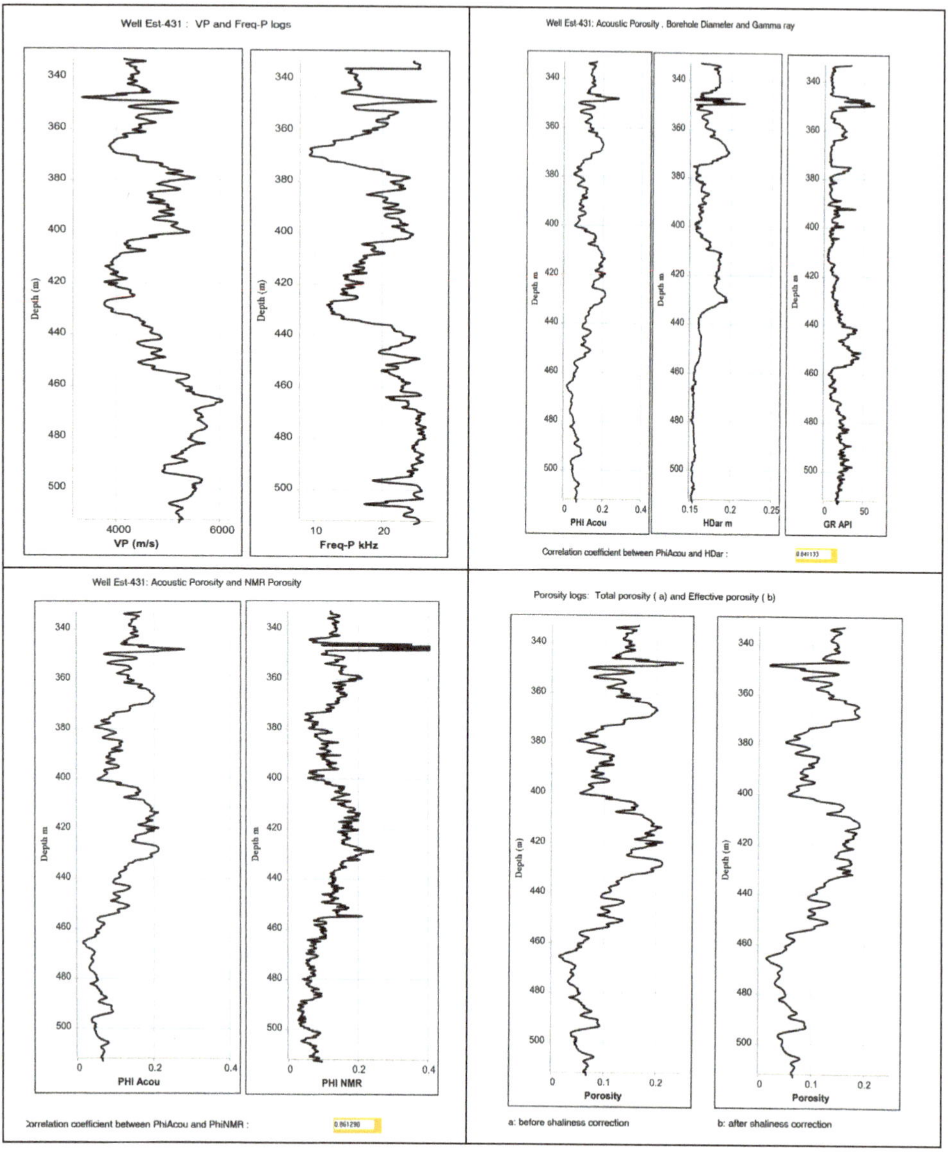

Fig. 2. Acoustic logs (P-wave velocity, P-wave frequency, acoustic porosity), borehole diameter, and gamma ray log . Comparison between acoustic porosity and NMR porosity. Total porosity and effective porosity.

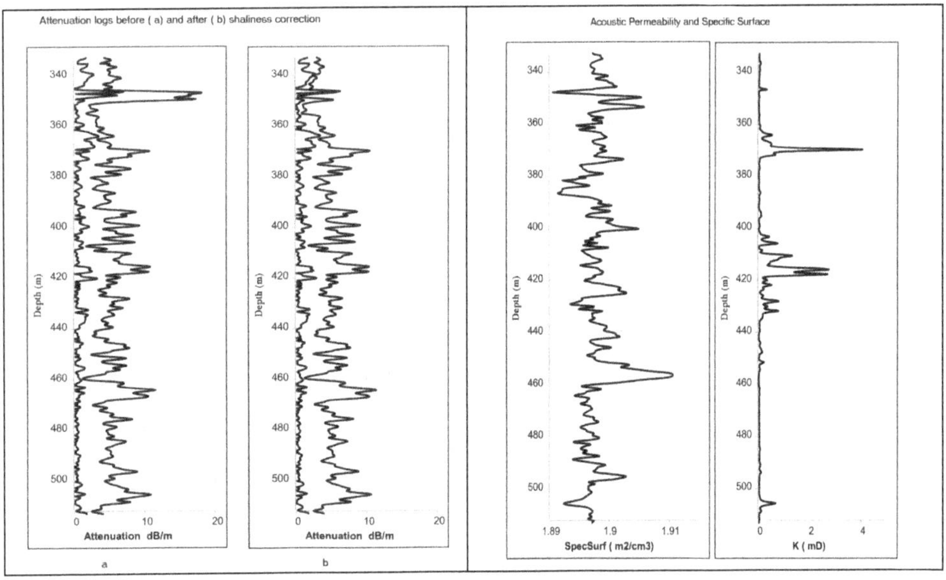

Fig. 3. Permeability estimation from acoustic logs (Specific surface, porosity, P-wave frequency). Left: Attenuation logs and their associated Std before (a) and after (b) shaliness correction, Right: Specific surface log and Predicted Permeability from acoustic logs.

3.3 Seismic acquisition and processing

The 3-D survey design is an orthogonal design. The active spread is composed of 8 receiver lines with 74 receivers per line. The distance between 2 adjacent receiver lines is 100 m. The distance between 2 receivers is 25 m. A single source line is fired per active spread. The distance between two source points is 25 m. The source is a vibroseis source generating a signal in the 14-140 Hz frequency bandwidth. The distance between 2 adjacent source lines is 100 m. The cell or bin size is 12.5 m x 12.5 m. The nominal fold is 37. The size of the area covered by the 3D is 4 km^2.

A conventional seismic sequence was applied to the data set. It includes amplitude recovery, deconvolution and wave separation, static corrections, velocity analysis, CMP-stacking and time migration. After migration, a model-based stratigraphic inversion (a priori impedance model obtained from well data) provides a 3D impedance model cube. Figure 4 is an example of an impedance section extracted from the 3D block. It also displays the porous layers (Hp1 to Hp4) associated with the Oxfordian carbonate.

3.4 From 3D impedance to 3D porosity

At well locations, the acoustic impedance log was calculated from the density and velocity logs. The porosity vs. impedance cross plot, displayed in figure 5 (top), was used to define a linear law between the two. The porosity is expressed in % and the acoustic impedance in (m/s).(g/cm^3). The cross plot was obtained by using density, acoustic velocity, and porosity (NMR) logs recorded in 6 wells. This empirical relationship between porosity and acoustic impedance was used to convert the 3D impedance into porosity within the Oxfordian layers (Figure 5, bottom). The porous layers are clearly visible.

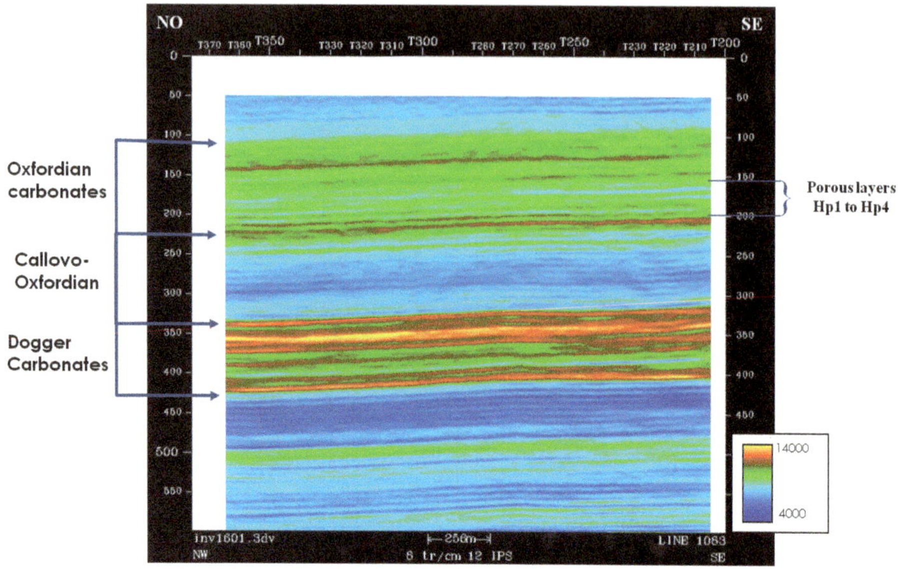

Fig. 4. Acoustic impedance section (Courtesy of Andra).

The 3D cube makes it possible to provide 3D imaging of the connectivity of the porous bodies. Core analysis was carried out to establish porosity vs. permeability laws and it demonstrated that the permeable bodies have porosity larger than 18 %. Consequently, a porosity cut-off of 18% was used to show the connectivity of the porous bodies (figure 6). A porosity cut-off of 21% was also applied to extract the porous bodies having the best hydraulic conductivity within the porous layers.

4. Second field example: Case study of a near surface heterogeneous aquifer

Many underground aquifers were developed as experimental sites during the past decade. These sites are designed for in-situ measurements and calibration of flow, transport and/or reactions in underground reservoirs that are heterogeneous by nature. The University of Poitiers (France) had a Hydrogeological Experimental Site (HES) built near the Campus for the sole purpose of providing facilities to develop long-term monitoring and experiments for a better understanding of flow and transfers in fractured rocks (Bernard et al., 2006; Kaczmaryk & Delay, 2007a,b ; Bourbiaux et al., 2007).

4.1 Geological context

The aquifer studied, 20 to 130 m in depth, consists of tight karstic carbonates of Middle Jurassic age. It lies on the borderline, named the "Poitou threshold", between the Paris and the Aquitaine sedimentary basins (Figure 7). The Hydrogeological Experimental Site (HES) covers an area of 12 hectares over which 35 wells were drilled to a depth of 120 m (Figure 7).

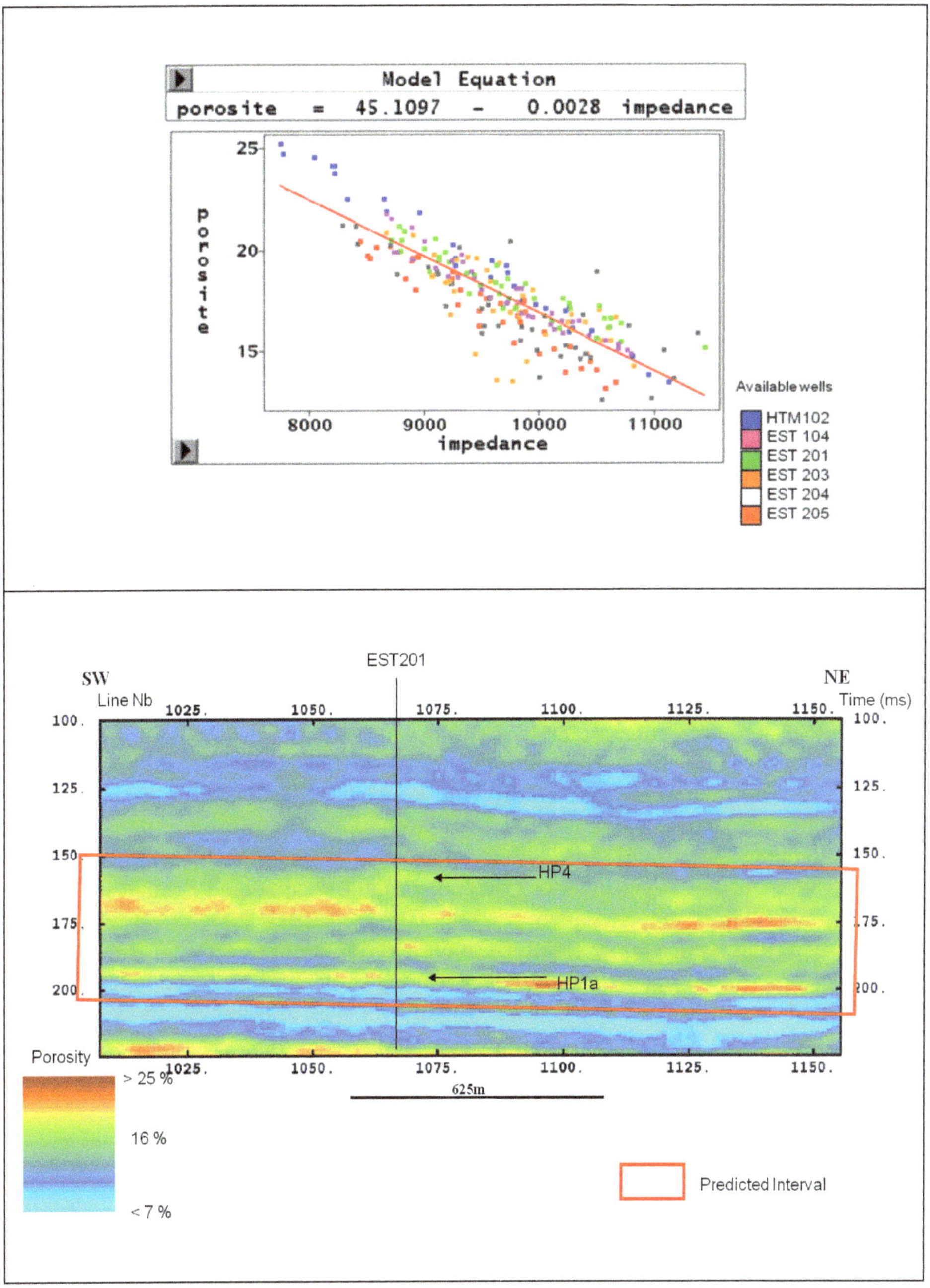

Fig. 5. Porosity vs. Acoustic Impedance law and Porosity section (Courtesy of Andra).

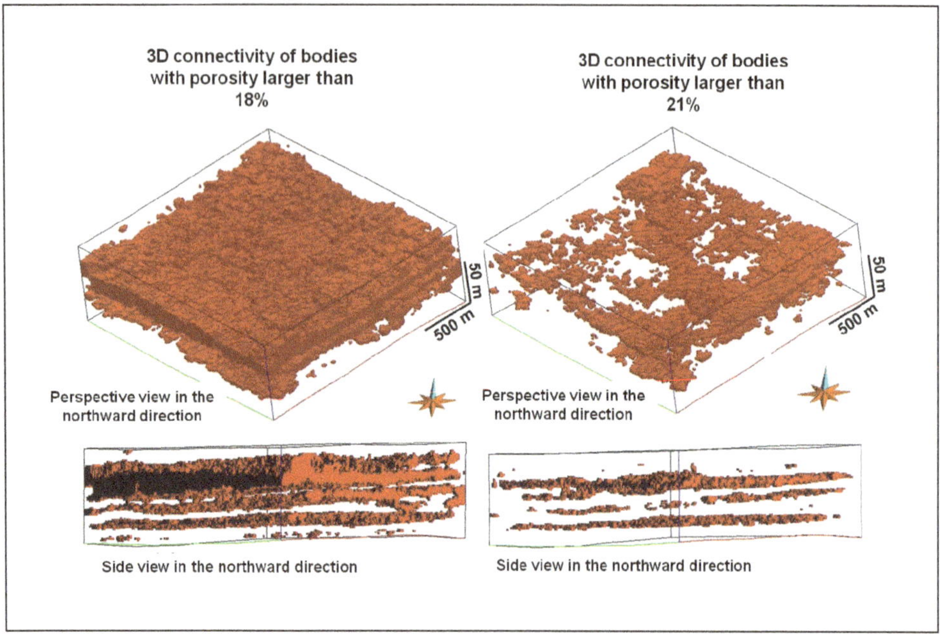

Fig. 6. 3D porosity blocks. 3D connectivity of bodies with porosity larger than 18 % (on the left) and larger than 21 % (on the right). (Courtesy of Andra).

Hydrogeological investigations show that maximum pumping rates vary from well to well and range from 5 to 150 m^3/h. The top of the reservoir was initially flat and horizontal, 150 millions years ago, but has been eroded and weathered since, during Cretaceous and Tertiary ages. It is shaped today as hollows and bumps with a magnitude reaching up to 20 m. The building phase started in 2002 and up to now, 35 wells have been bored over the whole thickness of the reservoir. Most wells dispose of documented drilling records and logs of various nature among which are gamma-ray, temperature, and acoustic logs. In addition, two wells were entirely cored. To sum up, the aquifer responds fairly evenly to the hydraulic stress of a pumped well, with pressure depletion curves merged together in time and amplitude whatever the distance from the pumped well. This merging is assumed to be the consequence of a local karstic flow in open conduits that have been unclogged by the drilling and pumping works. The presence of pervasive karstic drains is supported by recent logs in the wells using optic imaging. Almost all the wells have shown caves and conduits cut by the walls of the boreholes with sometimes mean apertures of 0.2 - 0.5 m. These conduits are mostly enclosed in three thin horizontal layers at a depth of 35, 88 and 110 m. Of course, these layers are intercepted by vertical wells and this potentially results in a good connection between wells and karstic drains. This connection is mainly controlled by the degree to which drains are re-opened in the vicinity of the well. In the end, it was considered crucial to better image the geometry of the reservoir with a resolution compatible with, on the one hand, the scale of a well and, on the other hand, the scale of the entire experimental disposal. High resolution geophysical tools seem well designed to undertake that kind of investigation.

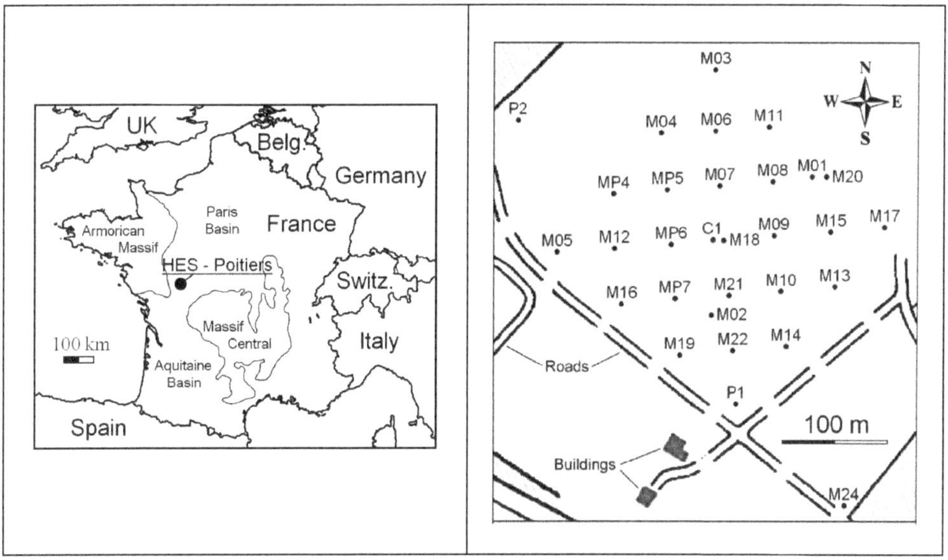

Fig. 7. Hydrogeological Experimental Site in Poitiers: site map and well locations.

4.2 3-D seismic acquisition and processing

Due to the limitations of the area, the length of the seismic line could not exceed 250 m in the in-line direction. In the cross-line direction, the extension of the area does not exceed 300 m. As a result, 20 receiver lines were implemented, with a 15 m distance between adjacent lines.

Figure 8 shows the map locating the seismic lines and the wells. In the acquisition of data a 48 channel recorder was used. An explosive source (25 g) was detonated and a single geophone (10 Hz) per trace was deployed. Such a source makes it easy to identify and pick first arrivals. A 5 m distance between two adjacent geophones was selected to avoid any spatial aliasing. A direct shot and a reverse shot were recorded per receiver line. Figure 8 shows an example of an in-line direct and reverse shot gather. Three shot points in the cross-line direction were fired at distances of 40, 50 and 60 m from the receiver line under consideration. Figure 8 shows an example of a cross-line shot gather. The range of offsets was selected to optimize the quality of the seismic image over the reservoir depth interval, between 40 and 130 m. The minimum offset distance was chosen equal to 40 m to reduce the influence of the surface waves. The time sampling interval is 0.25 ms and the recording length is 0.5 s.

4.2.1 Seismic refraction survey

To obtain the velocity of the refractor (top of the reservoir) and its depth, the Plus-Minus method was used. This method requires recordings where geophones are aligned with shot points. The arrival times of the direct and refracted waves were picked on all the in-line shots. The procedure was applied on each line independently. In order to obtain a map with a homogeneous sampling interval both in cross-line and in-line directions, the delay time curves were interpolated by kriging (Chilès & Delfiner, 1999) on a grid 2.5 m x 5m.

To perform the depth conversion, the velocity of the medium (Wz) situated above the refractor must be known. Here, it is given by the slope of the direct wave. In the area, the velocity V2 of the refractor was found to be 3350 m/s and the velocity of the Wz to be 850 m/s. On the Wz depth map (Figure 8), the arrow indicates the direction N 90° which corresponds to the main orientation of fracture corridors.

In order to add information in the inversion procedure, we used simultaneously in-line and cross-line cross shots with an offset of 60 m. The shots were selected to be sure to have the refracted wave as first arrival wave whatever the source receiver distance. The picked times of the first seismic arrivals on all the shots (in-line and cross-lines shots), the Wz depth map and the velocity model obtained by the Plus–Minus method are input data for the inversion procedure. The 3D *a priori* model is a homogeneous two layer model of V1 = 850 m/s and V2 = 3 350 m/s with the interface given by the Wz depth map.

For the inversion, we worked with 56 shots simultaneously: 40 shots acquired along the 2D-lines plus 16 shots acquired with a cross-offset equal to 60 m. A 400 Hz frequency was considered to compute the Fresnel volume. This value was estimated after tests with several frequencies to study the sensitivity of model velocity to the smoothing nature of the Fresnel volume method.

As we consider the 3D *a priori* model a valuable approach for the background velocity, we used it for the inversion SIRT procedure. Figure 9 shows the result of this process which was stopped when the *rms* error reached a value around σ = 0.87 ms, with an improvement of about 80% during the optimization procedure. Figure 9 shows the velocity distribution at 20 m in depth and the 2500 m/s iso-velocity depth map. We can notice a strong correlation between the 2500 m/s iso-velocity depth map (Figure 9, left) and the Wz depth map (Figure 8, bottom left). The correlation coefficient reaches 0.96. Figure 10 shows a 3D block with velocity sections located at a 0 m, 60 m and 180 m distance in the cross-line direction and velocity map located at 20 m in depth.

The velocity distribution versus depth can be represented by a two-layer model. The interface corresponds to the top of the karstic reservoir. Above the interface, the velocities are low (Vp ranging from about 850 m/s to about 2500 m/s) with no significant lateral variation but with a strong increase in depth associated with the 2500 m/s iso-velocity in the vicinity of the interface.

The inversion results obtained with 3D data emphasize the geological structures mentioned previously in Mari *et al.* (2009) allowing a better recognition of their alignments and shape (corridor of fractures). Furthermore, no cavities were detected near the surface.

The velocity model obtained by inversion of the first arrival picked times was used to extend the 3D velocity block obtained by the reflection survey (Mari *et al.* 2009) to the 0–35 m depth interval, emphasizing the necessity of employing 3D acquisitions for near surface studies.

4.2.2 Seismic reflection surveying

The processing sequence has been described in detail by Mari & Porel (2007). Each shot point (both in the cross-line direction and in the in-line direction) was processed independently to obtain a single-fold section with a sampling interval of 2.5 m (half the distance between 2 adjacent geophones) in the in-line direction and with a sampling interval of 5 m in the cross-line direction.

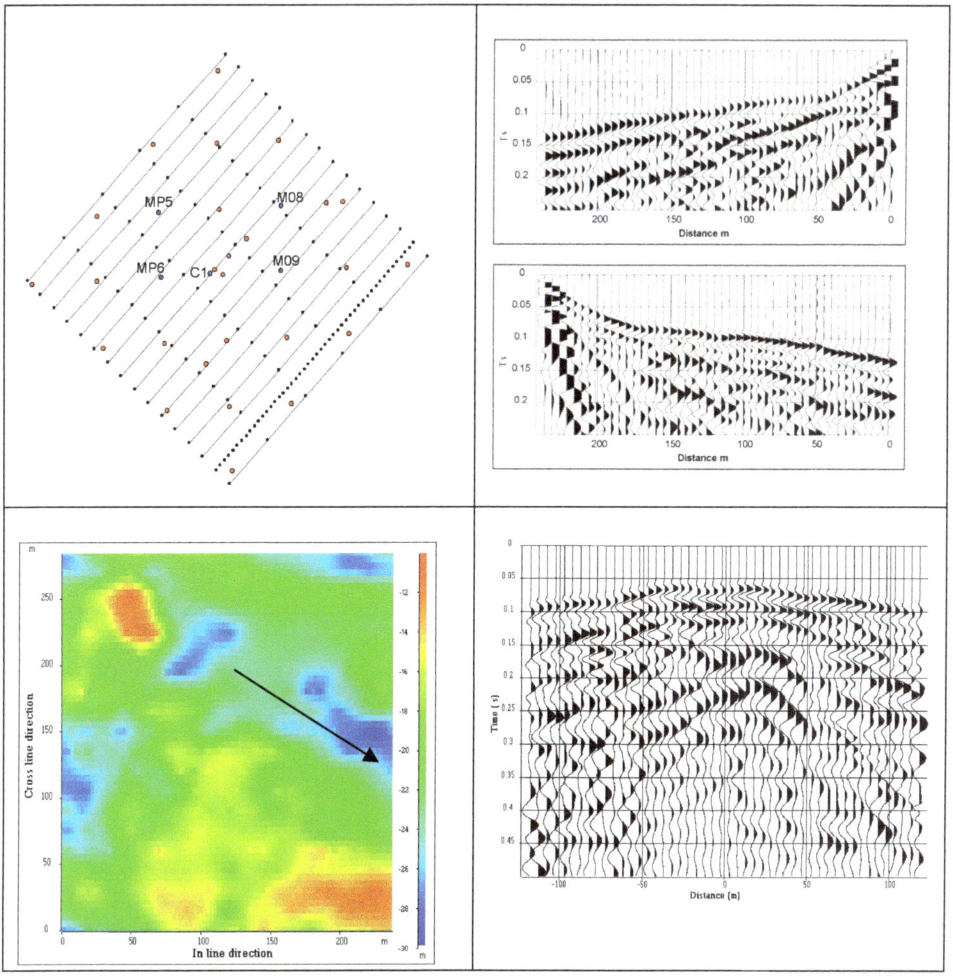

top left: Seismic line implementation and well location (red points),
top right: Example of direct and reverse in-line shot points,
bottom left: Wz depth map,
bottom right: example of cross-line shot point.

Fig. 8. Seismic acquisition: examples of shot points and Wz-depth map obtained by seismic refraction surveying.

For seismic imaging based on reflected waves, it is necessary to be able to separate weak reflected events from high-energy surface waves such as pseudo - Rayleigh waves. Wave separation is a crucial step in the processing sequence. This specific field case illustrates the benefit of combining two different wave-separation methods in order to remove all the energetic wave-field. The conventional F-K method was used to filter surface waves and converted refracted waves. The SVD method (Singular Value Decomposition) was then used to extract refracted waves.

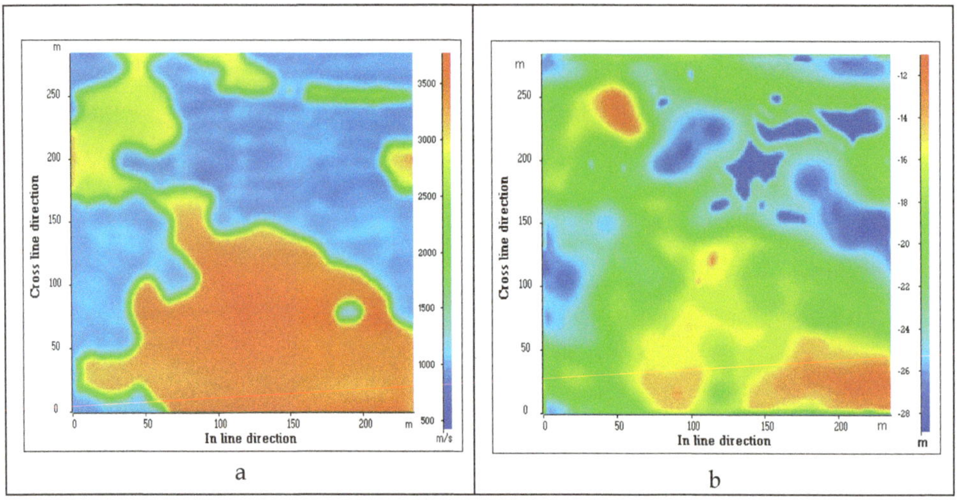

a: Velocity distribution at 20 m in depth, b: 2500 m/s iso-velocity depth map

Fig. 9. Results of 3D tomographic inversion .

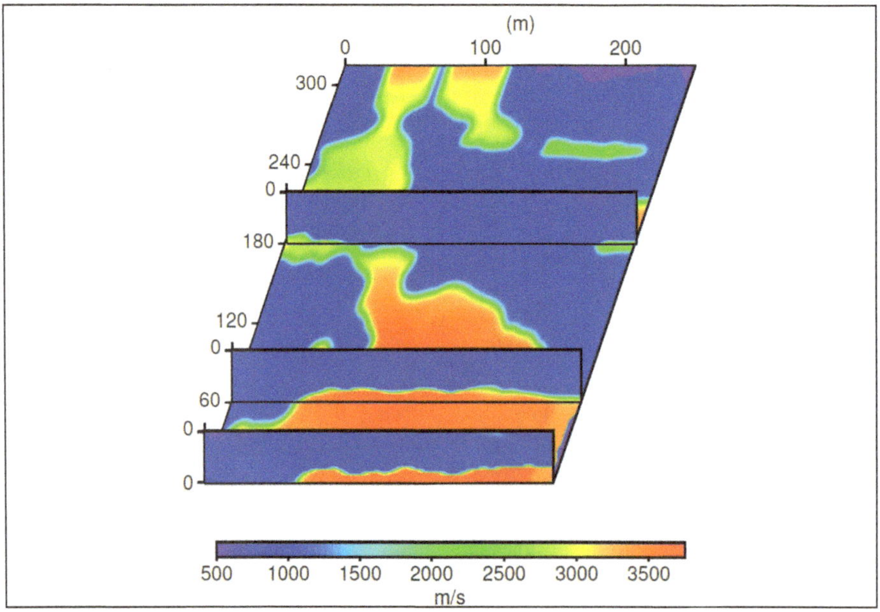

3D block with velocity sections located at a 0 m, 60 m and 180 m distance in the cross-line direction and velocity map located at 20 m in depth.

Fig. 10. Result s of 3D tomographic inversion.

After amplitude recovery, deconvolution and wave separation, normal move-out corrections were applied to the residual sections in order to obtain single-fold zero offset sections at normal incidence. In well C1 situated in the central part of the site, a Vertical Seismic Profile (VSP) was recorded. VSP data were processed to obtain a time versus depth relationship and a velocity model. The velocity model was used to apply the normal move - out corrections. The VSP time versus depth law was also used to convert the time sections into depth sections with a 0.5 m depth sampling interval. The single-fold depth sections were merged to create the 3D block. The width of the block in the in-line direction equals 240 m and 300 m in the cross-line direction. In the in-line direction, the abscissa zero indicates the location of the source line. The abscissa of the reflecting points varies between - 120 m and 120 m in the in-line direction, the distance between two reflecting points equals 2.5 m. In the cross-line direction, the distance between two reflecting points equals 5 m. The depth sections were deconvolved in order to increase the vertical resolution. They were then integrated with respect to depth to transform an amplitude block into a 3D pseudo velocity block in depth, using velocity functions (acoustic logs recorded at wells C1, MP5, MP6, M08, M09) as constraints. The pseudo velocity sections of the 3D block thus obtained were merged with those obtained by refraction tomography (Figure 10) to create a 3D extended velocity model from the surface. Figure 11 shows the in-line pseudo velocity sections extracted from the 3D extended velocity model and situated at cross-line distances of 0, 60 and 180 m. It also shows the velocity map at 87 m in depth. The 3D velocity model shows the large heterogeneity of the aquifer reservoir in the horizontal and vertical planes.

4.3 3D porous model building

In the area covered by the 3D seismic surveying, 23 wells (location shown in Figure 7) have been drilled. The wells are regularly spaced (~ 50 m) and are used to perform many hydraulic tests (interference pumping and slugs). In the wells, several logs have been recorded (electrical and gamma ray logs).

Due to the homogeneous spatial distribution of wells, in which resistivity logs have been recorded, we chose to use a method based on electrical measurements to quantify the 3D porosity distribution within that aquifer. The seismic interval velocity-to-porosity conversion was performed in two steps (Mari et al., 2009):

- first step: from 3D interval seismic velocity to 3D resistivity
- second step: from 3D resistivity to 3D porosity.

4.3.1 From 3D interval seismic velocity to 3D resistivity

Faust (1953) established an empirical relationship between seismic velocity V, depth Z, and electrical resistivity measurements Rt. For a formation of a given lithology, the velocity V can be written as a function of the depth Z and resistivity Rt as follows:

$$V = C.(Z.Rt)^{1/b} \tag{3}$$

with V the P-wave velocity of the formation in m/s, Z the depth in m, Rt the electrical resistivity in ohm.m, C and b the coefficients associated with the Faust's equation.

At each well where a long normal log was recorded, an interval velocity log was extracted from the 3D seismic interval velocity block. The two sets of data (resistivity and seismic velocity) were combined to calibrate an empirical Faust's law, which was then used as a local constraining function to transform the 3D pseudo-velocity block into a 3D pseudo-resistivity. For each well, the two coefficients, C (constant coefficient) and b (power law

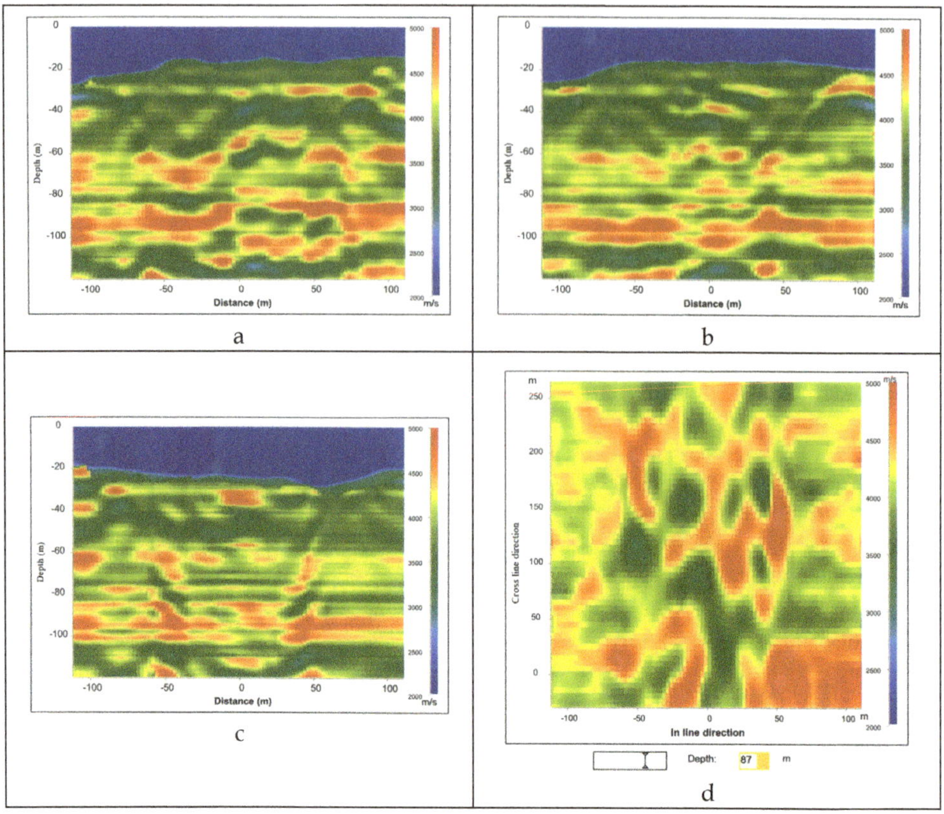

3D block with velocity sections located at a 0 m (a), 60 m (b) and 180 m (c) distance in the cross-line direction and velocity map located at 87 m in depth (d).

Fig. 11. Results of 3D seismic reflection processing.

exponent), of that empirical law were determined through a least-square minimization of the difference between the 3D-block-extracted seismic velocities and the velocities predicted from Faust's law using the long normal resistivity log data as input. 2D distribution maps of the C and b values over the site could then be built from the calibrated values in each of the 22 wells. These maps were used for the velocity - resistivity conversion of the 3D seismic block. Figure 12 shows the results obtained at well MP6 and the resistivity map at 87 m in depth.

4.3.2 From 3D resistivity to 3D porosity

Archie (1942) showed empirically that for water-saturated permeable formations, the relation between the true formation resistivity, Rt, and the resistivity, Rw, of the water impregnating the formation was given by :

$$Rt/Rw = F = \phi^{-m} \tag{4}$$

where F is the " resistivity formation factor ". ϕ is proportional to the formation porosity and m is a " cementation factor", that is a formation characteristic. The F value derived from the resistivity measurement, *Rt*, is unaffected by the mineralogical constituents of the formation matrix. Although the "cementation factor" value may vary between 1.3 and 3 according to the formation lithology, an approximate value of 2 is generally adopted for a well-cemented sedimentary log. Although applicability of Archie's law may be argued *a priori* for a karstic reservoir, two reasons motivated its choice. Firstly, the reservoir remains essentially a sedimentary carbonate formation at the seismic resolution scale. Actually, the size of the seismic bin (2.5 m in the in-line direction, 5 m in the cross-line direction), and the seismic vertical resolution ranging between 1 and 2 m, lead to an elementary seismic cell volume of 12 m^3 at least. Secondly, the volume of the karstic bodies represents only a small percentage (2 to 3 %) of the reservoir volume. This volume was estimated by analyzing borehole images (Audouin & Bodin, 2007).

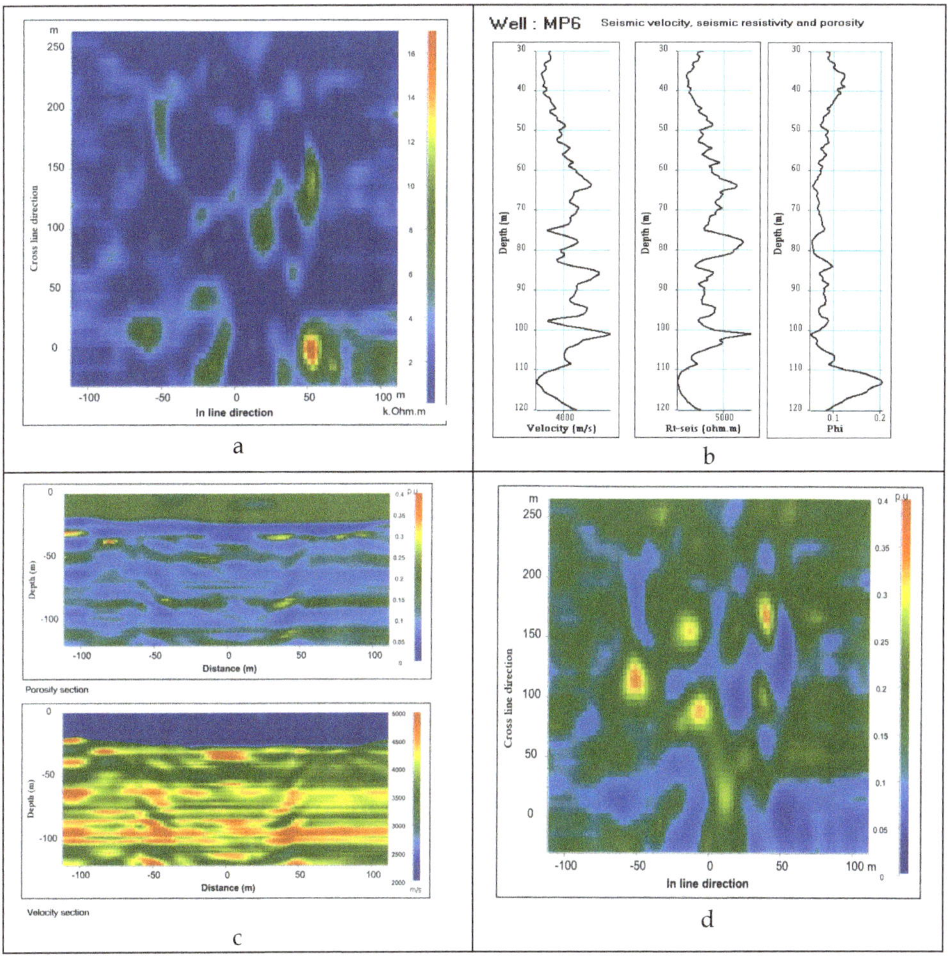

Resistivity and porosity maps located at 87 m in depth (a, d). Velocity, resistivity and porosity logs at well MP6 (b). Velocity and porosity sections located at a 180 m distance in the cross-line direction (c).

Fig. 12. From seismic velocities to porosity.

For the above two reasons, the previous seismic-derived 3D resistivity block (*Rt-seis*) was converted into a 3D pseudo-porosity block, by using the Archie-law given by eq. 5 with m= 2 and with the resistivity of the formation water, Rw, estimated at 20 ohm.m. Figure 12 shows the porosity log computed from the seismic-derived resistivity log, *Rt-seis*, at well MP6.

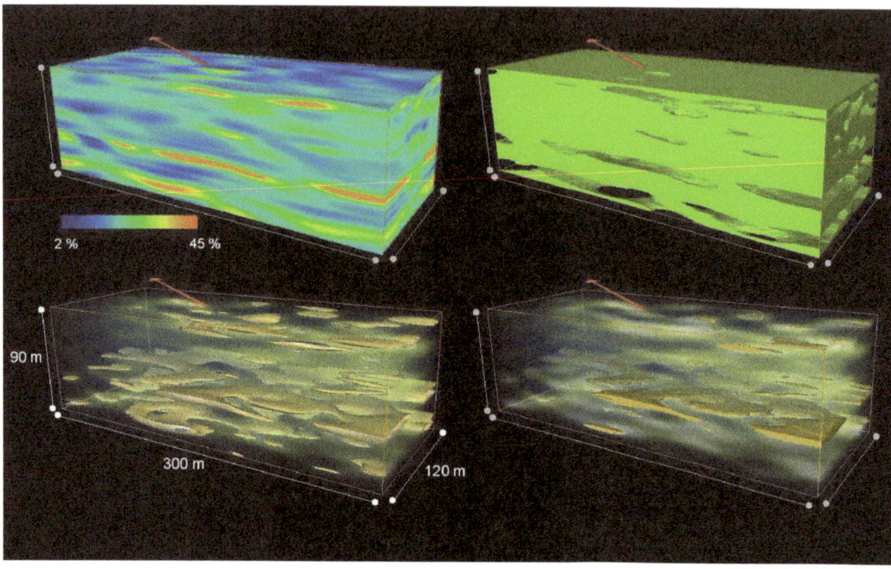

Fig. 13. 3D seismic pseudo porosity blocks.

Figure 12 also shows the in-line pseudo velocity and pseudo porosity sections (extracted from the 3D velocity model and from the 3D porosity block) situated at a cross-line distance of 180 m and the porosity map at 87 m in depth. The porosity section of Figure 12 clearly shows high-porosity layers located in the intervals of 85 - 87 m and 110 - 115 m depth. The production profiles measured in many wells revealed that these layers actually correspond to the major water feeding levels of wellbores. In order to further analyze the spatial distribution of porous bodies and of presumably-conductive flow paths, different cut-off values were applied to the 3D seismic porosity block. Figure 13 shows several 3D seismic pseudo-porosity blocks, associated with porosity cut-off values of 10 and 30%.The blocks were extracted from the central part of the area. They have a length of 120 m in the in-line direction and a length of 300 m in the cross-line direction. The extracted 3D reservoir volume with a porosity smaller than 10% (Figure 13, top right) actually represents the largest fraction of that aquifer, *i.e.* tight carbonates with a low permeability (less than a millidarcy).

This is consistent with the observation of very sparse and channeled flow paths within that aquifer. We may indeed assume that, within that extracted low-porosity volume, the density of conductive (karstic) bodies is too low to ensure a hydraulic communication between wells, because the velocity-to-porosity converted block is derived from a high-resolution 3D

seismic block, defined at a metric to pluri-metric scale that is significantly less than the well spacing.
The bodies with porosity higher than 10 % (Figure 13, bottom left) are mostly distributed within 3 layers, located at 30-35 m, 85-87 m and 110-115 m depth. Finally, a cut-off value of 30% was applied to the seismic porosity block to show up the most porous bodies of that aquifer. These highly-porous bodies (Figure 13, bottom right), mainly located in the intermediate porous layer situated in the 85 - 87 m depth interval, represent only 2 % of the total volume of aquifer investigated in this study.

4.3.3 3D porosity and hydraulic conductivity

The hydraulic slug tests, performed at the HES, show a very rapid propagation of the pressure wave over large distances, say 100 m on average. These observations allowed us to map a diffusivity distribution and the importance of connections between wells (Figure 14). Preferential connections are visible along the N90 direction (wells M13-M21-M22-M19 and wells M04-M06-M11). Incidentally, logging data show three water producing layers (35, 87, 115 m depth, respectively). Their presence is not systematic however at each well of the site. The 115 m-depth layer is located on the SW and NW borders of the site whereas the 87 m-layer is present everywhere. The variations of the seismic pseudo-porosity at different depths can be compared directly with hydrogeological data. The porosity map (Figure 14) drawn for the 87 m horizon confirms both hydrogeological data (flowmeters) and optical imaging obtained from the majority of wells. Actually, there is a significant connection between M13 and M21 which is also visible in the form of a high seismic porosity zone. It is therefore very likely that high porosity zones of geophysical maps do correspond to water productive areas. Interference testing gives information about hydrodynamic behaviour at the site scale but does not image the local hydraulic connection between wells, even though pressure transients may differ from one observed well to the other (Bernard et al., 2006; Kaczmaryk & Delay, 2007a).
It can be now wondered on the interest of a 3D porosity block for modelling flow in the aquifer. As stated above, the insurgence of karstic features is visible both onto the block and on responses to hydraulic stresses showing very rapid propagation of pressure wave depletions between distant wells. The question raised by these observations is how to model the static geometry of the porosity distribution and the dynamic hydraulic responses within a flow simulator that would keep inversion feasible. Stated otherwise, the point is to assign hydraulic parameters to the block by matching up numerical simulations onto flow data generally taken from hydraulic tests. In view of the spatial structure depicted by the porosity block, a continuous approach to flow seems well suited to fit a block of hydraulic conductivity (and specific storage capacity) that would satisfy the continuity of spatial porosity distribution drawn from seismic data. This makes us overlook discrete fracture networks and other object-based representations of fractured karstic aquifers, even though these interpretative depictions could probably be applied to the present case study.
Most of the hydraulic tests performed at the HES were based on the notion of interference between wells. It is recalled that interference testing monitors and interprets the system's response in terms of hydraulic head (or drawdowns) variations in an observation well located at some distance from another well where a pumping (injection) stress is applied.
Observations are usually representing a state of the system integrated over the whole screened intervals of the wells. Several variations of this basic configuration are then implemented.

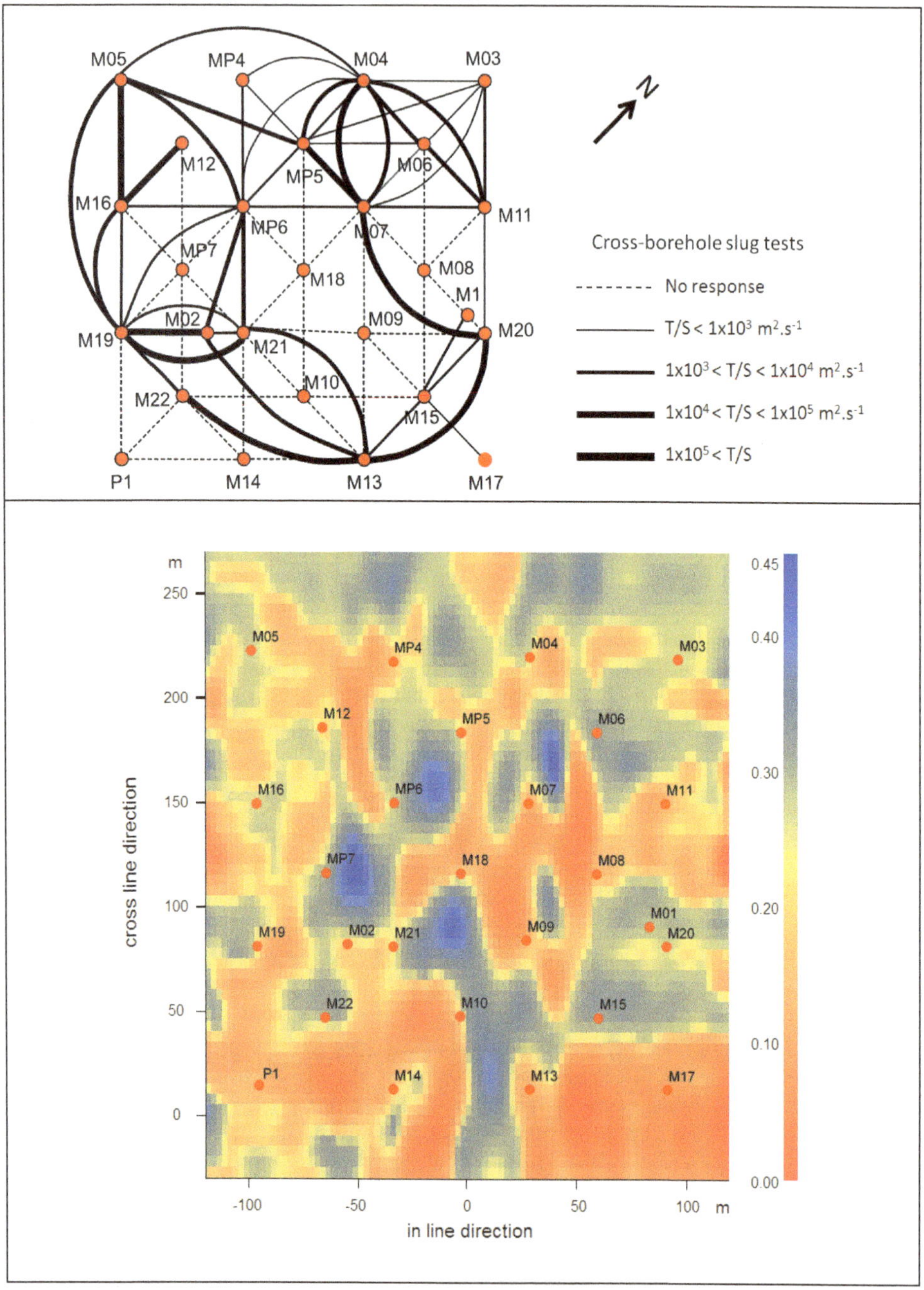

Fig. 14. Diffusivity map from slug tests and seismic porosity map at a depth of 87 m.

These include, for instance, dipole-tests, where groundwater is circulated between a pumping and an injection point, or cross-borehole flow logging, where the observation well is monitored at several depths. In general, the collected information is therefore partly integrated along the vertical direction but relevant to continuous approaches to flow. It can be sometimes dwelled on the capability of interference data to really conceal all the elements necessary for the assessment of 3D flow features (Delay et al., 2011), but this technical discussion is beyond the scope of the present contribution.

A classical dual-continuum approach was recently modified by Kaczmaryk & Delay (2007b) to account for the rapid propagation of pressure waves in a fractured karstic aquifer. This approach separates two overlapping continua, the fracture medium and the matrix medium, respectively, and adds hyperbolic wave propagation to the fracture medium. The local flow equations are written as

$$\begin{aligned} Ss_f(x)\frac{\partial h_f(x,t)}{\partial t} &= \nabla\left(K_f(x).\nabla h_f(x,t)\right) - u\nabla h_f(x,t) + \alpha(x)\left(h_m(x,t) - h_f(x,t)\right) \\ Ss_m(x)\frac{\partial h_m(x,t)}{\partial t} &= \alpha(x)\left(h_f(x,t) - h_m(x,t)\right) \end{aligned} \qquad (5)$$

Here, t [T] is time, x [L] is the vector of spatial coordinates, $h_f(x, t)$ and $h_m(x, t)$ are hydraulic head in the fracture and in the matrix continua, respectively; $K_f(x)$ [LT^{-1}] represents the hydraulic conductivity of the fracture medium, $Ss_f(x)$ and $Ss_m(x)$ [L^{-1}] respectively represent specific storage capacity in the fractures and in the matrix, $\alpha(x)$ [$L^{-1}T^{-1}$] is the exchange rate between fractures and matrix. From eq. 5, three basic types of models can be analyzed: (1) a single medium approach, *i.e.*, $u = 0$, $Ss_m = 0$ and $\alpha = 0$; (2) a classical dual porosity medium, *i.e.*, $u = 0$ and $Ss_m \neq 0$, $\alpha \neq 0$; and (3) a modified dual porosity medium, *i.e.*, $u \neq 0$ and $Ss_m \neq 0$, $\alpha \neq 0$.

These three types of models can be either taken homogeneous or heterogeneous in space. The first assumption is well suited to interpret drawdown curves one at a time (*i.e.*, the single relationship between the stress at the pumped well and the response at the observed well).By repeating inversion runs of a homogeneous model for all pairs pumped - observed wells available at the HES, one obtains a set of hydraulic parameters that bound the reasonable (or acceptable) values for the bulk hydraulic behavior of the system. The geometric structure of the aquifer revealed by the porosity block is obviously overlooked but this homogeneous approach may fix the ideas for further modeling tasks in the heterogeneous context. For instance, it could be considered that *i*- high porosity bodies are highly permeable with hydraulic conductivity values on the order of the upper bound of the homogeneous approach, *ii*- low porosity areas are affected with the small values of hydraulic conductivity from the lower bound of the homogeneous approach.

The exercise of inverting all drawdowns curves one at the time was carried out by Delay et al. (2007), Kaczmaryk & Delay (2007a, b) for the three main homogeneous configurations of eq. 5, *i.e.*, a single continuum, a dual fracture-matrix continuum, a dual continuum with wave propagation. In the end, it was shown by the authors that the relationships between local stresses and local responses were sensitive to the three assumptions prevailing to flow but always rendered quite homogeneous hydraulic conductivity values. Actually, the three homogeneous configurations differ by their capacity parameters (specific storage and, to some extent, the exchange rate between fractures and matrix) that may vary in any case over wide ranges.

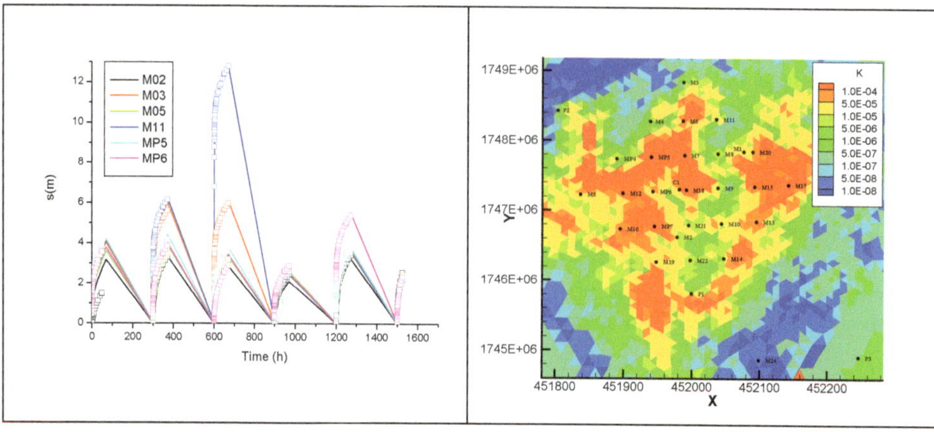

Fig. 15. Example of the fitting of fictitious scenario adding sequentially drawdown responses from different pumping tests (Left). Example of hydraulic conductivity field sought by inversion of a fictitious pumping scenario (Right).

Another attempt of drawdown inversion was carried out by Ackerer & Delay (2010). The authors considered flow in a single continuum model averaged over the vertical direction, thus yielding a diffusive 2D model widely used in groundwater engineering. They also inverted fictitious scenarios adding sequentially data obtained during various interference tests with the aim of providing to the calibration procedure several stress-response pairs evenly spread over the system. A result in terms of drawdown fitting and sought map of hydraulic conductivity is reported in Fig. 15. Again, there is no prior conditioning onto the structure given by the porosity block but the inverted hydraulic conductivity field is structured in space, obeying an exponential covariance not far from that used to filter seismic data. As shown in Fig. 15, inverting several drawdown curves results in variations in space of the hydraulic conductivity over four orders of magnitude, 10^{-8} - 10^{-4} ms^{-1}. This range could be considered as a good forecast of values that could be assigned to a 3D conductivity block inheriting from the structure provided by seismic data.

The exercise of inverting a 3D block of hydraulic conductivity conditioned onto the geometry of the porosity block is under investigations. The targeted objective is giving to the inverted fields of hydraulic parameters some consistency with the 3D spatial structure of the main flowing bodies. Several options can be proposed to deal with a tractable problem. The first option, which is quite inescapable, is that of thresholding the sought values of hydraulic conductivity. To make it short, it can be proposed to separate the system in *e.g.*, 3 types of conductivity values: high, mid and low values corresponding to similar cut-offs in the porosity values of the block. Without this simplification, it is likely that inverting a continuous distribution of hydraulic conductivity would become cumbersome or without tractable solution because of the too large number of freedom degrees given to the calibration procedure compared with available flow data. In the same vein, it can be discussed on the interest of proposing models of the dual-continuum type, eventually amended by adding wave propagations, in the case we dispose of a good image of the structural heterogeneity of the medium. In other words, is the picture given by the porosity block separating main flowing bodies from less permeable areas enough to represent the complexity of flow in a karstified aquifer, or is it needed to include locally more physics in the flow equations? Tackling the problem needs simply increasing the complexity of the local flow equations solved in a 3D block respecting the structure and then proceed with a sensitivity analysis of the models to parameters. The enhancement : single continuum → dual continuum → dual-continuum + hyperbolic wave propagation (see above) would inform on the relevance of adding or leaving out some mechanisms in the flow equations for a better understanding of measured data. Conversely, the sensitivity analysis while degrading the model would report on the capability of data to conceal information on the different mechanisms ruling flow in the fractured karstic aquifer. Notably, this topic remains a key question, unresolved for the moment, because very often the same set of data can be inverted with the same accuracy for different conceptualization of flow in the system. Incidentally another problem hampers for the moment the passage from a 3D block of seismic velocities to its equivalent in terms of hydraulic conductivity distributions. As stated above it is quite straightforward to build a prior estimate of the conductivity block by copying it on the distribution of seismic velocities. But it is not sure at all that this block would result in a reasonable fitting of interference data. A close look at interference tests performed over the HES shows that they mix reciprocal and non reciprocal responses. It is reminded that reciprocity, as defined by Lorenz with reference to electromagnetic fields, states that for a given process in a heterogeneous medium (e.g., Darcian flow in a porous medium) a stress at a location *A* yields a response at location *B* which coincides with the response in *A* due to an equivalent stress in *B*. In the present case of the HES, pumping a constant flow rate at well *A* may generate transient pressure drawdowns in well *B* that differ from the pressure response in *A* due to pumping in *B*. In these conditions, a 3D flow simulator for a single continuum based on the mass balance principle and Darcy's law should be discarded since it can be proved that this physical configuration always yields reciprocity irrespective the degree of heterogeneity in the medium and the boundary conditions (Delay et al., 2011). For 2D applications as discussed above, reciprocity gaps are damped by the integration of flow along the vertical direction making that hydraulic conductivity is integrated over the wetted thickness of the aquifer. For 3D flow with the aim of seeking accurately whether or not high-resolution seismic data coincide with hydraulic conductivity distribution, using non-reciprocal data for inverting a model which renders reciprocity would probably yield artifacts hardly graspable at this stage of our knowledge on the problem. Current studies (Delay et al., 2011) are analyzing the possible source of non reciprocity in the existence of inertial effects, non linear behavior as regard flow of the karstic features riddling the aquifer, complex flow as that in dual media, etc. These peculiar behaviors of the HES aquifer are incentive to first revisit the physics of a 3D flow model before proposing a block of hydraulic conductivity conditioned on both high-resolution seismic data and hydraulic tests.

5. Conclusion

We have shown with field examples how seismic and acoustic methods can be used to estimate porosity and condition inversion of hydraulic conductivity fields of geological formations for 3D models of reservoir.

The first field example concerns the study of the porous layers associated with the Oxfordian formation. A high resolution 3D seismic survey covering an area of about 4 km^2 was recorded. The processing gave rise to a time migrated 3D block. After migration, a model-based stratigraphic inversion provided a 3D cube of impedance after calibration at several wells. A porosity vs. impedance relationship, obtained by using density, acoustic velocity, and porosity (NMR) logs recorded in the wells, was used to convert the 3D impedance into porosity. The 3D cube makes it possible to provide 3D imaging of the connectivity of the porous bodies. A porosity cut-off of 21% was applied to extract the porous bodies with the best hydraulic conductivity within the porous layers. In order to complement the hydrogeological knowledge of the Oxfordian formation and to evaluate the power of the acoustic method to predict permeable zones, a full waveform acoustic log was recorded in an experimental well. The permeable zones detected by the acoustic logging were validated by hydraulic tests and conductivity measurements conducted later on . The absence of active hydraulic fracture in this borehole was confirmed by hydrogeological tests and acoustic logs.

The second field example concerns a near surface karstic reservoir. Both surface seismic refraction and seismic reflection data were recorded on the experimental hydrogeological site that has been developed for several years near Poitiers. Seismic reflection data enabled us to derive a 3D seismic pseudo-velocity block. Refraction tomography applied to the field data is conducive: (i) to obtain a complete velocity model in the first 35m; (ii) to map the top of the karstic reservoir; (iii) to detect the main corridor of fractures; (iv) to be sure that no cavities are present in the first 35 m. The velocity model obtained by refraction tomographic inversion was used to extend the 3D velocity block obtained by reflection survey to the 0-35 m depth interval. The 3D seismic pseudo-velocity block reveals a large heterogeneity of the aquifer reservoir in the horizontal and vertical planes. The low-velocity areas were found to correspond to the conductive levels and regions, as identified from well logging and flow interference tests. In order to quantify the porosity variations within that aquifer, the seismic-interval velocities were first converted into resistivity values. For that purpose, the empirical relationship between seismic velocity and true formation resistivity proposed by Faust (1953) was used. Resistivity values were then converted into porosity values, by using Archie's law (1942). The resulting 3D seismic pseudo-porosity block revealed three high-porosity, presumably-water-productive, layers, at depths of 35-40, 85-87 and 110-115 m. The 85-87 m-depth layer is the most porous one, with bodies of a porosity higher than 30%, that represent the karstic part of the reservoir. That seismic pseudo-porosity distribution appears to be consistent with the available hydrogeological data recorded at the site. It is therefore very likely that the high seismic-porosity zones (higher than 30 %) of the geophysical 3D block correspond to water producing layers. However, the conversion of that pseudo-porosity block into a dynamic-flow-property block is under investigations. In the case of the HES showing evidences of karstic features yielding complex flow and sometimes non-reciprocal behavior, the passage from seismic velocities to hydraulic conductivity is less straightforward as simply copying the block of conductivity on the block of velocities. However, we point out that high resolution seismic profiles inform with precision on the location of, *e.g.*, conduit flow compared with matrix flow. To conclude, the very high resolution seismic survey of that near-surface aquifer makes it possible the construction of a 3D seismic-porosity block providing a new insight to a deterministic high-resolution model of reservoir.

6. Acknowledgment

We thank Andra for permission to use the data presented in the first field example. We thank Daniel Guillemot and Beatrice Yven for their valuable help and advice.

We thank the University of Poitiers and IFP Energies nouvelles for permission to use the data presented in the second field example. We thank Gilles Porel and Bernard Bourbiaux for very useful discussions on various occasions, specifically for their advice and help in the interpretation of hydro geological data.

7. References

Ackerer P.& Delay F. (2010). Inversion of a set of well-test interferences in a fractured limestone aquifer by using an automatic downscaling parameterization technique. *Journal of Hydrology* 389:42-56. doi 10.1016/j.jhydrol.2010.05.020.

Audouin, O. & Bodin, J. (2007). Analysis of slug-tests with high-frequency oscillations. *Journal of Hydrology*, 334, 282-289.

Archie, G. E. (1942). The electrical resistivity log as an aid in determining some reservoir characteristics, *Petroleum Technology*, 146, 54-62.

Bernard, S.; Delay, F. & Porel, G. (2006). A new method of data inversion for the identification of fractal characteristics and homogenization scale from hydraulic pumping tests in fractured aquifers. *Journal of Hydrology* 328, 647-658.

Bourbiaux, B.; Callot, J.P.; Doligez, B.; Fleury, M.; Gaumet, F.; Guiton , M.; Lenormand, R.; Mari, J.L. & Pourpak, H. (2007). Multi-Scale Characterization of a Heterogeneous Aquifer Through the Integration of Geological, Geophysical and Flow Data: A Case Study, *Oil and Gas Science and Technology, Rev IFP* 61, 347-373.

Chaouch, A. & Mari, J.L. (2006). 3-D land seismic surveys: Definition of geophysical parameters, *Oil and Gas Science and Technology, Rev IFP* 61, 611-630. doi: 10.2516/ogst/2006002.

Chilès, J.P. & Delfiner, P. (1999). *Geostatistics modeling spatial incertainty*. Wiley series in probability and statistics, Wiley and Sons, New York City.

Delay, F; Kaczmaryk A & Ackerer, P. (2007). Inversion of interference hydraulic pumping tests in both homogeneous and fractal dual media. *Adv Water Resour*, 30(3):314-34. doi: 10.1016/j.advwatres.2006.06.008.

Delay, F.; Ackerer, P. & Guadagnini A. (2011). Theoretical analysis and field evidences of reciprocity gaps during interference pumping tests. *Adv Water Resour* , 34, 592 - 606

Delay, J.; Rebours, H.; Vinsot, A. & Robin P. (2007). Scientific investigation in deep wells for Nuclear waste Disposal studies at the Meuse/Haute-Marne underground research laboratory, North Eastern France[J]. *Physics and Chemistry of the Earth* , 32 : 42-57.

Fabricius, I. L.; Baechle ,G.; Eberli, G.P. & Weger, R. (2007). Estimating permeability of carbonate rocks from porosity and Vp/Vs, *Geophysics*, 72, 5, 185-191,doi:10.1190/1.2756081.

Faust, L.Y. (1953). A velocity function including lithologic variation, *Geophysics*, 18, 271-288.

Ferry, S.; Pellenard, P.; Collin, P.-Y.; Thierry, J.; Marchand, D.; Deconinck, J.-F.; Robin, C.; Carpentier, C.; Durlet, C. & Curial, A. (2007). Synthesis of recent stratigraphic data on Bathonian to Oxfordian deposits of the eastern Paris Basin. - *Mémoire de la Société Géologique Française, special issue*, 178, 37-57.

Glangeaud, F. & Mari, J.L. (2000). *Signal Processing in Geosciences*, CD, Editions Technip, ISBN 2-7108-0768-8, Paris, France.

Hagedoorn, G.J. (1959). The Plus-Minus method of interpreting seismic refraction sections, *Geophysical Prospecting* 7, 158- 182.

Kaczmaryk, A. & Delay, F.(2007a). Interpretation of interference pumping tests in fractured limestone by means of dual-medium approaches. *Journal of Hydrology* 337, 133-146.doi: 10.1016/j.jhydrol.2007.01.004.

Kaczmaryk, A. & Delay, F.(2007b).Improving dual-porosity-medium approaches to account for karstic flow in a fractured limestone. Application to the automatic inversion of hydraulic interference tests. *Journal of Hydrology* 347, 391-403.doi: 10.1016/j.jhydrol.2007.09.037.

Mari, J.L. ; Arens, G.; Chapellier, D.; Gaudiani, P. (1999). *Geophysics of Reservoir and Civil Engineering*, Editions Technip, ISBN 2-7108-0757-2, Paris, France.

Mari, J.L. & Porel, G. (2007). 3D seismic imaging of a near - surface heterogeneous aquifer: a case study, *Oil and Gas Science and Technology, Rev IFP* 63, 179-201. doi: 10.2516/ogst/2007077.

Mari, J.L.; Porel, G. & Bourbiaux, B. (2009). From 3D seismic to 3D reservoir deterministic model thanks to logging data: the case study of a near - surface heterogeneous aquifer, *Oil and Gas Science and Technology, Rev IFP* 64, 119-131. doi: 10.2516/ogst/2008049.

Mendes, M. (2009). A hybrid fast algorithm for first arrivals tomography, *Geophysical Prospecting*, 57, 803-809.

Morlier P., Sarda J.P. (1971). Atténuation des ondes élastiques dans les roches poreuses saturées, Revue de l'Institut Français du Pétrole, Vol 26, N° 9, p. 731-755.

Robein, E. (2003). *Velocities, Time-Imaging and Depth-imaging in Reflection Seismics. Principles and Methods*, EAGE publications, ISBN 90-73781-28-0, Houten, The Netherlands.

Part 5

Modelling and Hydraulic Conductivity

17

The Role of Hydraulic Conductivity in Modeling the Movement of Water and Solutes in Soil Under Drip Irrigation

René Chipana Rivera
University Mayor de San Andrés
La Paz
Bolivia

1. Introduction

Due of de need to manage more rational irrigation water in the world drip irrigation is one of the technologies that are expanding more rapidly in modern irrigated agriculture in view of its great potential to improve the economics of water use. In this method the water is provided in soil in a timely high frequency, and thus the dimension parameters such as percentage of the root zone moist, spacing and location of emitters, application rates, frequency and irrigation, are governed by standard distribution of moisture in the soil profile, which in turn depends on soil hydraulic conductivity. There is difficulty in pinpointing the relationships between factors that affect the movement of soil water in response to surface point sources, this fact is due to the complex nature of the surface boundary conditions, which can result in problems of irrigation management. Due to these facts, mathematical models were developed to analyze three-dimensional movement of water in soil under drip irrigation, using methods of numerical analysis, finite elements, flow in two and three dimensions under dynamic equilibrium, volume control, theory of capillary tubes, among others.

Moreover, in recent years, the technique of application of chemicals through irrigation water (chemigation) is gaining acceptance, particularly in drip irrigation systems, due to the advantages it provides the possibility of manipulating and controlling plant nutrition irrigated. However, this technique requires that the irrigation is carried out under high efficiencies, otherwise it may cause economic problems and environmental damage. The chemicals applied through irrigation water processes suffer from spatial and temporal changes in soil, varying the distribution of solutes in the profile, resulting in different distribution patterns. Some solutes react with the soil matrix, others move through it, may result in dissolution and precipitation, within or outside the soil solution. The understanding of the simultaneous transport of water and solutes in two or three dimensions, from a point source, allows to develop efficient strategies in the implementation of water and mineral fertilizers (fertigation). Although fertigation has widespread use, little information related to the simultaneous movement of water and chemicals from point sources is available.

The high costs involved in field research are making mathematical models a very viable tool, enabling a prediction and assessment of the fate and behavior of water and solutes in drip irrigation. However, the development of mathematical models to accurately describe the transport of water and solutes under real field conditions presents great complexity due to the large number of intervening factors, as well as the variability of hydraulic conductivity.

The hydraulic conductivity expressed the "ease" with which certain fluid flows in a way, dependent on the medium and fluid. The main characteristics of the soil matrix that interfere with the hydraulic conductivity are the distribution of the diameters of the pores, the tortuosity, specific surface area and porosity. In the fluid, the main features are the density and dynamic viscosity. The methods for determining the hydraulic conductivity can be classified into permanent, variable and computing. The direct measurement of hydraulic conductivity is theoretically simple, but experimentally difficult, because of their high variability in the field.

2. Model water and solute movement

The modeling was based on the numerical solution of two partial differential equations of second order applied to point sources under transient flow, ie the equation of motion of water in the soil also known as the Richards equation (equation 1) and the transport equation solutes (equation 2, considering the existence of linear equilibrium sorption), allowing to determine the distribution of water and solutes in soil as a function of both space and time

$$\frac{\partial\theta}{\partial t}=\frac{\partial}{\partial x}\left[Kx(\theta)\frac{\partial H}{\partial x}\right]+\frac{\partial}{\partial y}\left[Ky(\theta)\frac{\partial H}{\partial y}\right]+\frac{\partial}{\partial z}\left[Kz(\theta)\frac{\partial H}{\partial z}\right] \quad (1)$$

where θ is the soil moisture (L^3L^{-3}), t time (T), K (θ) is the hydraulic conductivity of unsaturated soil ($L\ T^{-1}$), H is the total hydraulic potential of water in the soil (L), x, y and z position coordinates

$$Fr\frac{\partial\theta C}{\partial t}=-\frac{\partial(qC)}{\partial z}+\frac{\partial}{\partial z}\left[D\theta\frac{\partial C}{\partial z}\right]-\frac{\partial(qC)}{\partial x}+\frac{\partial}{\partial x}\left[D\theta\frac{\partial C}{\partial x}\right]-\frac{\partial(qC)}{\partial y}+\frac{\partial}{\partial y}\left[D\theta\frac{\partial C}{\partial y}\right] \quad (2)$$

Where C is the mass of solute per unit volume of solution ($M\ L^{-3}$), q the volume of liquid (solution) through unit area of soil in unit time, also known as flux density (LT^{-1}), D is the diffusion coefficient of dispersion, also known as hydrodynamic dispersion coefficient, Fr is the retardation factor, dimensionless, defined by:

$$Fr=1+\frac{\rho Kd}{\theta} \quad (3)$$

where ρ is the density of the soil ($M\ L^{-3}$), Kd distribution coefficient or partition ($L^3\ M^{-1}$).

The retardation factor can be defined as the ability to retain or effect "buffer" of a particular element, or as the velocity of the solute in the speed of solution in the pore (Matos et al., 1999). If there are no interactions between the solute and soil, the Kd value is zero and the retardation factor becomes unity. If Fr is less than one means that only a fraction of the liquid phase participates in the transport process, as occurs for example in dense clusters containing very small pore diameter (Van Genuchten & Wierenga, 1986). According Toride et al. (1999) there is a negative correlation between speed of solution in the pore (ϑ, with dimensions LT-1) and distribution coefficient or partition (Kd), implying that Fr and ϑ are also inversely correlated.

Thus, the parameters of the soil dispersivity (λ, with dimensions L) and Fr, determine the way forward or a solute is retained in the soil (Alvarez et al., 1995). The most direct way to determine the parameters λ and Fr is from experiments using soil columns, in which a solution is applied on top of the column and the values of solute concentration are collected in the output. Nielsen & Biggar (1962) indicate that the standard form of presentation of the concentration data collected at the exit is called the breakthrough curve, and one should be careful about using the correct boundary conditions, otherwise its use can lead to serious errors when experimental data are extrapolated to the field.

These equations were solved considering a system of control volumes, two-dimensionally characterized by radial and vertical dimensions (Δr and Δz, respectively) and performing successive increments of time (Δt). In developing the model, we took into account the basic assumptions described by Botrel (1988) and Cruz (2000), under which the study region (wet bulb) can be described in a cylindrical space and imaginatively composed of rings concentric, each with a width (Δr) and depth (Δz). The rings have a center vertical line as part of that point source at the soil surface and vary in size according to their position (i) and (j), where (i) represents the radial axis and (j) axis vertical, as shown in figure 1.

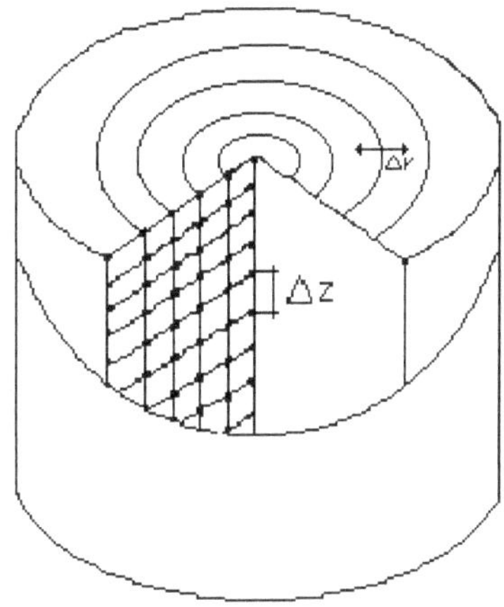

Fig. 1. Diagram of concentric rings considered for the numerical solution of the problem.

2.1 Determination of hydraulic conductivity and matric potential

The determination of hydraulic conductivity for unsaturated soils K (θ) is initially carried out considering values of relative hydraulic conductivity (Kr, with dimensions L T^{-1}) for conditions of no saturation calculated as a function of time and space.

Mualem (1976) and van Genuchten (1980) developed equations and methodologies for determining the hydraulic conductivity for unsaturated soils, from the relative hydraulic conductivity (Kr), using parameters of water retention of soil water, as shown in equations 4 to 7.

$$\Theta = \frac{\theta - \theta r}{\theta s - \theta r} \tag{4}$$

Where Θ is the effective saturation, so reflecting on the water content in soil (dimensionless), θ is the volumetric soil water content in the current condition ($L^3\ L^{-3}$), θr is the residual soil water content ($L^3\ L^{-3}$), θs is the volumetric soil water content at saturation point ($L^3\ L^{-3}$).
Θ can also be expressed by the equation 5:

$$\Theta = \left[\frac{1}{1+(\alpha\psi)^n}\right]^m \tag{5}$$

where m and n are regression parameters of equation (dimensionless), α is the parameter with dimension equal to the reversal potential (L-1), ψ is the matrix water potential in soil (L). Equation 6 expresses the value of relative hydraulic conductivity:

$$Kr = \Theta^{1/2}\left[\frac{\int_0^{\Theta}\frac{1}{\psi(x)}dx}{\int_0^{1}\frac{1}{\psi(x)}dx}\right]^2 \tag{6}$$

known value of Kr can be determined the value of hydraulic conductivity of unsaturated soil (K (θ)) as the product of relative hydraulic conductivity on the conductivity of saturated soil (Ko, with dimensions L T^{-1}):

$$K(\theta) = Kr\,Ko \tag{7}$$

Kr can also be calculated using equation (8), which is a consequence of equations (4, 5 and 6):

$$Kr = \Theta^{0,5}\left[1-(1-\Theta^{1/m})^m\right]^2 \tag{8}$$

On the other hand the matrix potential of soil water (ψ) was determined by transforming the equation (5), as presented in equation (9):

$$\psi = \frac{\left[\frac{1}{\Theta^{1/m}}-1\right]^2}{\alpha} \tag{9}$$

2.2 Transport of water in the soil profile

2.2.1 Calculating the volume of soil

When considering the soil divided into concentric rings, the axis of symmetry is formed by the vertical line passing through the wmitter. In this case, the first ring has a radius equal to half the increase in the horizontal axis (Δr/2).
The calculation of the volume of soil was done by considering the volumes of the rings in the radial and vertical axes. For example, the volume of the ring (i, j) is the volume of the cylinder defined by the larger radius of the ring (R_{ij}), less the volume of the cylinder defined by the smaller radius of the ring ($R_{i,j-1}$), considering a high Δz, as shown in equation (10).

$$Vs_{(i,j)} = \pi(R_{i,j}^{2} - R_{i,j-1}^{2})\Delta z \tag{10}$$

where Vs is the volume of soil in each ring, L^3, R is the radius of the ring, L and Δz is the depth of the ring, L. In vertical direction the volumes of the rings vary in function of their distance from the axis, but are constant for each position (j), whatever the depth. The volume of water stored in each ring (Va) is determined by multiplying the volume of soil by the value of volumetric soil water content (θ):

$$Va = Vs\theta \tag{11}$$

this calculation was performed for each time increment (ΔT) either in the radial direction and vertical

2.2.2 Determination of flux density (q)

The calculation of flux density (q) from one cell to another was made following considerations advocated by van der Ploeg and Benecke (1974) and van Genuchten (1980), whereby it is assumed that the flow inside the rings or cells occurs in two directions, both for entry and exit for water and solutes (figure 2).

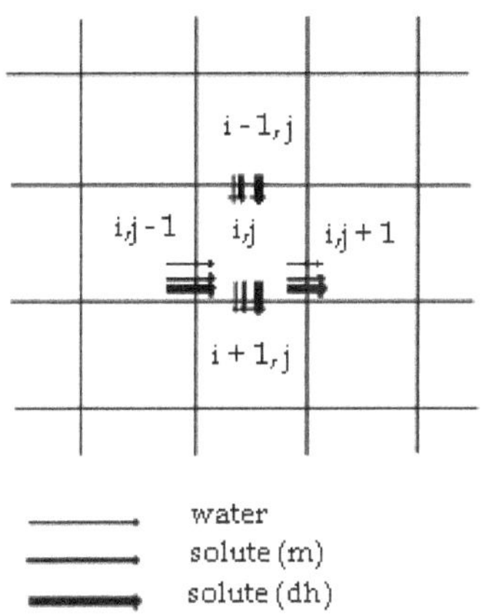

Fig. 2. Coordinates of adjacent cells required to determine the movement of water and solutes by convection (m) and hydrodynamic dispersion (dh) in the radial and vertical direction

a)Radial flux density

The movement of water from one ring to another in the radial axis is calculated according to equation (12).

$$q = -K(\theta)\frac{dH}{ds} \tag{12}$$

Considering that the flow occurs between two adjacent rings, it was assumed that the unsaturated hydraulic conductivity of the region where the flow takes place, can be represented by the unsaturated hydraulic conductivity of the medium adjacent rings.

$$K_{(i,j-1;\,i,j)} = \frac{K_{(i,j-1)} + K_{(i,j)}}{2} \tag{13}$$

where $K_{(i,j-1;\,i,j)}$ is the average unsaturated hydraulic conductivity cell (i, j-1) and (i, j); $K_{(i,j-1)}$ is the unsaturated conductivity cell (i, j-1) and $K_{(i,j)}$ is the unsaturated hydraulic conductivity of the cell (i,j). Therefore the equation (12) for determining the water flow rewritten as can be follows:

$$q_{(i,j-1;\,i,j)} = -K_{(i,j-1;\,i,j)} \frac{H_{(i,j)} - H_{(i,j-1)}}{\Delta r} \tag{14}$$

where $q_{(i,j-1;\,i,j)}$ is the flux density between the cell (i, j-1) and cell (i, j); $H_{i,j}$ is the total hydraulic potential of water in cell (i, j); $H_{i,j-1}$ is the total hydraulic potential of water in cell (i, j-1), and Δr is the increment in radius, ie the difference between the average radius of two cells.

Likewise, the radial flux density between the cell (i, j) and cell (i, j +1) is calculated on the basis of equations (15) and (16):

$$K_{(i,j;\,i,j+1)} = \frac{K_{(i,j)} + K_{(i,j+1)}}{2} \tag{15}$$

$$q_{(i,j;\,i,j+1)} = -K_{(i,j;\,i,j+1)} \frac{H_{(i,j+1)} - H_{(i,j)}}{\Delta r} \tag{16}$$

b) Vertical flux density

Similarly to the case of radial transport, the vertical flux density of cells (i-1, j) (i, j) and (i, j) (i +1, j), was determined by the following equations:

$$K_{(i-1,j;\,i,j)} = \frac{K_{(i-1,j)} + K_{(i,j)}}{2} \tag{17}$$

$$q_{(i-1,j;\,i,j)} = -K_{(i-1,j;\,i,j)} \frac{H_{(i,j)} - H_{(i-1,j)}}{\Delta z} \tag{18}$$

$$K_{(i,j;\,i+1,j)} = \frac{K_{(i,j)} + K_{(i+1,j)}}{2} \tag{19}$$

$$q_{(i,j;\,i+1,j)} = -K_{(i,j;\,i+1,j)} \frac{H_{(i+1,j)} - H_{(i,j)}}{\Delta z} \tag{20}$$

2.2.3 Moisture balance in each ring

Before and after application of water by a point source estimated the moisture at depth (i) and radial position (j) for each increment of time (Δt). The volume of water flowing horizontally from cell (i, j-1) (i, j) is determined from the lateral area of each ring and the flux density:

$$Va_{(i,j-1;\,i,j)} = q_{(i,j-1;\,i,j)} 2\pi(R_{(i,j-1)})\Delta z\,\Delta t \quad (21)$$

however the volume passing through the cell (i, j) (i, j +1) is represented by:

$$Va_{(i,j;\,i,j+1)} = q_{(i,j;\,i,j+1)} 2\pi(R_{(i,j)})\Delta z\,\Delta t \quad (22)$$

Similarly, the amount the volume of water flowing vertically between the rings (i-1, j) to (i, j) and (i, j) to (i +1, j) is calculated by:

$$Va_{(i-1,j;\,i,j)} = q_{(i-1,j;\,i,j)}\pi({R_{(i,j)}}^2 - {R_{(i,j-1)}}^2)\Delta t \quad (23)$$

$$Va_{(i,j;\,i+1,j)} = q_{(i,j;\,i+1,j)}\pi({R_{(i,j)}}^2 - {R_{(i,j-1)}}^2)\Delta t \quad (24)$$

The amount of water that remained in a particular ring after the increment Δt is represented by the change in volume of water in the ring ($\Delta Va_{(i,j)}$) due to water gain ($Va_{(i,j-1\ a\ i,j)}$ and $Va_{(i-1,j\ a\ i,j)}$) and loss ($Va_{(i,j\ a\ i,j+1)}$ e $Va_{(i,j\ a\ i+1,j)}$), so:

$$\Delta Va_{(i,j)} = Va_{(i,j-1;\ i,j)} + Va_{(i-1,j;\ i,j)} - Va_{(i,j;\ i,j+1)} - Va_{(i,j;\ i+1,j)} \quad (25)$$

Thus, the new moisture ring at the end of Δt is calculated by equation (26):

$$\theta_{(i,j)} = \theta_{(i,j,ini)} + \frac{\Delta Va_{(i,j)}}{Vs_{(i,j)}} \quad (26)$$

where $\theta_{(i,j,\ ini)}$ is the soil moisture at the beginning of the interval Δt, L^3L^{-3}.
Therefore, when considering the control volume discretization, equation (1) to two dimensions is transformed into:

$$\frac{\Delta\theta}{\Delta t} = \frac{\Delta qx}{\Delta x} + \frac{\Delta qz}{\Delta z} \quad (27)$$

or

$$\frac{\Delta\theta}{\Delta t} = -\overline{K}x(\theta)\frac{\Delta Hx}{\Delta x} - \overline{K}z(\theta)\frac{\Delta Hz}{\Delta z} \quad (28)$$

One important hypothesis was considered for resolving cases in which the humidity inside the ring became greater than the moisture content under saturation ($\theta_{(i,j)} > \theta s_{(i,j)}$). In this case the rings are the volume of surface water did not infiltrate transfer the water at outer rings (forming a circle saturated surface) and in the rings of the surface, the volume of water that can not be transported is held in the upper rings.

2.3 Solute transport in the bulb

After calculating the values of the volume of water flow and changes in moisture rings, starts determining the concentration of solute in each ring. Again the technique is employed to control volumes to calculate the mass flow, the flow by hydrodynamic dispersion and the variation of solute concentration in soil (ΔC).

2.3.1 Mass flow of solute

The mass flow of solute is determined for both the radial direction and for the vertical direction.

Radial mass flow

$$Jc_{(i,j-1;\ i,j)} = q_{(i,j-1;\ i,j)}\, C_{(i,j-1)} \tag{29}$$

where $Jc_{(i,j-1;\ i,j)}$ is the mass flow of the solute in cell (i,j-1) to cell (i,j) , $ML^{-2}T^{-1}$.

$$Jc_{(i,j;\ i,j+1)} = q_{(i,j;\ i,j+1)}\, C_{(i,j)} \tag{30}$$

where $Jc_{(i,j-1;\ i,j)}$ is the mass flow of the solute in cell (i,j) to cell (i,j+1).

Vertical mass flow

The mass flow of the solute in cell (i-1, j) to cell (ij) and cell (i, j) to cell (i +1, j) is determined by:

$$Jc_{(i-1,j;\ i,j)} = q_{(i-1,j;\ i,j)}\, C_{(i-1,j)} \tag{31}$$

$$Jc_{(i,j;\ i+1,j)} = q_{(i,j;\ i+1,j)}\, C_{(i,j)} \tag{32}$$

Mass transport of solute between adjacent rings

The mass (M) of solute that passes from one ring to another (E) in both the horizontal and vertical direction, is calculated by the following mathematical relationships:

Radial direction

$$Ec_{(i,j-1;\,i,j)} = Jc_{(i,j-1;\,i,j)} 2\pi R_{(i,j-1)}\, \Delta z\, \Delta t \tag{33}$$

where $Ec_{(i,j-1;\ i,j)}$ is the mass of solute passing through the cell (i, j-1) to cell (i,j), M.

$$Ec_{(i,j;\,i,j+1)} = Jc_{(i,j;\,i,j+1)} 2\pi R_{(i,j)}\, \Delta z\, \Delta t \tag{34}$$

where $Ec_{(i,j;\ i,j+1)}$ is the mass of solute passing through the cell (i,j) to cell (i,j+1), M.

Vertical direction

$$Ec_{(i-1,j;\,i,j)} = Jc_{(i-1,j;\,i,j)} \pi (R_{(i,j)}{}^2 - R_{(i,j-1)}{}^2) \Delta t \tag{35}$$

$$Ec_{(i,j;\,i+1,j)} = Jc_{(i,j;\,i+1,j)} \pi (R_{(i,j)}{}^2 - R_{(i,j-1)}{}^2) \Delta t \tag{36}$$

2.3.2 Flow by hydrodynamic dispersion

Before the calculation of flow by hydrodynamic dispersion, we determined the parameters of the diffusion coefficient or molecular ion (Dm, with dimensions L^2T^{-1}) and hydrodynamic dispersion coefficient (D, with dimensions L^2T^{-1}) for each increment Δt between consecutive cells:

$$D = Dm + \left(\frac{\lambda \bar{q}}{\bar{\theta}} \right) \tag{37}$$

where $\overline{\theta}$ é is the the average humidity between rings or consecutive cells.
Then the flow equation of hydrodynamic dispersion of solutes in soil can be written as:

$$Jdh = -\overline{\theta} D \frac{\Delta C}{\Delta s} \tag{38}$$

where Jdh is the hydrodynamic dispersion flow ($ML^{-2}T^{-1}$).
From equation (38) calculates the flow by hydrodynamic dispersion, both horizontally and vertically, with the number of equations 39 to 42:

Radial flow by hydrodynamic dispersion

$$Jdh_{(i,j-1;\,i,j)} = -\left(\frac{\theta_{(i,j-1)} + \theta_{(i,j)}}{2}\right) D_{(i,j-1;\,i,j)} \left[\frac{C_{(i,j)} - C_{(i,j-1)}}{\Delta r}\right] \tag{39}$$

$$Jdh_{(i,j;i,j+1)} = -\left(\frac{\theta_{(i,j)} + \theta_{(i,j+1)}}{2}\right) D_{(i,j;\,i,j+1)} \left[\frac{C_{(i,j+1)} - C_{(i,j)}}{\Delta r}\right] \tag{40}$$

Vertical flow by hydrodynamic dispersion

$$Jdh_{(i-1,j;i,j)} = -\left(\frac{\theta_{(i-1,j)} + \theta_{(i,j)}}{2}\right) D_{(i-1,j;\,i,j)} \left[\frac{C_{(i,j)} - C_{(i-1,j)}}{\Delta z}\right] \tag{41}$$

$$Jdh_{(i,j;\,i+1,j)} = -\left(\frac{\theta_{(i,j)} + \theta_{(i+1,j)}}{2}\right) D_{(i,j;\,i+1,j)} \left[\frac{C_{(i+1,j)} - C_{(i,j)}}{\Delta z}\right] \tag{42}$$

Solute transport by dispersion hydrodynamics between rings

Analogous to the convective motion of solutes was calculated solute mass (E) which passes from one ring to another in the radial and vertical directions.

Radial direction

$$Edh_{(i,j-1;\,i,j)} = Jh_{(i,j-1;\,i,j)} 2\pi R_{(i,j-1)} \Delta z \,\Delta t \tag{43}$$

$$Edh_{(i,j;\,i,j+1)} = Jh_{(i,j;\,i,j+1)} 2\pi R_{(i,j)} \Delta z \,\Delta t \tag{44}$$

Vertical direction

$$Edh_{(i-1,j;\,i,j)} = Jh_{(i-1,j;\,i,j)} \pi (R_{(i,j)}{}^2 - R_{(i,j-1)}{}^2) \Delta t \tag{45}$$

$$Edh_{(i,j;\,i+1,j)} = Jh_{(i,j;\,i+1,j)} \pi (R_{(i,j)}{}^2 - R_{(i,j-1)}{}^2) \Delta t \tag{46}$$

2.3.3 Solute balance in each ring

Similarly to the case of transport of water after applying the solution by a point source, we calculated the concentration of solute in the ring located at the depth (i) and horizontal (j) for each increment of time (Δt) . The amount of solute that was remaining in a ring after the increment Δt is represented by the change in solute mass in the ring ($\Delta E_{(i,j)}$) due to mass gain by:

Convection (Ecg):

$$Ecg_{(i,j)} = Ec_{(i,j-1;\ i,j)} + Ec_{(i-1,j;\ i,j)} \tag{47}$$

Hydrodynamic dispersion (Ehg)

$$Ehg_{(i,j)} = Eh_{(i,j-1;\ i,j)} + Eh_{(i-1,j;\ i,j)} \tag{48}$$

and mass loss by:

Convection (Ecp):

$$Ecp_{(i,j)} = Ecp_{(i,j;\ i,j+1)} + Ecp_{(i,j;\ i+1,j)} \tag{49}$$

Hydrodynamic dispersion (Ehp)

$$Ehp_{(i,j)} = Ehp_{(i,j;\ i,j+1)} + Ehp_{(i,j;\ i+1,j)} \tag{50}$$

obtaining finally:

$$\Delta E_{(i,j)} = \frac{Ecg_{(i,j)} + Ehg_{(i,j)} - Ecp_{(i,j)} - Ehp_{(i,j)}}{Fr} \tag{51}$$

Consequently, the new concentration of solute in the ring at the end of Δt is determined by the following equation:

$$C_{(i,j)} = \left[\frac{\dfrac{C_{(i,jini)}}{\left(\theta_{(i,jini)} Vs_{(i,j)}\right)^{-1}} + \Delta E_{(i,j)}}{\theta_{(i,j)} Vs_{(i,j)}} \right] \tag{52}$$

where $C_{(i,jini)}$ is the solute concentration at the beginning of the interval Δt, ML^{-3}.
When applying the technique to control volume discretization, equation (2) to two dimensions can be rewritten as:

$$Fr\frac{\Delta \theta C}{\Delta t} = -\frac{\Delta(qC)}{\Delta z} + \frac{\Delta}{\Delta z}\left[D\bar{\theta}\frac{\Delta C}{\Delta z}\right] - \frac{\Delta(qC)}{\Delta x} + \frac{\Delta}{\Delta x}\left[D\bar{\theta}\frac{\Delta C}{\Delta x}\right] \tag{53}$$

3. Model validation

The soil used in the validation of the model was derived from a profile classified as Oxisol, sandy phase, called Series "Sertaozinho". Soil moisture was determined by gravimetric method to a depth of 100 cm. For sampling in layers of 10 cm was used cup hole coupled to a jacketed system using a PVC tube to prevent contamination of the lower layers.

3.1 Breakthrough curve and the solute transport parameters

In the validation of the model was used potassium ion. With the aim of obtaining the transport parameters of potassium in the wet bulb from the breakthrough curve, was

Parameters of water retention curve				
θr (cm^3cm^{-3})	θs (cm^3cm^{-3})	α (cm^{-1})	n	m
0,113	0,482	0,029428	1,828069	0,452975

Table 1. Parameters of water retention, according to the model (van Genuchten 1980)

mounted an experiment which was used as solute potassium chloride, with a potassium concentration of 500 mgL^{-1}.

The values of potassium concentration, pore volume and cumulative time were used as input data CXTFIT program developed by U.S. Salinity Laboratory, USDA, Riverside, CA, version 2.1, written in FORTRAN. The CXTFIT determines the values of the parameters of solute transport in soil that is the dispersion coefficient (D), the pore water velocity (V), the dispersivity (λ) and retardation factor (Fr), through attempts to maximize the coefficient of determination of regression between the averaged concentration on the solution (C/Co) and pore volume. Thus, we obtained the values of transport parameters that are shown in Figure 3 (breakthrough curve) and Table 2

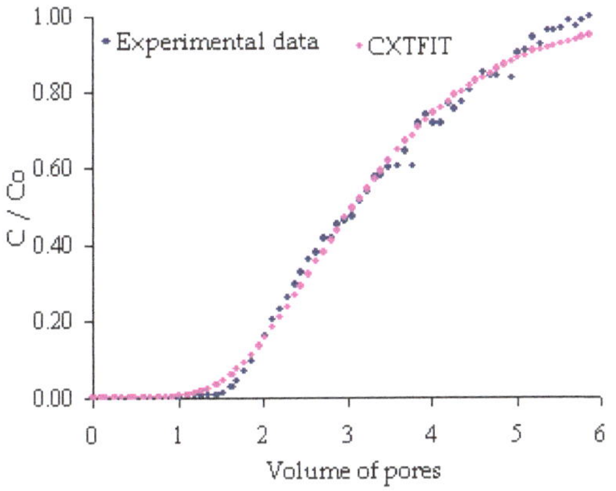

Fig. 3. Breakthrough curve of potassium solution in the soil column.

Transport parameters of potassium			
V ($cm\ min^{-1}$)	D ($cm^2\ min^{-1}$)	Fr (adimensional)	λ (cm)
1,489	2,705	4,730	1,816655

Table 2. Transport parameters of potassium in the soil obtained from the model CXTFIT: pore water velocity (V), dispersion-diffusion coefficient (D), retardation factor (Fr) and dispersivity (λ)

3.2 Soil moisture after irrigation

Data of soil moisture after 24 hours after irrigation generated by the model and the observed are presented through moisture isolines in Figure 4. Inside the bulb, ie a radius of 26 cm and depth of 30 cm from the emitter soil moisture varies in a range from 0.17 to 0.20 $cm^3\ cm^{-3}$, decreasing both in the radial or vertical direction, as it away from the emitter, the bulb acquiring a hemispherical shape. Observe that the radius of the bulb was slightly greater than its depth. In this respect Hachum et al. (1976) indicate that the force of gravity has a limited effect on clay and loam soils and, where capillary forces dominate the effects on water flow.

Outside the region of the bulb, there was practically no change in moisture content, which allowed the sampling was done only up to a radius of 50 cm and 70 cm deep, including therefore a safety margin.

The water content in the center of concentric rings (where it was located the point source) was approximately 0.20 cm^3 cm^{-3} in both cases (observed and simulated). This represents a little less than half the moisture saturation (0.43 cm^3 cm^{-3}), indicating that the time interval elapsed since the completion of irrigation until the moment of sampling (24 hours) there was a marked redistribution of the solution .

The soil moisture obtained (simulated an observed) were similar, being a value for the standard error of 0.011 cm^3 cm^{-3}. The central axis, the wet bulb has reached 35 cm with a moisture content of 0.176 and 0.172 cm^3 cm^{-3} for the model and test, respectively. As for the periphery (at 35 cm radius) the bulb has reached 20 cm depth for the model and about 15 cm depth for the test with moisture content of 0.170 and 0.163 cm^3 cm^{-3}, respectivily.

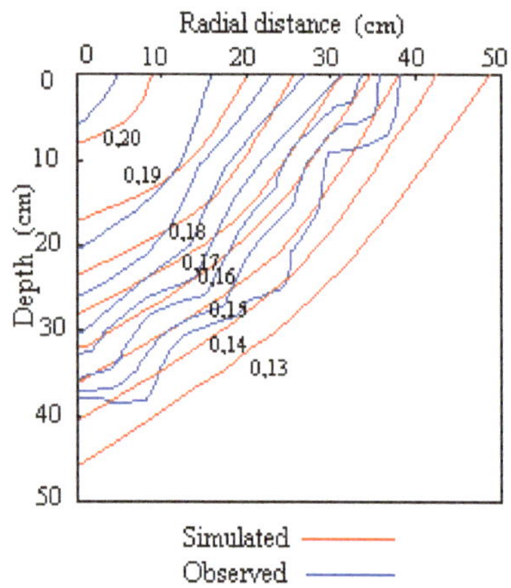

Fig. 4. Volumetric soil water content (cm^3 cm^{-3}) simulated by the model and observed 24 hours after the end of irrigation.

For a time of redistribution of 48 hours after irrigation, the dimensions of the bulb remained constant when compared with the time of 24 hours of redistribution, but there was a decrease in humidity, especially in cells near the point source. On the ground surface, a radial distance of 10 cm, volumetric soil water content was approximately 0.18 cm^3 cm^{-3} for the two cases (simulated and observed), suffering a decrease of approximately 7.5% moisture when compared with a time of 24 hours. The simulated moisture values were slightly higher than those observed, but the trend of the data in different cells was the same, determining a value for the standard error of 0.015 cm^3 cm^{-3}.

Elapsed time of 72 hours after irrigation, in both conditions (simulated and observed) cells close to the point source continued losing water to the adjacent cells, leaving a soil moisture around the 0.178 $cm^3\ cm^{-3}$, which represents a decrease of 5% over the 48 hours time, but still, this did not contribute significantly to changes in the dimensions of the bulb. Both humidity (simulated and observed) had an approximate behavior with regard to the distribution of moisture, resulting in a value for the standard error of 0.008 $cm^3\ cm^{-3}$. In this type of soil, one can say that 24 hours after the redistribution of water within the wetted practically ends with the largest changes observed in cells near the sender, ie the water flow that occurred after this time was small, so that the bulb dimensions have to remain virtually unchanged.

In general, the experimental values were similar to the model simulated. Small differences should possibly be due to limitations of the model and accuracy of measurements. It is important to emphasize that the assumptions made in developing the model were that the soil is homogeneous, there is no evaporation, the initial water content in soil is uniform, there is no occurrence of hysteresis, and the water applied by the issuer is distributed uniformly in all directions. Moreover, the model assumed average values for hydraulic conductivity between two adjacent cells, whereas in reality the change from one point to another is gradual. Already Nogueira et al. (2000) used the harmonic mean of three layers in the determination of hydraulic conductivity. In relation to homogeneous and isotropic soil, there was possibly a variation of physical and hydraulic properties in soil, thus filling the box with soil was done in layers of 10 cm, a fact that probably involves a change in soil density. With respect to evaporation during the test there was a slight condensation of water at the bottom of the plastic covering the box. Regarding the hysteresis, according to Levin et al. (1979), during the redistribution due to the low levels of soil moisture on the periphery of the bulb, the hysteresis has a pronounced influence, which may have occurred in the test.

Another fact during testing is the formation of a thin crust under the emitter, possibly by the dispersion of particles during application of the solution, which probably led to a reduction in porosity and hydraulic conductivity in this region, which was also observed by Lafolie et al. (1989). Also do not rule out a possible contamination of soil layers has been made when sampling with the auger, but has since been made the hole when the jacketing of sampling, there may have been the contact of small amounts of soil.

3.3 Concentration of potassium in the soil after irrigation

Potassium is adsorbed by clay particles, this fact can be seen in Table 2, reveals that to obtain a value for the retardation factor (Fr) equal to 4.73 (similar to that found by Miranda (2001) for the same soil type) indicating that one part of potassium applied to the solution is retained by the soil.

The results obtained in the concentration of potassium in the soil solution in different cells 24 hours after irrigation for both simulated and observed, are shown in Figure 5. This cation was retained in the surface layers. For both simulated and observed highest concentrations were around the emitter in a radius of 10 cm and 20 cm depth, where concentrations of potassium ranged from 62 to 817 mg L^{-1}.

The volume of soil containing considerable amounts of potassium (potash bulb) was lower than the wet bulb. The displacement of this cation is not fully followed the displacement of water (mass or convective flow). This can be explained by potassium in the soil solution interact with the cationic exchange complex of soil (expressed by the retardation factor), and therefore this element retained in the soil in the region closer to the emitter, so that the solution that went into the outer regions of the bulb possibly possessed a lower concentration of potassium. The values of potassium concentration simulated by the model were similar to those observed in the test (with a value for the standard error of 30.07 mg L^{-1}), following the same pattern of distribution. However, in general, the experimental values were slightly lower than those generated by the model.

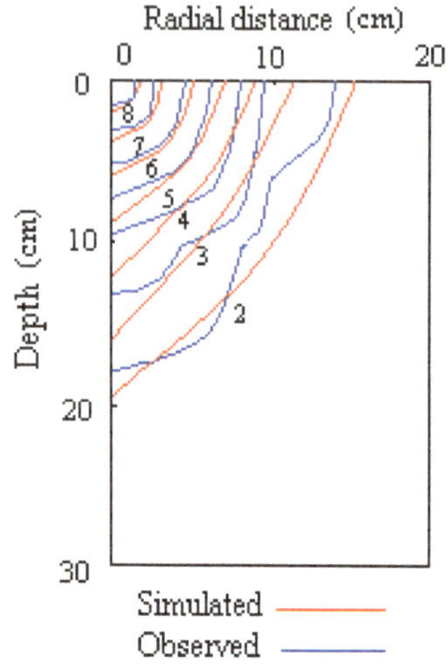

Fig. 5. Potassium concentration in soil solution (mg L^{-1}x100) simulated and observed 24 hours after irrigation.

For a time of redistribution of 48 hours after irrigation the content of potassium in the bulb cells remained in an almost unchanged, although the concentration of this element in general have risen as a result of decreased water content in the cells. The potassium present in soil solution in the rings located between 10 and 20 cm shaft showed concentrations ranging from 46 to 250 mg L^{-1} for both cases, showed similar trend and, with a standard error equal to 12.67 mg L^{-1} (in this case for less than 24 hours), the difference between simulated and observed values was in the range of 10%.

The content of potassium into cells in the soil 72 hours after irrigation has remained virtually unchanged with respect to that shown for 48 hours, and the concentration of this element on the soil surface within a radius of 10 to 20 cm from the emitter, ranged between 48 and 258 mg L^{-1}, with differences between predicted and observed roughly 12% with a value for the standard error of 11.30 mg L^{-1}, slightly lower than the value obtained for 48 hours.

Overall, the three times of redistribution considered, the concentration of potassium in the soil solution simulated by the model was similar to that seen in experimental conditions, and the displacement of this element partially followed the displacement of water, focusing primarily on cells more internal bulb. As explained in the case of moisture, the differences found between the concentrations of potassium possibly simulated and observed were mainly due to limitations of the model, because during development were not considered the process of hysteresis, considering only movement and adsorption process. Also soil physical-hydric variation, the errors in sampling and errors in determining the averaged concentration of potassium in solution in the laboratory.

It was also observed for the three times of redistribution, the values of potassium concentration obtained experimentally were lower than those generated by the model, which may be due to variations in bulk density in filling the box soil, the interaction of potassium soil solution with the other cations in the exchange complex of soil, as Ca, Na, Mg and Al. In addition the fact that the determination of the retardation factor was made in a soil washed with distilled water under conditions saturation practically free of cations in solution, which might not exactly represent the delay of potassium that occurs under real conditions, when cells of the bulb are under different levels of moisture (unsaturated condition) and when there are other cations in soil solution.

Moreover, the concentration of potassium observed in the three study time showed very low rates of redistribution and concentrated on the soil surface in the vicinity of emitter, which shows that besides being trapped in the inner layers to interact with the soil matrix, potassium is transported mainly by convective flow with the water (also called mass flow) in proportion to its flow and concentration. Therefore, the general pattern of distribution of potassium in the soil considered by the model, is dependent on the initial concentration of soil solution, the concentration in irrigation water, the emitter flow and physical-chemical properties of soil.

3.4 Sensitivity analysis of the moisture distribution in the bulb

As seen in Figure 6 respect to the saturated hydraulic conductivity, it appears that this has a important effect on the moisture of the bulb, especially when varied negatively, however for positive increments of the a standard error curve had lower slopes, and the model therefore less sensitive to increases in hydraulic conductivity than the reductions in this parameter.

The moisture content of the bulb is very sensitive to moisture changes of soil saturation, however, when this parameter begins to decrease (due to a reduction in porosity), the cells become saturated near the emitter, and begins to form a water depth at the soil surface, an aspect that is no longer considered by the model.

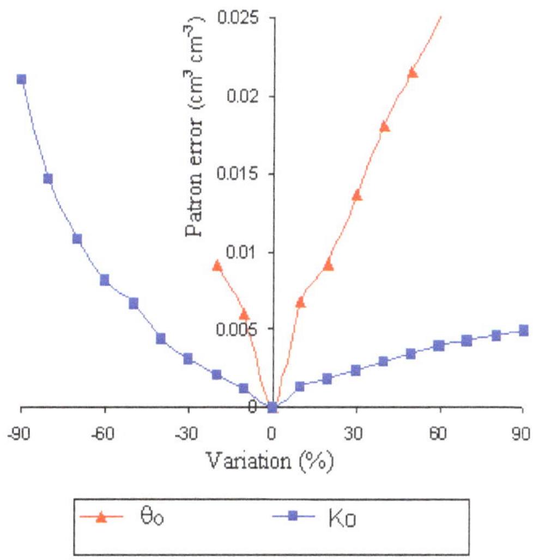

Fig. 6. Representation of the sensitivity analysis of the model with the profile of soil moisture on the bulb, applying from -90% to + 90% variation in soil moisture saturation (θo) and saturated hydraulic conductivity (Ko).

3.5 Sensitivity analysis of the potassium distribution in the bulb

As can be seen in Figure 7 referred to the saturated hydraulic conductivity, the model is sensitive to this parameter decreases, especially for very low values. For positive changes, the model is little sensitive, and be almost insensitive to high values of Ko.

With regard to soil moisture at saturation, the model is sensitivity to small increments of θo. For decreases of θo behavior is similar, but to a limited extent, as the soil begins to soak causing the impossibilities already discussed. The sensitivity of the model on this parameter was lower than that in the case of distribution of moisture in the bulb, because the potassium does not follow exactly the same dynamics of soil water.

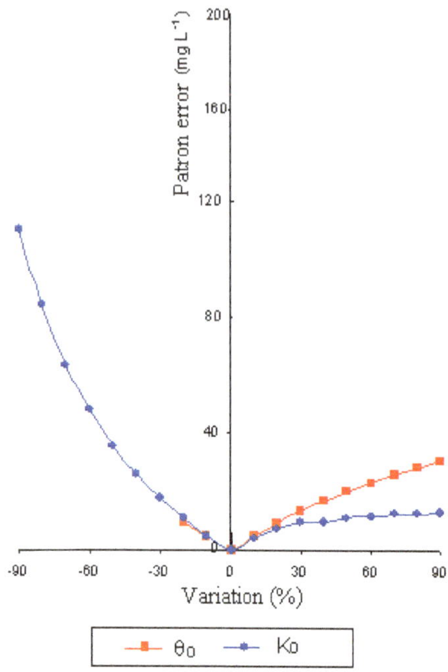

Fig. 7. Representation of the sensitivity analysis of the model relative to the concentration of potassium in the bulb, applying from -90% to + 90% variation in soil moisture saturation (θo) and saturated hydraulic conductivity (Ko).

4. Conclusions

The movement and distribution of water and potassium applied through drip irrigation can be modeled mathematically by the solution of the equations of transient flow by using the method of control volumes. There was a good fit in the values of the distribution of water and potassium in the bulb when compared to data simulated by the model with experimental data, yielding average values for the standard error equal to 0.0114 $cm^3\ cm^{-3}$ and 18.013 mg L^{-1} to moisture and potassium, respectively. The potassium distribution was limited to the inner layers of the bulb, which delayed the shift of this cation to interact with the soil matrix. The model, with respect to the distribution of soil water, is very sensitive to moisture saturation variations and moderately sensitive to hydraulic conductivity for saturated soil, on the other hand potassium distribution is affected mainly by the physical-chemical soil properties, therefore the model is very sensitive to negative variations of hydraulic conductivity for saturated soil.

5. References

Alvarez, J.; Herguedas, A; Atienza, J. (1995). *Modelización numerica y estimación de parámetros para la descripción del transporte de solutos en columnas de suelo en laboratorio.* Madrid: INIA. 69p.

Botrel, T. A. (1988). *Simulação da distribuição espacial da água em solo irrigado com gotejador.* Piracicaba. 61p. Tese (Doutorado) - Escola Superior de Agricultura "Luiz de Queiroz", Universidade de São Paulo.

Cruz, R. L. (2000). Modelização do balanço hídrico de uma cultura irrigada por um sistema de irrigação localizada. Botucatu. 80 p. Tese (Livre Docente) - Faculdade de Ciências Agronômicas, Universidade Estadual Paulista "Júlio de Mesquita Filho".

Hachum, A. Y.; Alfaro, J. F.; Willardson, L. S. (1976). Water movement in soil from trickle source. *Journal of Irrigation and Drainage Engineering,* v.102, n.2, p.179-192.

Lafolie, F.; Guennelon, R.; van Genuchten, M. Th. (1989). Analysis of water flow under trickle irrigation: I Theory and numerical solution. *Soil Science Society American Journal,* n. 53, p. 1310-1318.

Levin, I.; van Rooyen, P. C.; van Rooyen, F. C. (1979). The effect of discharge rate and intermitent water application by point-source irrigation on the soil moisture distribution pattern. *Soil Science Society American Journal,* n. 43, p. 8-16.

Matos, A. T.; Costa, L. M.; Fontes, M. P.; Martinez, M. A. (1999). Retardation factors and the dispersion-diffusion coeficients of Zn, Cd, Cu and Pb in soils from Viiçosa-MG, Brazil. *Transactions of the ASAE,* v. 42, n.4, p.903-910.

Miranda, J. H. (2001). Modelo para simulação da dinâmica de nitrato em colunas verticais de solo não saturado. Piracicaba. 79 p. Tese (Doutorado) - Escola Superior de Agricultura "Luiz de Queiroz", Universidade de São Paulo.

Mualem, Y. A. (1976). A new model for predicting the hidraulic conductivity of unsaturated porous media. *Water Resources Research,* v. 12, n.3, p.513-522.

Nielsen, D. R.; Biggar, J. W. (1962). Miscible displacement: III. Theorical considerations. *Soil Science Society of American Proceedings,* v. 26, n. 2, p.216-221.

Nogueira, C. P.; Coelho, E. F.; Leão M. C. S. (2000). Caracteristicas e dimensões do volume de um solo molhado sob gotejamento superficial e subsuperficial. *Revista Brasileira de Engenharia Agrícola e Ambiental,* v. 4, n. 3, p.315-320.

Toride, N.; Leij, F. van Genuchten, M. Th. (1999). *The CXTFIT code for estimating parameters from laboratory or field tracer experiments, verion 2.1.* Califórnia: Research Report U. S.; Salinity Laboratory Agricultural Research Service; U. S. Departament of Agriculture. 85p.

van Der Ploeg, R. R.; Benecke, P. (1974). Unsteady unsatured, n-dimensional moisture flow in soil: a computer simulation program. *Soil Science Society of America Proceedings,* v.38, p. 881-885.

van Genuchten, M. Th. (1980). A closed form equation for predicting the hydraulic conductivity of unsatured Soils. *Soil Science Society American Journal,* v. 44, p. 892-898.

van Genuchten, M. Th.; Wierenga, P. J. (1986). Solute dispersion coefficients and retardation factors. In: KLUTE, A. *Methods of Soil Analysis. I-Physical and Mineralogical Methods.* Madison: Soil Science Society of America. p.1025-1054.

18

Effects of Model Layer Simplification Using Composite Hydraulic Properties

Nicasio Sepúlveda and Eve L. Kuniansky
U. S. Geological Survey
United States

1. Introduction

Groundwater provides much of the fresh drinking water to more than 1.5 billion people in the world (Clarke et al., 1996) and in the United States more that 50 percent of citizens rely on groundwater for drinking water (Solley et al., 1998). As aquifer systems are developed for water supply, the hydrologic system is changed. Water pumped from the aquifer system initially can come from some combination of inducing more recharge, water permanently removed from storage, and decreased groundwater discharge. Once a new equilibrium is achieved, all of the pumpage must come from induced recharge and decreased discharge (Alley et al., 1999). Further development of groundwater resources may result in reductions of surface water runoff and base flows. Competing demands for groundwater resources require good management. Adequate data to characterize the aquifers and confining units of the system, like hydrologic boundaries, groundwater levels, streamflow, and groundwater pumping and climatic data for recharge estimation are to be collected in order to quantify the effects of groundwater withdrawals on wetlands, streams, and lakes. Once collected, three-dimensional (3D) groundwater flow models can be developed and calibrated and used as a tool for groundwater management. The main hydraulic parameters that comprise a regional or subregional model of an aquifer system are the hydraulic conductivity and storage properties of the aquifers and confining units (hydrogeologic units) that confine the system.

Many 3D groundwater flow models used to help assess groundwater/surface-water interactions require calculating "effective" or composite hydraulic properties of multi-layered lithologic units within a hydrogeologic unit. The calculation of composite hydraulic properties stems from the need to characterize groundwater flow using coarse model layering in order to reduce simulation times while still representing the flow through the system accurately. The accuracy of flow models with simplified layering and hydraulic properties will depend on the effectiveness of the methods used to determine composite hydraulic properties from a number of lithologic units.

The computation of the composite transmissivity for heterogeneous aquifers was presented for steady flow by Copty et al. (2006) by using a weighted average of the log transforms of the locally measured transmissivity within the aquifers. This probabilistic approach leads to a large-scale transmissivity that is approximately equal to the geometric mean of local transmissivity. The large-scale transmissivity has been shown to be larger than the geometric mean of local tests at site-specific studies (Sánchez-Vila et al., 1996). Simplification of model layers with composite hydraulic properties is not always feasible (Wen & Gómez-Hernández, 1996). Calculation of composite hydraulic properties has been mostly limited in the literature to hydraulic conductivity and transmissivity. Feng et al. (2007) computed composite hydraulic conductivity using the constant-flux assumption of Darcy's law within a finite-difference block by integrating, over the thickness of the block, the effective pressure-head dependent hydraulic conductivity divided by the head difference within the block. Parameter estimation of hydraulic properties through inverse modeling was used by Ward et al. (2006) to derive the composite hydraulic conductivity in the unsaturated zone.

This chapter evaluates the effects of simplifying hydraulic property layering within an unconfined aquifer and confining unit system that overlies a deeper confined aquifer, which is the principal water supply. This goal is accomplished in three steps. First, reference values of hydraulic properties are derived for all of the lithologic units identified in the unconfined aquifer and confining unit at two well fields in central Florida (USA) through inverse modeling of aquifer-test data. Second, simplified representations of the referenced lithologic units at the well fields are constructed by computing composite hydraulic properties for several aggregated units from the reanalysis of the aquifer-test data. Finally, the effects of lithologic simplification and the performance of the methods used to compute composite hydraulic properties are evaluated using fully 3D and quasi-3D flow models in which the number of lithologic units is gradually reduced. The most effective composite flow model is identified from flow residuals. A shallow and a deep well field in the surficial aquifer system (SAS) were selected to conduct aquifer tests to define a hydraulic conductivity field with the highest resolution possible.

Four multi-observation-well aquifer tests at two well fields were conducted and analyzed. The head data and the hydraulic properties determined for the two well fields from the analysis of the aquifer-test data were used to develop several conceptual 3D groundwater flow models by gradually using fewer layers to determine acceptable levels of layer simplification for simulation of water levels and leakage rates at the two well fields. Additionally, quasi-3D flow models were compared to fully 3D models (McDonald & Harbaugh, 1988) to assess the effects of assuming only a vertical flow component (quasi-3D) versus allowing both vertical and horizontal flow components in the leaky confining unit (fully 3D). The reader is referred to Harbaugh et al. (2000) for further details establishing the difference between these two conceptual flows.

Results of simulated flows and heads with a simplified number of layers are compared to those simulated with the most discretized set of hydraulic properties derived from the aquifer-test data analyses to assess the feasibility of preserving the main flow features with a simplified layering model. Multiple flow simulations generated from the methods of calculating composite hydraulic properties are compared to identify potentially unrealistic conceptual flows. This comparison allows a quantitative assessment of the simulated flow residuals introduced by the simplification.

2. Derivation of preliminary hydraulic parameters through slug-test data analyses

The SAS, composed mostly of sand, silt, clay, and gravel, is the uppermost water-bearing unit in most of central Florida and overlays the Floridan aquifer system (FAS). Slug tests were performed in two well fields in the SAS; one named Carrot Barn located in northwest Lake County and another one named Lyonia Preserve located in southwest Volusia County,

Florida (Fig. 1). The analyses of the slug test data resulted in the derivation of horizontal hydraulic conductivity from water-level displacement data recorded after introducing or withdrawing a closed cylinder to or from the wellbore. The stress exerted on the aquifer by the removal or injection of a closed cylinder is considered small as it involves no pumping and thus, the hydraulic conductivity values obtained from the analysis of slug test data apply to a maximum of 1 to 2 meters of open interval to the aquifer. The length of this open interval is considered to be of limited spatial range when compared to an aquifer test where the production well stresses a larger percentage of the aquifer thickness; a percentage that depends on the design of the production well.

Fig. 1. Locations of the surficial aquifer system well fields in (a) Carrot Barn, Lake County and Lyonia Preserve, Volusia County, Florida.

Aquifer responses to slug tests are classified depending on the shape of the graph of the water levels as they return to equilibrium after the injection or removal of the closed cylinder. Responses are classified as overdamped (Bouwer & Rice, 1976) when the water levels return to equilibrium following the shape of a slowly decaying exponential, or as underdamped (Van der Kamp, 1976) when the water levels vary sinusoidally while returning to equilibrium. Hydraulic conductivity values resulting from the analyses of underdamped aquifer responses are greater than those from overdamped responses. While an overdamped aquifer response to a slug test is typical for a clay, silt, fine silty sand, or low porosity limestone, an underdamped aquifer response can occur for gravel, well-sorted sand, or karstified limestone (media with large diffusivity, defined as transmissivity divided by storage). Underdamped water-level responses were recorded at wells L-0892, L-0894, and L-0901 (Fig. 2a) and in wells V-0876, V-0877, V-0878, V-0881, V-0882, and V-0883 (Fig. 2b). Overdamped responses were recorded at the remaining wells shown in figure 2.

Results of the horizontal hydraulic conductivities from slug test data are shown in the cross sections in figure 2. These hydraulic conductivity values derived from the slug tests analysis correspond to the immediate vicinity of the well open to the aquifer. These values are compared to those obtained from the drawdown analysis of aquifer test data, which correspond to a much larger spatial scale than that reflected by slug tests. In turn, these aquifer test results are then compared to regional hydraulic conductivity values derived from groundwater flow simulation to provide a perspective of scale variability under regional flow.

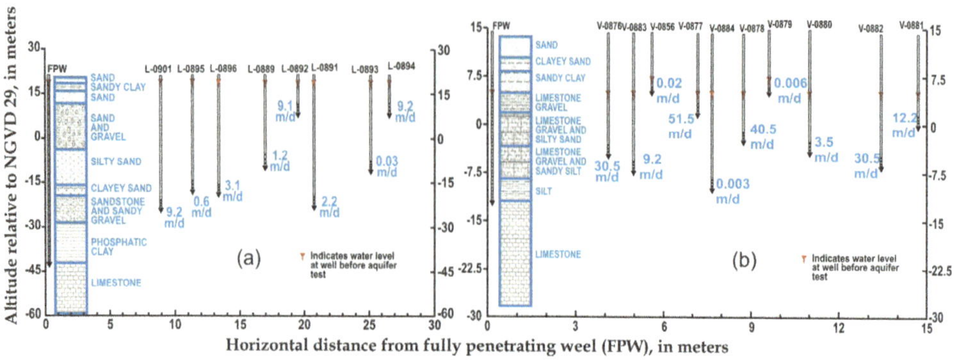

Fig. 2. Lithologic units identified in the fully penetrating well (FPW) and hydraulic conductivity values derived from slug tests performed at open intervals of (a) Carrot Barn SAS wells and (b) Lyonia Preserve SAS wells, in meters per day. [Altitudes refer to the National Geodetic Vertical Datum (NGVD) of 1929 (Snyder, 1987)]

The complexity of the intricate layering shown in figure 2 is exemplified by the relatively wide range of hydraulic conductivity values derived from slug tests. Analysis of drawdown data derived from larger scale aquifer tests can result in hydraulic conductivity values that depict a more regional perspective than the highly discretized layering properties shown in figure 2.

3. Derivation of reference hydraulic properties through aquifer-test data analyses

Multi-observation well aquifer tests were conducted at the Carrot Barn and the Lyonia Preserve well fields (Fig. 1). Wells open to the SAS, an unconfined aquifer, were pumped to determine the hydraulic properties of all lithologic units identified above the generally confined Upper Floridan aquifer (UFA). These lithologic units were identified while drilling fully (FPW) and partially penetrating production wells (PPW). Aquifer tests for both FPW and PPW were conducted to generate the drawdown data that allow the computation of most hydraulic properties. When the entire aquifer thickness is stressed by pumping from the FPW, a predominantly radial flow is generated towards the well. Such flow generally allows the calculation of the hydraulic properties of all the lithologic units identified in the SAS, with the exception of the vertical hydraulic conductivity. The calculation of the vertical to horizontal hydraulic conductivity ratio of the lithologic units in the SAS can be obtained from the aquifer response to pumping from a PPW (Hantush, 1964) because such stress generates a non-negligible vertical flow component towards the well.
The SAS overlies a leaky confining unit called the intermediate confining unit (ICU). Observation wells tapping the SAS in a shallow water-table setting (Carrot Barn well field, Fig. 1a) and in a deep water-table setting (Lyonia Preserve well field, Fig. 1b), were screened to the upper, middle, and lower depths of the SAS and to the ICU to isolate the water-level response at these depths. Flow in the SAS and ICU at each one of these well fields should be regarded as independent flow subsystems of the larger SAS, ICU, and UFA flow system in east-central Florida because of the different number of lithologic units within the SAS and ICU identified at these two well fields.

Although the SAS is generally unconfined with discontinuous clay lenses that make this aquifer semiconfined in some areas. The SAS consists predominantly of fine-to-coarse grained quartz sand with varying amounts of silt and clay. The sediments in the SAS generally grade into less permeable silty or clayey sands with increasing depth. The ICU is composed primarily of phosphatic clays, silts, and sands. The ICU separates the SAS from the UFA, which is composed of fractured carbonate rocks (Miller, 1986). The UFA is the major aquifer used for water supply in Florida.

Four aquifer tests were conducted, two at each well field, by pumping the FPW and PPW during separate tests. Each aquifer test began after water levels had reached static equilibrium. At the Carrot Barn well field, both FPW and PPW were pumped at 5.8 x 10^{-3} m^3/s for 70 hours. At the Lyonia Preserve well field, the FPW was pumped at 6.2 x 10^{-3} m^3/s for 67 hours and the PPW was pumped at 4.5 x 10^{-3} m^3/s for 30 hours. Pumping rates for each test were constant. Horizontal and vertical hydraulic conductivity, specific yield, and specific storage at various depths in the SAS and ICU were determined at each well field.

Walton (2008) contends that multiple observation well aquifer tests should be analyzed with more sophisticated techniques, such as the use of numerical models with regression-based parameter estimation techniques. The four tests were analyzed by developing radial, axisymmetric groundwater flow models. Radial axisymmetric models are typically used to approximately simulate 3D flow to wells (Reilly & Harbaugh, 1993; Langevin, 2008). The computer code MODOPTIM (Halford, 2006), based on MODFLOW-96 (Harbaugh & McDonald, 1996), was used to derive the hydraulic properties for each lithologic unit using regression-based parameter estimation techniques to minimize the sum of square residuals (SSR) between measured and simulated drawdown (Figs. 3, 4). Analysis of the lithologic data from drilling the FPW and PPW at the Carrot Barn well field identified five individual lithologic units in the SAS and one lithologic unit in the ICU (Fig. 5a). A similar analysis of data from the Lyonia Preserve well field indicated the presence of three lithologic units in the SAS and one lithologic unit in the ICU (Fig. 5b).

The Carrot Barn well field has an average land surface altitude of about 20 m above the NGVD of 1929 with the farthest well about 27 m away from the FPW. The Lyonia Preserve well field has an average land surface altitude of 14 m above the NGVD of 1929 with the farthest well about 15 m away from the FPW. The aquifer tests were designed to derive hydraulic properties at the local scale of the well fields and the numerical grids used to simulate flow in these well fields were spaced radially to consider the same local scale.

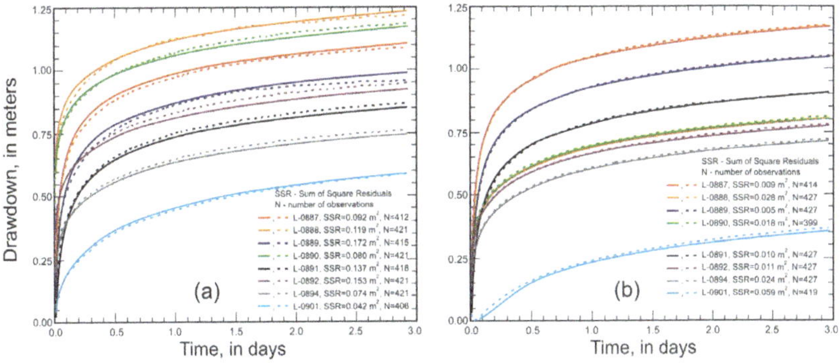

Fig. 3. Simulated (solid lines) and measured (dashed lines) water levels at observation wells in the Carrot Barn well field after pumping from the (a) FPW and (b) PPW.

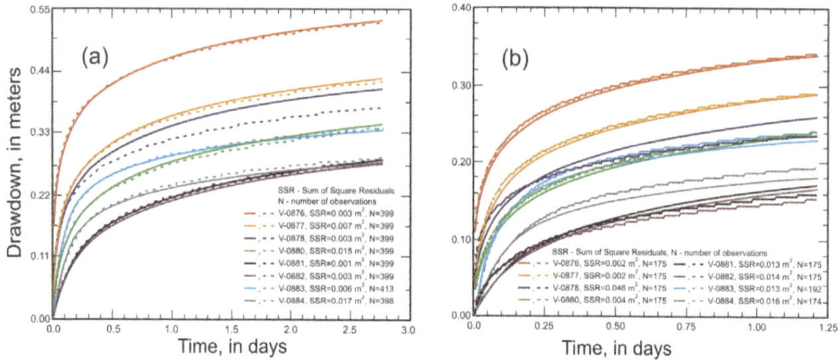

Fig. 4. Simulated (solid lines) and measured (dashed lines) water levels at observation wells in the Lyonia Preserve well field after pumping from the (a) FPW and (b) PPW.

The model grid used for the analysis of the aquifer-test data consisted of rows of uniform thickness of 0.305 m in the SAS, ICU, and UFA (vertical discretization) with a uniform hydraulic property assigned to each model layer (row) representing the lithologic unit identified in figure 5. The number of rows in each model grid can be calculated by dividing the thicknesses of the lithologic units shown in figure 5 by 0.305 m. A high resolution vertical grid was used to minimize truncation errors in the simulated heads in each layer. The column spacing (radial distance from the well) consisted of 150 columns that gradually increased in width from 0.12 to 2,593 m for a total distance of 40,234 m, extending beyond the radius of influence for the pumping rates applied in each test. There was no rainfall during any of the aquifer tests and thus, no-flow boundaries were imposed at the top of the SAS and at the lateral boundary away from the well. Models were calibrated to measured drawdown. The high transmissivity of the UFA and the fact that only the SAS was pumped were the reasons for choosing a constant-head boundary at the bottom of the UFA layer.

The computation of the hydraulic properties for all lithologic units identified in the SAS represents the highest vertical discretization of layers for the Carrot Barn (Fig. 5a) and Lyonia Preserve (Fig. 5b) well fields. Specific yield, computed only for the uppermost layer of the SAS, was 0.118 and 0.266 for the Carrot Barn and Lyonia Preserve well fields, respectively. These specific yield values were applied to the uppermost layer of the SAS in all subsequent simplified conceptual flow models. Specific yield values determined from the aquifer-test data for the uppermost simulated layer were kept constant from one simplified model to another as the specific yield used for the uppermost simulated layer.

Pumping from the SAS did not cause measurable drawdown in the UFA, thus, hydraulic properties for the UFA could not be estimated from the aquifer-test drawdown data. The simulated horizontal hydraulic conductivity for the UFA at the Carrot Barn (30.1 m/d) and Lyonia Preserve (82.5 m/d) well fields were obtained from model-simulated values reported by McGurk & Presley (2002). These hydraulic properties for the UFA are not used in the simulated layer simplification process because composite hydraulic properties are derived only for the lithologic units in the SAS and ICU. Vertical hydraulic conductivities were obtained by assuming a vertical to horizontal hydraulic conductivity anisotropy ratio of 0.01 because such a ratio is the most prevailing in the UFA flow simulations in east-central Florida (McGurk & Presley, 2002; Sepúlveda, 2002). Specific storage for the UFA (Fig. 5) was assigned based on water and limestone compressibility values (Domenico, 1972).

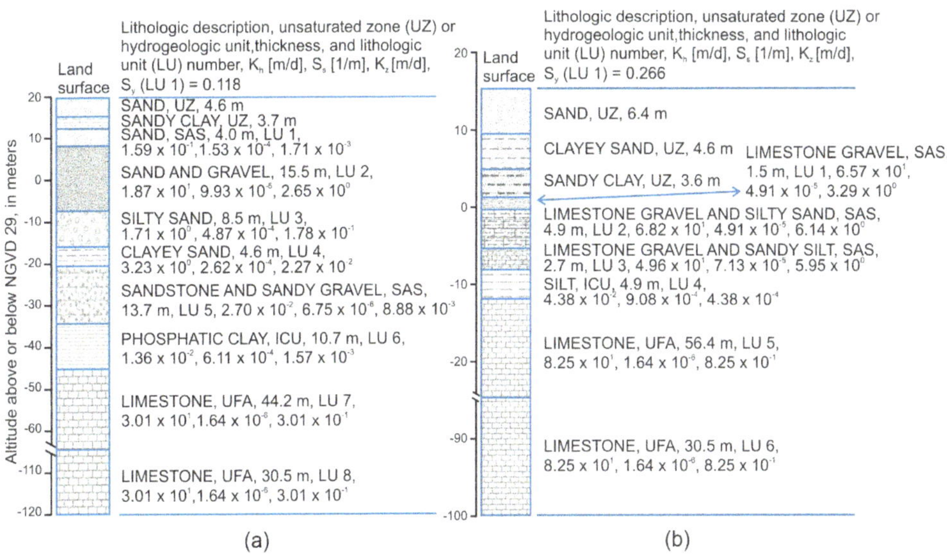

Fig. 5. Highest possible lithologic unit discretization within the SAS at (a) Carrot Barn well field and at (b) Lyonia Preserve well field and hydraulic properties derived from the analysis of aquifer-test data. The abbreviated vertical axis indicates relatively thick layers.

A comparison of the horizontal hydraulic conductivities derived from the analysis of slug test data for the cross sections in figure 2 and those hydraulic conductivity values derived from the analysis of aquifer test data for the appropriate lithologic units in figure 5 indicate a trend in the relation between these two sets of values. At the Carrot Barn well field, average hydraulic conductivity values from slug test data for wells with open intervals to lithologic units 2 to 5 in the SAS were 9.15, 0.62, 1.85, and 5.7 meters per day, respectively. The corresponding hydraulic conductivity values derived from the aquifer tests for the same lithologic units were 18.7, 1.71, 3.23, and 13.7 meters per day (Fig. 5a). At the Lyonia Preserve well field, average hydraulic conductivity values from slug test data for lithologic units 1 to 3 in the SAS were 51.5, 26.4, and 18.4 meters per day, respectively. The corresponding hydraulic conductivity values derived from the aquifer tests for the same lithologic units were 65.7, 68.2, and 49.6 meters per day (Fig. 5b). The consistently higher hydraulic conductivity values derived from the aquifer tests indicate a slightly larger scale of flow is accounted for over a larger area to better represent flow more areally extensive. The different magnitudes and scales of aquifer stresses generated by slug and aquifer tests result in different hydraulic conductivity values. The identification of this variability in hydraulic conductivity resulting from slug tests and aquifer tests is a result of the high degree of discretization used to identify the vertical extent of uniform soil grains of each lithologic unit as shown in figures 2 and 5.

4. Derivation of composite hydraulic properties for simplified model layering

Composite hydraulic properties for simplified model layering can be derived by computing equivalent hydraulic properties for aggregated layers using several methods. The first method presented in this chapter is to compute the thickness weighted average horizontal hydraulic conductivity and specific storage and the thickness weighted harmonic mean vertical hydraulic conductivity for a reduced number of layers in the unconfined aquifer and the leaky confining unit. The second method is to compute hydraulic properties by using the observed head from aquifer tests and applying the regression-based parameter estimation (PE) techniques (Poeter et al., 2005; Halford, 2006; Hill & Tiedeman, 2007). As the number of simulated layers in the unconfined aquifer is decreased, the performance of each model is assessed by analyzing changes in simulated heads under aquifer test conditions and by simulating flow exchanges between the hydrogeologic units under hypothetical but realistic recharge rates to the unconfined aquifer and withdrawals from the confined aquifer.

Composite horizontal hydraulic conductivity K_{hc}, and vertical hydraulic conductivity K_{zc}, and specific storage S_{sc}, from any set of L lithologic units (LUs) are calculated from the equations:

$$K_{hc} = \frac{K_{h1} b_1 + K_{h2} b_2 + \cdots + K_{hN} b_L}{b_1 + b_2 + \cdots + b_L}, \tag{1}$$

$$K_{zc} = \frac{b_1 + b_2 + \cdots + b_L}{\frac{b_1}{K_{z1}} + \frac{b_2}{K_{z2}} + \cdots + \frac{b_L}{K_{zL}}}, \text{ and} \tag{2}$$

$$S_{sc} = \frac{S_{s1} b_1 + S_{s2} b_2 + \cdots + S_{sL} b_L}{b_1 + b_2 + \cdots + b_L}, \tag{3}$$

where $b_1, b_2, \cdots$, and b_L are the thicknesses; $K_{h1}, K_{h2}, \cdots$, and K_{hL} are the horizontal hydraulic conductivity; $S_{s1}, S_{s2}, \cdots$, and S_{sL} are the specific storage; and $K_{z1}, K_{z2}, \cdots$, and K_{zL} are the vertical hydraulic conductivity of each LU. Eqs. (1) and (2) are derived in Bouwer (1978). Assuming that the porosity and water density remain nearly constant among layers, Eq. (3) states that aquifer matrix compressibility can be averaged over layer thickness, thus it mimics the form of Eq. (1). Note that Eq. (3) is not used to estimate the specific yield of the uppermost layer of the unconfined aquifer because specific yield, being several orders of magnitude greater than specific storage multiplied by layer thickness, was used for the uppermost layer of the SAS in all conceptual flow models.

Composite hydraulic properties from an arbitrary number of layers using PE techniques are derived in this chapter by minimizing the normalized residuals between the measured drawdown during the aquifer tests and the simulated drawdown by models where several LUs become a single composite layer. The head residuals in the unconfined aquifer are used to determine if the main flow features in this aquifer are preserved by each simplified conceptual flow model with the use of a radial, axisymmetric 2D flow model. The feasibility of each simplified model to preserve the main flow features in both the unconfined and confined aquifers requires a 3D groundwater flow model that simulates recharge to the unconfined aquifer and withdrawals from the confined aquifer.

The heads in the SAS, in the ICU, in the UFA, and the leakage rates to the UFA are the observations used to assess the accuracy of the simplified conceptual flow models by comparing these flow characteristics with those simulated by the highly discretized model. If *Di* and *Si* are subscripts used to denote flow parameters in the highly discretized model and the simplified conceptual flow model with a decreased number of layers, respectively, then dimensionless head and leakage rate residuals can be calculated from:

$$\sum_{i=1}^{N}\left(\frac{h_{SAS_{D_i}}-h_{SAS_{S_i}}}{h_{SAS_{D_i}}}\right)^2, \sum_{i=1}^{N}\left(\frac{h_{ICU_{D_i}}-h_{ICU_{S_i}}}{h_{ICU_{D_i}}}\right)^2, \sum_{i=1}^{N}\left(\frac{h_{UFA_{D_i}}-h_{UFA_{S_i}}}{h_{UFA_{D_i}}}\right)^2, \text{and} \sum_{i=1}^{N}\left(\frac{L_{D_i}-L_{S_i}}{L_{D_i}}\right)^2, \quad (4)$$

where the heads in the SAS are $h_{SAS_{D_i}}$ and $h_{SAS_{S_i}}$, the heads in the ICU are $h_{ICU_{D_i}}$ and $h_{ICU_{S_i}}$, the heads in the UFA are $h_{UFA_{D_i}}$ and $h_{UFA_{S_i}}$, the leakage rates to the UFA are L_{D_i} and L_{S_i}, and N is the number of simulated or measured head or leakage rate points. The leakage rates to the UFA were computed from the application of Darcy's law:

$$L_{D_i} = \frac{K_{zD}\left(h_{ICU_{D_i}} - h_{UFA_{D_i}}\right)}{b_{ICU,UFA}} \quad \text{and} \quad L_{S_i} = \frac{K_{zS}\left(h_{ICU_{S_i}} - h_{UFA_{S_i}}\right)}{b_{ICU,UFA}}, \quad (5)$$

where $b_{ICU,UFA}$ is the vertical distance between the center of the ICU and the UFA layers, and K_{zD} and K_{zS} are the equivalent vertical hydraulic conductivity between the lower half of the ICU and the upper half of the UFA layer, computed from Eq. (2) using the thicknesses of these two units.

5. Composite hydraulic properties for model layer simplification of the unconfined aquifer and confining unit

Eqs. (1)-(3) use lithologic thicknesses to calculate composite hydraulic properties to simplify model layering. However, the use of regression-based PE techniques (Halford, 2006) requires using the radial axisymmetric 2D flow models similar to the ones used to analyze the aquifer-test data, but with simplified model layers and calibrated to the same aquifer-test data. The models that simulated the combined flow in the SAS and ICU as one lithologic unit (quasi-2D flow model) were developed assuming an ICU layer with zero specific storage and zero horizontal hydraulic conductivity in Eqs. (1) and (3), such that only vertical flow can occur in the ICU.

The most simplified flow model used a quasi-2D approach to represent the effect of only vertical flow in the ICU. In this approach, the head and storage changes in the ICU are not simulated. The performance of the two methods of computing composite hydraulic properties for the various simplified models was assessed by examining the aquifer test residuals between measured water levels in the SAS and the simulated water levels by the 2D radial axisymmetric flow models. The computed root-mean square residuals (RMSRs) for each well field were used to characterize the accuracy of each simplified model to simulate flow.

A comparison of all root-mean square residuals (RMSRs) between measured and simulated aquifer-test drawdown in the SAS at the Carrot Barn well field shows that the highest increase in RMSRs occurred from the highest layering discretization available, or the 8 LU radial axisymmetric flow model, to the 3 LU or quasi-2D flow model (Table 1), whether using average hydraulic properties from Eqs. (1)-(3), or from a regression-based PE technique. Similarly, the highest increase in RMSRs at the Lyonia Preserve well field occurred from the highest lithologic unit discretization available, or the 6 LU radial axisymmetric flow model, to the 3 LU or quasi-2D flow model (Table 2). The largest changes in flow characteristics occurred from fully 2D to quasi-2D models. In both cases, the regression-based PE technique resulted in lower RMSR values when compared to the methods that used Eqs. (1)-(3) to calculate composite hydraulic properties for the simplified layers in each model.

Conceptual Flow Model	RMSRs, in meters	
	Using Eqs. (1)-(3)	Using regression-based PE technique
8 LU (FPW)	Not applicable	1.60×10^{-2}
6 LU (FPW)	1.00×10^{-1}	3.65×10^{-2}
4 LU (FPW)	3.18×10^{-1}	1.08×10^{-1}
3 LU (FPW)	3.95×10^{-1}	1.98×10^{-1}
8 LU (PPW)	Not applicable	7.00×10^{-3}
6 LU (PPW)	1.72×10^{-1}	3.72×10^{-2}
4 LU (PPW)	3.48×10^{-1}	1.98×10^{-1}
3 LU (PPW)	3.96×10^{-1}	2.12×10^{-1}

Table 1. Root-mean-square residuals (RMSRs) between measured and simulated drawdown at observation wells in the surficial aquifer system (SAS) of the Carrot Barn well field due to pumping from the fully (FPW) or partially penetrating well (PPW) tapping the SAS. [RMSRs were computed from the radial axisymmetric 2D flow models based on 3,335 and 3,367 measured drawdown points for the FPW and PPW tests, respectively, for the original 8 lithologic unit (LU) model and the simplified 6 LU, 4 LU, and 3 LU models; PE, parameter estimation. "Not applicable" indicates, for the 8 LU (FPW) and 8 LU (PPW) conceptual flow models, that these two are the highest layering discretization models available from aquifer tests, or not simplified models. For these, Eqs. (1)-(3) did not have to be used to generate parameters for simplified models]

Composite hydraulic properties for the simplified models derived from Eqs. (1)-(3) or from regression-based PE techniques at each well field showed that the largest variations occurred in the vertical hydraulic conductivity (Tables 3 and 4). The main difference between composite hydraulic properties for the simplified conceptual flow models derived from Eqs. (1)-(3) and from the regression-based PE technique was the composite vertical hydraulic conductivity derived for the 3 LU models for Lyonia Preserve, where the difference was one order of magnitude (Table 4). This difference could be explained by the contrasts in derived vertical hydraulic conductivity between the ICU and its overlying lithologic unit. Such a large contrast in vertical hydraulic conductivity between the ICU and its overlying unit was not observed in the Carrot Barn well field (Table 3). The largest increases in RMSRs from one conceptual model to another coincided with the largest contrast in vertical hydraulic conductivity between two contacting LUs. In particular, a large contrast in vertical hydraulic conductivity between the lowermost LU of the SAS and the ICU for the Lyonia Preserve models caused large increases in RMSRs. These large increases did not occur for the Carrot Barn models due to a smaller contrast in vertical hydraulic conductivity between the lowermost LU of the SAS and the ICU.

Conceptual Flow Model	RMSRs, in meters	
	Using Eqs. (1)-(3)	Using regression-based PE technique
6 LU (FPW)	Not applicable	4.09×10^{-3}
5 LU (FPW)	6.78×10^{-3}	4.18×10^{-3}
4 LU (FPW)	5.11×10^{-3}	4.19×10^{-3}
3 LU (FPW)	2.58×10^{-1}	1.86×10^{-1}
6 LU (PPW)	Not applicable	8.74×10^{-3}
5 LU (PPW)	9.18×10^{-3}	8.76×10^{-3}
4 LU (PPW)	8.85×10^{-3}	8.77×10^{-3}
3 LU (PPW)	1.69×10^{-1}	1.33×10^{-1}

Table 2. Root-mean-square residuals (RMSRs) between measured and simulated drawdown at observation wells in the surficial aquifer system (SAS) of the Lyonia Preserve well field due to pumping from the fully (FPW) or partially penetrating well (PPW) tapping the SAS. [RMSRs were computed from the radial axisymmetric 2D flow models based on 3,205 and 1,416 measured drawdown points for the FPW and PPW tests, respectively, for the original 6 lithologic unit (LU) model and the simplified 5 LU, 4 LU, and 3 LU models; PE, parameter estimation. "Not applicable" indicates, for the 6 LU (FPW) and 6 LU (PPW) conceptual flow models, that these two are the highest layering discretization models available from aquifer tests, or not simplified models. For these, Eqs. (1)-(3) did not have to be used to generate parameters for simplified models]

A substantially thicker UFA than the SAS or the ICU results in a larger transmissivity of the UFA than the SAS or ICU. This applies to both the Carrot Barn and the Lyonia Preserve sites. Hydraulic properties of the UFA for layer numbers 7 and 8 remained unchanged throughout the calculation of composite hydraulic properties because the hydraulic conductivity and thickness of the UFA is much greater than those of the ICU. Thus, pumping from the SAS does not generate a large enough stress to cause drawdown in the UFA.

<table>
<tr><th>LU number</th><th>6 LU model Eqs. (1)-(3)</th><th>6 LU model PE</th><th>4 LU model Eqs. (1)-(3)</th><th>4 LU model PE</th><th>3 LU model Eqs. (1)-(3)</th><th>3 LU model PE</th></tr>
<tr><td>SAS - 1</td><td>1.59 x 10^{-1}
1.53 x 10^{-4}
1.71 x 10^{-3}</td><td>1.59 x 10^{-1}
1.53 x 10^{-4}
1.71 x 10^{-3}</td><td rowspan="5">6.93 x 10^{0}
1.64 x 10^{-4}
1.13 x 10^{-2}</td><td rowspan="5">7.94 x 10^{0}
1.54 x 10^{-4}
8.08 x 10^{-2}</td><td rowspan="6">6.93 x 10^{0}
1.64 x 10^{-4}
5.22 x 10^{-3}</td><td rowspan="6">5.67 x 10^{0}
7.19 x 10^{-5}
4.71 x 10^{-3}</td></tr>
<tr><td>SAS - 2</td><td>1.87 x 10^{1}
9.93 x 10^{-5}
2.65 x 10^{0}</td><td>1.87 x 10^{1}
9.93 x 10^{-5}
2.65 x 10^{0}</td></tr>
<tr><td>SAS - 3</td><td>1.11 x 10^{0}</td><td>6.60 x 10^{-1}</td></tr>
<tr><td>SAS - 4</td><td>2.03 x 10^{-4}</td><td>3.20 x 10^{-4}</td></tr>
<tr><td>SAS - 5</td><td>1.50 x 10^{-2}</td><td>8.05 x 10^{-2}</td></tr>
<tr><td>ICU - 6</td><td>1.36 x 10^{-2}
6.11 x 10^{-4}
1.57 x 10^{-3}</td><td>1.36 x 10^{-2}
6.11 x 10^{-4}
1.57 x 10^{-3}</td><td>1.36 x 10^{-2}
6.11 x 10^{-4}
1.57 x 10^{-3}</td><td>1.36 x 10^{-2}
6.11 x 10^{-4}
1.57 x 10^{-3}</td></tr>
<tr><td>UFA - 7
UFA - 8</td><td>3.01 x 10^{1}
1.64 x 10^{-6}
3.01 x 10^{-1}</td><td>3.01 x 10^{1}
1.64 x 10^{-6}
3.01 x 10^{-1}</td><td>3.01 x 10^{1}
1.64 x 10^{-6}
3.01 x 10^{-1}</td><td>3.01 x 10^{1}
1.64 x 10^{-6}
3.01 x 10^{-1}</td><td>3.01 x 10^{1}
1.64 x 10^{-6}
3.01 x 10^{-1}</td><td>3.01 x 10^{1}
1.64 x 10^{-6}
3.01 x 10^{-1}</td></tr>
</table>

Table 3. Hydraulic properties of the SAS, ICU, and UFA, for the simplified 6 LU, 4 LU, and 3 LU conceptual 2D radial axisymmetric flow models in the Carrot Barn well field, derived from the 8 LU model properties either using Eqs. (1)-(3) or regression-based parameter estimation (PE) with MODOPTIM. [Listed properties are: horizontal hydraulic conductivity (m/d), specific storage (1/m), and vertical hydraulic conductivity (m/d)]

<table>
<tr><th>LU number</th><th>5 LU model Eqs. (1)-(3)</th><th>5 LU model PE</th><th>4 LU model Eqs. (1)-(3)</th><th>4 LU model PE</th><th>3 LU model Eqs. (1)-(3)</th><th>3 LU model PE</th></tr>
<tr><td>SAS - 1</td><td>6.57 x 10^{1}
4.91 x 10^{-5}
3.29 x 10^{0}</td><td>6.57 x 10^{1}
4.91 x 10^{-5}
3.29 x 10^{0}</td><td rowspan="3">6.22 x 10^{1}
5.57 x 10^{-5}
5.32 x 10^{0}</td><td rowspan="3">6.29 x 10^{1}
6.24 x 10^{-5}
5.66 x 10^{0}</td><td rowspan="4">6.22 x 10^{1}
5.57 x 10^{-5}
1.26 x 10^{-3}</td><td rowspan="4">1.44 x 10^{1}
1.19 x 10^{-4}
4.39 x 10^{-4}</td></tr>
<tr><td>SAS - 2</td><td rowspan="2">6.15 x 10^{1}
5.71 x 10^{-5}
6.07 x 10^{0}</td><td rowspan="2">6.28 x 10^{1}
6.13 x 10^{-5}
5.65 x 10^{0}</td></tr>
<tr><td>SAS - 3</td></tr>
<tr><td>ICU - 4</td><td>4.38 x 10^{-2}
9.08 x 10^{-4}
4.38 x 10^{-4}</td><td>4.38 x 10^{-2}
9.08 x 10^{-4}
4.38 x 10^{-4}</td><td>4.38 x 10^{-2}
9.08 x 10^{-4}
4.38 x 10^{-4}</td><td>4.38 x 10^{-2}
9.08 x 10^{-4}
4.38 x 10^{-4}</td></tr>
<tr><td>UFA - 5
UFA - 6</td><td>8.25 x 10^{1}
1.64 x 10^{-6}
8.25 x 10^{-1}</td><td>8.25 x 10^{1}
1.64 x 10^{-6}
8.25 x 10^{-1}</td><td>8.25 x 10^{1}
1.64 x 10^{-6}
8.25 x 10^{-1}</td><td>8.25 x 10^{1}
1.64 x 10^{-6}
8.25 x 10^{-1}</td><td>8.25 x 10^{1}
1.64 x 10^{-6}
8.25 x 10^{-1}</td><td>8.25 x 10^{1}
1.64 x 10^{-6}
8.25 x 10^{-1}</td></tr>
</table>

Table 4. Hydraulic properties of the SAS, ICU, and UFA, for the simplified 5 LU, 4 LU, and 3 LU conceptual 2D radial axisymmetric flow models in the Lyonia Preserve well field, derived from the 6 LU model properties either using Eqs. (1)-(3) or regression-based parameter estimation (PE) with MODOPTIM. [Listed properties are: horizontal hydraulic conductivity (m/d), specific storage (1/m), and vertical hydraulic conductivity (m/d)].

6. Evaluation of 3D models with simplified layers and hypothetical pumping of the confined aquifer

Although water was pumped out of the SAS during the aquifer tests, transient 3D groundwater flow models simulating pumping from the UFA were needed to assess the effects of model layer simplification in the SAS and ICU on heads and flows in this two-aquifer system, because the UFA is the principal water supply. These 3D groundwater flow models, regional in nature, were developed to simulate temporal variations in recharge to the SAS and pumping from the UFA, and to evaluate how simulated flow changes in the SAS and UFA might be affected by model layering simplification through the computation of composite hydraulic properties for the SAS and ICU.

The maximum number of model layers simulated for each well field varied according to site lithology (Fig. 5). Each model had 400 rows and 400 columns, with no-flow conditions imposed along the lateral boundaries, either a constant-head or no-flow boundary at the bottom of the UFA, and a specified recharge rate imposed at the top of the SAS. Square cells 80.5 m long were used to span an area of about 1,037 km^2 in models for each well field. Flow models used uniform hydraulic properties as shown in figure 5 for each simulated layer. Composite hydraulic properties for several lithologic units were calculated for the Carrot Barn and Lyonia Preserve models, as shown by the hydraulic properties applied to more than one unit in Tables 3 and 4. The composite hydraulic properties for the full thicknesses of the SAS and ICU units were calculated only for the 3 LU models, leaving the hydraulic properties of the UFA unchanged for all models.

Changes in SAS heads, ICU heads, UFA heads, and leakage rates to the UFA were analyzed with the transient 3D models using MODFLOW-2000 (Harbaugh et al. 2000). The combined flow of the SAS and the ICU was simulated by using a quasi-3D flow approximation (McDonald and Harbaugh 1988) where the composite hydraulic properties for these two units were calculated assuming zero horizontal hydraulic conductivity and zero specific storage in the ICU. The vertical hydraulic conductivity of the SAS and of the ICU are used in Eq. (2) in the quasi-3D flow approximation to compute the composite vertical hydraulic conductivity of the two, which results in a value closer to the less permeable lithologic unit. The assumption of no horizontal flow in the ICU in a quasi-3D flow approximation causes the SAS and ICU to be represented with a lower composite horizontal hydraulic conductivity from Eq. (1) than in a fully 3D flow simulation. The flow residuals simulated with a quasi-3D flow approximation are quantified and compared in this section with those of a fully 3D flow.

Flow in the UFA for all models was simulated with two layers with equal hydraulic properties. Hypothetical pumping wells tapped only the upper UFA lithologic unit (Fig. 5) to isolate the effects of pumping from the UFA on the induced flow from the SAS in each well field, to simulate a partial penetration of many pumping UFA wells in the area, and to better assess the effect of the boundary condition applied at the bottom of the lower UFA lithologic unit. The number of each LU in figure 5 refers to the vertical zones within the SAS, ICU, and UFA that have distinctive hydraulic properties. Each vertical zone or LU within the SAS, ICU, and UFA was further divided into two model layers with half the thickness and the same hydraulic properties for each LU in Tables 3 and 4 to increase the resolution of the simulated vertical hydraulic gradients.

Variable monthly recharge rates (Fig. 6a) were simulated, during a one-year period, in the top 3D flow model layer of the SAS. Total simulated recharge to the SAS was 50.8 cm/year, a reasonable average estimate for the area. This reasonable average recharge rate to the SAS was used to discard recharge as a variable that could contribute to water level residuals at observations wells in the SAS. The top 3D flow model simulates the effects of recharge, hydraulic conductivity, pumping rates, and initial heads on simulated heads in the SAS, ICU, and UFA. Variable monthly pumping rates were applied to six hypothetical UFA wells distributed near the center of the grid (Fig. 6b). Measured heads in the SAS, ICU, and UFA from nearby observation wells were used to specify initial heads for the 3D flow models. The main hydraulic parameter that causes variations in head residuals and changes in leakage rates to the ICU is the horizontal and vertical hydraulic conductivity.

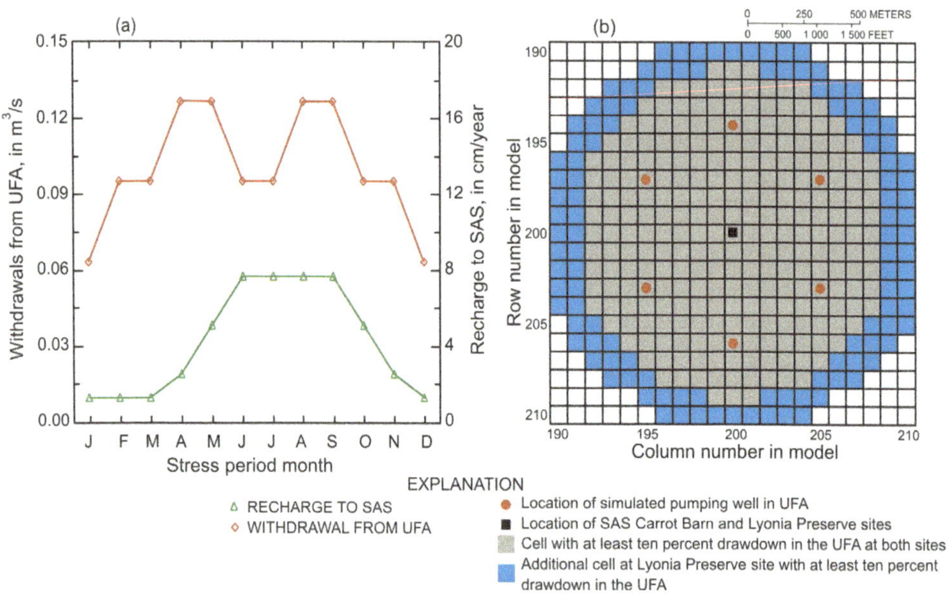

Fig. 6. (a) Simulated monthly recharge and pumping rates and (b) locations of simulated pumping wells from the Upper Floridan aquifer in each model.

The simulated heads in the SAS, ICU, and UFA, and the simulated leakage rates to the UFA with the 8 LU model for the Carrot Barn well field or with the 6 LU model for the Lyonia Preserve well field were used as the "observed" heads and flows to calculate residuals resulting from model layer simplification. The accuracy of the methods used to compute composite hydraulic properties (Eqs. (1)-(3) and PE) for the simplified 3D flow models was assessed by examining the RMSRs between observed and simulated SAS heads, ICU heads, UFA heads, and leakage rates to the UFA from the overlying ICU unit. The largest RMSRs were calculated when the composite hydraulic properties for the combined SAS and ICU lithologic units were used to simulate a single layer (Tables 5, 6). The increase in SAS, ICU, and UFA head RMSRs from a fully 3D to a quasi-3D model was the largest for both Lyonia Preserve and Carrot Barn simulations than any other single-step transition in conceptual flow models. The difference in vertical hydraulic conductivity between the lowermost LU of the SAS and the ICU was the main reason for the increases in residuals.

Leakage rates from the SAS to the UFA, in these areas characterized as recharging the UFA, were simulated using Eq. (5), the assigned hydraulic conductivity to the UFA by McGurk and Presley (2002), and a vertical to horizontal hydraulic conductivity anisotropy of 0.01 in the vicinity of the Carrot Barn and Lyonia Preserve well fields. RMSRs for the leakage rates to the UFA and heads in the ICU simulated by the quasi-3D models were among the largest from all conceptual flow models (Tables 5, 6). Head and leakage RMSR rates were calculated relative to the highest vertically discretized models (6 LU model for Carrot Barn and 5 LU model for Lyonia Preserve). The largest RMSR percentages were simulated for the heads in the UFA by the quasi-3D model for the Lyonia Preserve well field when the no-flow boundary was imposed at the bottom of the UFA lower layer (Table 6B). The largest contrasts in vertical hydraulic conductivity between the ICU and the LUs lying above or below occurred at the Lyonia Preserve well field (Table 4, Fig. 5b).

Model	(A) Root-mean-square residuals, constant-head boundary applied at the bottom of the UFA							
	SAS heads (m)		ICU heads (m)		UFA heads (m)		Leakage rates to UFA (cm/year)	
	Eqs. (1)-(3)	PE	Eqs. (1)-(3)	PE	Eqs. (1)-(3)	PE	Eqs. (1)-(3)	PE
6 LU	0.05 (0.26)	0.26 (1.40)	0.02 (0.11)	0.07 (0.45)	0.00 (0.00)	0.00 (0.00)	0.41 (2.84)	1.48 (11.61)
4 LU	0.05 (0.24)	0.35 (1.87)	0.02 (0.13)	0.11 (0.68)	0.00 (0.00)	0.01 (0.05)	0.43 (3.23)	2.15 (16.06)
3 LU	0.15 (0.83)	0.24 (1.30)	1.43 (8.71)	1.43 (8.71)	0.04 (0.27)	0.03 (0.22)	13.36 (100.2)	10.97 (82.19)
Model	(B) Root-mean-square residuals, no-flow boundary applied at the bottom of the UFA							
	SAS heads (m)		ICU heads (m)		UFA heads (m)		Leakage rates to UFA (cm/year)	
	Eqs. (1)-(3)	PE	Eqs. (1)-(3)	PE	Eqs. (1)-(3)	PE	Eqs. (1)-(3)	PE
6 LU	0.02 (0.13)	0.09 (0.57)	0.06 (0.65)	0.34 (3.78)	0.04 (0.48)	0.17 (2.30)	0.57 (1.75)	4.09 (16.13)
4 LU	0.08 (0.52)	0.30 (1.97)	0.03 (0.28)	0.44 (4.82)	0.01 (0.19)	0.22 (2.82)	0.65 (2.30)	4.74 (16.79)
3 LU	0.30 (1.92)	1.63 (10.72)	2.81 (29.90)	3.81 (38.57)	0.63 (7.29)	2.29 (24.95)	20.21 (78.22)	13.37 (57.81)

Table 5. Root-mean-square residuals (percentages between parenthesis) between simulated SAS heads, ICU heads, UFA heads, and leakage rates to the UFA by the three-dimensional 8 LU flow model and those simulated by the 6 LU, 4 LU, and 3 LU simplified flow models for the aquifer stresses shown in figure 6a applied to the Carrot Barn well field. [Hydraulic properties were derived from either Eqs. (1)-(3) or from regression-based parameter estimation (PE) in Table 3]

Model	(A) Root-mean-square residuals, constant-head boundary applied at the bottom of the UFA							
	SAS heads (m)		ICU heads (m)		UFA heads (m)		Leakage rates to UFA (cm/year)	
	Eqs. (1)-(3)	PE	Eqs. (1)-(3)	PE	Eqs. (1)-(3)	PE	Eqs. (1)-(3)	PE
5 LU	0.00 (0.00)	0.00 (0.00)	0.00 (0.00)	0.00 (0.00)	0.00 (0.00)	0.00 (0.00)	0.00 (0.00)	0.01 (0.22)
4 LU	0.00 (0.00)	0.01 (0.11)	0.00 (0.00)	0.00 (0.00)	0.00 (0.00)	0.00 (0.00)	0.00 (0.00)	0.02 (0.41)
3 LU	0.35 (5.49)	1.71 (25.86)	0.44 (9.24)	1.02 (22.13)	0.01 (0.19)	0.00 (0.00)	5.70 (103.5)	0.22 (2.21)
Model	(B) Root-mean-square residuals, no-flow boundary applied at the bottom of the UFA							
	SAS heads (m)		ICU heads (m)		UFA heads (m)		Leakage rates to UFA (cm/year)	
	Eqs. (1)-(3)	PE	Eqs. (1)-(3)	PE	Eqs. (1)-(3)	PE	Eqs. (1)-(3)	PE
5 LU	0.00 (0.00)	0.02 (0.24)	0.00 (0.00)	0.02 (0.32)	0.00 (0.00)	0.02 (0.41)	0.01 (0.07)	0.03 (0.32)
4 LU	0.00 (0.00)	0.04 (0.49)	0.00 (0.00)	0.04 (0.66)	0.00 (0.00)	0.04 (0.80)	0.01 (0.06)	0.02 (0.25)
3 LU	12.91 (157.8)	5.28 (64.80)	13.73 (247.3)	5.39 (95.09)	13.14 (269.0)	4.75 (91.70)	10.57 (129.1)	1.34 (14.88)

Table 6. Root-mean-square residuals (percentages between parenthesis) between simulated SAS heads, ICU heads, UFA heads, and leakage rates to the UFA by the three-dimensional 6 LU flow model and those simulated by the 5 LU, 4 LU, and 3 LU simplified flow models for the aquifer stresses shown in figure 6a applied to the Lyonia Preserve well field. [Hydraulic properties were derived from either Eqs. (1)-(3) or from regression-based parameter estimation (PE) in Table 4]

Results from the calculated composite hydraulic properties suggest that the net effect on flow accuracy of transitioning from a fully 3D to a quasi-3D model lies somewhere between the residuals in Tables 5A and 5B and in Tables 6A and 6B. The flow differences between the fully 3D and quasi-3D models increase when the no-flow boundary is simulated at the bottom of the UFA lower layer instead of the constant-head boundary case because downward flow to the UFA is induced from the overlying SAS with a more contrasting vertical hydraulic conductivity than the value used for the fully 3D flow model. A constant-head boundary at the bottom of the UFA lower layer induces increased flows through the constant-head boundary because the hydraulic conductivity of the UFA is higher than that of the ICU.

Leaky confining unit thicknesses at Carrot Barn and Lyonia Preserve of 10.7 and 4.9 m, respectively (Fig. 5), were not thick enough to constitute a significant source of water from storage to cause a change in UFA heads. Flow simulations, assuming a hypothetical increase in the thickness of the ICU unit at the Carrot Barn and Lyonia Preserve well fields by one order of magnitude, were conducted to assess the effects of these thicknesses on changes in heads in the UFA. Simulation results showed the RMSRs in heads in the SAS, ICU, and UFA, and leakage rates to the UFA were reduced from 1.63, 3.81, and 2.29 m, and 13.37 cm/year (Table 5B) for the Carrot Barn well field to 0.93, 2.01, and 1.40 m, and 5.29 cm/year, respectively, when using the composite hydraulic properties derived from the regression-based PE technique for the quasi-3D models after hypothetically increasing the thickness of the ICU by one order of magnitude. Corresponding flow simulations for the Lyonia Preserve well field resulted in RMSR changes from 5.28, 5.39, and 4.75 m, and 1.34 cm/year to 4.22, 2.19, 1.46 m, and 1.04 cm/year. These head and leakage rate residuals suggest that for a thick confining unit where storage effects may be important, application of the quasi-3D approach may mask conceptual error because residual error may be reduced through the adjustment of composite hydraulic properties.

7. Conclusions

Conceptual flow models with decreasing numbers of model layers in an unconfined aquifer were developed to determine the effects on flow of calculating composite hydraulic properties from two or more lithologic units at a time. The thickness-weighted arithmetic mean was used to compute the horizontal hydraulic conductivity and the specific storage, and the thickness-weighted harmonic mean was used to compute the vertical hydraulic conductivity. A regression-based parameter estimation technique was also used to compute composite hydraulic properties. Flow in the unconfined aquifer can be simulated with reasonable accuracy if all lithologic units within the aquifer are combined into a single model layer, composite hydraulic properties are calculated from thickness-weighted averages, and the contrast in vertical hydraulic conductivity among the lithologic units is not large. The same level of accuracy cannot be achieved with composite properties when only vertical flow is simulated within the confining unit because of the increases in head residuals in the unconfined aquifer, confining unit, and confined aquifer, and increases in leakage rate residuals to the confined aquifer that occurred with the quasi-3D approach.

These residuals, which illustrate the main difference between the quasi-3D and fully 3D conceptual flow models, become larger as the contrast in vertical hydraulic conductivity between the lower lithologic units of the unconfined aquifer and the confining unit increases.

The derivation of composite hydraulic properties for several lithologic units through regression-based parameter estimation techniques resulted in lower drawdown residuals in the 2D flow simulations when compared to the residuals derived from composite hydraulic properties derived from thickness-weighted averages. The same composite hydraulic properties derived from the parameter-estimation techniques resulted in higher simulated head and leakage rate residuals than those from thickness-weighted averages in the 3D flow simulations, which considered recharge to the unconfined aquifer and pumping from the confined aquifer. The lack of accuracy in heads and leakage rates can be explained by the contrasts in vertical hydraulic conductivity between the unconfined aquifer and the confining unit. Differences in specific storage were a small factor in the head and leakage rate residuals because the confining units of both well fields analyzed in this study were thin. As the thickness of the confining unit increases, larger water volumes from storage would be neglected by the quasi-3D flow simulations and differences in simulated heads in the confined aquifer could be larger than those presented in this chapter.

The derivation of composite hydraulic properties using regression-based parameter estimation techniques for the combined flow properties of the unconfined aquifer and the leaky confining unit showed that residuals for the water levels in the unconfined aquifer, heads in the confined aquifer, and leakage rates to the confined aquifer could be reasonably reduced when all lithologic units in the unconfined aquifer were simulated as one. These residuals increase when the composite hydraulic properties of the unconfined aquifer and the leaky confining unit are calculated, namely when the quasi-2D or quasi-3D flow assumptions are made. Given the flow residuals presented in this chapter, inaccuracies in simulated flow increase when a transition in conceptual flow models is made from a fully 3D to a quasi-3D flow simulation. When composite hydraulic properties are calculated for the simplified conceptual flow models, the head and leakage rate residuals increase as the contrast in hydraulic properties between the unconfined aquifer and the leaky confining unit increase, in particular, as the contrast in vertical hydraulic conductivity increases between these two lithologic units.

8. References

Alley, W.M.; Reilly, T.E. & Franke, O.L. (1999). Sustainability of Groundwater Resources. *US Geological Survey Circular 1186*, Washington, DC.

Bouwer, H. (1978). *Groundwater hydrology*. McGraw Hill, ISBN 0070067155, New York.

Bouwer, H. & Rice, R.C. (1976). A slug test for determining hydraulic conductivity of unconfined aquifers with completely or partially penetrating wells. *Water Resources Research*, Vol. 12, No. 3, pp. 423-428.

Clarke, R.; Lawrence, A.; & Foster, S. (1996). Groundwater: A Threatened Resource (United Nations Environment Programme Environment Library No. 15, Nairobi, Kenya.

Copty, N.K.; Sarioglu, M.S. & Findikakis, A.N. (2006). Equivalent transmissivity of heterogeneous leaky aquifers for steady state radial flow. *Water Resources Research*, Vol. 42, No. 4, W04416, doi: 10.1029/2005WR004673.

Domenico, P.A. (1972). *Concepts and models in groundwater hydrology*. McGraw Hill, ISBN 0070175357, New York.

Feng, C.Y.; Lee T.H.; Lee W.S. & Chen C.H. (2007). Estimating finite difference block equivalent hydraulic conductivity for numerically solving the Richards' equation. *Hydrological Processes* Vol. 21, pp. 3587-3600.

Halford, K.J. (2006). MODOPTIM: A general optimization program for ground water flow model calibration and ground water management with MODFLOW. *US Geological Survey Scientific Investigations Report* 2006-5009.

Hantush, M.S. (1964). Hydraulics of wells. *Advances in Hydroscience 1*, V.T. Chow, pp. 281-432, Academic Press, New York.

Harbaugh, A.W. & McDonald, M.G. (1996). User's documentation for MODFLOW-96, an update to the U.S. Geological Survey modular finite-difference ground-water flow model. *US Geological Survey Open-File Report* 96-485.

Harbaugh, A.W.; Banta, E.R.; Hill, M.C.; & McDonald, M.G. (2000). MODFLOW-2000, the U.S. Geological Survey modular ground-water model - User guide to modularization concepts and the ground-water flow process. *US Geological Survey Open-File Report* 02-92.

Hill, M.C. & Tiedeman C.R. (2007). *Effective groundwater model calibration - with analysis of data, sensitivities, predictions, and uncertainty.* Wiley, ISBN: 978-0-471-77636-9, New Jersey.

Langevin, C.D. (2008). Modeling axisymmetric flow and transport. *Ground Water*, Vol. 46, No. 4, pp. 579-590.

McDonald, M.G. & Harbaugh, A.W. (1988). A modular 3D finite-difference ground-water flow model. *US Geological Survey Techniques of Water-Resources Investigation Book* 6, Chap. A1.

McGurk, B. & Presley, P. (2002). Simulation of the effects of groundwater withdrawals on the Floridan aquifer system in east-central Florida: model expansion and revision. *St. Johns River Water Management District Technical Paper* 2002-3.

Miller, J.A. (1986). Hydrogeologic framework of the Floridan aquifer system in Florida and in parts of Georgia, Alabama, and South Carolina. *US Geological Survey Professional Paper* 1403-B.

Poeter, E.P.; Hill, M.C.; Banta, E.R.; Mehl, S. & Christensen, S. (2005). UCODE_2005 and Six Other Computer Codes for Universal Sensitivity Analysis, Calibration, and Uncertainty Evaluation. *US Geological Survey Techniques and Methods* 6-A11, Reston, Virginia.

Reilly, T.E. & Harbaugh, A.W. (1993). Computer note: Simulation of cylindrical flow to a well using the U.S. Geological Survey modular finite-difference ground-water flow model. *Ground Water*, Vol. 31, No. 3, pp. 489-494.

Sánchez-Vila, X.; Carrera, J. & Girardi, J.P. (1996). Scale effects in transmissivity. *J Hydrology*, Vol. 183, pp. 1-22.

Sepúlveda, N. (2002). Simulation of ground-water flow in the intermediate and Floridan aquifer systems in peninsular Florida: *US Geological Survey Scientific Investigations Report* 2002-4009.

Solley, W.B.; Pierce, R.R.; & Perlman, H.A. (1998). Estimated Use of Water in the United States in 1995. *US Geological Survey Circular 1200*, Washington, DC.

Snyder, J.P. (1987). Map projections-A working manual, 3rd ed. *US Geological Survey Professional Paper* 1395, Washington, DC.

Van der Kamp, G. (1976). Determining aquifer transmissivity by means of well response tests: the underdamped case: *Water Resources Research*, Vol. 12, No. 1, pp. 71-77.

Walton, W.C. (2008). Upgrading aquifer test analysis. *Ground Water*, Vol. 46, No. 5, pp. 660-662.

Ward, A.L.; Zhang, Z.F. & Gee, G.W. (2006). Upscaling unsaturated hydraulic parameters for flow through heterogeneous anisotropic sediments. *Advances in Water Resources*, Vo. 29, No. 2, pp. 268-280.

Wen, X.H. & Gómez-Hernández, J.J. (1996). Upscaling hydraulic conductivities in heterogeneous media: An overview. *J Hydrology*, Vol. 183, pp. ix-xxxii.

19

Simulation of Water and Contaminant Transport Through Vadose Zone - Redistribution System

Thidarat Bunsri[1], Muttucumaru Sivakumar[2] and Dharmappa Hagare[3]
[1]*Department of Environmental Engineering, Faculty of Engineering, King Mongkut's University of Technology Thonburi, Bangkok,*
[2]*School of Civil, Mining and Environmental Engineering, Faculty of Engineering, University of Wollongong, NSW,*
[3]*School of Engineering, University of Western Sydney, NSW,*
[1]*Thailand*
[2,3]*Australia*

1. Introduction

Movement of water in vadose zone, mainly focusing on infiltration and percolation that involves percolation of water under gravity from soil surface and redistribution which is the capillary rise of water movement upwards, is presented. In the global hydrologic cycle, 76% of the precipitating water enters the soil via percolation-infiltration, which leads to the downward movement of water (L'vovich 1974). The water used by natural processes, can move downwards due to infiltration and lift from groundwater table during natural redistribution process. The forecasting of water movement in unsaturated infiltration-redistribution system is linked between soil hydraulic properties and hydrologic condition of natural surface water system. The understanding of water movement processes associated with infiltration and redistribution has a number of practical applications. One such application is to predict the fate and transport of materials through soil including nutrients, organic carbon and microbes under natural processes, which in turn will help in developing appropriate management plans for irrigation, fertiliser application and waste disposal on land.

2. Infiltration and redistribution system

The prevailing weather condition can potentially affect the amount of water input to soil. The infiltration rate consists of (1) water input rate, which is the rate of water that arrives at soil surface due to rain, natural and artificial applications and (2) infiltration capacity, which is the maximum rate at which percolating water migrates through the soil pore. In general, water input rate responds to seasonal climatic variations and to any recharge from natural or artificial conditions. The infiltration capacity rate depends on the soil texture and soil hydraulic properties. The infiltration process can be separated into three categories (1) no ponding (2) saturation from above and (3) saturation from below. The "no ponding" refers to the condition, when water input rate is less than or equals to the infiltration capacity. The "saturation from above" presents the condition, when water input rate exceeds the

infiltration capacity rate. The "saturated from below" reveals the condition, when groundwater table has risen above the original saturation layer resulting in saturated soil, hence the infiltration rate becomes zero (Iwata et al. 1995, Dingman 2002).
The redistribution rate can be determined by the upward flow rate of water, which involves both the exfiltration and capillary rise. The exfiltration deals with evaporation or evaportranspiration. The exfiltration can reduce the soil pore water at the upper layer. The exfiltration rate of soil also depends on relative humidity of air and solar radiation. Furthermore, soil is a porous material and can behave as a series of capillary tubes. The surface tension force can abstract the water from aquifer into pore space above the water table. The capillary rise refers to the water movement from saturated zone to unsaturated zone owing to surface tension. The capillary rise acts opposite to the direction of gravitational force. The height of capillary rise relates to pore size and soil moisture content. Consequently, the height of capillary rise in fine grained soil is higher than in the case of coarse grained soil and the height of capillary rise into dry soil is lower than that of relatively wet soil. The water movement in unsaturated zone near the region of saturated zone or capillary fringe is mentioned because the water in the saturated and unsaturated zones is connected together and oscillates with fluctuations of groundwater table. The flucutation of groundwater level during wet-to-dry season, causes the movement of capillary fringe region that causes of smearing of contaminants to soil above groundwater table. The critical point of predicting the water movement in the capillary fringe is the non-uniform capillary flow. In order to simplify this infiltration-redistribution system that occurrs in the capillary fringe, it can be assumed that the redistribution rate is associated with the capillary height, which is normally considered under the equilibrium of capillary force (Fredlund & Rahardjo 1993, Dingman 2002).
The vadose zone is the entire zone of negative water pressure above the water table, so the pressure head at the deepest level of this zone is saturated or nearly saturated as a result of capillary rise. Almost all water in the vadose zone is available for plants and exfiltration (drainage). The plant available water is observed at pressure head ranged from -150 to -3.4 m H_2O and the drainable water will be at pressure head ranged from -3.4 m H_2O to saturation (Dingman 2002). The water infiltrating into vadose zone is influenced by both capillary and gravitational forces, and these are associated with upward and downward-directed pressure gradients, respectively. The movement of water due to infiltration capacity is described by the Richards equation and the movement of drainage is satisfactorily modeled with Darcy's law. The capillary gradient is determined by static capillarity equilibrium height using the capillary models (Bunsri et al. 2009).

3. Governing equation for unsaturated flow

Movement of water due to infiltration is assumed that water percolates under gravitational force. The vertical flow through interconnected intergrain pores, which are randomly distributed in a mix grained soil. The movement of water through the effective soil pores can be simplified so that water moves only in the liquid state (excluding, soil-water freezing or thawing) and there is no impact of airflow in soil pores. The 1D-vertical unsaturated flow in porous media can be described using Richards equation (Richards 1931).
The pressure head based equation is expressed as:

$$\frac{\partial}{\partial z}\left[K_z k_{rw}\left(\frac{\partial \psi}{\partial z}+1\right)\right]=S\frac{\partial \psi}{\partial t} \tag{1}$$

The volumetric moisture content based equation (Warrick, Islas & Lomen 1991) is:

$$\frac{\partial}{\partial z}\left(D_z\frac{\partial \theta}{\partial z}\right)-\frac{\partial K_z k_{rw}}{\partial z}=\frac{\partial \theta}{\partial t} \tag{2}$$

where k_{rw} is the relative hydraulic conductivity [unitless], K_z is the fully saturated hydraulic conductivity [L T^{-1}], S is the specific moisture capacity $(=\partial\theta/\partial\psi)$, θ is the volumetric moisture content [unitless], ψ is the pressure head [L] and D_z is the soil water diffusivity $=(\partial K_z k_{rw}/\partial S)$. Equations (1) and (2) contains Darcy's velocity, q_z, which is given by (the negative sign means downward flow) (Huyakorn et al. 1984):

$$q_z=-K_z k_{rw}\left(\frac{\partial[\psi+z]}{\partial z}\right) \text{ or } \frac{\partial \theta}{\partial t}=-\frac{\partial q_z}{\partial z} \tag{3}$$

where q_z is Darcy's velocity in vertical direction [LT^{-1}].
By inserting a series of tensiometers in different parts of a drainage field, the profiles of pressure head can be observed as shown in Figure 1. The negative pressure head determined in the unsaturated soil layer is due to suction head, zero pressure head occurred at the interface of groundwater table and the positive pressure head in the saturated soil layer is due to hydraulic head (gravitational head plus pore-water pressure head) (Fredlund and Rahardjo 1993).

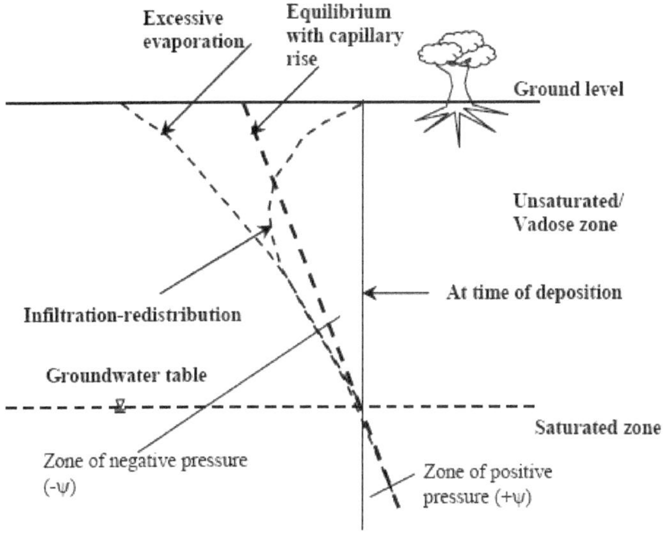

Fig. 1. Profiles of pore pressure under a steady state condition (Adapted from Fredlund & Rahardjo 1993)

Figure 2 presents possible pressure head profiles at varying Darcy's velocities (q_z). If the velocity is constant $(\partial\psi/\partial z = 0)$, the simplest pressure head profiles (case 1) are obtained. When $\partial\psi/\partial z$ is negative with $-K_z k_{rw} < q_z < 0$, this will lead to vertical downward flow (case 2). If $\partial\psi/\partial z$ is positive with $q_z < -K_z k_{rw}$, water moves downward with suction head (case 3). The upward flow is yielded (case 4), if Darcy's velocity is greater than zero. However, in most unsaturated zone cases, the water will experience a downward flow, these are either case 2 for evaporation or case 3 for irrigation (Warrick et al. 1991).

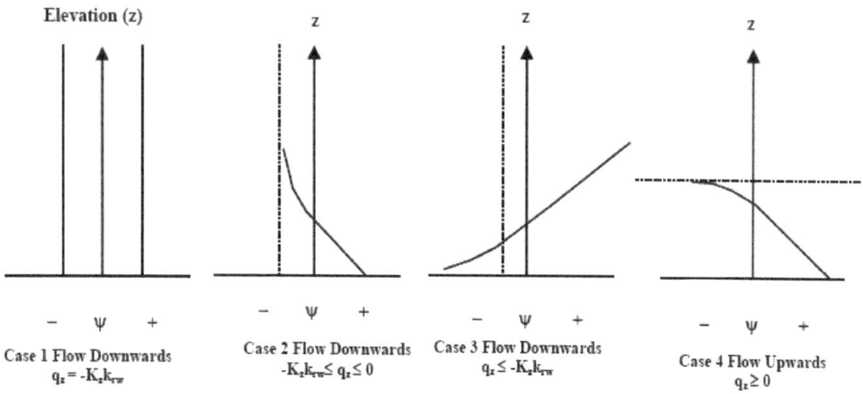

Fig. 2. Pressure head profiles at varying Darcy's velocities (Adapted from Warrick et al. 1991)

The relationship between relative hydraulic conductivity ($k_{rw} K_z$), pressure head (ψ) and volumetric moisture content (θ) is defined as the hydraulic properties function, which is highly nonlinear. Typically used hydraulic properties equations in this research include those of Brooks and Corey (1964), Haverkamp et al. (1977), van Genuchten (1980) and Saxton et al. (1986). Brooks & Corey (1964) hydraulic properties function was firstly derived by modifying the statistical porewater interaction models of Childs and Collis-George (1950) (cited in Fredlund & Rahadjo 1993). The Brooks & Corey (1964) model is:

$$\theta = \left[\left(\frac{\psi_a}{\psi}\right)^{\tilde{N}} (\theta_S - \theta_r)\right] + \theta_r \tag{4a}$$

$$K_z k_{rw}(\psi) = K_z \left(\frac{\psi_a}{\psi}\right)^{\tilde{M}} \tag{4b}$$

$$K_z k_{rw}(\theta) = K_z \left(\frac{\theta - \theta_r}{\theta_S - \theta_r}\right)^{\tilde{M}/\tilde{N}} \tag{4c}$$

where ψ_a is the air entry suction pressure head [L], θ_S and θ_r are the saturated and residual volumetric water content, respectively [unitless]. $\tilde{M}$ and $\tilde{N}$ are empirical constants.

Haverkamp et al. (1977) fitted the properties of homogeneous soil in unsaturated conditions by the least square method and proposed the following empirical equations (HV equations):

$$\theta = \frac{\alpha(\theta_S - \theta_r)}{\alpha + |\psi|^{\beta}} + \theta_r \tag{5a}$$

$$K_z k_{rw} = K_z \left(\frac{A}{A + |\psi|^{\gamma}} \right) \tag{5b}$$

where A, α, β and γ are the curve fitting coefficients [unitless].
van Genuchten (1980) derived the hydraulic properties equations based on the equation of Brooks & Coley (1964) and proposed the following empirical equations for hydraulic properties (VG equations):

$$\theta = \theta_r + \frac{\theta_S - \theta_r}{\left[1 + a|\psi|^{p}\right]^{q}} \tag{6a}$$

$$K_z k_{rw} = K_z \frac{\left[1 - \left(a|\psi|^{p-1}\right)\left(1 + \left(a|\psi|^{p}\right)\right)^{-q}\right]^2}{\left[1 + a|\psi|^{p}\right]^{q/2}} \tag{6b}$$

where a is the soil water retention function [L^{-1}], q and p are the empirical parameters, $(q = 1 - [1/p])$ [unitless].
The possible values for the coefficients presented in VG equations are given in Table 1. The coefficients are sorted by soil textures in accordance with USDA textural classes (Carsel et al. 1988).

Soil Type	θ_S	θ_r	a (cm^{-1})	p
Clay*	0.38	0.068	0.008	1.09
Clay loam	0.41	0.095	0.019	1.31
Loam	0.43	0.078	0.036	1.56
Loam sand	0.41	0.057	0.124	2.28
Silt	0.46	0.034	0.106	1.37
Silt loam	0.45	0.067	0.020	1.41
Silty clay	0.36	0.070	0.005	1.09
Silty clay loam	0.43	0.089	0.010	1.23
Sand	0.43	0.045	0.145	2.68
Sandy clay	0.38	0.100	0.027	1.23
Sandy clay loam	0.39	0.100	0.059	1.48
Sandy loam	0.41	0.065	0.075	1.89

Note: *Agricultural soil, less than 60% clay.

Table 1. Recommended empirical coefficients for VG equations (Carsel et al. 1988).

The hydraulic properties are estimated from the soil texture using a method generalised by Saxton et al. (1986). The textural class is assessed according to the USDA system. The water retention curve is fitted with linear regression and the formulations are in S.I. unit as follows:

1) When the hydraulic pressure is between 10 and 1500 kPa (or 1.02 to 15.3 m H_2O), then the expression for ψ is given by:

$$\psi = \psi_a \left(\frac{\theta - \theta_r}{\theta_S - \theta_r} \right)^H \tag{7a}$$

By assuming $\theta_r = 0$; will give

$$\psi = J\theta^H \tag{7b}$$

with $J = \psi_a \theta_S^{-H}$

where ψ and ψ_a are the hydraulic pressure and the air entry pressure, respectively [kPa]. H and J are obtained by the statistical curve fitting of 44 soil samples with an R^2 =0.99.

$$H = -3.140 - 0.00222 \cdot [\%\ clay]^2 - 3.484 \times 10^{-5} [\%\ sand]^2 [\%\ clay] \tag{8a}$$

$$J = \exp\begin{pmatrix} -4.396 - 0.0715 \cdot [\%\ clay] - 4.880 \times 10^{-4} \cdot [\%\ sand]^2 \\ -4.285 \times 10^{-5} \cdot [\%\ sand]^2 \cdot [\%\ clay] \end{pmatrix} \times 100 \tag{8b}$$

2) When the hydraulic pressure is between ψ_a and 10 kPa (or 1.02 H_2O), then:

$$\psi = 10.0 + \left(\frac{(\theta - \theta_{10})(10.0 - \psi_a)}{\theta_S - \theta_{10}} \right) \tag{9}$$

where θ_{10} is the volumetric moisture content at 10 kPa, $\left(= \exp\left[(2.303 - LnJ)/H\right]\right)$ [cm³/cm³]. $\psi_a = 100.0(-0.108 + 0.341\theta_s)$ and $\theta_S = 0.332 - 7.251 \times 10^{-4} [\%\ sand] + 0.1276(Log[\%\ clay])$

3) when the hydraulic pressure is between 0.0 kPa and ψ_a (or 0.0 cm H_2O to ψ_a), then:

$$\theta = \theta_S \tag{10a}$$

Variable $K_z k_{rw}$ is estimated as follows:

$$K_z k_{rw} = 2.778 \times 10^{-6} \left(\exp\begin{pmatrix} 12.012 - 0.0755 \cdot [\%\ sand] \\ +(1/\theta)(-3.8950 + 0.0364 \cdot [\%\ sand] \\ -0.113 \cdot [\%\ clay] + 8.7546 \times 10^{-4} \cdot [\%\ clay]^2) \end{pmatrix} \right) \tag{10b}$$

where $K_z k_{rw}$ is the unsaturated hydraulic conductivity [m s^{-1}].

Kunze et al. (1968) established a relative permeability function based on Poiseuille's equation. An example of estimating relative permeability using Kunze's equation is illustrated in Figure 3.

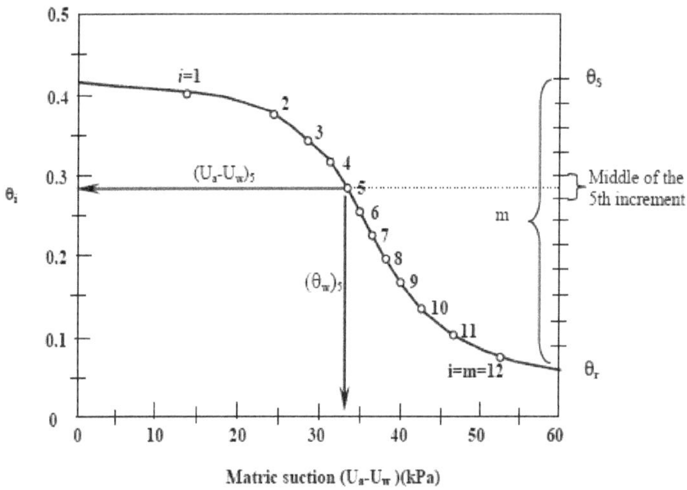

Fig. 3. Graphical estimation of relative permeability (Adapted from Fredlund & Rahardjo 1993)

The coefficient of relative permeability was obtained from the relationship between the matrix suction and volumetric moisture content. The equation was given as follows:

$$k_{rw}(\theta_i) = \frac{(K_z)_{mea}}{(K_z)_{cal}} \frac{\sigma_S^2 \rho_w g \theta_S^P}{2\mu_w N^2} \sum_{j=1}^{m} \left[(2j+1-2i) |\psi|_j^{-2} \right]$$

$$i = 1,2,....m \qquad (11)$$

where $k_{rw}(\theta)_i$ is the calculated coefficient of permeability for a specified volumetric moisture content; θ_i, corresponding to the ith interval [unitless]. i is the interval on the water retention curve [unitless] and j is the counter number from i to m [unitless]. $(K_z)_{mea}$ and $(K_z)_{cal}$ are the measured saturated coefficient of permeability and the calculated coefficient of permeability, respectively [L T^{-1}]. σ_S is the surface tension of water [M T^{-2}], ρ_w is the water density [M L^{-3}] and μ_w is the absolute viscosity of water [M T^{-1}]. P is a constant which accounts for the interaction of pores of various sizes, usually assumed to be 2.0 (Green & Corey 1971 cited in Fredlund & Rahardjo 1993) [unitless]. m is the total number of intervals between the saturation volumetric water content and the residual volumetric water content [unitless] and N is the total number of intervals $(N = m\theta_S / [\theta_S - \theta_r])$ [unitless].

4. Numerical solution for the unsaturated flow equation

The finite discretionary scheme is given in Figure 4. The number of nodes in the system is assigned sequentially to the flow direction.

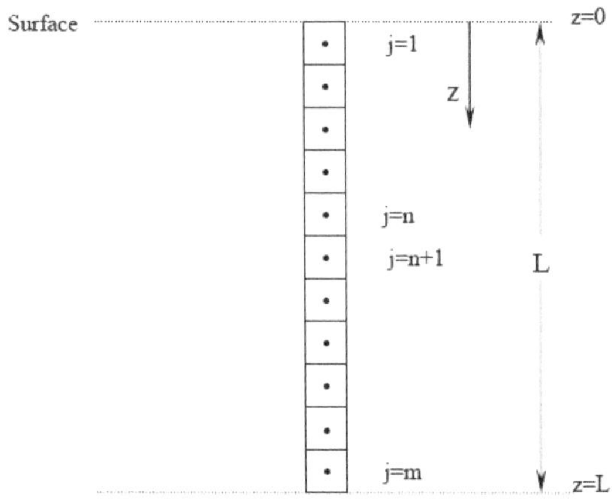

Fig. 4. Scheme of finite discretisation.

As the hydraulic properties model can be applied to modify Richards' equation between the θ based to ψ based formulations. Generally, ψ based Richards' equation is preferred over θ based equation as it is possible to accurately measure the pressure head using tensiometers. Several previous works (Wang & Anderson 1982, Coley 1983, Segerlind 1984, Paniconi et al.1991) have assumed the approximate solution to Richards' equation as:

$$\psi(z,t)=\sum_{j=1}^{m}N_j(z)\psi_j(t) \tag{12}$$

where $N_j(z)$ is the shape function [unitless], $\psi_j(t)$ is the unknown coefficient with corresponding to the value of nodal pressure head [L] and the subscript "*j*" is defined to denote a nodal sequence.
Richards' equation (Eq. 1) is now written in the form of $L[\psi]=0$ as follows.

$$L[\psi]=S\frac{\partial\psi}{\partial t}-\frac{\partial}{\partial z}\left[K_z k_{rw}\left(\frac{\partial\psi}{\partial z}+1\right)\right] \tag{13}$$

The weighting function, N_i was assigned the same value as N_j. Applying Galerkin's finite element method yields (Bunsri et al. 2009):

$$\int_{z=0}^{z=L} L[\psi]N_i dz=0 \tag{14}$$

Substituting Eq. 13 into Eq. 14, yields:

$$\int_{z=0}^{z=L} N_i S\frac{\partial\psi}{\partial t}dz-\int_{z=0}^{z=L} N_i\left[K_z k_{rw}\left(\frac{\partial\psi}{\partial z}\right)+1\right]dz=0 \tag{15}$$

The numerical solution of Richards' equation is (Bunsri et al. 2009):

$$\int_{z=0}^{z=L} N_i S \frac{\partial \psi}{\partial t} dz - N_i K_z k_{rw} \left(\frac{\partial \psi}{\partial z} + 1 \right) \Bigg|_{z=0}^{z=1} + \int_{z=0}^{z=L} K_z k_{rw} \psi_j \left(\frac{\partial N_i}{\partial z} \frac{\partial N_j}{\partial z} \right) dz + \int_{z=0}^{z=L} K_z k_{rw} \frac{\partial N_i}{\partial z} dz = 0 \quad (16)$$

The algebraic matrix systems were defined as follows (Bunsri et al. 2009):

$$\left[A_{ij} \right] \cdot \psi_j + \left[B_{ij} \right] \cdot \frac{\partial \psi_j}{\partial t} = \left\{ E_{ij} \right\} \quad (17)$$

where

$$\left[A_{ij} \right] = \sum_e \int_{z=0}^{z=L} K_z k_{rw} \left(\frac{\partial N_i}{\partial z} \frac{\partial N_j}{\partial z} \right) dz, \quad \left[B_{ij} \right] = \sum_e \int_{z=0}^{z=L} N_i S dz \text{ , and}$$

$$\left\{ E_i \right\} = N_i K_z k_{rw} \left(\frac{\partial \psi}{\partial z} + 1 \right) \Bigg|_{z=0}^{z=L} - \sum_e \int_{z=0}^{z=L} K_z k_{rw} \frac{\partial N_i}{\partial z} dz$$

Using Eq. 3, the vector matrix $\{E_i\}$ could be written in the form of Darcy's flux boundary condition as:

$$\left\{ E_i \right\} = N_i q_z \Big|_{z=0}^{z=L} - \sum_e \int_{z=0}^{z=L} K_z k_{rw} \frac{\partial N_i}{\partial z} dz \quad (18)$$

The time derivative approximation at a particular node " j " can be explained by (Wang & Anderson 1982, Paniconi et al. 1991, Ségol 1993):

$$\frac{\partial \psi_j}{\partial t} = \frac{\psi_j^{t+\Delta t} - \psi_j^t}{\Delta t} \quad (19)$$

where $\{\partial \psi / \partial t\}$ is a column matrix, which refers to time-dependent hydraulic pressure head $(\psi_j(t))$. The term " $\{\partial \psi / \partial t\}$ " can be simplified with the vector symbol " $\{\psi\}$ ". The iterative scheme obtained using a single Picard's method is given by (Wang & Anderson 1982, Paniconi et al. 1991, Ségol 1993):

$$\left(\frac{1}{2} \left[A_{ij} \right]^t + \frac{1}{\Delta t} \left[B_{ij} \right]^t \right) \left(\{\psi\}^{t+\Delta t} - \{\psi\}^t \right) = \{E_i\} - \left[A_{ij} \right]^t \{\psi\}^t \quad (20)$$

Celie et al. (1990) estimated the pressure head profiles using Richards' equation. The numerical model is developed using both finite difference model (FDM) and finite element method (FEM). The model developed using FEM has given an oscillation of pressure head that is near to infiltration front. This oscillation pressure head could be reduced by applying a diagonal time matrix (mass lumping technique). The net water balance between inflow and outflow of soil pore water at any node and time step t (MB) is defined based on a diagonal time matrix. This aspect is investigated to evaluate the existence of any errors within the calculation process if existed. The water balance equation for the finite element technique is presented as follows (Celie et al. 1992, Ségol 1993):

$$MB^{t+1} = \frac{\sum_{i=1}^{E-1}\left(\theta_i^{t+1} - \theta_i^0\right)\Delta z + \left(\theta_0^{t+1} - \theta_0^0\right)\frac{\Delta z}{2} + \left(\theta_E^{t+1} - \theta_E^0\right)\frac{\Delta z}{2}}{\sum_{t=1}^{t+1}\left(q_0^t - q_E^t\right)\Delta t} \tag{21}$$

where q_0 and q_E are the boundary fluxes associated with z_0 and z_L, respectively [L T^{-1}]. Symbols "*i*" and "*t*" count for the sequence of nodes and time steps, respectively. Subscripts "*0*", "*E*" refer to the upper boundary of downward flow and the number of elements, respectively. Superscripts "*0*", "*t*" *and* "*t*+1" refer to the initial, previous and current iteration time step, respectively.

5. Boundary condition for infiltration system

The initial condition was defined as the starting condition of the system at *t=0*. The boundary conditions were classified into upper and lower boundaries that were located on the top and bottom of the considered system (Huyakorn & Pinder 1983). The upper boundary was the condition at the discharge point and the lower boundary was the condition at the water table or the column base. The boundary and initial conditions for the solute transport model and Richards' equation included the known concentration of contaminant and pressure head, respectively (Bear 1979, Huyakorn & Pinder 1983). The change of infiltration is mainly controlled by intensity of infiltration and surface soil properties. The infiltration rates are determined by: (1) Rate of water approaching above soil surface via rainfall, snowmelt, irrigation, natural or artificial recharge and depth of ponding on the surface; (2) Fully saturated hydraulic conductivity at soil surface; (3) Degree of saturation at soil surface when the infiltration begins; (4) Inclination and roughness of topsoil; and (5) Chemical characteristics of top soil and physical and chemical properties of water.

The infiltration rate; $f(t)$ is computed as (Nassif & Wilson 1975):

$$f(t) = i - q(t) \tag{22}$$

where i is natural precipitation or application rate and $q(t)$ is the rate of surface runoff. The natural infiltration recharge can be technically determined using Green-Ampt approach (Green & Ampt 1911). In the case of infiltration coming from runoff or rainfall recharge, the percolation rate is based on top soil properties. It can be classified into 2 categories as:

1) Water input rate is less than saturated hydraulic conductivity ($i<K_z$). At initial stage (t=0), the hydraulic conductivity is defined as K_{z0} and it increases to K_z at the specified time; t_w. At this stage, water will percolate and it is stored until the soil pore is fully filled.

$$f(t) = i \quad 0 < t < t_w \text{ and } f(t) = 0 \quad t >> t_\omega \tag{23}$$

2) Water input rate is greater than saturated hydraulic conductivity ($i>K_z$). This process is observed at an early stage of infiltration in which the excess water cannot be transmitted downwards. The maximum water content and the hydraulic conductivity are limited at θ_S and K_z, respectively. Therefore, when the soil surface reaches saturation, ponding will occur or in the case of a hilly area, overland flow will take place. However, the

corresponding equation for this case cannot give a valid value of *f(t)* as there is no well developed relationship for ponding.

The percolation of water during infiltration process with no ponding can be expressed as:

$$f_p = \frac{K_z \psi_w}{z_S} + K_z \tag{24a}$$

$$I = (\theta_S - \theta_0) z_S \tag{24b}$$

$$t = \frac{\theta_S - \theta_0}{K_z}\left\{ z_S - \psi_w Ln\left(1 + \frac{z_S}{\psi_S}\right)\right\} \tag{24c}$$

where θ_0 is the initial soil moisture content [unitless], f_p is the percolation rate, ψ_w is the pressure head at the wetting front, I is the cumulative infiltration and z_S is the wetting front distance for Green-Ampt model.

Schmid (1990) has modified Green-Ampt approach with a Taylor series expansion, and obtained the explicit approximation for a function of the infiltration rate.

$$f(t) = w\left[1 + 2\frac{(w - K_z)^2}{K_z |\psi_f| (\phi - \theta_0)^t}\right]^{-0.5} \tag{25}$$

where ψ_f is the effective tension at the wetting front $(= 0.76|\psi_S|)$ (Brakensiek 1977, Freyberg et al. 1980), ψ_S is the pressure head at saturation and term $(\phi - \theta_0)$ is the initial soil water deficit. The soil moisture content under infiltration process can be estimated using Wang et al. (2003) equation as follows.

$$\theta = \begin{cases} \left(1 - \frac{z}{z_f}\right)^{\delta} (\theta_S - \theta_0) + \theta_0, & z \le z_f \\ \theta_0 & z > z_f \end{cases} \tag{26a}$$

$$I = \frac{(\theta_S - \theta_0) z_f}{1 + \delta} \tag{26b}$$

$$f_P = \frac{K_z}{\varphi z_f} + K_z \tag{26c}$$

$$t = \frac{(\theta_S - \theta_0)}{(1 + \delta) K_z}\left(z_f - \frac{Ln\left[\varphi z_f + 1\right]}{\varphi}\right) \tag{26d}$$

where z_f is the wetting front distance that is located between soil surface and the bottom of the wetted soil layer. Variable δ is the shape coefficient of Brook and Corey (1964) hydraulic properties model, $(= \tilde{M} / \tilde{N})$. φ is the soil suction allocation coefficient $(= \tilde{M} / \xi)$ and ξ is a constant.

6. Boundary condition for redistribution system

The boundary conditions for a redistribution system can be evaluated according to the physical model of capillary rise. The zone of negative pressure is observed within the depth of capillary height. The capillary height is the height of the water level inside the capillary tube. The capillary height could be estimated using (Freudlune & Rahardjo):

$$U_w = -\rho_w h_C g \tag{27}$$

where U_w is the hydraulic pressure at the capillary height [M L^{-1} T^{-2}], ρ_w is water density [M L^{-3}], h_C is the capillary height [L] and g is the gravitational acceleration [L T^{-2}].
The physical model of capillary pressure force in unsaturated soil is presented in Figure 5. If matric suction (U_a-U_w) or capillary height (h_C) is plotted against pore radius (r) on a Log-Log plot, a linear relationship is expected. Further details can be found in Fredlune & Rahardjo (1993). The variables "σ_S", "r_S" and "α" refer to the surface tension, the radius of capillary tube and the contact angle, respectively. The variable "U_a" is the atmospheric pressure (guage) that is normally taken as 0 cm H_2O.

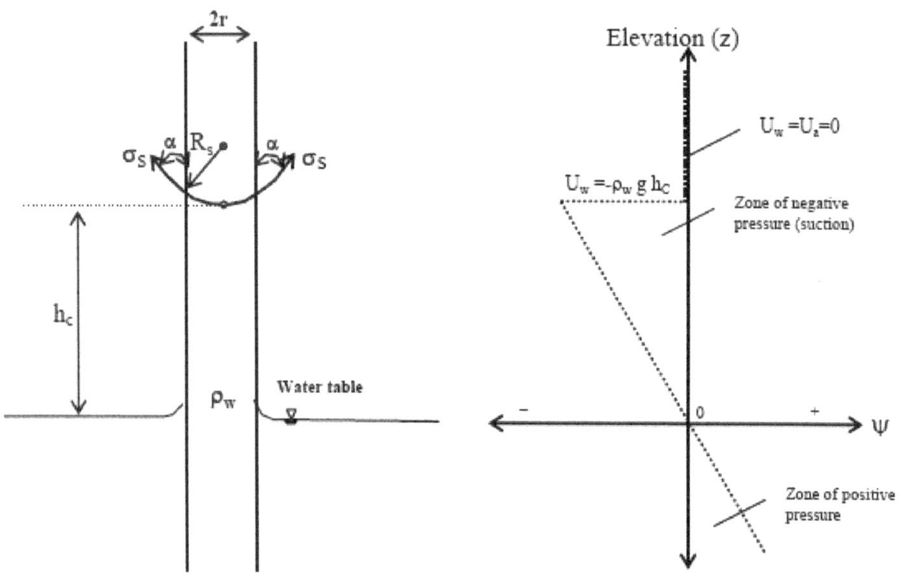

Fig. 5. Physical model of capillarity (Adapted from Fredlund & Rahardjo 1993)

7. Governing equation for solute transport

In practice, the objective is not only to predict the movement of water in the vadose zone, but also to determine the movement of reactive and non-reactive contaminants through the soil pores. The developed model can predict the fate and transport of reactive constituents such as phosphate, nitrate, organic carbon compounds and microbes.
The governing equation for multi-component transportation of contaminants in porous media under variable saturation conditions could be expressed in a general form as follows (Schnoor 1996).

$$\underset{(accumulate)}{\frac{\partial C}{\partial t}} = \underset{(dispersion)}{\frac{\partial}{\partial z}\left(D_z \frac{\partial C}{\partial z}\right)} - \underset{(advection)}{\frac{\partial}{\partial z}\left(q_z C\right)} - \underset{sorption}{\frac{\rho_B}{\theta}\left(\frac{\partial C^*}{\partial t}\right)} \pm \underset{(reaction)}{\sum_{x=1}^{n} r_x} \tag{28}$$

where C^* is the concentration of a considered constituent in sorbed phase [M L-3]; D_z is the dispersion coefficient; $(= D_m + D^*)$ [L2 T-1]; D_m and D^* are the mechanical and the molecular dispersion [L2T-1], respectively; and r_x is the rates of reaction "x" with x=1,2,..., n [M L-3 T-1].

The dispersion can be obtained from Fick's Law (Fetter 1992) as follows.

$$D_z = \omega v_i + D^* \tag{29}$$

where D^* is an effective molecular diffusion coefficient [L2 T-1], ω is a coefficient relating to tortuosity [unitless] and v_i is the average linear velocity in the vertical direction; $(= q_z / \theta)$ [L T-1].

7.1 Nitrogen and organic carbon compounds

The total rate of organic carbon compounds (substrate) utilisation, r_S was a combination of the substrate utilisation rate due to aerobic and nitrate respiration (Widdowson et al. 1988).

$$r_S = r_{SO} + r_{SN} \tag{30}$$

where r_S, r_{SO} and r_{SN} are the total substrate utilisation, substrate utilisation under aerobic respiration and substrate utilisation under nitrate respiration, respectively [T -1].

Using a modified Monod's equation, the substrate utilisation rates could be derived as follows (Widdowson et al. 1988):

$$r_{SO} = \frac{\mu_O}{Y_o}\left[\frac{C_S}{K_{SO} + C_S}\right]\left[\frac{C_O}{K_O + C_O}\right]\left[\frac{C_A}{K_{SA} + C_A}\right] \tag{31a}$$

$$r_{SN} = \frac{\mu_N}{Y_N}\left[\frac{C_S}{K_{SN} + C_S}\right]\left[\frac{C_O}{K_N + C_N}\right]\left[\frac{C_A}{K_{AN} + C_A}\right] I\left[C_O\right] \tag{31b}$$

where μ_O and μ_N are the maximum specific growth rate for aerobic and denitrifying bacteria, respectively [T-1]. Y_O and Y_N are a heterotrophic yield coefficient for aerobic and denitrifying bacteria, respectively [unitless]. C_S, C_O and C_A are the concentration in aqueous of organic carbon, oxygen and ammonia, respectively [M L-3]. K_{SO}, K_O and K_{AO} are half concentration of substrate, oxygen and ammonia nitrogen under aerobic respiration, respectively [M L-3]. K_{SN}, K_N and K_{AN} are half concentration of substrate, nitrate and ammonia nitrogen under nitrate respiration, respectively [M L-3]. $I[C_O]$ is the inhibition factor $I[C_0] = (1 + C_o / K_c)^{-1}$ [unitless] and K_C is the coefficient of inhibition [M L-3].

By referring to the rate of kinetic reaction (Eq. 31a and 31b) for organic carbon compounds biodegradation, the equation for transport of organic carbon compounds is written as (Schnoor 1996):

$$\frac{\partial C_S}{\partial t} = -\frac{\partial}{\partial z}(q_z C_S) + \frac{\partial}{\partial z}\left(D_z \frac{\partial C_S}{\partial z}\right) - r_{SO} C_S \theta \tag{32}$$

The nitrate transport equation can be formulated as follows (Schnoor 1996):

$$\frac{\partial C_N}{\partial t} = -\frac{\partial}{\partial z}(q_z C_N) + \frac{\partial}{\partial z}\left(D_z \frac{\partial C_N}{\partial z}\right) - r_{SN} C_S C_N \theta \tag{33}$$

Harvey et al. (1984) determined the kinetic coefficients for nitrifying and denitrifying bacteria. These coefficients are given in Table 2.

Parameter	Value	Parameter	Value
M_b (mg/cm³)	5.64x10⁻⁴	K_{SO} (mg/cm³)	0.040
μ_O (1/day)	3.1	K_{SN} (mg/cm³)	0.040
μ_N (1/day)	2.9	K_O (mg/cm³)	0.00077
Y_O	0.45	K_N (mg/cm³)	0.00260
Y_N	0.5	K_{AO} (mg/cm³)	0.0010
k_o (1/day)	0.02	K_{AN} (mg/cm³)	0.0010
k_N (1/day)	0.02		

Table 2. Kinetic parameters for organic carbon and nitrate retardation (Harvey et al. 1984 cited in Widdowson et al. 1988).

7.2 Phosphate phosphorus compounds

Phosphorus adsorption on soil was formulated as follows (Shah et al. 1975):

$$\rho_B \frac{\partial C_P^S}{\partial t} = K_t\left(C_P^w - C_P^{w*}\right) \tag{34}$$

where C_P^S is the concentration of adsorbed phosphorus [M M^{-1}] and K_t is the overall volumetric mass transfer coefficient [L T^{-1}]. C_P^{w*} is the concentration of phosphorus in liquid phase that is in equilibrium with the concentration of phosphorus in the solid phase [M M^{-1}] and $C_P^w - C_P^{w*}$ is a driving force for transferring phosphorus from liquid to solid phase [M L^{-3}].

The Langmuir isotherm equation is the best fit for phosphorus adsorption (Shah et al. 1975). Variable C_P^{w*} is modified into Langmuir isotherm equation, yielding (Schnoor 1996, Watts 1997):

$$C_P^{w*} = \frac{k_L k_M C_P^w}{1 + k_M C_P^w} \tag{35}$$

where k_L and k_M are the coefficients (Langmuir rate constant) [unitless] and the maximum phosphate adsorption capacity on soil [L^3 M^{-1}], respectively.

The soluble phosphate, C_P can be assumed as soil pore water, which can be presented as volumetric portion of moisture (Fetter 1992). The transport of soluble phosphate is obtained by:

$$\frac{\partial(\theta\kappa C_P)}{\partial t} = -\frac{\partial}{\partial z}(q_z C_P) + \frac{\partial}{\partial z}\left(D_z \frac{\partial C_P}{\partial z}\right) \tag{36}$$

where κ is a retardation factor $\left(=1+\left[\rho_B K_d / \theta\right]\right)$ and K_d is the distribution coefficient, $\left(=k_L k_M / \left(1+K_M C_P^w\right)^2\right)$.

7.3 *E. coli*

The overall reaction rates of microbial kinetics were the summation of production, maintenance, decay, adsorption and desorption. The retardation equation was given as (Zysset et al. 1994):

$$n\frac{d(C_b^f)}{dt} = n v_y \eta k_\mu C_S C_b^f - n k_d C_b^f + \theta k_c C_b^w - n k_s C_b^f \tag{37}$$

where n is the fraction of aqueous volume and biofilm in total volume (equals to porosity) [unitless]. C_b^f and C_b^w are the concentration of adhering microbes and free swimming microbes [M L^{-3}], respectively. C_S is the concentration of limiting substrate in aqueous compartment [N L^{-3}], v_y is the stoichiometric coefficient [M N^{-1}], η is an effectiveness of biofilm [unitless]. k_μ and k_d are Monod's constant for substrate utilisation in biomass [L^3 M^{-1} T^{-1}] and the constant of decay rate [T^{-1}], respectively. k_S and k_C are the constant desorption (detachment) and adsorption (attachment) rate [T^{-1}], respectively. The unit N is the quantity of microbes involved (cfu or MPN).

The concentration of *E. coli* relates to the substrates consumed. Substrate utilisation during metabolisation processes is defined using the first order Monod's kinetics equation as follows (Zysset et al. 1994).

$$\frac{\partial C_S}{\partial t} = -k_\mu C_S C_b^t - k_m C_b^t \tag{38}$$

where $k_\mu = \mu_b / Y_b(K_{Sb} + C_S)$, but $C_S << K_{Sb}$, so $k_\mu = \mu_b / Y_b K_{Sb}$. C_S and C_b^t are the carbonaceous substrate concentration [M L^{-3}] and the total microbial concentration [M L^{-3}], respectively. μ_b and Y_b are the microbial maximum specific growth rate [T^{-1}] and a heterotrophic microbial yield coefficient [unitless], respectively. K_{Sb} and k_m are the substrate concentration when the rate of utilisation of half of the maximum rate under aerobic condition [M L^{-3}] and a biomass maintenance rate [T^{-1}], respectively.

The *E. coli* transport equation was governed as follows (Zyset et al. 1994):

$$\frac{\partial(\kappa\theta C_b)}{\partial t} = \frac{\partial}{\partial z}\left(D_z \frac{\partial C_b}{\partial z}\right) - \frac{\partial}{\partial z}(q_z C_b) - \lambda_b \theta\kappa C_b \tag{39}$$

where $\lambda_b = (k_\mu - k_m) \cdot C_b^t$, which involves assimilation and dissimilation of microbes and $C_b^t = \kappa C_b$.

A general form of the mathematical model for contaminant transport coupling retardations could be written as follows (Huyakorn et al. 1985):

$$\underset{(dispersion)}{\frac{\partial}{\partial z}\left(D_z\frac{\partial C}{\partial z}\right)} \quad \underset{(advection)}{-\frac{\partial}{\partial z}(q_zC)} \quad = \quad \underset{(accumulation)}{\frac{\partial}{\partial t}(\theta\kappa C)} \quad + \quad \underset{(1^{st}\ order\ \ decay)}{\lambda\theta\kappa C} \tag{40}$$

The retardation factor, κ equals 1.0 for organic carbon and nitrate compound transport equations. The biodecay factor (involving growth of microbes), λ equals 0.0 for phosphate compounds transport equation. Only microbial transport contains all of these factors.

8. Numerical solution for solute transport equation

The approximate solution of contaminant concentration at any nodes and time "t" is defined as (Segerlind 1984, Huyakorn et al. 1985, Clement et al. 1998):

$$C(z,t)=\sum_{j=1}^{m}N_j(z)C_j(t) \tag{41}$$

where $N_j(z)$ is the shape function [unitless], $C_j(t)$ is the concentration of contaminant at time t [M L^{-3}].
The mathematical model presented in Eq. 40 can be modified as follows:

$$L[C]=\frac{\partial}{\partial z}\left(D_z\frac{\partial C}{\partial z}\right)-q_z\frac{\partial C}{\partial z}-\theta\kappa\left[\frac{\partial C}{\partial t}+\lambda C\right] \tag{42}$$

The integral form of Eq. 42 is given as follows:

$$\int_{z=0}^{z=L}N_iL[C]dz=0 \tag{43}$$

where $[0,L]$ is the extent of the vertical direction (one dimension) domain. The subscripts "i" and "j" denoted the sequence of elements in the domain as presented in Figure 4.
Substituting Eq. 42 into 43, gives:

$$\int_{z=0}^{z=L}N_i\frac{\partial}{\partial z}\left(D_z\frac{\partial C}{\partial z}\right)dz-\int_{z=0}^{z=L}N_iq_z\frac{\partial C}{\partial z}dz-\int_{z=0}^{z=L}N_i\theta\kappa\left[\frac{\partial C}{\partial t}+\lambda C\right]dz=0 \tag{44}$$

The numerical solution of Eq. 44 is governed as:

$$\int_{z=0}^{z=L}\left(D_z\frac{\partial N_i}{\partial z}\frac{\partial N_j}{\partial z}+N_i\frac{\partial N_j}{\partial z}q_z\right)C_jdz+\int_{z=0}^{z=L}N_iN_j\theta\kappa\left[\frac{\partial C_j}{\partial t}+\lambda C_j\right]dz-N_iD_z\left(\frac{\partial C}{\partial z}\right)\Bigg|_{z=0}^{z=L}=0 \tag{45}$$

The algebraic matrix systems are defined as follows:

$$\left\{\left[P_{ij}\right]+\left[R_{ij}\right]\right\}C_j+\left[Q_{ij}\right]\frac{\partial C_j}{\partial t}=\{S_i\} \tag{46}$$

where

$$\left[P_{ij}\right]=\sum_{e}\int_{z=0}^{z=L}\left(D\frac{\partial N_i}{\partial z}\frac{\partial N_j}{\partial z}+N_i\frac{\partial N_j}{\partial z}q_z\right)dz\ ,\ \left[Q_{ij}\right]=\sum_{e}\int_{z=0}^{z=L}N_iN_j\theta\kappa dz$$

$$\left[R_{ij}\right]=\sum_{e}\int_{z=0}^{z=L}N_iN_j\theta\kappa\lambda dz\ \text{ and } \left\{S_i\right\}=N_iD_z\left(\frac{\partial C}{\partial z}\right)\Bigg|_{z=0}^{z=L}$$

The initial concentration in the entire domain $\left[0,L\right]$ at time t = 0 is defined as follows (Bear 1979, Huyakorn & Pinder 1983, Huyakorn, et al. 1985, Ségol 1993, Clement et al. 1998):

$$C_j\left(z,0\right)=C_0\left(z\right) \tag{47}$$

where $C_0\left(z\right)$ is the known distribution of solute concentration at time t = 0 [ML^{-3}].

The boundary concentration on the edge of domain $\left[0,L\right]$ at time "t" is defined using the Dirichlet boundary condition (Bear 1979, Huyakorn & Pinder 1983, Huyakorn, et al. 1985, Ségol 1993, Clement et al. 1998).

$$C_j\left(z,t\right)=C\left(z,t\right)\ \ on\ \ z_1<z<z_2 \tag{48}$$

The specific dispersive flux on the edge of domain $\left[0,L\right]$ at time "t" is employed using Neumann boundary condition. The dispersive flux was defined as $\partial C/\partial z$, (Bear 1979, Huyakorn & Pinder 1983, Huyakorn, et al. 1985, Ségol 1993, Clement et al. 1998).

$$\text{For rainfall infiltration}\quad D_z\frac{\partial C}{\partial z}=q_C^D\qquad on\qquad 0\le z\le L \tag{49a}$$

$$\text{For seepage flow}\quad D_z\frac{\partial C}{\partial z}-q_zC=q_C^T\qquad on\qquad 0\le z\le L \tag{49b}$$

where q_C^D and q_C^T are the portion of the boundary flux attributable to concentration due to dispersion [M L^{-1} T^{-2}] and the portion attributable to total concentration [M $L^{-1}T^{-2}$], respectively.

9. Model applications

The simulation results of solute transport in unsaturated infiltration-redistribution system is an active research area where a variety of attempts are being made to determine the dynamics of vadose zone in relation to water and contaminant movement and quality of soil. The model presented above is used to simulate the experimental data obtained by Paniconi et al. (1991). The input parameters for this case study are given in Table 3.

The Neumann and the Dirichlet conditions were applied to the upper and lower boundary, respectively. Both the experimentally observed and simulated pressure head and moisture content profiles are presented in Figures 6 and 7, respectively. The developed models predicted the experimental observations of Paniconi et al. (1991) very well. This implied that the developed model is robust, and that it could effectively predicte the water movement in the infiltration-redistribution sytems. On the other hand, the simulation results from the solute transport model could not be compared with experimental observations due to lack of appropriate data. However, a series of simulations are carried out as shown in Figures 8 through 11. The kinetic rate constants of each contaminant are assumed based on the case study of contaminants movement in sandy soil near Perth (McArthur & Bettenay 1964, cited in Whelan & Barrow 1984). The input parameters used in model are presented in Table 4.

Parameters	Values
Domain	Column with depth of 10 m (saturation maintained at the base of the column).
Hydraulic properties model	van Genutchten model with saturated hydraulic conductivity; K_z is 5 m/h, Air entry pressure; ψ_a is -3.0 mH$_2$O. Saturated moisture content; θ_S is 0.45 and residual moisture content; θ_r is 0.08. q and p are 0.667 and 3.0, respectively.
Boundary and initial conditions	At the top, Darcy flux (q_z) is applied as a function of time, equalled to *t/64* m/h. Constant head at the column base, ψ_{bot} of 0 m. Constant head at the initial, ψ_i of 0 m.
Number of element; *nelem*	100.
Increment time interval; *dt*	Time step varies between 0.1 to 0.5 hour.

Table 3. Input parameters for water movement model (Paniconi et al. 1991).

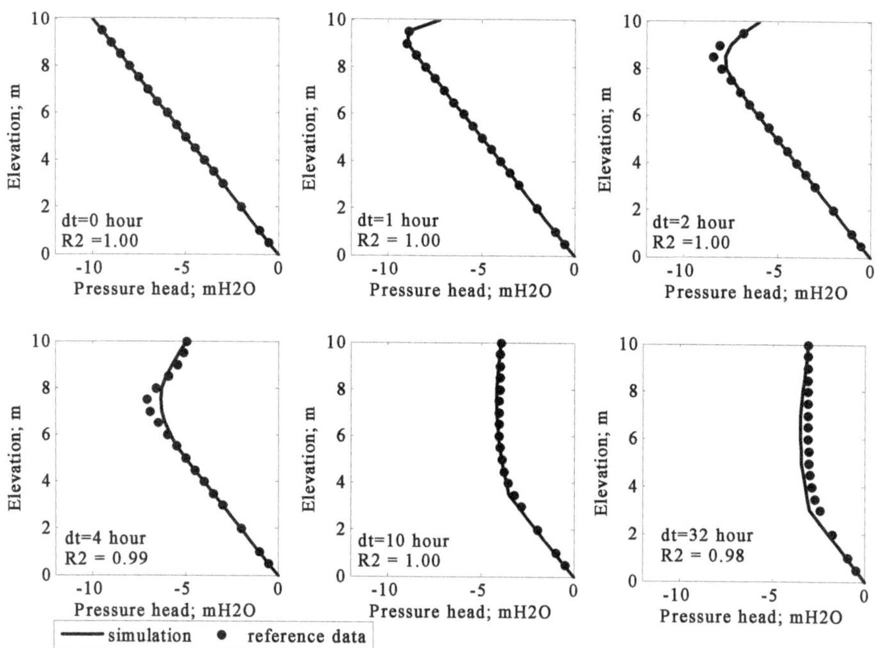

Fig. 6. Pressure head profiles (datum was at the groundwater table).

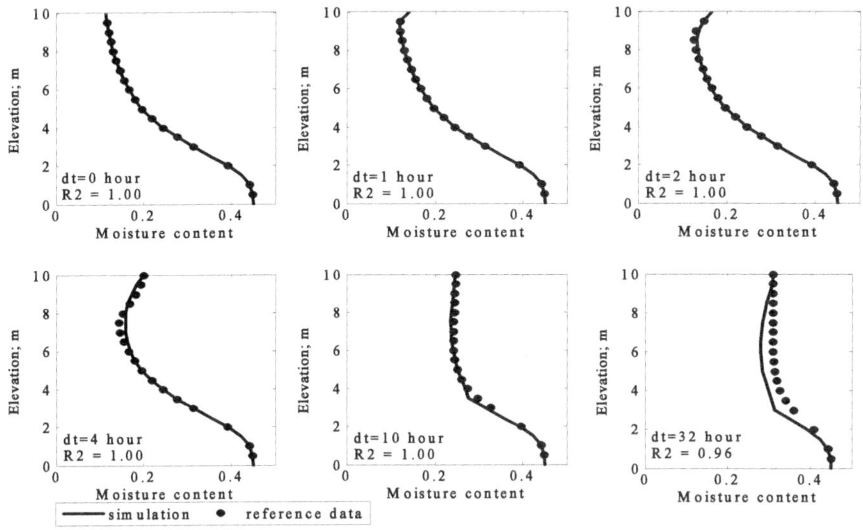

Fig. 7. Moisture content profiles (datum was at the groundwater table)

Parameters	Values
Contaminant transport	Organic carbon compounds: Constant concentration at the surface, C_{top} of 1250 mg/L and at the datum, C_{bot} of 39 mg/L. Initial concentration, C_{int} of 39 mg/L. Nitrate nitrogen compounds: Constant concentration at the surface, C_{top} of 2 mg/L and at the datum, C_{bot} of 83 mg/L. Initial concentration, C_{int} of 2 mg/L. Phosphate phosphorus compounds: Constant concentration at the surface, C_{top} of 15 mg/L and at the datum, C_{bot} of 0.72mg/L. Initial concentration, C_{int} of 0.72 mg/L. *E. coli*: The initial microbial concentration 1.03×10^{9} cfu/L. The microbial concentration at the column surface and base were 2.14×10^{10} and 1.318×10^{3}cfu/g (weight of E.Coli is 7×10^{-10} mg/cfu).
Dispersivity; α_T (cm)	3.25 (organic carbon), 7.54 (nitrate), 1.57 (phosphate) and 7.59 (*E.Coli*)
Molecular diffusion; D^* (cm²/h)	0.03924 (Organic C), 0.06840 (Nitrate), 0.07055 (Ammonia), 0.02304 (Phosphate) and 0.0479 (*E.coli*) (Kemper 1986, Stevik 1999).
Kinetic rate constant	Organic C and Nitrate: μ_o =74.4 h^{-1}, Y_O =0.45, K_{SO} =0.040 mg/cm^3, K_O = 0.00077 mg/cm^3, K_{AO} = 0.001 mg/cm^3, C_A =50 mg/L, C_O = 2 mg/L, μ_N = 69.6 h^{-1}, Y_N =0.5, K_{SN} =0.040 mg/cm^3, K_N =0.0026mg/cm^3, K_{AN} =0.001 mg/cm^3 and $I[C_O]$=0.0802 (Widdowson 1998). Phosphate: K_d =1.22×10^{-9} cm^3/g and ρ_B =1.7 kg/L. (McArthur & Bettenay 1964, cited in Whelan & Barrow 1984). *E.coli*: Adsorption rate= 0.015 h^{-1} , Growth-decay rate 56.8 h^{-1}, (Stevik et al 1999, Schnoor 1996).
Increment time interval; *dt* (hour)	Varies between 0.1-0.5.
Maximum simulation time; t (hour)	32

Table 4. Input parameters for solute transport model (adopted from McArthur & Bettenay 1964, cited in Whelan & Barrow 1984).

Figures 8, 9, 10 and 11 present the simulations of nitrate, organic carbon, phosphate and *E.coli* transport in infiltration-redistribution system. The simulation results reveal that the contaminants could reach the groundwater table over a longer period. Also, it appeares that the top soil can remove substantial amount of contaminants. Particularly, the organic carbon and *E.coli*, are removed within the few centimetres of the top soil layer. On the other hand, phosphate can move downwards to a depth of 2 metres. This might relate to phosphate adsorption capacity, which is relatively low in the sandy soil. However, to make proper assessment of contaminant transport and its potential contamination of groundwater longer periods of simulation are required. At this stage it can be concluded that the contaminant transport model presented earlier could be used to predict the contaminant transport within the unsaturated (vadose) zone.

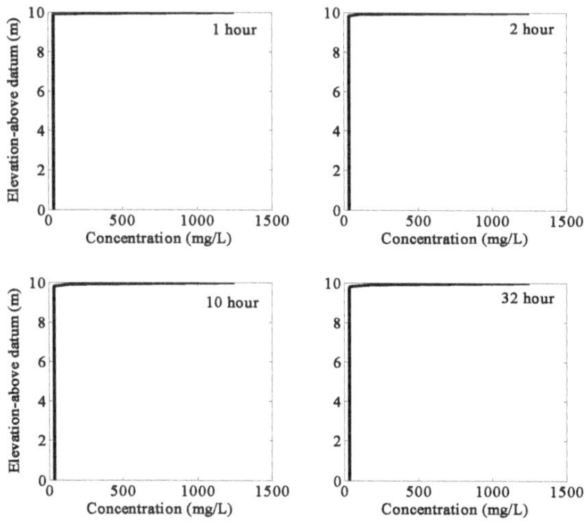

Fig. 8. Organic carbon concentration profile (datum was at the groundwater table).

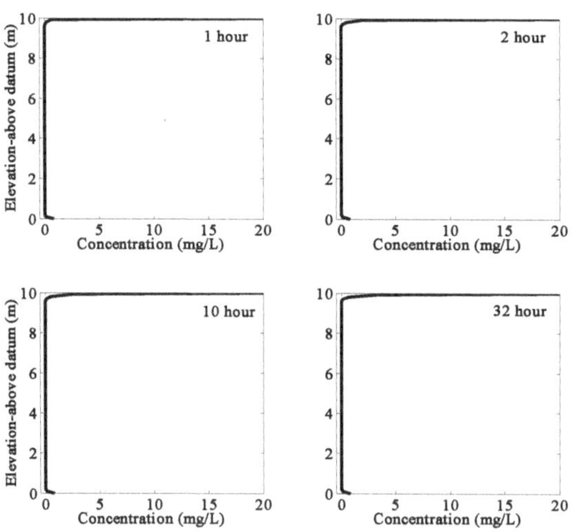

Fig. 9. Nitrate concentration profile (datum was at the groundwater table).

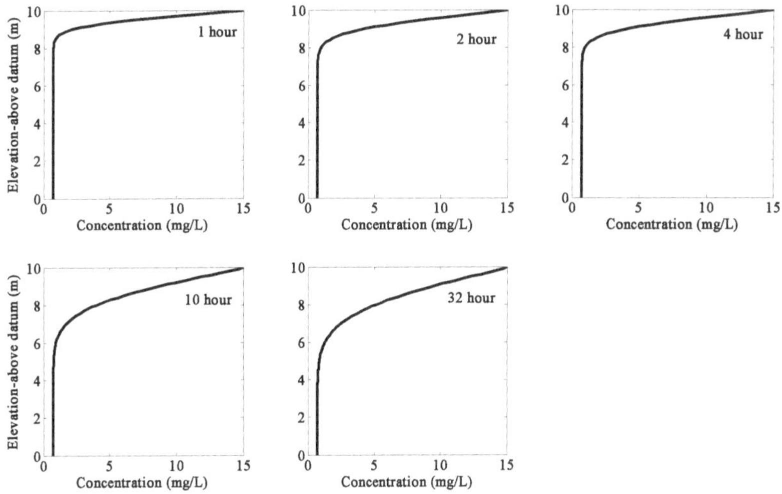

Fig. 10. Phosphate concentration profile (datum was at the groundwater table).

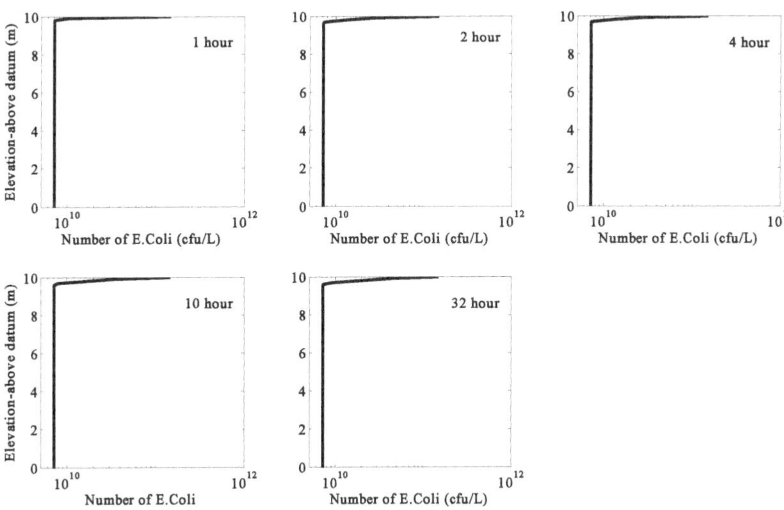

Fig. 11. *E.coli* distribution profile (datum was at the groundwater table).

10. Conclusion

A comprehensive model for predicting the movement of water and contaminants through unsaturated soil is presented. The models were developed based on Richards equation and mass balance relationships. Also, Finite Element Method (FEM) based solution techniques were developed for obtaining numerical solution to the models developed. The water movement simulation model predicted fairly well the water movement observed through column studies. Thus, the water movement model can be used for predicting the water movement through unsaturated zone with confidence. However, it is important to use site specific input parameters for reliable results. In the case of contaminant transport model, it could not be tested with experimental observations due to lack of appropriate data. However, some of the simulation results for organic carbon, nitrate, phosphate and *E.coli* appear to indicate that the model is able to predict the contaminant transport through soil. It is recommended that the model is further tested with experimental and/or field data. The water can carry contaminants during percolation, however there are some chemical and biological mechanisms, which can retard the migration of contaminants. The capillary force can extract water from aquifer, which is important for redistribution system. With the infiltration-redistribution system, the pore velocity of water may be reduced. This can lead to self protection of groundwater from contamination. Further, it is apparent that the relative hydraulic conductivity $K_z k_{rw}$ is one of the critical parameters that influence the water and contaminant transport through vadose zone. In this study the influence of relative hydraulic conductivity on the water and contaminant transport was not fully investigated. Evidently, further studies are required to fully understand the influence of relative hydraulic conductivity and hence identify critical soil types that may be readily affected by the on-site waste disposal systems.

11. Acknowledgement

The model was originally developed as part of the first author's PhD project at the University of Wollongong, Australia. Grant was provided by the National Centre of Excellence for Environmental and Hazardous Waste Management-KMUTT satellite centre. Funding was partially provided by the Higher Education Research Promotion and National Research University Project of Thailand, Office of the Higher Education Commission.

12. References

Bear, J. (1979). *Hydraulics of groundwater*, McGraw-Hill, ISBN 978-048-6453-552, New York, USA.

Brakensiek, D.L. (1977). Estimating the effective capillary pressure in the Green-and-Ampt infiltration equation, *Water Resources Research*, Vol. 13, pp. 680-682, ISSN 0043-1397.

Brooks, R.H. & Corey, A.J. (1964). *Hydraulic properties of porous media*, Hydrol. Paper 3, Colorado State University, ISBN 951-22-7195-8, Fort Collins, USA.

Brouwer, J., Willatt, S.T. & van der Graaff, R. (1979). The hydrology of on-site septic tank effluent disposal on a yellow duplex soil, *Hydrology and Water Resources Symposium, ACT IEAust.*, ISBN 0-642-712107, Perth, September 1979.

Bunsri, T., Sivakumar, M. & Hagare, D. (2009). Simulation of water movement through unsaturated infiltration-redistribution system, *Journal of Applied Fluid Mechanics*, Vol. 2, No.1, pp. 45-53, ISSN 1735-3645.

Carsel, R. F. & Parrish, R. S. (1988). Developing joint probability distributions of soil-water retention characteristics, *Water Resources Research*, Vol. 24, No.5, pp. 755-769, ISSN 0043-1397.

Celia, M.A., Boulotas, E.T. & Zarba, R.L. (1990). A general mass-conservative numerical solution for the unsaturated flow equation, *Water Resources Research*, Vol. 26, No. 7, pp. 1483-1496, ISSN 0043-1397.

Clement, T.P., Sun, Y., Hooker, B.S. & Petersen, J.N. (1998). Modeling multispecies reactive transport in ground water, *Groundwater Monitoring Remediation Journal*, Vol. 18, pp. 72-92, ISSN 1745-6592.

Dingman, S.L. (2002). *Physical Hydrology*, Macmillan, ISBN 978-002-3297-458, New York, USA.

Fetter, C.W. (1992). *Contaminant hydrogeology*, Macmillan, ISBN 978-002-3371-356, New York, USA.

Fredlund, D.G. & Rahardjo, H. (1993). *Soil mechanics for unsaturated soils*, John Wiley & Sons, ISBN 978-047-1850-083, New York, USA.

Freyberg, D.L., Reeder, J.W., Franzini, J.B. & Remson, I. (1980). Application of the Green-Ampt model to infiltration under time-dependent surface water depths, *Water Resources Research*, Vol. 16, pp. 517-528, ISSN 0043-1397.

Green, R.E. & Corey, J.C. (1971). Calculation of hydraulic conductivity a further evaluation of some predictive methods, *Soil Science Society of America Proceeding*, Vol. 35, pp. 3-8, ISSN: 0038-0776.

Green, W.H. & Ampt, G.A. (1911). Studies on soil physics, 1: The flow of air and water through soil, *Journal of Agricultural Science*, Vol. 4, No. 1, pp.1-24, ISSN 1916-9752.

Harvey, R.W., Smith, R.L. & George, L. (1984). Effect of organic contamination upon microbial distribution and heterotrophic uptake in Cape Cod, Mass., Aquifer, *Applied and Environmental Microbiology*, Vol. 48, pp. 1197-1202, ISSN 0099-2240.

Haverkamp, R., Vauclin, M., Touma, J., Wierenga, P.J. & Vachaud, G. (1977). A comparison of numerical simulation models for one-dimensional infiltration, *Soil Science Society of America Journal*, Vol. 41, pp. 285-294, ISSN 0361-5995.

Huyakorn, P.S. & Pinder, G.F. (1983). *Computational methods in subsurface flow*, Academic Press, ISBN 978-012-3634-818, New York, USA.

Huyakorn, P.S., Mercer, J.M. & Ward, D.S. (1985). Finite element matrix and mass balance computational schemes for transport in variably saturated porous media. *Water Resources Research*, Vol. 21, No.3, pp. 346-358, ISSN 0043-1397.

Huyakorn, P.S., Thomas, S.D. & Thompson, B.M. (1984). Technique for making finite element competitive in modeling flow in variably saturates porous media, *Water Resources Research*, Vol. 20, No. 8, pp. 1099-1115, ISSN 0043-1397.

Iwata, S., Tabuchi, T. & Warkentin, B.P. (1995). *Soil water interactions: mechanisms and applications*, Marcel Dekker, ISBN 0-8247-9293-9, New York, USA.

Kemper, W.D. (1986). Solute Diffusivity, In, *Methods of soil analysis part1 Physical and mineralogical methods, 2nd ed. Agronomy 9, Vol. 1*, Klute A. et al. (Eds.), 1007-1024, Academic Press, ISBN 978-0891-1188-117 , New York, USA.

Kunze, R.J., Uehara, G. & Graham, K. (1968). Factors important in the calculation of hydraulic conductivity, *Soil Science Society of America Proceedings*, Vol. 32, pp. 760-765, ISSN 0038-0776.

L'vovich, M.I. (1974). *World water resources and their future*, Translated by R.L. Nace, American Geophysical Union, ISBN 0875902243, Washington DC. USA.

McArthur, W.M. & Bettenay, E. (1964). *The development and distribution of the soils of Swan Coastal Plain, Western Australia*, CSIRO, ISBN 064-3001-085, Melbourne, Australia.

Nassif, S.H. & Wilson, E.M. (1975). The influence of slope and rain intensity on runoff and infiltration. *Hydrological Science Bulletin*, Vol. 20, pp. 539-553, ISSN 0020-6025.

Paniconi, C., Aldama, A.A. & Wood, E.F. (1991). Numerical evaluation of iterative and noniterative methods for solution of the nonlinear Richards equation, *Water Resources Research*, Vol. 27, No. 6, pp. 1147-1163, ISSN 0043-1397.

Richard, A.L. (1931). Capillary conduction of liquids through porous media. *Physics*, Vol. 1, pp. 316-333, ISSN 0148-6349.

Saxton, K.E., Rawls, W. J., Romberger, J.S. & Papendick, R.I. (1986). Estimating generalised soil water characteristics from texture, *Soil Science Society of America Journal*, Vol. 50, pp.1031-1036, ISSN 0361-5995.

Schmid, B. (1990). Derviation of an explicit equation for infiltration on the basis of the Mein-Larson model, *Hydrological Sciences Journal*, Vol. 35, pp. 197-208, ISSN 0262-6667.

Schnoor, J.L. (1996). *Environmental modeling: fate and transport of pollutants in water, air and soil*, John Wiley & Sons, ISBN 978-0-471-12436-8, New York, USA.

Segerlind, L.J. (1984). *Applied finite element analysis*, John Wiley & Sons, ISBN 978-047-1806-622, New York, USA.

Ségol, G. (1993). *Classical groundwater simulations: Proving and improving numerical models*, PTR Prentice Hall, ISBN 978-013-1379-930, New Jersey, USA.

Shah, D.B., Coulman, G.A., Novak, L.T. & Ellis, B.G. (1975). A mathematical model for phosphorus movement in soils, *Journal of Environmental Quality*, Vol. 4, No. 1, pp. 87-92, ISSN 0047-2425.

Stevik, T.K., Ausland, G., Hansseb, J.F. & Jessen, P.D. (1999). The influence of physical and chemical factors on the transport of *E.coli* through biological filters for wastewater purification, *Water Research*, Vol. 33, No. 18, pp. 3701-3706, ISSN 0043-1354.

van Genuchten, M.Th. (1980). A closed-form equation for predicting the hydraulic conductivity of unsaturated soils, *Soil Science Society of America Journal*, Vol. 44, pp. 892-898, ISSN 0361-5995.

Wang, H.F. & Anderson, M.P. (1982). *Introduction to groundwater modelling: Finite differences and finite element methods*, W.H. Freeman, ISBN 978-012-7345-857, San Francisco, USA.

Warrick, A.W., Islas, A. & Lomen, D.O. (1991). An analytical solution to Richards' Equation for time-varying Infiltration. *Water Resources Research*, Vol. 27, No.5, pp. 763-766, ISSN 0043-1397.

Watts, R. J. (1997). *Hazardous wastes: sources, pathways, receptors*, John Wiley & Sons, ISBN 978-047-1002-383, New York, USA.

Whelan, B.R. & Barrow, N.J. (1984). The movement of septic tank effluent through sandy soils near Perth. I. Movement of nitrogen. *Australian Journal Soil Research*, Vol. 22, pp. 283-292, ISSN 0004-9573.

Zysset, A., Stauffer, F. & Dracos, T. (1994). Modeling of reactive groundwater transport governed by biodegradation, *Water Resources Research*, Vol. 30, No. 8, pp. 2423-2434, ISSN 0043-1354.

20

Measurement and Modeling of Unsaturated Hydraulic Conductivity

Kim S. Perkins
United States Geological Survey
United States of America

1. Introduction

The unsaturated zone plays an extremely important hydrologic role that influences water quality and quantity, ecosystem function and health, the connection between atmospheric and terrestrial processes, nutrient cycling, soil development, and natural hazards such as flooding and landslides. Unsaturated hydraulic conductivity is one of the main properties considered to govern flow; however it is very difficult to measure accurately. Knowledge of the highly nonlinear relationship between unsaturated hydraulic conductivity (K) and volumetric water content (θ) is required for widely-used models of water flow and solute transport processes in the unsaturated zone. Measurement of unsaturated hydraulic conductivity of sediments is costly and time consuming, therefore use of models that estimate this property from more easily measured bulk-physical properties is common. In hydrologic studies, calculations based on property-transfer models informed by hydraulic property databases are often used in lieu of measured data from the site of interest. Reliance on database-informed predicted values with the use of neural networks has become increasingly common. Hydraulic properties predicted using databases may be adequate in some applications, but not others.

This chapter will discuss, by way of examples, various techniques used to measure and model hydraulic conductivity as a function of water content, $K(\theta)$. The parameters that describe the $K(\theta)$ curve obtained by different methods are used directly in Richards' equation-based numerical models, which have some degree of sensitivity to those parameters. This chapter will explore the complications of using laboratory measured or estimated properties for field scale investigations to shed light on how adequately the processes are represented. Additionally, some more recent concepts for representing unsaturated-zone flow processes will be discussed.

2. Hydraulic conductivity measurement

The most direct and most generally reliable measurements of unsaturated hydraulic conductivity are from steady-state flow methods. These methods are seldom applied, however. In the simple gravity-driven implementations they have serious drawbacks: limitation to the wettest soil conditions, and slowness, sometimes requiring months for a single K measurement. The Steady-State Centrifuge (SSC) method, used to measure $K(\theta)$ for the samples presented here, extends the range to lower steady-state K measurements by at least three orders of magnitude, and allows at least six points of the unsaturated K relation with θ to be characterized for a pair of samples in about five days (Nimmo and others, 2002). The steady state centrifuge (SSC) method used to measure $K(\theta)$ on 40 samples from the Idaho National Laboratory is the Unsaturated Flow Apparatus[1] (UFA) version (Conca and Wright, 1998; Nimmo et al., 2002) of the method originally developed by Nimmo et al. (1987). The core samples were sub-cored in the laboratory using a mechanical coring device into a 4.9-cm-long, 3.3-cm-diameter retainer designed specifically to fit into the buckets of the UFA centrifugal rotor.

The SSC method requires that steady-state conditions be established within a sample under centrifugal force. Steady-state conditions require application of a constant flow rate and a constant centrifugal force for sufficient time that both the water distribution and the water flux within the sample become constant. When these conditions are satisfied, Darcy's law relates K to θ and matric pressure (ψ) for the established conditions as follows:

$$q = -K(\theta)\left(\frac{d\psi}{dr} - C\rho\omega^2 r\right) \tag{1}$$

where q is the flux density (cm/s), C is a unit conversion factor of 1 cm-water/980.7 dyne/cm^2 (1 cm of water is equal to a pressure of 980.7 dyne/cm^2 and 1 dyne is equal to 1 g cm/s^2), ρ is the density of the applied fluid (g/cm^3), ω is the angular velocity (rad/s), and r is the radius of centrifugal rotation (cm). If the driving force is applied with the centrifuge rotation speed large enough to ensure that $d\psi/dr << \rho\omega^2 r$, i.e., matric-pressure gradients that develop in the sample during centrifugation are insignificant, the flow is essentially driven by centrifugal force alone. The flow equation then simplifies to

$$q \approx K(\theta)C\rho\omega^2 r\ . \tag{2}$$

The ω threshold for which the $d\psi/dr$ gradient can become negligible depends on the hydraulic properties of the medium of interest. Nimmo et al. (1987) presented model calculations showing that the $d\psi/dr$ gradient becomes negligible at relatively low speeds for a sandy medium and at higher speeds for a fine-textured medium. Speeds high enough for this purpose also normally result in fairly uniform water content throughout the sample, permitting the association of the sample-average θ and ψ values with the measured K.

After achieving steady flow at a given q, θ is measured by weight and ψ is measured by non-intrusive tensiometer (Young and Sisson, 2002) or the filter paper method (Fawcett and Collis-George, 1967) in cases where suctions exceed 800 cm. These measurements along with the computation of K, yields a triplet of data (K, θ, ψ) for the average water content within the sample. Repeat measurements with different q and in some cases different rotational speed give additional points needed to define the $K(\theta)$, and $\theta(\psi)$ characteristics. There are five samples for which $\theta(\psi)$ was not measured. Three of the samples had gravel at the top surface which prohibited sufficient contact with the tensiometer and the other two simply lack that measurement.

K_{sat} was measured using either the standard benchtop falling head method (Reynolds and others, 2002) or the falling head centrifuge method (Nimmo et al., 2002) in cases where samples had very low (less than about 10^{-6} cm/s) K_{sat} values. The increased driving force

[1] Use of brand names does not constitute endorsement by the US Geological Survey

allows for rapid measurement at low K_{sat} values using the equation from Nimmo et al. (2002)

$$K_{sat} = \frac{2aL}{A\rho g\left(t_f - t_i\right)} \ln\left[\frac{\left(gz_f + \omega^2 r_b^2\right)}{\left(gz_i + \omega^2 r_b^2\right)}\right] \quad (3)$$

where a is the cross sectional area of the inflow reservoir (cm^2), L is the sample length (cm), A is the cross sectional area of the sample (cm^2), t is time (initial and final), z is the height of water above the plane in which the sample rotates (initial and final height, cm), and r_b is the position of the bottom of the sample (cm).

3. Hydraulic conductivity estimation

When direct measurements of $K(\theta)$ are not obtainable it is possible to estimate using more easily measured properties such as particle size distributions. To estimate $K(\theta)$, $\theta(\psi)$ and K_{sat} parameters (measured and modeled) were combined with Mualem's (1976) capillary-bundle model, one of the most widely used $K(\theta)$ models available. Mualem's model infers a pore-size distribution for a soil from its $\theta(\psi)$ curve based on capillary theory, which assumes that a pore radius is proportional to the ψ value at which that pore drains. Mualem's model conceptualizes pores as pairs of capillary tubes whose lengths are proportional to their radii; the conductance of each capillary-tube pair is determined according to Poiseuille's law[2]. In this formulation, $K(\theta)$ is defined as

$$K(\theta) = K_r(\theta) K_{sat} \quad (4)$$

where $K_r(\theta)$ is relative hydraulic conductivity. To compute $K(\theta)$ for the whole medium, the conductance of all capillary-tube pairs is integrated as

$$K_r(\theta) = S_e^{\,L} \left[\frac{\int_0^{\theta} \frac{1}{\psi(\theta)} d\theta}{\int_0^{\theta_{sat}} \frac{1}{\psi(\theta)} d\theta}\right]^2 \quad (5)$$

Where L is a dimensionless parameter interpreted as representing the tortuosity and connectivity of pores with different sizes, usually given the value 0.5, $S_e = \frac{\theta - \theta_r}{\theta_{sat} - \theta_r}$ (degree of saturation), θ_r is residual water content and θ_{sat} is saturated water content and $\psi(\theta)$ is the retention curve with matric pressure expressed as a function of water content.

The parameters used in Mualem's model as described above were obtained in several ways for the data presented here (fig 1) measured water retention data were fit with the Rossi-Nimmo (1994) model, 2) measured water retention data were fit with the van Genuchten (1980) model, and 3) water retention data were estimated based on particle size distributions and bulk density using the Rosetta model (Schaap et al., 1998).

---

[2] By Poiseuille's law the flow rate per unit cross-sectional area of a capillary tube is proportional to the square of the radius.

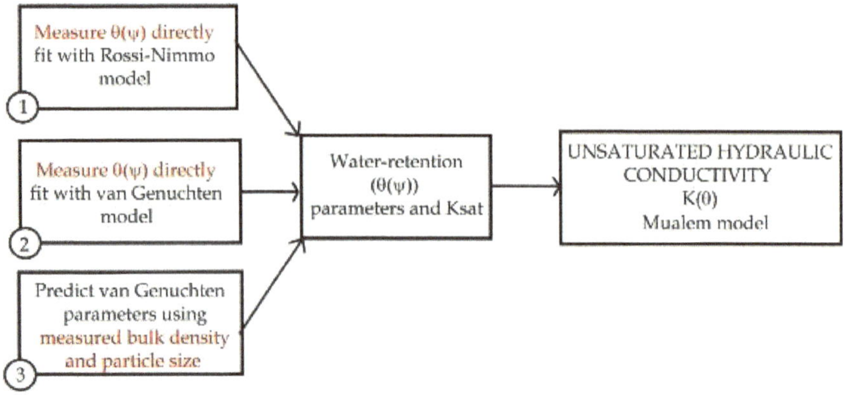

Fig. 1. Steps used in the estimation of unsaturated hydraulic conductivity.

4. Water retention models and hydraulic conductivity estimation

The choice of the water-retention model used to produce the parameters required for hydraulic conductivity estimation also has an effect on the resulting parameters. The Rossi-Nimmo junction (RNJ) model is one that was chosen here to fit the $\theta(\psi)$ measurements because this model is more physically realistic over the entire range of θ from saturation to oven dryness (ψ_d) than other parametric models (for example Brooks and Corey, 1964 and van Genuchten, 1980) that include the empirical, optimized residual water content (θ_r) parameter, which is not well defined. According to capillary theory the largest pores are associated with ψ values near zero and drain first, followed by drainage of successively smaller pores as θ approaches θ_r. With the Brooks and Corey and van Genuchten models there is an asymptotic approach to θ_r meaning that if it is taken to be >0, the number of small pores approaches infinity at a water content above zero, which is physically unrealistic. The $\theta(\psi)$ curve represented by the RNJ model does not have a parameter analogous to θ_r; the curve goes to zero θ at a fixed value of ψ calculated for the conditions of ψ_d. The RNJ model, like many other parametric water retention models, can be analytically combined with the capillary-bundle model of Mualem (1976) to estimate $K(\theta)$ (Fayer and others, 1992; Rossi and Nimmo, 1994; Andraski, 1996; Andraski and Jacobson, 2000).

The RNJ model consists of three functions joined at two points (defined as ψ_i and ψ_j, fig. 2): a parabolic function for the wet range of ψ ($\psi_i \le \psi \le 0$), a power law function (Brooks and Corey, 1964) for the middle range of ψ ($\psi_j \le \psi \le \psi_i$), and a logarithmic function for the dry range of ψ ($\psi_d \le \psi \le \psi_j$). This model has two independent parameters: (1) the scaling factor for ψ (ψ_o), and (2) the curve-shape parameter (λ). Sometimes, ψ_o is associated with ψ at which air first enters a porous material during desaturation (referred to as "air-entry pressure"), but, actually, air begins displacing water in the largest pores at a higher (less negative) ψ than ψ_o as evidenced by the departure of θ from saturation to the right of ψ_o on the $\theta(\psi)$ curve (fig. 2). Unlike the model of Brooks and Corey, which holds θ fixed between $\psi = 0$ and the air-entry pressure, the RNJ model produces a smooth curve near saturation, represented by a parabolic function, that allows the pore-size distribution (the first derivative of the $\theta(\psi)$ curve) to be represented more realistically. The curve-shape parameter λ indicates the relative steepness of the middle portion of the $\theta(\psi)$ curve, described by the power-law function. Larger λ values cause the drainage portion of the $\theta(\psi)$ curve to appear steeper.

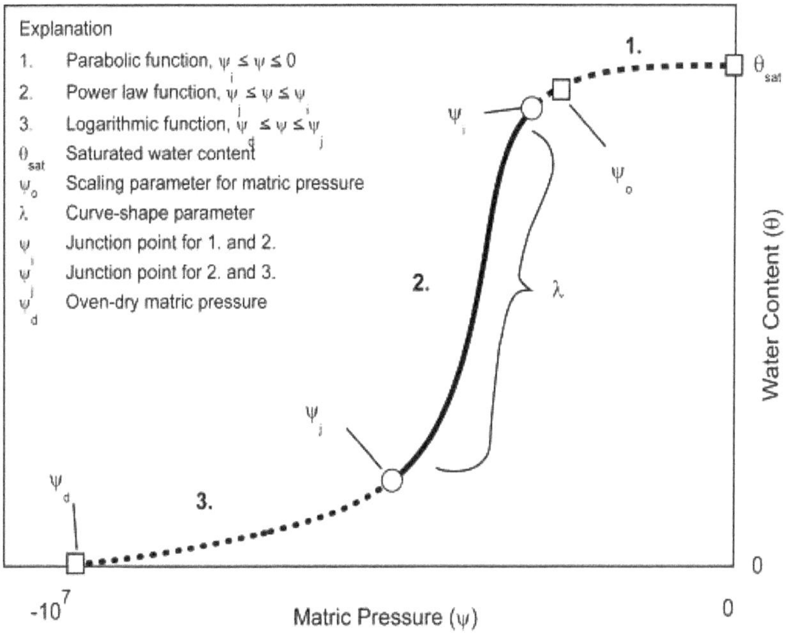

Fig. 2. Example of water-retention ($\theta(\psi)$) curve showing the components of the curve-fit model developed by Rossi and Nimmo (1994).

The parabolic function applies for $\psi_i \leq \psi \leq 0$, and is represented by:

$$\frac{\theta}{\theta_{sat}} = 1 - c\left(\frac{\psi}{\psi_o}\right)^2, \tag{6}$$

where θ_{sat} is saturated water content expressed volumetrically and c is a dimensionless constant calculated from an analytical function involving the parameter λ (described below), which also is dimensionless.
The power law function applies for $\psi_j \leq \psi \leq \psi_i$, and is represented by:

$$\frac{\theta}{\theta_{sat}} = \left(\frac{\psi_o}{\psi}\right)^{\lambda}. \tag{7}$$

The logarithmic function applies for $\psi_d \leq \psi \leq \psi_j$, and is represented by:

$$\frac{\theta}{\theta_{sat}} = \alpha \ln\left(\frac{\psi_d}{\psi}\right), \tag{8}$$

The dependent parameters are calculated as follows:

$$\alpha = \lambda e \left(\frac{\psi_o}{\psi_d} \right)^{\lambda},$$

$$\psi_i = \psi_o \left(\frac{2}{2+\lambda} \right)^{\frac{-1}{\lambda}}, \tag{9}$$

$$\psi_j = \psi_d e^{\frac{-1}{\lambda}},$$

and

$$c = 0.5\lambda \left(\frac{2}{2+\lambda} \right)^{\frac{\lambda+2}{\lambda}}.$$

For convenience, a ψ_d value of -1 × 10^7 cm-water (the pressure at which the curve goes to zero θ) was used in the model fits for all core samples. This is a reasonable ψ value for a soil dried in an oven at 105°-110°C under typical laboratory conditions (Ross and others, 1991; Rossi and Nimmo, 1994).

The RNJ model is integrable in closed form for use in the Mualem (1976) hydraulic-conductivity model as described below (Rossi and Nimmo, 1994). Relative hydraulic conductivity, the ratio between the unsaturated and saturated conductivity can be expressed as:

$$K_r(\theta) = \sqrt{\frac{\theta}{\theta_{sat}}} \frac{I^2(\theta)}{I^2(\theta_{sat})}, \tag{10}$$

where

$$I(\theta) = I_{III}(\theta) \quad \text{for } 0 \le \theta \le \theta_j,$$

$$I(\theta) = I_{II}(\theta) \quad \text{for } \theta_j \le \theta \le \theta_i, \tag{11}$$

$$I(\theta) = I_{I}(\theta) \quad \text{for } \theta_i \le \theta \le \theta_{sat},$$

and

$$I_{III}(\theta) = \frac{\alpha}{\psi_d} \left[\exp\left(\frac{1}{\alpha} \frac{\theta}{\theta_{sat}} \right) - 1 \right],$$

$$I_{II}(\theta) = I_{III}(\theta_j) + \frac{1}{\psi_o} \frac{\lambda}{\lambda+1} \left[\left(\frac{\theta}{\theta_{sat}} \right)^{\frac{(\lambda+1)}{\lambda}} - \left(\frac{\theta_j}{\theta_{sat}} \right)^{\frac{(\lambda+1)}{\lambda}} \right], \tag{12}$$

$$I_I(\theta) = I_{II}(\theta_i) + \frac{2c^{1/2}}{\psi_o}\left[\left(1-\frac{\theta_i}{\theta_{sat}}\right)^{1/2} - \left(1-\frac{\theta}{\theta_{sat}}\right)^{1/2}\right],$$

$$\theta_i = \theta(\psi_i) \qquad \theta j = \theta(\psi_j).$$

The measured water-retention data also were fit with the empirical formula of van Genuchten (1980) which has the form:

$$\theta(\psi) = \theta_r + \left\{(\theta_{sat} - \theta_r) / \left[1 + (\alpha\psi)^n\right]^m\right\}, \tag{13}$$

where α, n, and m are empirical, dimensionless fitting parameters. Using measured θ and ψ values, α and n parameters are optimized to achieve the best fit to the data. The parameter m is set equal to $1\text{-}1/n$ in order to reduce the number of independent parameters allowing for better model convergence and to permit convenient mathematical combination with Mualem's model (van Genuchten, 1980) as follows

$$K(\theta) = K_{sat} S_e^{\mathrm{L}} \{1 - [1 - S_e^{n/(n-1)}]^{1-(1/n)}\}^2, \tag{14}$$

Where K_{sat} is saturated hydraulic conductivity and L is a dimensionless curve-fitting parameter.
Most widely-used unsaturated flow and transport models use the van Genuchten model rather than the Rossi-Nimmo junction model to represent $\theta(\psi)$. The van Genuchten equation is parameterized by θ_{sat}, θ_r, α, and n, where the scaling parameter for ψ is α (analogous to ψ_o) and the curve-shape parameter is n (analogous to λ).

5. Hydraulic property databases

Property transfer models (PTMs) are another way to estimate unsaturated zone hydraulic property data such as $\theta(\psi)$ and $K(\theta)$. PTMs, which can be based on simple or complex relationships among variables of interest, serve the purpose of estimating hydraulic properties from more easily measured bulk properties such as particle size distribution and bulk density. Published databases of hydraulic properties, such as those of Holtan et al. (1968), Mualem (1976), Nemes et al. (1999), Wösten (1999), are often used in studies when direct measurements are not possible or when data for a large number of samples are required, such as in development and testing of new models and theories or in comparative or regionally extensive analyses. Some PTM development and testing studies use these published databases (Vereecken et al., 1989 and 1990; Schaap et al., 1998; Hwang and Powers, 2003), while others use unpublished data or data presented only in parameterized or graphical form (Gupta and Larson, 1979; Arya and Paris, 1981; Wagner et al., 2001). Schaap and Leij (1998) evaluated the effect of data accuracy on the uncertainty of PTMs and concluded that the performance of a PTM depends strongly on the data being used for calibration and testing, however, estimated properties may be sufficient depending on the application for which they are used.
Desirable features of a database include (1) high reliability and precision of measurements, (2) high quality, minimally disturbed samples, (3) a large and diverse sample population, (4)

consistency in measurement techniques across the data set, (5) a full suite of hydraulic and bulk property data for each sample, and (6) ease of use. The database of Perkins and Nimmo (2009) used for the examples in this chapter presents a data set that, though smaller in sample number than many published databases, is ideal in several ways. Sample collection and preparation techniques were selected to ensure minimal sample disturbance, measurements were performed by the same laboratory techniques under highly controlled conditions, and measurements were done by a limited number of researchers. Samples are from diverse geographic, climatic, and geomorphic environments, and the data were originally generated for various research purposes including recharge estimation, simulation of variably saturated flow and solute transport, theoretical studies of porous media, and property transfer model development. Samples were from various soil depths, including many from below the root zone. Other published data sets commonly include samples from shallow depths; about 80% of the samples in the data set of Nemes et al. (1999) come from depths less than 1 m. Additionally, published databases often contain measurements done on disturbed agricultural soils. The data used here for illustrative purposes include bulk density (ρ_b), particle density (ρ_p), particle-size distribution, saturated hydraulic conductivity (K_{sat}), hydraulic conductivity as a function of water content ($K(\theta)$), and water content as a function of matric potential ($\theta(\psi)$).

6. Data analysis

The data used to illustrate the effect of parameterization on $K(\theta)$ estimation and numerical flow simulations is from the database of Perkins and Nimmo (2009) described above. Specifically they are from a core sample from the unsaturated zone at the Idaho National Laboratory (INL) in eastern portion of the Snake River Plain. The medium is sandy in texture with a K_{sat} of 3.90 x 10^{-3} cm/s and a porosity of 0.42. Additionally, measurements from 40 INL samples were used to evaluate the error in $K(\theta)$ produced by each estimation technique.

6.1 Error calculations

The root-mean-square error (RMSE), also referred to as the standard error of the estimate, is used here as a goodness-of-fit indicator between measured and predicted values of $K(\theta)$. The parameters for predicting $K(\theta)$ were obtained using water retention data fit with the RNJ model, the van Genuchten model, and retention parameters predicted by the Rosetta model. The RMSE is calculated as:

$$RMSE = \sqrt{\frac{\sum_{j=1}^{n}(y_j - \hat{y}_j)^2}{n}}, \tag{15}$$

where y_j is the measured value, $\hat{y}_j$ is the predicted value of the dependent variable, and n is the number of observations. Smaller values of RMSE indicate that the predicted value is closer to the measured value of the variable. $K(\theta)$ values span several orders of magnitude which, in effect, unequally weights points in the RMSE calculation, therefore the values were logarithmically transformed prior to calculation. The number of $K(\theta)$ points measured for each sample was between three and ten.

6.2 Parameter testing with numerical simulation

Parameterized $\theta(\psi)$ and $K(\theta)$ curves, representative of the modeled media, are required input for numerical flow and solute transport simulations, therefore numerical simulations were run in order to assess the influence of the input parameters on modeled results. Utilizing parameterized unsaturated-hydraulic properties ($K(\theta)$ and $\theta(\psi)$) flow was simulated using the U.S. Geological Survey variably-saturated two-dimensional transport model (VS2DT) (Lappala and others, 1983; Healy, 1990; Hsieh and others, 1999) in order to assess the effect of the chosen input parameters. The model was modified to allow for the use of the Rossi-Nimmo water-retention parameters (Healy, personal communication 2006) in additional to the van Genuchten model parameters. VS2DT solves the finite difference approximation to Richards' equation (Richards, 1931) for flow and the advection-dispersion equation for transport. The flow equation is written with total hydraulic potential as the dependent variable to allow straightforward treatment of both saturated and unsaturated conditions. Several boundary conditions specific to unsaturated flow may be utilized including ponded infiltration, specified fluxes in or out, seepage faces, evaporation, and plant transpiration. As input, the model requires saturated hydraulic conductivity, porosity, parameterized unsaturated hydraulic conductivity and water-retention functions, grid delineation, and initial hydraulic conditions.

Three simulations based on soil core properties are presented here. Relations between pressure head and water content were represented by functions developed by Rossi and Nimmo and van Genuchten, in all cases using Mualem's model to calculate relative hydraulic conductivity. Parameters were also obtained with the Rosetta model, as described earlier. The 2- by 2-m domain was discretized into 1- by 1-centimeter (cm) grid blocks with a boundary condition chosen to simulate 60 minutes of infiltration at a constant rate of 0.01 cm/s over a 25 cm section at the top left of the domain. Initial hydraulic conditions were specified as uniform water content (10% volumetrically).

7. Discussion

Errors were calculated for $K(\theta)$ for the 40 samples and ranged from 0.21 to 9.45 with the best overall performance achieved using the RNJ model for fitting the measured water retention data where 47.5% of the samples has the lowest RMSE values. The maximum RMSE value for the RNJ model is at least a factor of 4 lower than the other parameterizations. On average, the Rosetta estimations were slightly better than those achieved by fitting the measured water retention data using the van Genuchten model (27.5% vs. 25.0% of the samples having the lowest RMSE values respectively). Table 1 shows the values for all 3 parameterizations for each sample.

For several of the van Genuchten $K(\theta)$ estimates, the RMSE values were unusually large (fig. 3). This occurred in cases where few data points were available and the data clustered within a small range in θ. The van Genuchten model is not physically realistic for the entire range of θ from saturation to ψ_d because the model uses θ_r as an optimized parameter. For the particular cases where no $\theta(\psi)$ data are available in the dry range and the measured points slope steeply within a small range in θ, the curve asymptotically approaches θ_r starting from near θ_{sat}. On the resulting $K(\theta)$ curve, K decreases sharply with little change in θ. The $\theta(\psi)$ curve represented by the RNJ model goes to zero θ at a fixed value of ψ calculated for the conditions of ψ_d; therefore, even when few data points are available, the relation remains somewhat realistic and, in turn, allows for a better estimate of $K(\theta)$. The

RNJ model has a much narrower range in values than the other models and also produces no extreme outliers.

Sample	RMSE Values for predicted $K(\theta)$		
	van Genuchten model	Rossi-Nimmo junction model	Rosetta model
INEEL UZ98-2 42.98 m	0.633	0.659	0.583
INEEL UZ98-2 43.09 m	0.763	0.796	0.427
INEEL UZ98-2 43.21 m	0.884	0.629	0.655
INEEL UZ98-2 43.31 m	1.760	0.979	8.062
INEEL UZ98-2 45.21 m	0.504	0.785	0.564
INEEL UZ98-2 48.16 m	0.445	0.314	1.571
INEEL UZ98-2 48.26 m	0.367	0.342	1.043
INEEL UZ98-2 48.44 m	1.051	0.358	0.439
INEEL UZ98-2 48.92 m	0.427	0.292	1.273
INEEL UZ98-2 49.02 m	0.596	0.266	1.160
INEEL UZ98-2 49.23 m	1.671	0.795	0.392
INEEL UZ98-2 49.30 m	0.949	0.359	0.878
INEEL UZ98-2 49.79 m	2.055	1.205	1.176
INEEL UZ98-2 49.89 m	1.503	0.779	0.977
INEEL UZ98-2 49.99 m	0.608	0.368	0.408
INEEL UZ98-2 50.06 m	0.711	0.214	0.315
INEEL UZ98-2 50.10 m	0.443	0.554	0.610
INEEL UZ98-2 50.30 m	0.345	0.622	0.649
ICPP-SCI-V-215 45.85 m	2.011	0.643	0.472
ICPP-SCI-V-215 46.10 m	3.142	0.598	0.530
ICPP-SCI-V-215 46.37 m	2.101	0.813	1.328
ICPP-SCI-V-215 46.45 m	2.310	0.823	1.599
ICPP-SCI-V-215 47.12 m	8.337	0.739	1.891
ICPP-SCI-V-215 47.42 m	9.451	0.564	2.103
ICPP-SCI-V-215 58.36 m	3.461	2.190	1.211
ICPP-SCI-V-215 58.55 m	2.598	1.293	0.863
ICPP-SCI-V-215 59.20 m	0.571	0.668	1.572
ICPP-SCI-V-215 59.48 m	0.743	0.596	1.365
ICPP-SCI-V-215 59.70 m	5.859	1.662	0.333
ICPP-SCI-V-215 59.92 m	0.841	0.568	1.959
ICPP-SCI-V-189 36.59 m	0.671	0.895	2.158
ICPP-SCI-V-198 42.91 m	0.776	1.068	0.415
ICPP-SCI-V-204 48.41 m	0.943	1.652	3.398
ICPP-SCI-V-205 42.30 m	0.610	1.085	3.802
ICPP-SCI-V-213 37.91 m	1.362	1.699	0.276
ICPP-SCI-V-213 56.08 m	0.787	1.620	7.624
ICPP-SCI-V-214 39.40 m	0.641	1.346	0.510
ICPP-SCI-V-214 43.85 m	0.650	1.057	0.753
ICPP-SCI-V-214 43.96 m	0.804	0.689	0.806
ICPP-SCI-V-214 56.24 m	0.575	1.354	0.619
Average	**1.624**	**0.848**	**1.419**
Maximum value	**9.451**	**2.190**	**8.062**
Minimum value	**0.345**	**0.214**	**0.276**
Number of best fit values	**10**	**19**	**11**

Table 1. RMSE vales for 40 samples from the INL with water retention parameters obtained from curve fitting with the van Genuchten and Rossi-Nimmo models, and estimated with the Rosetta model.

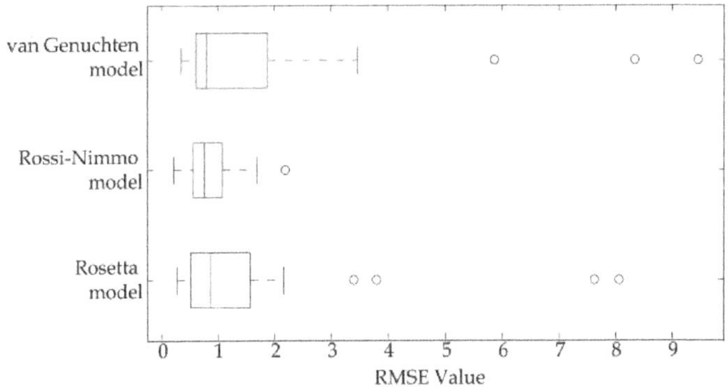

Fig. 3. Comparison of the RMSE values predicted hydraulic conductivity (log K) based on parameters from curves fits to water retention data with the van Genuchten and Rossi-Nimmo models, and the Rosetta model. The box indicates the interquartile range (25th to 75th percentile), the red line is the median value, the whiskers indicate the values that lie within 1.5 of the interquartile range, and the points indicate outliers.

Simulation results from the VS2DT model illustrate the effect of parameterization on model results. Figure 4 shows the hydraulic conductivity curves used in the model plotted with measured values.

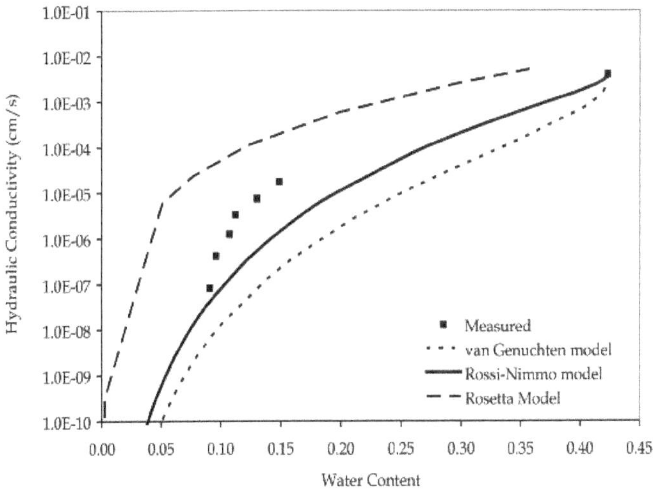

Fig. 4. Measured and estimated hydraulic conductivity curves used in the VS2DT numerical simulations.

The model domain, which was 2 m wide and 2 m deep, had uniform hydraulic properties (sandy textured material) and constant infiltration over a 50 cm section at the top left corner for 60 minutes at a rate of 0.01 cm/s. Changes in water content were output at the end of the infiltration period at 11 observation points (fig. 5) to assess lateral and vertical water movement.

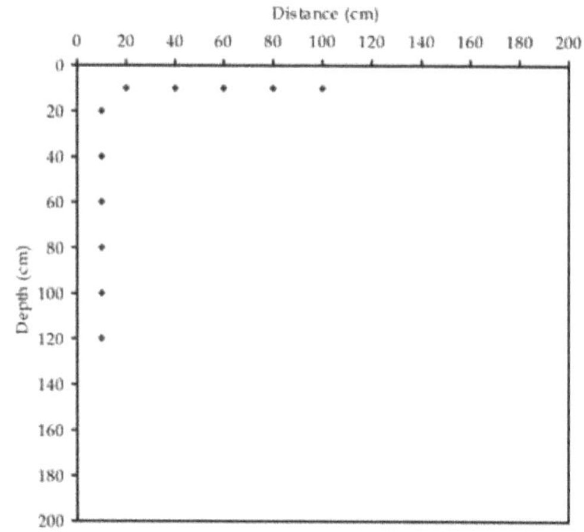

Fig. 5. Observation points for water content output at the end of the infiltration period.

The Rosetta parameterization, one of the most commonly used methods, had a $K(\theta)$ RMSE of 8.06, the highest of the 3 cases. The wetting front was very diffuse compared to the other models and reached the 100 m vertically and laterally to 30 cm beyond the infiltration boundary within the infiltration period (fig. 6). With the water retention curve fitting as the basis for obtaining parameters, the maximum water content is known from the curve. The Rosetta model only has textural information therefore the porosity is estimated and the water contents never reach the true known saturation value, which is slightly higher than the model predicts. The van Genuchten parameterization had a $K(\theta)$ RMSE of 1.76. The wetting front was very sharp and progressed vertically to 60 cm with some wetting laterally to 10 cm beyond the edge of the infiltration boundary (fig. 6). Saturation was reached more quickly than the other 2 cases. The Rossi-Nimmo parameterization had the lowest $K(\theta)$ RMSE of the 3 cases at 0.98. The wetting front progressed vertically about 60 cm and laterally to 10 cm beyond the edge of the infiltration boundary (fig. 4). The results using known retention curves with the Rossi-Nimmo and van Genuchten models to predict $K(\theta)$ are very similar and presumably closest to reality. This analysis shows variable simulation results due to parameterization even though the conceptual model is highly simplified and homogeneous. The effects of soil structure, layering, preferential flow, variable water inputs, nonuniform initial moisture conditions, and other mechanisms that are often found to dominate field situations are not considered here. These effects would further complicate the model uncertainty as influenced by the chosen parameter estimation method in ways that may be difficult to anticipate.

The extremely simplified type of analysis presented here assumes that flow through the unsaturated zone proceeds as described by the Richards' equation. Many studies show that there are processes, such as preferential flow, that are not adequately described by Richards' equation (Perkins et al., 2011; Tyner et al., 2006; Köhne and Gerke, 2005; Jaynes et al., 2001; Yoder, 2001; Iragavarapu et al., 1998, Flury et al., 1994). There are recent studies suggesting that hydraulic conductivity and water retention may be important for modeling diffuse matrix flow; however, they may not capture some of the dominant field-scale processes such as preferential flow. These processes may be better represented by the addition of parameters related to the nature of the water input source and preferential flow path density which will improve the capability of unsaturated zone flow models (Nimmo, 2007; Nimmo 2010). The examples given here show that model results are highly sensitive to the estimated parameters. It is also certain that the use of field-measured (small or large scale), laboratory-measured, and empirically-estimated data in numerical models will yield significantly different results. Another important aspect to consider in evaluating unsaturated zone flow is whether or not the numerical model can accurately represent all of the processes in the conceptual model which should be based on field observations to the extent possible (e.g. information about soil structure, macropores, anthropogenic modifications, etc.).

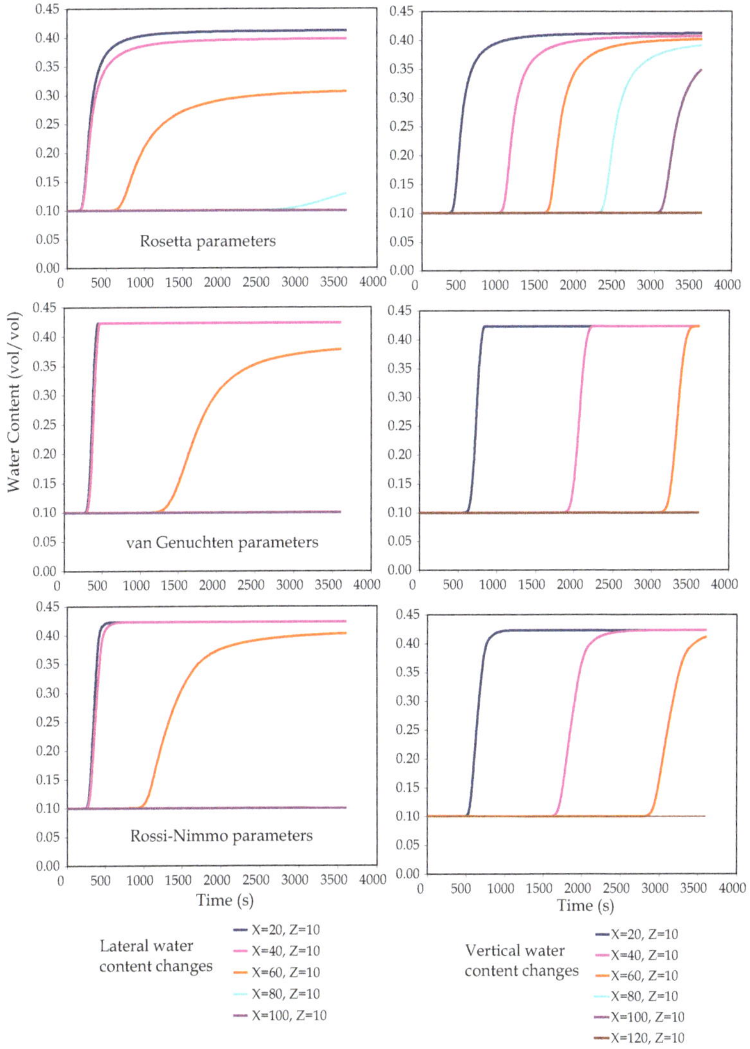

Fig. 6. Water contents siimulated for the simple infiltration model using the VS2DT model.

8. Summary and conclusions

Understanding the highly nonlinear relationship between water content and hydraulic conductivity is one element in predicting water flow and solute transport processes in the unsaturated zone. Measurement of unsaturated hydraulic conductivity of sediments is costly and time consuming, therefore use of models that estimate this property from more easily measured bulk-physical properties is common. In hydrologic studies, especially those using dynamic unsaturated zone moisture modeling, calculations based on property transfer models informed by hydraulic property databases are often used in lieu of measured data from the site of interest. The degree to which the unsaturated hydraulic conductivity curves estimated from property-transfer-modeled water-retention parameters and saturated hydraulic conductivity approximate the laboratory-measured data were evaluated. Results indicate that using the physically realistic water retention model of Rossi and Nimmo (1994) yields results that are closer to measured values especially where measured data are sparse. Using the Rosetta pedotransfer model or the van Genuchten water retention model produced about the same goodness of fit between the measured and modeled hydraulic conductivity data. Because numerical models of variably-saturated solute transport require parameterized hydraulic properties as input, simulation results were shown to illustrate the effect of the various parameters on model performance. It is clear that the model results vary widely for a highly simple conceptual model and it is likely that the addition of more physically realistic characteristics would further affect the model performance in complex ways.

9. References

Andraski, B.J., 1996, Properties and variability of soil and trench fill at an arid waste-burial site: Soil Science Society of America Journal, v. 60, p. 54–66.

Andraski, B.J., & E.A. Jacobson, 2000, Testing a full-range soil-water retention function in modeling water potential and temperature: Water Resources Research, v. 36, no. 10, p. 3081–3089.

Arya L.M., & J.F. Paris, 1981, A physicoempirical model to predict the soil moisture characteristic from particle-size distribution and bulk density data, Soil Sc.i Soc. Am. J. 45:6 1023-1030.

Brooks R.H., & A.T. Corey, 1964, Hydraulic properties of porous media, Colorado State University Hydrology Paper 3, 27 p.

Campbell G.S., & G.W. Gee, 1986, Water potential: miscellaneous methods, In Methods of soil analysis, part I—Physical and mineralogical methods (second edition), Soil Sci. Soc. Am. Book Series No. 9, Madison, Wisconsin, edited by A. Klute, pp 628-630.

Conca J.L., & J.V. Wright, 1998, The UFA method for rapid, direct measurement of unsaturated transport properties in soil, sediment, and rock, Aust. J. Soil Res. 36: 291-315.

Fawcett R.G., & N. Collis-George, 1967, A filter paper method for determining the moisture characteristics of soil. Aust. .J Exper. Ag. and Anim. Hu.s 7(24):162-167.

Fayer, M.J., M.L. Rockhold, & M.D. Campbell, 1992, Hydrologic modeling of protective barriers—Comparison of field data and simulation results: Soil Science Society of America Journal, v. 56, p. 690–700.

Flury, M., H. Flühler, W.A. Jury, & J. Leuenberger, 1994, Susceptibility of soils to preferential flow of water: a field study: Water Resour. Res. 30, 1945-1954.

Gupta S.C., & W.E. Larson, 1979, Estimating soil water retention characteristics from particle size distribution, organic matter content, and bulk density, Water Resour. Res. 15(6):1633-1635.

Healy, R.W., 1990, Simulation of solute transport in variably saturated porous media with supplemental information on modifications to the U.S. Geological Survey's computer program VS2DT: U.S. Geological Survey Water-Resources Investigations Report 90-4025, 125 p.

Holtan H.N., C.B. England, G.P. Lawless, & G.A. Schumacher, 1968, Moisture-tension data for selected soils on experimental watersheds, Rep. ARS 41-144, USDA-ARS, Beltsville, MD.

Hwang S., & S.E. Powers, 2003, Lognormal distribution model for estimating soil water retention curves for sandy soils, Soil Sci, 168(3):156-166.

Hsieh, P.A., Wingle, W., and Healy, R.W., 1999, VS2DTI—A graphical user interface for the variably saturated flow and transport computer program VS2DT: U.S. Geological Survey Water-Resources Investigations Report 99-4130, 13 p.

Iragavarapu, T.K., Posner, J.L., & Bubenzer, G.D., 1998. The effect of various crops on bromide leaching to shallow groundwater under natural rainfall conditions. J. Soil Water Conserv. 53, 146–151.

Jaynes, D.B., Ahmed, S.I., Kung, K.J.S., & Kanwar, R.S., 2001. Temporal dynamic of preferential flow to a subsurface drain. Soil Sci. Soc. Am. J. 65, 1368–1376.

Köhne, J.M., Gerke, H.M., 2005. Spatial and temporal dynamics of preferential bromide movement towards a tile drain. Vadose Zone J. 4, 78–88.

Lappala, E.G., Healy, R.W., and Weeks, E.P., 1983, Documentation of the computer program VS2D to solve the equations of fluid flow in variably saturated porous media: U.S. Geological Survey Water-Resources Investigations Report 83–4099, 184 p.

Mualem Y., 1976, A Catalogue of the Hydraulic Properties of Unsaturated Soils. Haifa, Israel, Technion, Israel Institute of Technology, 118 p.

Nemes A., M.G. Schaap, & F.J. Leij, 1999, UNSODA model version 2.0, see http://www.ars.usda.gov/Services/docs.htm?docid=8967.

Nimmo, J.R., 2010, Theory for Source-Responsive and Free-Surface Film Modeling of Unsaturated Flow: Vadose Zone Journal, v. 9, no. 2, p. 295–306.

Nimmo, J.R., 2007, Simple Predictions of Maximum Transport Rate in Unsaturated Soil and Rock: Water Resources Research, v. 43, no. 5.

Nimmo J.R., J. Rubin , & D.P. Hammermeister, 1987, Unsaturated flow in a centrifugal field: measurement of hydraulic conductivity and testing of Darcy's law, Water Resour. Res. 23:124-134.

Nimmo J.R., K.S. Perkins, & A.M. Lewis, 2002, Steady state centrifuge, In Methods of soil analysis, Part 4—Physical methods, Soil Sci. Soc. Am. Book Series No. 5, Madison, Wisconsin, edited by J.H. Dane and G.C. Topp, pp 903-916.

Perkins, K. S., J.R. Nimmo, C.E. Rose, & R.H. Coupe, 2011, Field Tracer Investigation of Unsaturated Zone Flow Paths and Mechanisms in Agricultural Soils of Northwestern Mississippi, USA, Journal of Hydrology, v. 369, p. 1-11.

Perkins, K.S., & J.R. Nimmo, 2009, High-quality unsaturated zone hydraulic property data for hydrologic applications: Water Resour. Res., 45, W07417, doi:10.1029/2008WR007497.

Reynolds, W.D., D.E. Elrick, E.G. Youngs, A. Amoozegar, H.W.G. Bootlink, & J. Bouma, 2002, Saturated and field-saturated water flow parameters, *In* J.H. Dane and G.C. Topp (ed.), Methods of Soil Analysis, Part 4- Physical Methods, SSSA, Madison, WI, P. 797-878.

Richards, L.A., 1931, Capillary conduction of liquids through porous media: Physics, v. 1, p. 318-333.

Ross P.J., J. Williams, & K.L Bristow, 1991, Equation for extending water-retention curves to dryness, Soil Sci. Soc. Am. J. 55:923–927.

Rossi C., & J.R. Nimmo, 1994, Modeling of soil water retention from saturation to oven dryness, Water Resour. Res. 30(3):701–708.

Schaap M.G., and F.J. Leij, 1998, Database related accuracy and uncertainty of pedotransfer functions., Soil Sci. 163:765–779.

Schaap M.G., F.J. Leij, & M.T. van Genuchten, 1998, Neural network analysis for hierarchical prediction of soil hydraulic properties, Soil Sci. Soc. Am. J. 62(4):847-855.

Tyner, J.S., Wright, W.C., & Yoder, R.E., 2006. Identifying long-term preferential and matrix flow recharge at the field scale. Trans. ASABE 50, 2001–2006.

van Genuchten M.T., 1980, A closed-form equation for predicting the hydraulic conductivity of unsaturated soils, Soil Sci. Soc. Am. J. 44(5):892-898.

Vereecken H., J. Maes, J. Feyen, & P. Darius, 1989, Estimating the soil moisture retention characteristic from texture, bulk density, and carbon content, Soil Sci. 148(6):389-403.

Vereecken H., J. Maes, & J. Feyen, 1990, Estimating unsaturated hydraulic conductivity from easily measured properties, Soil Sci. 149(1):1-12.

Wagner B., V.R. Tarnawski, V. Hennings, U. Müller, G. Wessolek, & R. Plagge, 2001, Evaluation of pedo-transfer functions for unsaturated hydraulic conductivity using an independent data set, Geoderma 102:275-297.

Wösten J.H.M., A. Lilly, A. Nemes, & C. Le Bas, 1999, Development and use of a database of hydraulic properties of European Soils, Geoderma 90:169-185.

Yoder, R.E. 2001. Field-scale preferential flow at textural discontinuities. In: Proceedings 2nd International Symposium of Preferential Flow, St Joseph, Mich. ASAE, 65–68.

Young M.H., & J.B. Sisson, 2002, Tensiometry, In Methods of soil analysis, Part 4—Physical methods, Soil Sc. Soc. Am. Book Series No. 5, Madison, Wisconsin, edited by J.H. Dane and G.C. Topp, pp 575-606.

Permissions

The contributors of this book come from diverse backgrounds, making this book a truly international effort. This book will bring forth new frontiers with its revolutionizing research information and detailed analysis of the nascent developments around the world.

We would like to thank Lakshmanan Elango, for lending his expertise to make the book truly unique. He has played a crucial role in the development of this book. Without his invaluable contribution this book wouldn't have been possible. He has made vital efforts to compile up to date information on the varied aspects of this subject to make this book a valuable addition to the collection of many professionals and students.

This book was conceptualized with the vision of imparting up-to-date information and advanced data in this field. To ensure the same, a matchless editorial board was set up. Every individual on the board went through rigorous rounds of assessment to prove their worth. After which they invested a large part of their time researching and compiling the most relevant data for our readers. Conferences and sessions were held from time to time between the editorial board and the contributing authors to present the data in the most comprehensible form. The editorial team has worked tirelessly to provide valuable and valid information to help people across the globe.

Every chapter published in this book has been scrutinized by our experts. Their significance has been extensively debated. The topics covered herein carry significant findings which will fuel the growth of the discipline. They may even be implemented as practical applications or may be referred to as a beginning point for another development. Chapters in this book were first published by InTech; hereby published with permission under the Creative Commons Attribution License or equivalent.

The editorial board has been involved in producing this book since its inception. They have spent rigorous hours researching and exploring the diverse topics which have resulted in the successful publishing of this book. They have passed on their knowledge of decades through this book. To expedite this challenging task, the publisher supported the team at every step. A small team of assistant editors was also appointed to further simplify the editing procedure and attain best results for the readers.

Our editorial team has been hand-picked from every corner of the world. Their multi-ethnicity adds dynamic inputs to the discussions which result in innovative outcomes. These outcomes are then further discussed with the researchers and contributors who give their valuable feedback and opinion regarding the same. The feedback is then collaborated with the researches and they are edited in a comprehensive manner to aid the understanding of the subject.

Apart from the editorial board, the designing team has also invested a significant amount of their time in understanding the subject and creating the most relevant covers. They scrutinized every image to scout for the most suitable representation of the subject and create an appropriate cover for the book.

The publishing team has been involved in this book since its early stages. They were actively engaged in every process, be it collecting the data, connecting with the contributors or procuring relevant information. The team has been an ardent support to the editorial, designing and production team. Their endless efforts to recruit the best for this project, has resulted in the accomplishment of this book. They are a veteran in the field of academics and their pool of knowledge is as vast as their experience in printing. Their expertise and guidance has proved useful at every step. Their uncompromising quality standards have made this book an exceptional effort. Their encouragement from time to time has been an inspiration for everyone.

The publisher and the editorial board hope that this book will prove to be a valuable piece of knowledge for researchers, students, practitioners and scholars across the globe.

List of Contributors

K. B. Singh
Krishi Vigyan Kendra, Moga, Punjab Agricultural University, Ludhiana, India

Cevza Melek Kazezyılmaz-Alhan
Istanbul University, Civil Engineering Department, Turkey

Rejane Nascentes
Universidade Federal de Viçosa, Campus de Rio Paranaíba, Brazil

Ernani Lopes Possato
Universidade Federal de Lavras, Campus Universitário, Brazil

Izabel Christina Duarte de Azevedo
Universidade Federal de Viçosa, Campus Universitário, Brazil

Francis D. Udoh
Department of Chemical & Petroleum Engineering,University of Uyo, Uyo - Akwa Ibom State, Nigeria

María del Carmen Martínez-Ballesta, María del Carmen Rodríguez- Hernández, Carlos Alcaraz-López, César Mota-Cadenas, Beatriz Muries and Micaela Carvajal
Dpto. Nutrición Vegetal, Centro de Edafología y Biología, Aplicada del Segura (CEBAS-CSIC) Campus de Espinardo, Murcia, Spain

Mirela Tulik and Katarzyna Marciszewska
Warsaw University of Life Science - SGGW (WULS-SGGW), Poland

Daniel G. Neary
USDA Forest Service, Rocky Mountain Research Station, United States of America

Marie-France Jutras and Paul A. Arp
Faculty of Forestry and Environmental Management, University of New Brunswick, Fredericton, NB, Canada

Rainer Schuhmann, Franz Königer, Katja Emmerich, Eduard Stefanescu and Markus Stacheder
Karlsruhe Institute of Technology (KIT), Germany
Competence Centre for Material Moisture (CMM), Karlsruhe, Germany

Ali Rasoulzadeh
Water Engineering Dept., College of Agriculture, University of Mohaghegh Ardabili, Ardabil, Iran

Fred T. Tracy
U.S. Army Engineer Research and Development Center, USA

Maria Clementina Caputo and Lorenzo De Carlo
Water Research Institute (IRSA), National Research Council (CNR), Via F. De Blasio, Bari, Italy

Rafael Segovia Rosales, Eugenio Sanz Pérez and Ignacio Menéndez Pidal
Laboratorio de Geología Aplicada, Escuela Técnica, Superior de Ingenieros de Caminos, Canales y Puertos, Universidad Politécnica de Madrid, Spain

Cheng-Yu Ku
National Taiwan Ocean University, Taiwan

Shih-Meng Hsu
Sinotech Engineering Consultants, Inc, Taiwan

Jose Antonio Gutierrez Gnecchi
Instituto Tecnológico de Morelia, Departamento de Ingeniería Electrónica, México

Adriana Téllez Anguiano, Arturo Méndez Patiño, Gerardo Marx Chávez Campos and Fernando Landeros Paramo
Instituto Tecnológico de Morelia, Departamento de Ingeniería Electrónica, México

Alberto Gómez-Tagle (Jr)
Instituto de Investigaciones Sobre Los Recursos Naturales, Laboratorio de Suelos, Michoacán, México.

Philippe Lobit
Instituto De Investigaciones Agropecuarias y Forestales, México

Jean Luc Mari and Frederick Delay
IFP School and University of Strasbourg, France

René Chipana Rivera
University Mayor de San Andrés, La Paz, Bolivia

Nicasio Sepúlveda and Eve L. Kuniansky
U. S. Geological Survey, United States

Thidarat Bunsri
Department of Environmental Engineering, Faculty of Engineering, King Mongkut's, University of Technology Thonburi, Bangkok, Thailand

Dharmappa Hagare
School of Engineering, University of Western Sydney, NSW, Australia

Muttucumaru Sivakumar
School of Civil, Mining and Environmental Engineering, Faculty of Engineering, University of Wollongong, NSW, Australia

Kim S. Perkins
United States Geological Survey, United States of America